Part I

Overview and Materials

This course is unlike many that you may have taken in that the material that will be covered is primarily a number of unit processes that are quite distinct from each other. The book then has the flavor of a survey of topics that will be covered rather than a linear progression. This part of the book will lay the foundations that will be needed to later understand the various fabrication processes.

Just as the Industrial Revolution enabled man to apply and control greater physical power than his own muscle could provide, so electronics has extended his intellectual power. [1]

The first chapter will provide a roadmap of the course and an introduction to integrated circuit fabrication. The processes that are covered are briefly described in a qualitative manner and the relationships between the various topics are discussed. A simple example of a semiconductor technology, the fabrication of integrated resistors, is used to demonstrate a flow of these processes which we will call a technology. Extensions of the technology to include capacitors and MOSFETs are also discussed.

The second chapter will introduce the topic of crystal growth and wafer production. The chapter contains basic materials information that will be used throughout the rest of the book. This includes crystal structure and crystal defects, phase diagrams, and the concept of solid solubility. Unlike the other unit processes that will be covered in the later chapters, very few integrated circuit fabrication facilities actually grow their own wafers. The topic of wafer production however demonstrates some of the important properties of semiconductor materials that will be important both during the fabrication process and to the eventual yield and performance of the integrated circuit. The differences in the production of silicon and GaAs wafers are discussed.

---

[1]Dr. Robert Noyce (1977) Co-founder, Fairchild Semiconductor, Intel

Chapter 1

An Introduction to Microelectronic Fabrication

The electronics industry has grown rapidly in the last three decades. This growth has been driven by a revolution in microelectronics. In the early 1960s, putting more than one transistor on a piece of semiconductor was considered cutting edge. Integrated circuits (ICs) containing tens of devices were unheard of. Digital computers were large, slow, and extremely costly. Bell Labs, which had invented the transistor a decade earlier, rejected the concept of ICs. They reasoned that in order to achieve a working circuit all of the devices must work. Therefore, to have a 50% probability of functionality for a 20 transistor circuit, the probability of device functionality must be $(0.5)^{1/20} = 0.966$, or 96.6%. This was considered to be ridiculously optimistic at the time, yet today integrated circuits are built with tens of millions of transistors.

Early transistors were made from germanium, but most circuits are now made on silicon substrates. We will therefore emphasize silicon in this book. The second most popular material for building ICs is gallium arsenide (GaAs). Where appropriate, the book will discuss the processes required for GaAs ICs. While GaAs has a higher electron mobility than silicon, it also has several severe limitations including low hole mobility, less stability during thermal processing, a poor thermal oxide, high cost, and perhaps most importantly, much higher defect densities. Silicon has therefore become the material of choice for highly integrated circuits, and GaAs is reserved for circuits that operate at very high speeds but with low to moderate levels of integration. Currently the most common application of GaAs is analog circuits operating at speeds in excess of a gigahertz (10^9 Hz).

To chart the progress of silicon microelectronics it is easiest to follow one type of chip. Memory chips have had essentially the same function for many years, making this type of analysis meaningful. Furthermore, they are extremely regular and can be sold in large volumes, making technology customization for the chip design economical. As a result, memory chips have the highest density of all ICs. Figure 1.1 shows the density of dynamic random access memories (DRAMs) as a function of time. The vertical axis is logarithmic. The density of these circuits increase by increments of 4×. Each of these increments takes three to four years. One of the most fundamental changes in the fabrication process that allows this technology evolution is the minimim feature size that can be printed on the chip. Not only does this increase IC density, the shorter distances that electrons and holes have to travel improve the transistor speed. Part of the IC performance improvement comes

3

Figure 1-1   Memory and minimum feature sizes for dynamic random access memories as a function of time.

from this increased transistor performance, and part of it comes from being able to pack the transistors closer together, decreasing the parasitic capacitance. The right-hand side of Figure 1.1 shows that ICs have progressed from 10 microns (μm) (1 μm = 10^{-4} cm) to under 1 μm. For sake of reference, Figure 1.2 shows an electron micrograph of a silicon based IC, along with a human hair. The vertical and horizontal lines are metal wires used to interconnect the transistors. The transistors themselves are below the metal and are not visible in the micrograph. At this rate of progress, gigabit chips with 0.25 μm features will be seen before the year 2000.

At first glance, these incredible densities and the associated design complexity would seem extremely daunting. This book, however, will focus on how these circuits are built rather than how they are designed or how the transistors operate. The fabrication process is

Figure 1-2   Scanning electron micrograph (SEM) of an IC circa mid 1980s. The visible lines correspond to metal wires connecting the transistors.

similar no matter how many transistors are on the chip. The first half of the book will cover the basic operations required to build an IC. Using mechanical construction as an analogy, these would include steps such as forging, cutting, bending, drilling, and welding. These steps will be called unit processes in this text. If one knows how to do each of these steps for a certain material (e.g., steel), and if the machines and material required are available, they could be used to make a ladder, or a high pressure cylinder, or a small ship. The required number and order of the steps will clearly depend on what is being built, but the basic unit processes remain the same. Furthermore, once a sequence that produces a good ship has been worked out, other ships of similar design could probably be built with the same process. The design of the ship, that is, what goes where, is a separate task. The shipbuilder is handed a set of blueprints to which he or she must build.

The collection and ordering of these unit processes for making a useful product will be called a *technology*. Part V of the book will cover some of the basic semiconductor technologies. Whether the technology is used to make microprocessors, I/O controllers, or any other digital function is largely immaterial to the fabrication process. Even many analog designs can be built using a technology very similar to that used to build most digital circuits. An IC, then, starts with a need for some sort of electronic device. A designer or group of designers translates the requirements into a circuit design; that is, how many transistors, resistors, and capacitors must be used, what values they must have, and how they must be interconnected. The designer must have some input from the fabricator. In the shipbuilding example, the blueprints must somehow reflect the limitation that the shipbuilder cannot put rivets over weld joints or use small rivets and still expect them to hold very high pressures. The builder must therefore give the designer a document that says what can and cannot be done. In microelectronics this document is called the *design rules* or *layout rules*. They specify how small or large some features can be or how close two different features can be. If the design conforms to these rules, the chip can be built with the given technology.

1.1 Microelectronic Technologies—A Simple Example

Instead of blueprints, the circuit designer hands the IC fabricator a set of photomasks. The photomasks are a physical representation of the design that has been produced in accordance with the layout rules. As an example of this interface, assume that a need exists for an IC consisting of a simple voltage divider as shown in Figure 1.3. The technology to build this design is shown in Figure 1.4. Silicon wafers will be used as the substrate since they are flat, reasonably inexpensive, and most IC processing equipment is set up to handle them. The production of these substrates will be discussed in the second chapter. Since the wafer is at least somewhat conductive, an insulating layer must first be deposited to prevent leakage between adjacent resistors. Alternatively, a thermal oxide of silicon could be grown, since it is an excellent insulator. The thermal oxidation of silicon is covered in Chapter 4. Next a conducting layer is deposited that will be used for the resistors. Several techniques for depositing both insulating and conducting layers will be discussed in Chapters 12–14.

This conducting layer must be divided up into individual resistors. This can be done by removing portions of the conducting layer, leaving rectangles of the film that are isolated from each other. The resistor value is given by

$$R = \rho \frac{L}{W \cdot t}$$

where ρ is the material resistivity, L is the resistor length, W is the resistor width, and t is the thickness of the layer. The designer can therefore select different values of resistors by choosing the width to

(A)

(1) Starting wafer   **(2) Grow oxide**

(3) Deposit resistor material   **(4) Pattern resistor material**

(5) Deposit insulator   **(6) Pattern insulator**

(7) Deposit metal   **(8) Pattern metal**

(B)

Figure 1-3 A simple resistor voltage divider. At left is a circuit representation; at right is a physical layout. The layers shown at right are resistor, contact, and low resistance metal.

Figure 1-4 The technology flow for fabricating the resistor IC shown in Figure 1.3.

length ratio, subject to the limits specified by the layout rules. The technologist chooses the film thickness and the material (and therefore ρ) to give the designer an appropriate range of resistivities without forcing him to resort to extreme geometries. Since ρ and t are determined during the fabrication and are approximately constant across a wafer, the ratio ρ/t is more often specified than ρ or t. This ratio is called the *sheet resistance*, ρ_s. It has units of $\Omega/\square$ where the number of squares is the ratio of the length to width of the resistor line.

The resistor information from the design, namely L and W for each resistor, must be transferred from the photomask to the wafer. This is done using a process called *photolithography*. The most commonly used type of photolithography is *optical lithography*. In this process, a photosensitive layer called *photoresist* is first spread on the wafer (Figure 1.5). Light shining through the mask exposes the resist in the regions of the wafer where some of the metal resistor layer must be removed. In these exposed regions, a photochemical reaction occurs in the resist that causes it to be easily dissolved in a developer solution. After the develop step, the photoresist remains only in the areas where a resistor is desired. The wafer is then immersed in an acid that dissolves the exposed metal layer but does not significantly attack the resist. When the etch is complete, the wafers are removed from the acid bath, rinsed, and the photoresist is removed. The photolithographic process will be covered in Chapters 7 through 9. Chapter 11 will cover etching.

While the resistors have now been formed, they still need to be interconnected and metal lines must be brought to the edge of the chip, where they can later be attached to metal wires for contact to the external world. This latter operation, called *packaging*, will not be covered in this text. If the metal lines have to cross over the resistors, another insulating layer must be deposited. To make electrical contact to the resistors, one can open up holes in the insulating layer using the same photolithographic and etch processes we had used for patterning the resistors, although the composition of the acid bath may be different. Finally, the fabrication sequence can be completed by depositing a highly conductive metal layer, applying a third mask, and etching this metal interconnect layer.

(1) Starting wafer with layer
to be patterned

(2) Coat with photoresist

(3) Bake the resist to set its
dissolution properties

(4) Expose resist by shining light
through a photomask

(5) Immerse exposed wafer
in developer

Developer

(6) Etch the film

Etchant

Figure 1-5   Steps required for a pattern transfer using optical lithography.

The technology consists of four layers: the lower insulator, the resistor film, the upper insulator, and the interconnect metal. Photolithography is used to selectively remove some of these layers in certain regions, but any point on the wafer must be made of some subset of these layers in the same order that they were built up. Except for the edges of the patterns, the thickness of these films is constant. The technology uses three photolithography steps, three etch steps, and four thin film deposition (including oxide growth) steps. A very similar set of steps could be used to fabricate a capacitor. With only a few more steps, simple transistors can be built. Notice that the comparison with shipbuilding breaks down in one critical respect. The effort required to build the IC is independent of the number of resistors. The photomask might define one resistor or 1,000,000 resistors (assuming a 1,000,000 resistor circuit could be useful). In fact, two different sets of photomasks, one that defines only a few resistors and one that defines thousands of resistors, could be used interchangeably with exactly the same technology and would require the same amount of work. This is because most of the unit processes described in this book operate in the whole wafer at the same time instead of one rivet at a time.

1.2 Unit Processes and Technologies

Some of the basic steps used in building an IC have already been discussed: photolithography, thin film deposition, and etching. Unit processes for thin film deposition include the physical processes of sputtering and evaporation. These are physical processes in that they do not generally depend on a chemical reaction. Sputtering is done by using charged atoms of argon called ions (Ar^+) to bombard a target containing the material for deposition. The target erodes under this bombardment, and some of the material falls onto the wafers, coating them with a thin film of material. Evaporation involves heating the material to be deposited to a high temperature so that a vapor stream is created. The wafers are placed in this stream for coating. The third thin film deposition process that will be discussed is chemical vapor deposition. In this technique one or more gasses are made to flow into a chamber that contain the wafers to be coated. In many cases the wafers are also heated. A chemical reaction occurs that leaves the desired solid product on the surface of the wafer.

A resistor was chosen to simplify the first example. Most semiconductor devices require the formation of doped regions. For example, consider the n-channel MOSFET shown in Figure 1.6. Some familiar layers from the resistor example are recognizable: a blanket insulator and a patterned metal layer. One can also selectively dope the source and drain regions, putting together a technology to make the transistor. Dopants are either donors (n-type) or acceptors (p-type). For silicon, the most common n-type dopants are arsenic, phosphorus, and antimony, and the most common p-type dopant is boron. For gallium arsenide the most common n-type dopants are silicon, sulphur, and selenium, and the most common p-type dopants are carbon, beryllium, and zinc. In early semiconductor technologies, impurities were introduced by exposing heated wafers to a dopant containing gas. For example, a hydrogen/phosphine (H_2/PH_3) mixture can be used to introduce phosphorus into silicon.

Figure 1-6 Cross section of an MOS transistor showing gate, source, drain, and substrate electrodes. The "+" and "−" indicate very heavy and very light dopings, respectively.

The introduction of dopant using this technique and the subsequent movement of the impurities when the wafer is heated is called *diffusion*. This type of dopant introduction method allows the impurities to diffuse deep into the wafer. As a result, it is not desirable for the small devices required in modern fabrication technologies. Ion implantation, which has replaced it, uses a beam of ionized atoms or molecules electrostatically accelerated toward the wafer. This method allows the process technologist to control the amount of impurity introduced (dose) and the depth of the impurity in the wafer (range). To limit the diffusion of impurities, a new class of processes have been developed that allow the rapid heating (to high temperatures) and cooling of the wafer. This type of process is called rapid thermal processing (RTP).

A number of processes that allow the growth of thin layers of semiconductor on top of the wafer will be discussed. These processes are called *epitaxial growth*. They allow the production of patterned dopant regions below the surface of the wafer. The book will first discuss the more traditional techniques of growing silicon on silicon and gallium arsenide on gallium arsenide (homoepitaxy). It will then cover more advanced techniques that allow the growth of extremely thin layers for the fabrication of advanced device structures.

The unit processes can be assembled into functional process modules. These modules are designed to carry out specific tasks such as the electrical isolation of adjacent transistors, low resistance contacts to transistors, and multiple layers of high density interconnect. Clear trade-offs exist among the various modules in terms of process complexity, circuit density, planarity, and performance. These modules and the basic transistor fabrication are assembled into technologies. Three of the most popular technologies that represent a reasonable cross section of the microelectronics industry will be reviewed. Finally, techniques required for high volume manufacturing of ICs will be discussed.

1.3 A Roadmap for the Course

The various unit processes for fabrication are fairly independent. Each of the next 13 chapters will cover a different unit process. To keep the book to a manageable size, each process can be only briefly introduced. In many cases, the chapters themselves can be expanded into books. The material in the chapter will provide references that will allow you to further investigate each topic if you have an interest. The result of this approach is sometimes called a *survey course*.

Figure 1.7 shows a map of the course chapters. The order that your instructor chooses to follow is completely arbitrary as long as the necessary introductory material is covered. These chapters and sections are marked with a °. The chapters and sections marked with a + are additional material somewhat beyond the basic processes needed to describe simple semiconductor technologies.

The last five chapters of the book are dedicated to semiconductor technologies. The basic unit processes discussed earlier are brought together to form ICs made from silicon CMOS, bipolar transistors, and GaAs field effect transistors. These technology examples have been chosen because they are popular and because they are representative of many other common technologies. As you might infer from the previous discussion, the same unit processes could be used to fabricate flash memories, charge coupled devices (CCDs), lasers, solar cells, light emitting diodes (LEDs), and even micro-

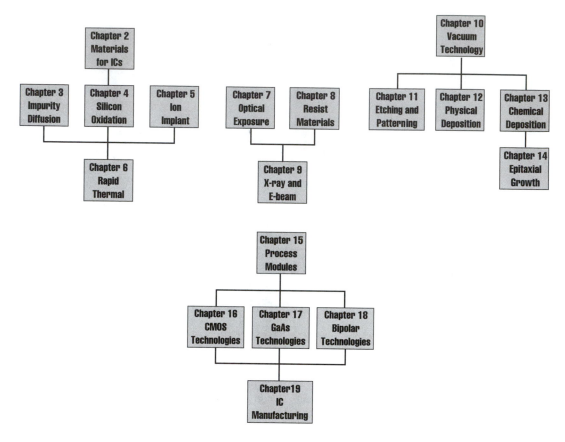

Figure 1-7 A roadmap for the course indicating the relationships between the chapters.

mechanical devices like transducers and actuators. The only differences are the number, type, and sequence of the processes used to form the technology. You are encouraged to look into one of these other fabrication technologies after you complete the course to see some of the other ways that these unit processes are applied.

1.4 Summary

Integrated circuits have developed with incredible levels of complexity, rapidly approaching 1,000,000,000 transistors per chip. Transistor density, as measured by dynamic random access memories (DRAMs), quadruples about every three years, as it has since 1968. This book will introduce the technologies used to fabricate the ICs. The building blocks of these technologies are the unit processes of photolithography, oxidation, diffusion, ion implantation, etching, thin film deposition, and epitaxial growth. The unit processes can be assembled in different order and number, depending on the circuit to be built.

Chapter 2

Semiconductor Substrates

This section of the book will deal with unit processes that depend strongly on the properties of the semiconductor wafers themselves. Diffusion for example, depends on the crystalline perfection in the wafers, which in turn depends on the process temperature. We will begin with a description of phase diagrams. This material is particularly useful for understanding the formation of alloys that will be used later in the book. This topic also leads naturally into a discussion of solid solubility and the doping of semiconductor crystals. Following that, the chapter will concentrate on crystal structures and defects in crystalline materials. The second half of the chapter will discuss the techniques used to fabricate semiconductor wafers. These wafers, which vary in diameter from 1 in. for some compound semiconductors up to 300 mm for some silicon wafers, are the basic starting point for device fabrication. Although few fabrication facilities still make their own wafers, the study of the semiconductor substrate makes a good place to begin to develop an understanding of semiconductor processing.

The materials used for microelectronics can be divided into three classifications, depending on the amount of atomic order they possess. In single crystal materials, almost all of the atoms in the crystal occupy well defined and regular positions known as lattice sites. Most of the semiconductor substrates in which the active devices are actually made are single crystal. Amorphous materials, such as SiO_2, are at the opposite extreme. The atoms in an amorphous material have no long range order. Instead, the chemical bonds have a range of lengths and orientations. The third class of materials is polycrystalline. These materials are a collection of small single crystals randomly oriented with respect to each other. The size and orientation of these crystals often change during processing and sometimes even during circuit operation.

2.1 Phase Diagrams and Solid Solubility°

Most of the materials of interest to us in this text are not elemental; rather, they are mixtures of materials. Even silicon is not very useful in a pure state. Instead, it is mixed with impurities that affect its electrical properties. A very convenient way to present the properties of mixtures of materials is a *phase diagram*. Binary phase diagrams can be thought of as maps that show the regions of stability for mixtures of two materials as a function of percent composition and temperature. Phase diagrams may also have a pressure dependence, but all of the diagrams of interest in semiconductor device fabrication will be at 1 atm.

Figure 2.1 shows the phase diagram for Ge-Si, an example of the simplest type of system [1]. There are two solid lines on the graph. The upper, or liquidus, curve describes the temperature at which a given mixture will be in a completely liquid state. The lower, or solidus, curve describes the temperature at which the mixture will completely freeze. Between these two curves is a region containing both liquid and solid mixtures. The composition of the melt can be readily determined from the diagram. If a solid mixture with equal atomic concentrations of Si and Ge is heated from room temperature, the material will begin to melt at 1108°C. Assume that the heating proceeds slowly enough so that the system remains in thermodynamic equilibrium. In the region between the two solid lines, the concentration of the melt will be determined by the composition at which the liquidus line crosses the temperature. For example, at 1150°C, the composition of the melt will be 22 atomic percent silicon. The concentration of the solid can be read off of the graph as the concentration at which the solidus line crosses the temperature. In this example, the remaining solid will be 58 atomic percent silicon. Notice that the amount of the solid that will melt can be calculated using these values. As the temperature is increased, the composition of the melt moves toward the original value, while the concentration of the solid moves closer to pure silicon. When the temperature reaches the liquidus line the entire charge melts. For the 50% starting mixture this occurs at about 1272°C. Upon cooling,

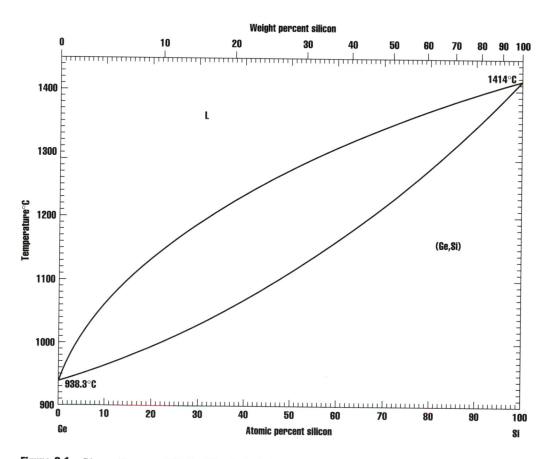

Figure 2-1   Phase diagram of Si-Ge. The dashed lines correspond to a heating process that remains in thermodynamic equilibrium *(courtesy of ASM International)*.

the same processes occur. In either process, maintaining thermodynamic equilibrium in the solid is much more difficult than it is in the melt.

Figure 2.2 shows the phase diagram for GaAs [2]. Material systems like GaAs that have two solid phases that melt to form a single liquid phase are called *intermetallics*. To examine the phase diagram, start in the lower right-hand corner. This is a solid phase, since it is below the solidus line. The vertical line at the center of the diagram indicates that the compound GaAs will form in this material system. The lower left region is a solid mixture of GaAs and Ga, while the region in the lower right is a solid mixture of GaAs and As. If this As-rich solid is heated to 810°C, the solution will begin to melt. Between this temperature and the liquidus line, the concentration of the melt can be determined as before. For Ga-rich charges, the mixed state begins at about 30°C, only slightly above room temperature. This gives rise to problems associated with the growth of GaAs layers that will be described both later in this chapter as well as in a later chapter on epitaxial growth.

As a final example consider Figure 2.3, which shows the phase diagram for the As-Si system [3]. While the structure looks quite complicated with several different solid phases mapped out, microelectronic applications are primarily interested in the low arsenic concentration limit. Even very heavily doped silicon is normally less than 5% arsenic. Notice that there is only a very small region in which As will dissolve in silicon as a dopant without forming a compound. The maximum concentration of an impurity that can be dissolved in another material under equilibrium conditions is

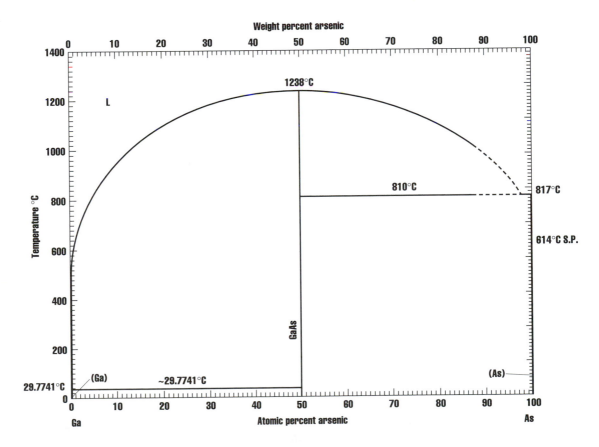

Figure 2-2 Phase diagram for GaAs *(courtesy of ASM International).*

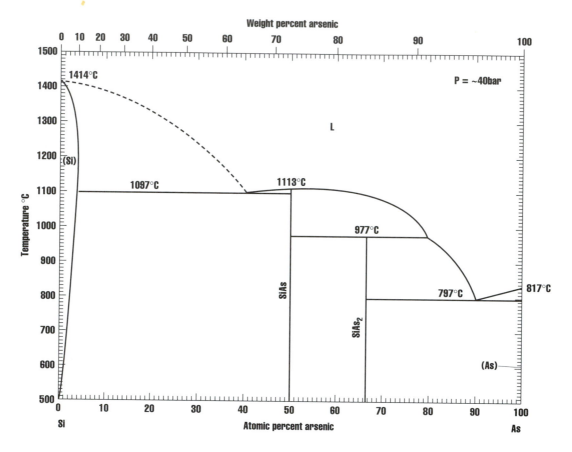

Figure 2-3   Phase diagram for As-Si *(courtesy of ASM International).*

called the solid solubility. Notice that the solid solubility increases as the temperature approaches 1097°C, where it is about 4 atomic percent, and that in this region a vertical line will actually intersect three curves. The portion of the line that goes from 0 atomic percent Si at 500°C to 4 atomic percent Si at 1097°C is called the *solvus curve*, as it represents a solubility. The remaining two lines are the solidus and liquidus, respectively. The solid solubility of As in Si is a comparatively large number, and it means that As can be used to form very heavily doped and therefore low resistance regions such as source and drain contacts for MOS transistors (Chapter 16) and emitter and collector contacts to bipolar transistors (Chapter 18). Since it is the solid solubility from the phase diagram that is of primary interest for dopant impurities, and the solid solubility varies by orders of magnitudes for different impurities in silicon, the phase diagram information for a number of common dopants in silicon have been combined [4] on a semilog plot in Figure 2.4.

Consider what happens in the following situation. A silicon wafer is heated to 1097°C and doped to 3.5% atomic with As. Doping methods will begin to be discussed in the next chapter. Impurity concentrations are normally expressed in number per unit volume. A 3.5% atomic concentration in silicon corresponds to 1.75×10^{20} cm^{-3}. The phase diagram indicates that as the wafer is cooled, it will eventually exceed the maximum concentration that can be in solution. If it is to maintain thermodynamic equilibrium, the excess As must condense out, either by coming out of the surface or more likely, by forming solid precipitates in the silicon crystal. In order for this to happen, the As

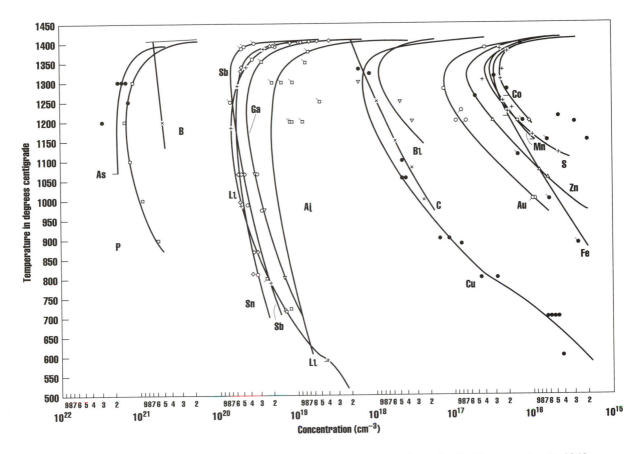

Figure 2-4   Solid solubility of common silicon impurities.

atoms must be mobile in the crystal. If the wafer is cooled rapidly enough, the precipitates can not form, and a higher concentration of impurities than is thermodynamically allowed can be frozen in. Metallurgists refer to this process as *quenching*.

2.2 Crystallography and Crystal Structure°

Crystals are described by their most basic structural element: the unit cell. A crystal is simply an array of these cells, repeated in a very regular manner over three dimensions. The unit cells of interest have cubic symmetry with each edge of the unit cell being the same length. Figure 2.5a shows three common types of cubic crystals. The directions in a crystal are identified using a Cartesian coordinate system as [x,y,z]. For a cubic crystal, the faces of the cell form planes that are perpendicular to the axes of the coordinate system. The symbol (x,y,z) is used to denote a particular plane that is perpendicular to the vector that points from the origin along the [x,y,z] direction. Figure 2.5b shows several common crystal directions. The set of numbers x, y, and z that are used to describe planes in this manner are called the *Miller indices* of a plane. They are found for a given plane by taking the inverse of the points at which the plane in question crosses the three coordinate axes, then multiplying by the smallest possible factor to make x, y, and z integers. The notation {x,y,z} is also used to represent

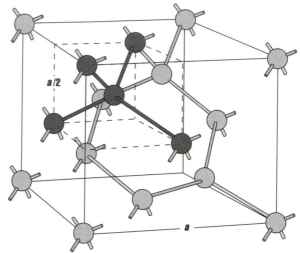

Figure 2-5 (a) Common cubic crystals: simple cubic, body centered cubic, and face centered cubic. (b) Crystal orientations in the cubic system.

Figure 2-6 The diamond structure.

crystal planes. This representation is meant to include not only the given plane, but also all equivalent planes. For example, in a crystal with cubic symmetry, the (100) plane will have exactly the same properties as the (010) and (001) planes. The only difference is an arbitrary choice of coordinate system. The notation {100} refers to all three.

Silicon and germanium are both Group IV elements. They have four valence electrons and need four more to complete their valence shell. In crystals, this is done by forming covalent bonds with four nearest neighbor atoms. None of the basic cubic structures in Figure 2.5 would therefore be appropriate. The simple cubic crystal has six nearest neighbors, the body centered cubic (BCC) has eight, and the face centered cubic (FCC) has 12. Instead, Group IV semiconductors form in the diamond structure shown in Figure 2.6. The unit cell can be constructed by starting with an FCC cell and adding four additional atoms. If the length of each side is a, the four additional atoms are located at $(a/4, a/4, a/4)$, $(3a/4, 3a/4, a/4)$, $(3a/4, a/4, 3a/4)$, and $(a/4, 3a/4, 3a/4)$. This crystal structure can also be thought of as two interlocking FCC lattices. GaAs also forms in this same arrangement; however, when two elements are present, the crystal has a reduced level of symmetry. The structure is then called Zinc-Blende.

2.3 Crystal Defects

Semiconductor wafers are highly perfect single crystals. Nevertheless, crystal defects play an important role in semiconductor fabrication. Semiconductor defects or imperfections, can be divided into four types, depending on their dimensionality. Point defects do not extend in any direction. Line defects extend in one direction through the crystal. Area and volume defects are 2- and 3-D defects, respectively. Each type of defect impacts different areas of the fabrication process. Point defects are extremely important to the understanding of doping and diffusion. The prevention of line defects is important to any thermal processing, particularly in rapid thermal processing. Volume defects can play a useful role in yield engineering. Figure 2.7 shows a few of the most important semiconductor defects.

One of the most common types of point defect is a lattice site without an atom. This defect is

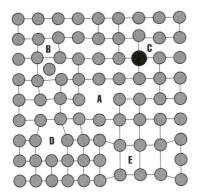

Figure 2-7 Simple 0- and 1-D semiconductor defects include: (a) vacancies, (b) self-interstitials, (c) substitutional impurities, (d) edge dislocations, and (e) dislocation loops.

a vacancy. A closely related point defect is an atom that resides not on a lattice site, but in the spaces between the lattice positions. It is referred to as an interstitial. If the interstitial atom is of the same material as the atoms in the lattice, it is a self-interstitial. In some cases, the interstitial comes from a nearby vacancy. Such a vacancy interstitial combination is called a Frenkel defect. The interstitial or vacancy may not remain at the site at which it was created. Both types of defects can move through the crystal, particularly under the high temperatures typically found during processing conditions. Either defect might also migrate to the surface of the wafer where it is annihilated.

Vacancies and self-interstitials are intrinsic defects. Just as one talks about intrinsic carriers in semiconductors, at nonzero temperatures intrinsic defects will tend to occur in an otherwise perfect crystal. Thermal excitation creates a very small percentage of electrons and holes in a semiconductor. It will also remove a small number of atoms from their lattice sites, leaving behind a vacancy. Generally, the vacancy concentration is given by an Arrhenius function, which is an equation of the form

$$N_v^o = N_o \exp^{-E_a/kT} \tag{2.1}$$

where N_o is the number density of atoms in the crystal lattice (5.02×10^{22} cm^{-3} for silicon), E_a is the activation energy associated with the formation of a vacancy, and k is Boltzmann's constant. For silicon, E_a is of order 2.6 eV, so at room temperature only one in 10^{44} lattice sites would be vacant in an otherwise perfect crystal. At 1000°C, however, the number of defects rises to 1 in 10^{10}. An identical equation can be used to describe the equilibrium concentration of interstitials in a material. In silicon, the activation energy for interstitials is higher than that for vacancies (about 4.5 eV). Notice from these two equations that, unlike intrinsic electrons and holes, the equilibrium vacancy concentration is not generally the same as the equilibrium interstitial concentration. Several important point defect sources and sinks exist, including extended defects and the surface of the wafer that allow this discrepancy.

GaAs has two types of lattice sites. Gallium sites have four arsenic nearest neighbors. Arsenic sites have four gallium nearest neighbors. The concentration of vacancies in GaAs is given by

$$N_{v,Ge}^o = 3.3 \times 10^{18} \text{ cm}^{-3} e^{-0.4/kT} \tag{2.2}$$

for gallium and

$$N_{v,As}^o = 2.2 \times 10^{20} \text{ cm}^{-3} e^{-0.7/kT} \tag{2.3}$$

for arsenic [5].

Of course, this picture has considerably simplified the subject of intrinsic point defects. This chapter began by pointing out that in single crystal silicon, each atom has four nearest neighbors to which it covalently bonds. The simplest thing that could happen when a vacancy is created is that all four of the bonds break. This would leave all of the atoms electrically neutral, but it also leaves four unsatisfied valence shells. Alternatively, an electron may remain behind when a vacancy was created. In doing so, it completes the valence shell of one of the neighboring atoms, leaving it with a net one

negative charge. This situation is described by saying that a -1 vacancy has been created (along with a $+1$ interstitial). The activation energy of this vacancy will be considerably different than that of a neutral vacancy. Furthermore, $-2, -3, -4, +1, +2, +3$, and $+4$ vacancies are also possible, although triply and quadrupally ionized species are rarely important in practice.

The concentration of charged vacancies at equilibrium is given by

$$N_v^- = N_v^o \frac{n}{n_i} e^{(E_i - E_v^-)/kT} \tag{2.4}$$

where n is the carrier concentration of free electrons in the semiconductor, n_i is the intrinsic carrier concentration (see Figure 3.4), E_i is the intrinsic energy level (the center of the bandgap), and E_v^- is the energy level associated with the negatively charged vacancy. The difference in the energy levels between the vacancy and the intrinsic level just plays the role of a new activation energy, E_a^{v-}. Similarly, the concentration of positively charged vacancies is given by

$$N_v^+ = N_v^o \frac{p}{n_i} e^{E_v^+ E_i/kT} = N_v^o \frac{p}{n_i} e^{-E_a^{v+}/kT} \tag{2.5}$$

The concentration of multiply charged vacancies is similarly proportional to the ratio of the charge density to the intrinsic carrier concentration, raised to the appropriate power. For example, the concentration of $v^=$ is given by

$$N_v^- = N_v^o \left[\frac{n}{n_i} \right]^2 e^{-E_a^{v-}/kT} \tag{2.6}$$

The second type of point defect that may exist in a semiconductor is an extrinsic defect. This is caused either by an impurity atom at an interstitial site or at a lattice site. In the latter case, it is referred to as a *substitution impurity*. The word ''defect'' should not necessarily carry with it a negative connotation. For example, dopant atoms that are required to modulate semiconductor conductivity are substitutional defects. While this text will concentrate on substitutional dopant impurities, interstitial impurities can also have a significant impact on device performance. Some impurities that tend to occupy interstitial sites have electronic states near the center of the bandgap. As a result, they are efficient sites for the recombination of electrons and holes. These recombination centers reduce the gain of bipolar transistors and can cause p-n diodes to leak.

Line defects extend in one dimension. The most common example is a dislocation. In this defect, an extra plane of atoms is inserted between two other planes of atoms. The simplest type of dislocation is an edge dislocation. Here, the extra plane is terminated on one end by the edge of the crystal. If the extra plane is completely contained in the crystal, the defect is referred to as a dislocation loop. The presence of a dislocation in the crystal is a sign of stress. The bonds just before the insertion of the extra plane are stretched, and bonds just after the plane are compressed. Dislocations are often formed by the agglomeration of point defects. Any defect in a crystal has associated with it a surface energy. The higher the surface area of the defect, the higher the energy stored in the defect. A high concentration of point defects therefore costs more energy than a dislocation, since the point defects have a larger total surface area. Point defects moving randomly through the crystal will tend to stick together until a dislocation or other higher dimensionality defect results. The process is called *agglomeration*.

There are various ways that defect inducing stress can occur during processing. If there is a substantial temperature difference across the wafer, the wafer will attempt to expand nonuniformly, and thermoplastic stress will occur in the wafer. Similar stresses can result if the wafer is rigidly clamped and heated or if layers with different thermal expansion coefficients are present when the wafer is heated. The equations governing thermoplastic stress will be covered in Section 6.4, since this problem is particularly important in rapid thermal processing. A second method of inducing dislocations is the presence of high concentrations of substitutional impurities in the crystal. Since these atoms are not the same size as the host atom, stresses will result. The effect of the stress is to lower the energy required to break bonds and form vacancies. The third type of dislocation formation mechanism, which will be reviewed in detail later, is physical damage of the crystal. During some processes, the surface of the wafer is bombarded with other atoms. These atoms transfer enough energy to the lattice to break bonds. Once again the process creates a high concentration of vacancies and interstitials that will tend to agglomerate into dislocations and other higher order defects.

Dislocations move by two primary mechanisms, as illustrated in Figure 2.8: *climb* and *glide*. Glide is the movement of a dislocation line in a direction other than along the line. It is the result of shear stress. In this situation, one of the planes of lattice atoms breaks to form a continuous plane with the previous extra plane leaving behind a new defect. This process, called *slip*, can continue until the entire width of the wafer has moved by one lattice site. Climb, on the other hand, is simply the growth (or shrinkage) of a dislocation line. Vacancies and/or interstitials are created in response to stress. These point defects then become part of the line defect.

The third type of defect is a *2-D* or *area defect*. The most obvious type of area defect is a polycrystalline grain boundary. An important defect from a device standpoint is a *stacking fault*. Similar to a dislocation line, a stacking fault is an extra plane of atoms. In this case, the pattern is disrupted in two dimensions and is only regular in the third. If, for example, an extra plane of atoms is inserted in the diamond lattice, as shown in Figure 2.9, a stacking fault results. Stacking faults are terminated either by the edge of the crystal or by dislocation lines. Bulk defects are irregular in all three dimensions. A common example of a bulk defect is a *precipitate*. An important class of such precipitates are impurity precipitates, discussed in Section 2.1.

The change in the Gibbs free energy associated with the impurity condensation can be expressed as

$$\Delta G = -V \cdot \Delta G_v + A\gamma + V\varepsilon + R_e \tag{2.7}$$

where V is the volume of the precipitate, ΔG_v is the change in free energy per unit volume associated with the transformation, A is the surface area of the precipitate, γ is the free energy per unit area of

Figure 2-8   Movement of an edge dislocation (center figure) by (a) climb and (b) glide.

the precipitate, ε is the strain energy per unit volume of the precipitate, and R_e is a strain relation term. If the concentration of point defects exceeds the equilibrium values given by Equations 2.2–2.4, the crystal is said to be supersaturated, and there will be a strong driving force toward the formation of extended defects. In the case at hand, as the temperature is lowered the degree of supersaturation increases, and the first term of Equation 2.7 becomes large enough to make ΔG negative, providing the thermodynamic force necessary to drive the precipitation. When the volume change upon precipitation is large, the precipitate causes strain that may be relaxed through the formation of self-interstitials and/or dislocations [6],[7].

It would seem that defects are undesirable. More precisely, 2-D and 3-D defects are undesirable in active regions (i.e., where the transistors are located). Defects in inactive regions have a beneficial effect as gettering sites. Gettering is a process by which impurities and defects diffuse through the crystal, becoming trapped at the gettering site. Beneficial gettering can be accomplished by providing highly strained or damaged regions far from the active devices, such as the backside of the wafer. This type of process is called *extrinsic gettering*.

Another common application of gettering in silicon technology is the use of oxygen precipitates in the bulk of the wafer. Point defects and residual impurities such as heavy metals become trapped at the precipitate sites, reducing their concentration near the active device regions. Since this process uses oxygen that is intrinsic to the wafer, it is called *intrinsic gettering*. Figure 2.10 shows the solubility of oxygen in silicon [8],[9]. Unlike the dopant curves in Figure 2.4, the solid solubility of oxygen shows a simple Arrhenius behavior, obeying the equation

$$C_{\text{ox}} = 2 \times 10^{21} \, \frac{\text{atoms}}{\text{cm}^3} \exp \frac{-1.032 \text{eV}}{kT} \qquad \textbf{(2.8)}$$

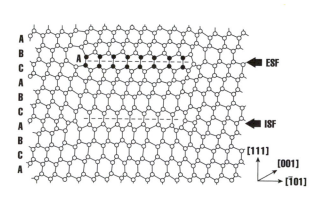

Figure 2-9 An intrinsic stacking fault is the removal of part of a plane of atoms in the {111} directions. An extrinsic stacking fault is the addition of a partial plane of atoms in the {111} directions. The labels A, B, and C correspond to the three different (111) planes in the diamond lattice *(after Shimura)*.

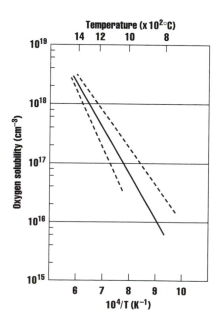

Figure 2-10 The solubility of oxygen in silicon. The dashed lines correspond to the highest and lowest variations *(after Shimura)*.

Figure 2-11   Cross section view of denuded wafer with bipolar transistor *(reprinted by permission, Academic Press, after Shimura).*

As will be discussed later in this chapter, dissolved oxygen results from the process by which the single crystal silicon is grown. Typical oxygen concentrations in silicon wafers are 10 to 40 ppm or about 10^{18} cm^{-3}. If the wafer temperature is less than about 1150°C, oxygen will tend to precipitate from the crystal, leading to 3-D defects. The shape of the precipitates depends on the temperature of the wafer. If the wafer is cooled rapidly, the excess oxygen is immobile and so cannot agglomerate to condense out. The oxygen is held in solid solution in a supersaturated (i.e., greater than would be allowed if the wafer were in thermodynamic equilibrium) condition. If the wafer is annealed at about 650°C, the precipitates are shaped like rods and lie along the $\langle 110 \rangle$ directions in the (100) planes [10]. When the wafer is annealed at about 800°C, square precipitates form on the (100) planes with [111] rounded edges [11]. For wafers annealed at 1000°C, the precipitates are shaped like octahydra [12].

In order to use an intrinsic gettering process, the wafer should have an oxygen concentration of 15 to 20 ppm (Figure 2.11). If the concentration is much smaller than that the oxygen impurities are too far apart to form agglomerates. Larger oxygen concentrations will precipitate, but they will also lead to wafer warpage, and other extended defects such as slip lines may thread through the active regions. A typical intrinsic gettering process consists of three steps: outdiffusion, nucleation, and precipitation [13]. The purpose of the outdiffusion step is to reduce the concentration of dissolved oxygen in a denuded zone near the surface of the wafer. The depth of the denuded zone must exceed the deepest junction depth in the technology, plus the depletion width of the junction at maximum reverse bias. Since this process has considerable uncertainty, it is common to use denuded zone widths of two or three times the required depth. The denuded zone is formed by annealing the wafer at high temperature [14],[15] in an inert ambient. The temperature of the step must be high enough to allow the oxygen to diffuse out of the surface, but low enough to reduce the concentration of the oxygen to less than 15 ppm. The width of the denuded zone in silicon can be approximated by [16]

$$L_d = \sqrt{0.091 \frac{\text{cm}^2}{\text{sec}} \, t \, \exp \frac{-1.2\text{eV}}{kT}} \qquad \textbf{(2.9)}$$

By comparing Equations 2.8 and 2.9 one can find process times and temperatures that satisfy both conditions: $C_{\text{ox}} < 15$ ppm and $L_d > 10$ μm. For example, an anneal at 1200°C for 3 hr gives an oxygen concentration of 6×10^{17} cm^{-3} (12 ppm) and an approximate denuded zone depth of 25 μm. Finally, one needs to observe that the carbon concentration must also be controlled in wafers intended for intrinsic gettering use, since the oxygen seems to precipitate at carbon impurities. The concentration of carbon in the wafer, therefore, is a factor in determining the size and shape of the oxygen precipitates.

2.4 Czochralski Growth

The technique used to produce most of the crystals from which semiconductor wafers are cut is called *Czochralski growth*. The process was first developed by Teal [17] in the early 1950s. Since silicon is a single component system, it is easiest to start by studying its growth. Once this is complete some of the complications associated with compound semiconductor growth will be discussed. The production of the high purity polycrystalline materials that are melted in the Czochralski furnace will not be discussed. While this is an interesting exercise in distillation, it is not very relevant to the IC fabrication process.

Czochralski growth involves the solidification of a crystal from a melt. The material used in single crystal silicon growth is electronic grade polycrystalline silicon (polysilicon) which has been refined from quartzite (SiO_2) until it is 99.999999999% pure. The poly is loaded into a fused silica crucible that is contained in an evacuated chamber (Figure 2.12) [18]. The chamber is back filled with an inert gas, and the crucible is heated to approximately 1500°C. Next, a small seed crystal (about 0.5 cm in diameter and 10 cm long) is lowered into contact with the melt. This crystal must be carefully oriented since it will serve as the template for the growth of the much larger crystal, called the *boule*. Modern boules of silicon can reach a diameter of over 300 mm and are 1 to 2 meters long.

Since both the liquid and the solid are at about the same pressure and have approximately the same composition, solidification must be accomplished by a reduction in temperature. As shown in the figure, temperature is lost by the increased surface area of the solid. Both natural convection and gray body radiation will cause the crystal to give off substantial heat and will give rise to a strong thermal gradient across the liquid solid interface. At the interface, additional energy must be lost to accommodate the latent heat of fusion for the solid. Balancing the energy flow in a unit volume at the interface with a simple 1-D analysis,

$$\left(-k_l A \frac{dT}{dx}\Big|_l \right) - \left(-k_s A \frac{dT}{dx}\Big|_s \right) = L \frac{dm}{dt} \qquad (2.10)$$

where the k's are the thermal conductivities of liquid and solid silicon at the melting point, A is the cross-sectional area of the boule, T is the temperature, and L is the latent heat of fusion (approximately 340 cal/g for silicon).

Both of the two thermal diffusion terms are positive under normal Czochralski growth conditions, with the first term larger than the second. This implies that there is a maximum rate at which the crystal can be pulled (Figure 2.13). This would occur if all of the heat diffusing up the solid is produced by the latent heat of fusion at the interface (i.e., the first term in Equation 2.10 is 0). Then there would be no temperature gradient in the liquid and

$$V_{max} = \frac{kA}{L}\frac{dT}{dm} = \frac{k}{\rho L}\frac{dT}{dx}\Big|_s \qquad (2.11)$$

If an attempt is made to pull the crystal from the melt faster than this, the solid cannot conduct the heat away and the material will not solidify in a single crystal [19]. Typical values for the temperature gradient in silicon Czochralski are about 100°C/cm, although for pull rates near the maximum, the temperature gradient will vary inversely with the diameter of the crystal (Figure 2.14). To minimize the temperature gradient in the melt, the boule and melt are typically rotated in opposite directions during the growth.

In reality, the maximum pull rate is not normally used. It has been found that the crystalline

Figure 2-12 Schematic and photograph *(courtesy of Ferrofluidics Corporation)* of Czochralski growth system.

Figure 2-13 Time lapse sequence of boule being pulled from the melt in a Czochralski growth *(reprinted with permission of Lattice Press).*

(A)

(B)

Figure 2-14 (a) 200 mm silicon growth facility, (b) prototype 300 mm wafers *(photographs courtesy of MEMC Electronic Materials, Inc.).*

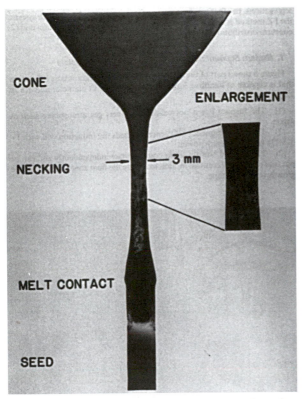

CONE

ENLARGEMENT

NECKING ← → 3 mm

MELT CONTACT

SEED

Figure 2-15   X-ray topograph of seed neck showing edge terminated dislocations *(reprinted by permission, Academic Press).*

quality is a sensitive function of the pull rate. The material near the melt has a very high density of point defects. It would be desirable to cool the solid quickly enough to prevent these defects from agglomerating. On the other hand, such rapid cooling means that large thermal gradients (and therefore large stresses) will occur in the crystal, particularly for large diameter wafers. Czochralski growth processes use this effect to minimize dislocations in the boule by rapidly beginning the pull. This produces a narrow, highly perfect region [20]–[22] just below the seed crystal, called a *tang*. Any dislocations in the seed crystal, whether there initially or caused by contact with the molten silicon, can be prevented from propagating into the boule in this manner [23]. An example of such an edge terminated dislocation can be seen in Figure 2.15. The melt temperature is then lowered and the pull rate reduced to shoulder out the boule to the desired diameter. Finally, the pull rate and furnace temperature are stabilized using feedback control from an optical detector set to measure the width of the boule.

Generally, as the diameter of the crystal is increased, the pull rate must be decreased. This is because the heat loss is proportional to the surface area, which is proportional to the diameter of the crystal, while the energy produced by fusion is proportional to the volume, which in turn is proportional to the square of the diameter. If the pull rate is too low, however, point defects will agglomerate. The type of defects formed most commonly are called *dislocation loops*. In semiconductor substrates these loops are called swirl because they are often distributed in swirls about the center of the wafer.

During the process of Czochralski growth several impurities will incorporate into the crystal. We have already discussed the importance of oxygen in the substrate. We are now in a position to understand its origin. Crucibles used to hold the molten silicon during the Czochralski process are usually fused silica (SiO_2, often mistakenly called quartz). At 1500°C, silica will release a considerable amount of oxygen into the molten silicon. Some of the oxygen will be incorporated into the growing crystal. Since this supply of oxygen is constantly replenished, it will be approximately constant along the length of the boule. The oxygen concentration can be controlled using the temperature of the melt. For reduced concentrations of oxygen, the boule can be grown under magnetic confinement. The first commercially significant magnetically confined Czochralski growths were reported in the early 1980s [24],[25]. Figure 2.16 shows a magnetically confined crystal growth system. The magnetic field may be directed along the length of the boule (axial); however, modern systems are nearly all perpendicular to the growth direction (transverse). In either case, the purpose of the field is to create a Lorentz force ($qv \times B$), which will change the motion of the ionized impurities in the melt in such a manner as to keep them away from the liquid/solid interface and therefore decrease the impurity incorporation in the crystal. In this arrangement, oxygen concentrations as low as 2 ppm have been reported [26]. This process also has the effect of minimizing resistivity variations across the wafer [27].

Figure 2-16   Photograph of commercial magnetically confined Czochralski system *(Thomas et al.)*.

It is also common to introduce dopant atoms into the melt so that a particular resistivity wafer can be made. To do this, one can simply weigh the melt, determine the number of impurity atoms that would be needed, and add that weight of impurity. The process is complicated, however, by the fact that impurities tend to segregate at solid/liquid interfaces. That is, the solid may be more or less likely to contain an impurity than the liquid. A segregation coefficient k can be defined as

$$k = \frac{C_s}{C_l} \tag{2.12}$$

where C_s and C_l are the impurity concentrations at the solid and liquid sides of the solid/liquid interface. Table 2.1 summarizes the segregation coefficients of common impurities in silicon.

To understand how dopant segregation impacts crystal uniformity, consider what happens if k > 1. In that case, the concentration of impurity in the solid is greater than that in the melt. To achieve

Table 2.1   Segregation coefficients for common impurities in silicon

Al	As	B	O	P	Sb
0.002	0.3	0.8	0.25	0.35	0.023

From Sze [28].

this, a higher proportion of dopant is pulled from the melt than is contained in the liquid. Consequently, the impurity concentration in the melt must decrease as the boule is pulled. Referring to the phase diagrams or the solid solubility curves, the concentration shift of the melt also shifts the concentration of the solid. If we define X as the fraction of the melt that has solidified, and assume that the solution is well mixed, it can be shown that

$$C_s = kC_0(1 - X)^{k-1} \qquad \textbf{(2.13)}$$

where C_0 is the initial melt concentration.

The well mixed approximation is not very good due to the existence of thermal gradients in the melt. The basic effect is shown in Figure 2.17. The hot walls of the crucible cause the melt near the walls to expand. The lower density of this heated material will cause it to rise. The cooler region of the melt near the boule will tend to sink. Collectively, the process is known as *natural convection* and the flow patterns that it produces are known as *buoyancy driven recirculation cells*. This effect is present to some extent whenever temperature gradients exist in a system that contains a liquid or gas with a finite, nonzero viscosity. The rotation of the crucible and the boule, along with the pulling action of the boule and the surface tension of the melt, all contribute to the flow patterns in the melt [29]. At the liquid/solid interface, the melt cannot flow, in the same way the water at the banks of a river cannot flow. In fact, because of the finite viscosity of the melt, there will be a region near the liquid/solid interface over which little flow of material will be present. This region of the melt is called the *boundary layer*. The impurities that are taken up by the solid must diffuse across this region. To take this effect into account, the segregation coefficient k can be replaced with an effective segregation coefficient k_e, where

$$k_e = \frac{k}{k + (1 - k)e^{-Vb/D}} \qquad \textbf{(2.14)}$$

b is an effective boundary layer thickness, V is the pull velocity, and D is the impurity diffusivity in the molten semiconductor.

The growth of GaAs from the melt is significantly more difficult than the growth of silicon. One reason is the difference in the vapor pressure of the two materials. Stoichiometric GaAs melts at 1238°C (see Figure 2.2). At that temperature, the vapor pressure of gallium is less than 0.001 atm, while the vapor pressure of arsenic is about 10^4 larger. It is obvious that maintaining stoichiometry through the boule will be challenging. A variety of systems have been designed to overcome this obstacle. This and the next section will present the two most popular choices: *liquid encapsulated Czochralski growth*, commonly called LEC, and the *Bridgman growth technique*. Bridgman wafers have the lowest dislocation densities (of order 10^3 cm^{-2}) and are commonly used for fabricating optoelectronic devices such as lasers. LEC grown wafers can be made with larger diameters, they are round, and can be made semi-insulating with resistivities of nearly 100 MΩ cm.

Figure 2-17   Formation of recirculation cells in a melt.

Figure 2-18 Schematic of a liquid encapsulated Czochralski growth system. The labeled parts include: (1) quartz crucible, (2) heat flux control system, (3) graphite shield, (4) cavity thermocouple, (5) radiation shield, (6) heater, (7) temperature control thermocouple, (8) water cooled support, (9) insulation support, (10) graphite crucible support, and (11) tubing support *(after Kelly et al.)*.

A disadvantage of LEC wafers is that they typically have defect densities greater than 10^4 cm^{-2}. The production and use of these semi-insulating substrates will be discussed in Chapter 15. As a result of this property, nearly all electronic GaAs devices are fabricated on LEC material.

Pyrolytic boron nitride (pBN) crucibles are used instead of quartz in LEC growth to avoid silicon doping of the GaAs boule from the quartz. To prevent outdiffusion of As from the melt, the LEC process uses a tightly fitting disk as shown in Figure 2.18. The most common sealant material is boric oxide (B_2O_3). A slight excess of arsenic is added to the charge to compensate for the arsenic loss that occurs until the cap becomes molten at about 400°C and seals the melt. Once the charge is molten, the seed crystal can be lowered through the boric oxide until it contacts the charge. The growth vessel is maintained at high pressure during the growth. For that reason, the process is sometimes called *high pressure LEC* or HP LEC. The initial argon pressure is about 40 atm, but it is slowly lowered throughout the growth. Typical pull rates are about 1 cm/hr [27].

The second problem encountered in LEC growth relates to differences in the material properties of silicon and GaAs. Table 2.2 summarizes the relevant data. The thermal conductivity of GaAs is about a third that of silicon. As a result, a GaAs boule is not able to dissipate the latent heat of fusion as rapidly as a silicon boule. Furthermore, the shear stress required to nucleate a dislocation at the melting point is about a fourth that of silicon. Not only is the material less able to dissipate heat, a smaller thermoplastic strain will induce defects. It is not surprising, therefore, that the Czochralski growth of GaAs is primarily limited to wafer diameters of 4 in. or less and that defect densities of Czochralski grown material are many orders of magnitude larger than comparable silicon wafers.

If the dislocation density is kept low enough, it does not present an insurmountable barrier to the fabrication of ICs with moderate to high levels of integration; however, when it exceeds 10^4 cm^{-2}, the dislocations can have a significant impact on transistor performance [30]. It has been found that thermally annealing the GaAs ingot after Czochralski growth somewhat reduces the dislocation den-

Table 2.2   Thermal and mechanical properties of semiconductors

	Melting Point (C)	**Thermal Conductivity (W/cm-K)**	**Critical Resolved Shear Stress at MP (MPa)**
Si	1420	0.21	1.85
Ge	960	0.17	0.70
GaAs	1238	0.07	0.40

Thomas et al. [27].

sity [31]. Alloying about 0.1 atomic percent indium into the wafer has been found to minimize the impact of dislocations in Czochralski grown GaAs. Ehrenreich and Hirth [32] have suggested that the replacement of a gallium atom with a much larger indium atom creates a strain field that traps the dislocations, effectively gettering them from any active layers grown on top of the wafer. It is now generally believed that the use of indium increases the critical resolved shear stress through a process known as *solid solution hardening* [33]. As a result, the dislocation density itself can be 10^3 cm^{-2} or less in indium doped wafers. Because of the hardening, indium doped wafers are even more brittle than pure GaAs and so more prone to chipping and breakage. This fact, combined with concerns of indium diffusion during the process and the material improvements made possible by boule annealing has caused the popularity of indium doped GaAs to decline in recent years.

2.5 Bridgman Growth of GaAs

Roughly three fourths of the market for growing GaAs is controlled by the horizontal Bridgman method and its variants. The basic process is shown in Figure 2.19 [34]. The solid Ga and As components are loaded into a fused silica ampoule, which is then sealed shut. In many cases, the ampoule includes a separate solid arsenic chamber with a restricted orifice to the main chamber. The arsenic only chamber provides the arsenic overpressure necessary to maintain stoichiometry. The ampoule is loaded into a SiC tube, which rests on a semicircular trough, typically also made from SiC. A tube furnace is then slowly rolled past the charge. Although the charge may be passed through the furnace, the opposite is normally done to minimize any disturbance of the crystal as it solidifies. The temperature of the furnace is set to melt the charge when it is completely inside. As the furnace rolls past the ampoule the molten GaAs charge in the bottom of the ampoule recrystallizes in a characteristic ''D'' shape. If desired, a seed crystal can be mounted so as to contact the melt. Typical crystal diameters produced this way are 1 to 2 in. The growth of larger crystals requires very accurate control of the stoichiometry of the axial and radial temperature gradients to control the dislocation density [35]. The important feature of the Bridgman approach is that the molten charge is held by the

Figure 2-19   Schematic of a horizontal Bridgman growth system *(after Sell)*.

ampoule. This allows the process to be run with very small thermal gradients and, therefore, routinely produces wafers with dislocation densities of less than 10^3 cm^{-2}.

The difficulty with standard Bridgman growth is that it is unable to produce high resistivity substrates because of the large intimate contact between the ampoule and the melt. A variety of techniques have been proposed to overcome this obstacle. Two of the most commonly cited are the vertical Bridgman [36] and vertical gradient freeze [37] methods. The basic idea of each approach is to turn a horizontal Bridgman apparatus on its side. The charge is held in a boat that typically is boron nitride, which is then sealed in a silica ampoule in such a way so as to minimize the dead volume. A small additional arsenic charge may be added to maintain stoichiometry. The ampoule is loaded into a furnace where it is raised to a temperature just above the melting point. The furnace is then slowly raised, cooling the ampoule and freezing the charge. Temperature gradients are less than 10°C/cm and the growth rate is limited to a few mm/hr [38]. Generally, these techniques produce results that are intermediate between the two extremes with resistivities of about 10 MΩ cm and dislocation densities of 2 to 5 × 10^3 cm^{-2}.

2.6 Float Zone Growth

When extremely high purity silicon is required the growth technique of choice is *float zone refining*. This technique is not generally used for GaAs. Figure 2.20 shows a plot of the crystal purity reported with both Czochralski growth and float zone refining. Carrier concentrations as low as 10^{11} cm^{-3} have been achieved with the float zone method. The basic feature of this growth technique is that the molten part of the sample is supported entirely by the solid part. There is no need for a crucible. The concept of the float zone method is shown in Figure 2.21. A rod of high purity polycrystalline material is held in a chuck while a metal coil driven by a high power radio frequency (RF) signal is slowly

Figure 2-20   Minimum achievable carrier concentration for various growth technologies *(Thomas et al.).*

Figure 2-21   Schematic of a float zone refining system.

Supply rod

RF Coil

Melt

Single crystal

Bottle neck

Supporting balls

Seed

Bolt

Figure 2-22 Schematic of a float zone system for growing large diameter boules (*reprinted from Keller and Mühlbauer by courtesy of Marcel Dekker*).

passed along its length. The field set up by the RF power induces eddy currents in the rod that lead to joule heating. The coil power is set such that the part of the rod closest to the coil is melted. Alternatively a focused electron beam can be used to locally melt the rod. The entire apparatus can be encased in a vacuum system or in an enclosed chamber with an inert gas purge.

To ensure growth along a preferred orientation, a seed crystal may also be used. In a method very similar to that of Czochralski growth, Dash injected a very thin seed crystal into the top of the molten rod [22]. In this technique, a thin neck about 3 mm in diameter and about 10 to 20 mm long is pulled and the pull rate and temperature lowered to shoulder the crystal out to a larger diameter. The difficulty of the top seeded technique is that the molten region must be able to support the weight of the entire rod, unlike Czochralski where the boule has not yet been formed. As a result, this technique is limited to boules of no more than a few kilograms. Figure 2.22 shows an arrangement that can be used for the float zone growth of large diameter crystals [39]. In this case the crystal is bottom seeded. After a sufficient length of dislocation free material has been produced, a funnel filled with small spheres is slid upward until it supports the weight of the boule.

A disadvantage of float zone growth is the difficulty of introducing a uniform concentration of dopants. There are four methods that can be used: *core doping*, *pill doping*, *gas doping*, and *neutron transmutation*. Core doping refers to the use of a doped polysilicon rod for the starting material. On top of this rod, additional undoped polysilicon is deposited until the average desired concentration is reached. The process can be repeated through several generations if necessary. Core doping is the preferred process for boron due to its high diffusivity and the fact that it does not tend to evaporate from the surface of the rod. Then the dopant concentration of the rod is given by

$$C(z) = C_c \left[\frac{r_d}{r_f} \right]^2 [1 - (1 - k)e^{-k\,z/l}] \qquad \textbf{(2.15)}$$

where C_c is the dopant concentration in the core rod, r_d is the radius of the dopant rod, r_f is the radius of the final boule, l is the length of the floating zone, z is the distance from the start of the boule, and k is the effective distribution coefficient that is similar to the segregation coefficient. For float zone growth, k(boron) = 0.9, k(phosphorus) = 0.5, and k(antimony) = 0.07. Neglecting the first few melt lengths, the concentration of boron in a boule is quite uniform. Gas doping is accomplished through the use of gasses such as PH_3, $AsCl_3$, or BCl_3. The gas may be injected as the polysilicon rod is deposited, or it may be injected at the molten ring during the float zone refining. Pill doping is accomplished by drilling a small hole in the top of the rod and inserting the dopant in the hole. If the dopant has a small segregation coefficient, most of it will be carried with the melt as it passes the length of the boule resulting in only a modest nonuniformity. Gallium and indium doping work well in this manner. Finally, for light n-type doping, float zone silicon can be doped through a process

known as transmutation doping. In this process, the boule is exposed to a high brightness neutron source. Approximately 3.1% of the silicon is the mass 30 isotope. Under neutron flux that isotope can be changed through the nuclear process [40]

$$.^{30}\text{Si}(\text{n}^{\cdot}, \gamma') \rightarrow .^{31}\text{Si} \xrightarrow{3.6 \text{ h}} - ^{31}\text{P} + \beta' \qquad (2.16)$$

Of course, a disadvantage of this process is that it is not suitable for forming p-type silicon.

2.7 Wafer Preparation and Specifications

After the boule has been grown the wafers must be made. The boule is first characterized for resistivity and crystal perfection. Then the seed and tail are cut off and the boule is mechanically trimmed to the proper diameter. The diameter at this point is slightly larger than the final wafer diameter, since additional etching will still be done. Flats are ground the entire length of the boule to denote the crystal orientation and the doping type and to provide a method for coarsely aligning the rotation of the wafer during subsequent photolithography steps. The largest flat, called the *primary*, is oriented perpendicular to the $\langle 110 \rangle$ direction. One or more minor flats will also be ground. Figure 2.23 shows the flat orientations for different wafer types.

After grinding the flats, the wafer is dipped in a chemical etchant to remove the damage caused by the mechanical grinding. Each manufacturer has a proprietary mixture, but they are usually based on the HF-HNO$_3$ system. After etching, the boule is sliced into wafers. This is a critical step in the process as it will determine the wafer bow and flatness. Typically, a steel blade impregnated with diamond particles is used. The wafers may then be edge rounded in another mechanical grinding process. It has been found that edge rounded wafers are less susceptible to defects created by me-

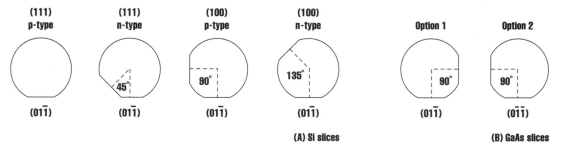

Figure 2-23   Standard flat orientations for different semiconductor wafers.

Table 2.3   Typical specifications for 2 in. GaAs wafers

	LEC	Vertical Gradient Freeze
Resistivity (Ω cm)	$>2 \times 10^7$	$>10^7$
Mobility (cm^2-V/s)	>4000	>5000
Carrier concentration (cm^{-3})	$<10^8$	—
Etch pit density (cm^{-2})	7×10^4 (typical)	$<10^3$
Flatness (μm)	<4	<5
Thickness (μm)	450	500
Cost ($/cm^2)	10	15

Table 2.4   Typical specifications for state of the art silicon wafer

Cleanliness (particle/cm^2)	<0.03
Oxygen concentration (cm^{-3})	Specified $+-3\%$
Carbon concentration (cm^{-3})	<1.5 × 10^{17}
Metal contaminants bulk (ppb)	<0.001
Grown in dislocations (cm^{-2})	<0.1
Oxidation induced stacking faults (cm^{-2})	<3
Diameter (mm)	≥150
Thickness (μm)	625 or 675
Bow (μm)	10
Global flatness (μm)	3
Cost ($/cm^2)	0.2

From Shimura [41].

chanical handling during the subsequent processing. It also prevents the pile up of liquids that will coat the wafer when using spin on processes.

Finally, a series of steps are performed to remove any residual mechanical damage and to prepare the wafers for device fabrication. First, the wafers are mechanically lapped in a slurry of alumina and glycerine, then etched as before to reduce the damage. Finally, one or both sides receive a polish in an electrochemical process involving a slurry of NaOH and very fine silica particles followed by a chemical clean to remove any residual contaminants. Tables 2.3 and 2.4 show typical wafer specifications.

2.8 Summary and Future Trends

This chapter reviewed some of the most basic properties of semiconductor materials. An introduction to phase diagrams was given and the idea of solid solubility was presented. Next, a basic description of point, line, area, and volume defects was given. In the second half of the chapter, crystal growth methods were presented. Czochralski growth is the most common method for preparing silicon wafers. Liquid encapsulated Czochralski is the dominant method for growing GaAs for electronic applications, but high defect densities are a problem. For optoelectronic applications the preferred method is the Bridgman technique.

Problems

1. A GaAs crystal is on a coordinate system such that an arsenic atom sits at the position 0,0,0 and a gallium atom sits at $a/4$, $a/4$, $a/4$. Find the x,y,z coordinates of the other three nearest neighbor gallium atoms to the arsenic atom at 0,0,0. What is the distance between these atoms? Compare the distance to the atomic radii given in Figure 1.6. Explain any differences (note: $a = 5.65$ Å).

2. For the crystal in Problem 1, find the three other nearest arsenic atoms to the cited gallium atom.

3. A mixture of 30% silicon and 70% germanium is heated to 1100°C. If the material is in thermal equilibrium, what is the concentration of silicon in the melt? At what temperature will the entire charge melt? The sample temperature is raised to 1300°C, then slowly cooled back down to 1100°C. What is the concentration of silicon in the solid?

4. During the molecular beam epitaxial growth (MBE) of GaAs layers, there is a strong tendency to form droplets of gallium on the surface of the wafer. These oval defects are a serious problem for MBE fabrication. Avoiding them requires a large arsenic to gallium flux ratio. Referring to the phase diagram, explain why thermodynamics would favor the formation of these droplets.

5. A process technology that has gained a great deal of interest in the last few years is rapid thermal annealing (see Chapter 6). The process allows wafers that have high concentrations of dopant atoms in them to be heated to high temperatures very rapidly, minimizing dopant diffusion. Explain the desirability of such a process based on the discussion of phase diagrams and solid solubility.

6. A silicon wafer that has 10^{16} cm^{-3} of boron is found to have a neutral vacancy concentration of 2×10^{10} cm^{-3} at some processing temperature and a singly ionized vacancy concentration of 10^9 cm^{-3} at the same temperature. Determine the temperature and the activation energy of the charged vacancy with respect to the intrinsic level, E_i.

7. Repeat Problem 6 if the wafer is doped with 2×10^{19} cm^{-3} of boron.

8. It is desired to form a denuded zone 10 μm thick using an 1100°C anneal. How long will the anneal need to be? What will the oxygen concentration be in the denuded zone?

9. If the temperature gradient in Czochralski silicon is 100°C/cm, calculate the maximum pull rate.

10. Assuming that the melt is at a uniform temperature and loses no heat except to the boule and that the boule is a perfect black body, set up the differential equations and boundary conditions in two dimensions that one would need to solve to find $T(r, z)$ in the boule.

11. A melt contains 0.1 atomic percent phosphorus in silicon. Assume the well mixed approximation and calculate the dopant concentration when 10% of the crystal is pulled, when 50% of the crystal is pulled, and when 90% of the crystal is pulled.

12. Why does Bridgman growth tend to have higher impurity concentrations than LEC?

References

1. *Binary Alloy Phase Diagrams*, 2nd ed., vol. 2, ASM Int., Materials Park, OH, 1990, p. 2001.
2. *Binary Alloy Phase Diagrams*, 2nd ed., vol. 1, ASM Int., Materials Park, OH, 1990, p. 283.
3. Ibid, p. 319.
4. F. A. Trumbore, ''Solid Solubilities of Impurities in Germanium and Silicon,'' Bell Systems Tech. J. **39**:210 (1960).
5. S. Ghandhi, *VLSI Fabrication Principles*, John Wiley and Sons, New York, 1983.
6. S. M. Hu, *Materials Res. Soc. Symp. Proc.* **59**:249 (1986).
7. S. Mahajan, G. A. Rozgonyi, and D. Brasen, ''A Model for the Formation of Stacking Faults in Silicon,'' *Appl. Phys. Lett.* **30**:73 (1977).
8. R. A. Craven, *Semiconductor Silicon 1981*, p. 254.
9. J. C. Mikkelson, Jr., S. J. Pearton, J. W. Corbett, and S. J. Pennycook, eds., *Oxygen, Carbon, Hydrogen and Nitrogen in Crystalline Silicone*, MRS, Pittsburg, 1986.
10. A. Bourret, J. Thibault-Desseaux, and D. N. Seidmann, ''Early Stages of Oxygen Segregation and Precipitation in Silicon,'' *J. Appl. Phys.* **55**:825 (1985).
11. K. Wada, N. Inoue, and K. Kohra, ''Diffusion Limited Growth of Oxygen Precipitation in Czochralski Silicon,'' *J. Cryst. Growth* **49**:749 (1980).
12. K. H. Yang, H. F. Kappert, and G. H. Schwuttke, ''Minority Carrier Lifetime in Annealed Silicon Crystals Containing Oxygen,'' *Phys. Stat. Sol.* **A50**:221 (1978).

13. D. C. Gupta and R. B. Swaroop, "Effects of Oxygen and Internal Gettering on Donor Formation," *Solid State Technol.*, p. 113 (Aug. 1984).

14. R. B. Swaroop, "Advances in Silicon Technology for the Semiconductor Industry," *Solid State Technol.,* p. 101 (July 1983).

15. D. Huber and J. Reffle, "Precipitation Process Design for Denuded Zone Formation in Czochralski Silicon Wafers," *Solid State Technol.,* p. 137 (Aug. 1983).

16. M. Stavola, J. R. Patel, L. C. Kimerling, and P. E. Freeland, "Diffusivity of Oxygen in Silicon at the Donor Formation Temperature," *Appl. Phys. Lett.* **42:**73 (1983).

17. G. K. Teal, "Single Crystals of Germanium and Silicon—Basic to the Transistor and the Integrated Circuit," *IEEE Trans. Electron Dev.* **ED-23:**621 (1976).

18. W. Zuhlehner and D. Huber, "Czochralski Grown Silicon," in *Crystals 8*, Springer-Verlag, Berlin 1982.

19. S. N. Rea, "Czochralski Silicon Pull Rates," *J. Crystal Growth* **54:**267 (1981).

20. W. C. Dash, "Evidence of Dislocation Jogs in Deformed Silicon," *J. Appl. Phys.* **29:**705 (1958).

21. W. C. Dash, "Silicon Crystals Free of Dislocations," *J. Appl. Phys.* **29:**736 (1958).

22. W. C. Dash, "Growth of Silicon Crystals Free from Dislocations," *J. Appl. Phys.* **30:**459 (1959).

23. T. Abe, "Crystal Fabrication," in *VLSI Electron.—Microstructure Sci.* **12,** N. G. Einspruch and H. Huff, eds., Academic Press, Orlando, 1985.

24. K. Hoshi, T. Suzuki, Y. Okubo, N. Isawa, *Electrochem. Soc. Ext. Abs.* St. Louis Meet., May 1980, p. 811.

25. K. Hoshi, N. Isawa, T. Suzuki, Y. Okubo, "Czochralski Silicon Crystals Grown in a Transverse Magnetic Field," *J. Electrochem. Soc.* **132:**693 (1985).

26. T. Suzuki, N. Izawa, Y. Okubo, and K. Hoshi, *Semiconductor Silicon 1981*, p. 90 (1981).

27. R. N. Thomas, H. M. Hobgood, P. S. Ravishankar, and T. T. Braggins, "Melt Growth of Large Diameter Semiconductors: Part I," *Solid State Technol.*, p. 163 (Apr. 1990).

28. S. Sze, *VLSI Technology*, McGraw-Hill, New York, 1988.

29. N. Kobayashi, "Convection in Melt Growth—Theory and Experiments," *Proc. 84th Meet. Cryst. Eng.* (Jap. Soc. Appl. Phys.), p. 1 (1984).

30. S. Miyazawa and F. Hyuga, "Proximity Effects of Dislocations on GaAs MESFET V_t," *IEEE Trans. Electron. Dev.*, **ED-3:**227 (1986).

31. R. Rumsby, R. M. Ware, B. Smith, M. Tyjberg, M. R. Brozel, and E. J. Foulkes, *Tech. Dig. GaAs IC Symp.*, Phoenix, p. 34 (1983).

32. H. Ehrenreich and J. P. Hirth, "Mechanism for Dislocation Density Reduction in GaAs Crystals by Indium Addition," *Appl. Phys. Lett.* **46:**668 (1985).

33. G. Jacob, *Proc. Semi-Insulating III-V Materials*, Shiva Publishing, Nantwick, U.K., 1982, p. 2.

34. H.-J. Sell, "Melt Growth Processes For Semiconductors," *Key Eng. Materials*, **58:**169 (1991).

35. T. P. Chen, T. S. Huang, L. J. Chen, and Y. D. Guo, "The Growth and Characterization of GaAs Single Crystals By a Modified Horizontal Bridgman Technique," *J. Cryst. Growth* **106:**367 (1990).

36. R. E. Kremer, D. Francomano, G. H. Beckhart, K. M. Burke, and T. Miller, *Materials Res. Soc. Symp. Proc.*, **144:**15 (1989).

37. C. E. Chang, V. F. S. Kip, and W. R. Wilcox, "Vertical Gradient Freeze Growth of GaAs and Naphthalene: Theory and Practice," *J. Cryst. Growth* **22:**247 (1974).

38. R. E. Kremer, D. Francomano, B. Freidenreich, H. Marshall, and K. M. Burke, in *Semi-Insulating Materials 1990*, A. G. Milnes and C. J. Miner, eds., Adam-Hilger 1990.
39. W. Keller and A. Mühlbauer, *Float-Zone Silicon*, Marcel Dekker, New York, 1981.
40. C. N. Klahr and M. S. Cohen, *Nucleonics* **22:**62 (1964).
41. F. Shimura, *Semiconductor Silicon Crystal Technology*, Academic Press, San Diego 1989.

Part II

Unit Processes I: Hot Processing and Ion Implantation

Until now only the semiconductor substrate itself has been discussed. This section will begin a discussion of unit processes. These are the individual process steps carried out in typical fabrication technologies. A later section will discuss how these unit processes are put together to form functional blocks (known as process modules) and ultimately, a technology. This first section on unit processes will discuss those processes related to dopant introduction and movement as well as the growth of thermal oxides. Since dopants are necessary for all types of devices, they are some of the first processes developed for fabrication. For the device to operate properly, the doped regions must have the right concentrations and sizes. This section will therefore first discuss the movement of dopant impurity atoms through diffusion. Early technologies used gaseous or liquid vapor sources in high temperature ovens to introduce the impurities into the wafer. As device size was reduced however, ion implantation was developed to better control the position and amount of impurity in the wafer. As standard implantation and high temperature annealing steps have begun to prove inadequate, special methods have also been developed to minimize the redistribution of impurities through diffusion. One of the most important of these is rapid thermal processing, which will be discussed in Chapter 6. Chapter 4 will also cover the thermal oxidation of silicon. Unlike the other chapters, only silicon oxidation will be discussed since this process is not used in compound semiconductor technologies.

> *To render a problem tractable, we must concentrate on its most important features.*[1]

---

[1]A. S. Grove, *Physics and Technology of Semiconductor Devices*, Wiley, New York, 1967.

Chapter 3

Diffusion

Every semiconductor device relies on the ability to fabricate well controlled, locally doped regions of the wafer. The chemical impurities must therefore first be introduced into some sections of the wafer, they must be active so that they contribute the desired carrier, and they must be the concentration desired by the device designer. Frequently, concentration profiles will be described. As shown in Figure 3.1, the impurity concentration or the carrier concentration is plotted on the vertical axis. The depth into the wafer is plotted on the horizontal axis. Typically, the y variable will vary over many orders of magnitude. For that reason, the concentration is normally given on a logarithmic scale. Recall that the number density of silicon is 5×10^{22} atoms/cm^3, so that typical impurity concentrations (10^{17} atoms/cm^3) for active device regions are doped as lightly as a few parts per million.

After the impurities are introduced they may redistribute in the wafer. This may be intentional or it may be a parasitic effect of some other thermal process. In either event, it must be controlled and monitored. The motion of impurity atoms in the wafer occurs primarily by diffusion, the net movement of a material that occurs near a concentration gradient as a result of random thermal motion. This chapter will introduce the differential equations that describe diffusion, solve the equation in closed form for two sets of boundary conditions, describe the physics involved in the diffusion coefficient, present models that describe the diffusion behavior of typical impurities in silicon and GaAs, and introduce SUPREM III, a program that calculates diffusion profiles under a wide variety of conditions.

3.1 Fick's Diffusion Equation in One Dimension

Any material that is free to move will experience a net redistribution in response to a concentration gradient. The movement will tend to reduce the size of the gradient. The source of this movement is the random motion of the material. Since the high concentration region has more impurity atoms, there is a net movement of impurities away from the concentration maximum. This is an effect not limited to impurities in semiconductors by any means. The basic laws of diffusion introduced here are used to describe heat transfer, the motion of electrons, gaseous impurities such as air pollution, and even animal population statistics.

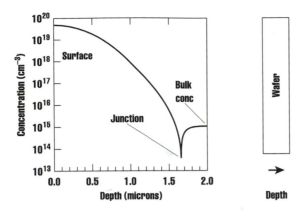

Figure 3-1   Typical concentration plot of impurities or carriers as a function of depth into the wafer. Note that these profiles are typically much less than 1% of the total wafer thickness.

Figure 3-2   An differential volume element in a bar of cross-sectional area A. J_1 and J_2 are the flux of an impurity into and out of the volume element.

The basic equation that describes diffusion is Fick's first law:

$$J = -D \, \frac{\partial C(x, t)}{\partial x} \tag{3.1}$$

where C is the impurity concentration, D is the coefficient of diffusion, and J is the net flux of material. The units of J are number per unit time per unit area. The negative sign expresses the fact that there is net movement in the direction of decreasing concentration.

While Fick's first law accurately describes the diffusion process, in this application there is no convenient way to measure the current density of the impurity. Unlike electrical current, the diffusing material is usually poorly confined and not easily detected. Therefore, a second expression for Fick's law has been developed that describes the same concept, but with more readily measurable quantities. In developing this expression it is easiest to start with a long bar of material with a uniform cross section A (Figure 3.2). Consider a small volume of length dx, then

$$\frac{J_2 - J_1}{\Delta x} = \frac{\partial J}{\partial x} \tag{3.2}$$

where J_2 is the flux leaving the volume and J_1 is the flux entering the volume. If these two fluxes are not the same, the concentration of the diffusing species in the volume must change. Recall that the number of impurities in this volume element is just the product of the concentration and the differential volume element $(A \cdot dx)$. Then the continuity equation is expressed as

$$-A(J_2 - J_1) = -A \, dx \, \frac{\partial J}{\partial x} = A \, dx \, \frac{\partial C}{\partial t}$$

or $\tag{3.3}$

$$\frac{\partial C(x, t)}{\partial t} = -\frac{\partial J}{\partial x}$$

From Fick's first law this can be written

$$\frac{\partial C(x,\ t)}{\partial t} = \frac{\partial}{\partial x}\left(D\,\frac{\partial C}{\partial x}\right)$$

(3.4)

Equation 3.4 is the most general representation of Fick's second law. If the diffusion coefficient is assumed to be independent of position, this reduces to the simpler form

$$\frac{\partial C(z,\ t)}{\partial t} = D\,\frac{\partial^2 C(z,\ t)}{\partial z^2}$$

(3.5)

where the position variable has been changed to z to suggest that the direction into the wafer (depth) is the one of primary interest. Finally, in three dimensions for an isotropic medium, Fick's second law is expressed as

$$\frac{\partial C}{\partial t} = D\nabla^2 C$$

(3.6)

One is left, then, with the solution of a second order differential equation. This requires the knowledge of at least two independent boundary conditions. The solution of the differential equation will be discussed later in the chapter. First we will focus on the application of Fick's second law to the problem of diffusion in semiconductors and discuss the factors that determine the diffusion coefficient D.

3.2 Atomistic Models of Diffusion

This section will discuss the physical mechanisms that determine the diffusion coefficient D. The argument begins by assuming that the crystal is isotropic. Without this approximation Fick's second law cannot be applied. While it allows us to derive solutions to Equation 3.6, it breaks down when the concentration of the dopant is large. Then the diffusivity becomes a function of the doping concentration and therefore depth.

In a crystal, the lattice sites are represented as the minima of parabolic potential wells. Each atom is at rest only in the limit of 0 K. At nonzero temperatures the atoms oscillate about their equilibrium position. Now insert an impurity atom into this crystal. The atom may sit between lattice sites in an interstitial position. Typically, atoms that do not bond readily with the matrix material are interstitial impurities. These impurities diffuse rapidly, but they do not directly contribute to doping. A second type of impurity is one that replaces the silicon atom on the lattice site. These substitutional impurities will be the primary focus of the chapter. Table 3.1 lists a number of silicon impurities and categorizes them into substitutional and interstitial.

Table 3.1	Silicon impurities	
	Substitutional	P, B, As, Al, Ga, Sb, Ge
	Interstitial	O, Au, Fe, Cu, Ni, Zn, Mg

Table 3.1 lists some impurities in silicon and characterizes them as tending to reside primarily on lattice sites (substitutional impurities) or in the spaces between the lattice sites (interstitial impurities).

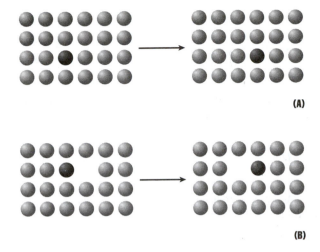

(A)

(B)

Figure 3-3   Diffusion of an impurity atom by direct exchange (b) and by vacancy exchange (a). The latter is much more likely due to the lower energy required.

Assume that the impurity atom in Figure 3.3a moves one lattice site to the right. By symmetry, no net energy was expended. Yet, in order for a substitutional atom to move in the crystal, it must have sufficient energy to surmount the potential well in which it rests. For the direct exchange shown in Figure 3.3a, at least six bonds must be broken for the host atom and the impurity to exchange positions. This is considerably easier, however, if the adjacent lattice site is occupied by a vacancy (Section 2.4). Then only three bonds must be broken. Vacancy exchange (Figure 3.3b) is therefore one of the dominant diffusion mechanisms for substitutional impurities.

Fair's vacancy model can be used to successfully describe the diffusion of many impurities in low and moderate concentrations at temperatures below 1000°C. It takes the simple picture from Figure 3.3b and adds one additional detail: vacancy charge. Recall that each atom in the silicon matrix must form a covalent bond with its four nearest neighbors in order to fill its valence shell. In the presence of a neutral vacancy, these four atoms are left with an unsatisfied shell. If the vacancy captures an electron, it satisfies one atom's valence, but becomes negatively charged. Similarly, an adjacent atom can lose an electron, and the vacancy appears to be positively charged.

Since vacancies are very dilute in a semiconductor at typical processing conditions, each of the possible charged states can be treated as independent entities. The diffusion coefficient then becomes the sum of all possible diffusion coefficients, weighted by their probability of existence. If we assume that the probability of charge capture is a constant, then the number of charged vacancies is proportional to the ratio $[C(z)/n_i]^j$, where $C(z)$ is the carrier concentration, n_i is the intrinsic carrier concentration, and j is the order of the charge state. Then, the most general expression for the total diffusion coefficient in the vacancy model is given by

$$D = D_i + \frac{n}{n_i}D^- + \left[\frac{n}{n_i}\right]^2 D^{2-} + \left[\frac{n}{n_i}\right]^3 D^{3-} + \left[\frac{n}{n_i}\right]^4 D^{4-}$$
$$+ \frac{p}{n_i}D^+ + \left[\frac{p}{n_i}\right]^2 D^{2+} + \left[\frac{p}{n_i}\right]^3 D^{3+} + \left[\frac{p}{n_i}\right]^4 D^{4+}$$

(3.7)

The intrinsic carrier concentration for silicon can be found from [1]

$$n_i(\text{cm}^{-3}) = n_{io}T(\text{K})^{3/2}e^{E_g/2kT}$$

(3.8)

where $n_{io} = 1.4*10^{15}$ cm^{-3} for silicon and $n_{io} = 4.2*10^{14}$ cm^{-3} for GaAs. The bandgap can be determined by

$$E_g = E_{g0} - \frac{\alpha T(\text{C})^2}{\beta + T(\text{C})}$$

(3.9)

where E_{g0}, α, and β are 1.17 eV, 0.000473 eV-K, and 636 K for silicon, and 1.52 eV, 0.000541 eV-K, and 204 K for GaAs. The intrinsic carrier concentration for silicon and GaAs are shown in

Figure 3-4 Intrinsic carrier concentration of silicon and GaAs as a function of temperature.

Figure 3.4. In heavily doped silicon the bandgap is also reduced by the bandgap narrowing effect

$$\Delta E_g = -7.1 \times 10^{-10}\text{eV} \sqrt{\frac{n_i(\text{cm}^{-3})}{T(\text{K})}} \quad \textbf{(3.10)}$$

For heavily doped diffusions ($C \gg n_i$) the electron or hole concentration is just the impurity concentration. For low concentration diffusions ($C(z) \ll n_i$) $n \approx n_i$. For substrates with excess free electrons, the positive terms in Equation 3.7 can be neglected and for substrates with excess free holes the negative terms can be neglected. Furthermore, the contributions of the third and fourth power terms are typically very small and are almost universally neglected. If charged vacancies must be considered, the electron or hole concentration, and therefore the diffusivity, is a function of position. In that case the simple form of Equation 3.5 cannot be used. Instead Equation 3.4 must be solved numerically.

If very dilute impurity profiles are measured before and after diffusion, a diffusion coefficient can be determined. If this procedure is repeated for several temperatures and the logarithm of the diffusivity is plotted against reciprocal temperature in Kelvin, an Arrhenius plot will result. The intrinsic diffusivity is of the form

$$D_i = D_{io}e^{-E_{ia}/kT} \quad \textbf{(3.11)}$$

where E_{ia} is the activation energy of the intrinsic diffusivity, and D_{io} is a nearly temperature independent term that depends on the vibrational frequency and geometry of the lattice. Table 3.2 summarizes the activation energies and pre-exponentials for some common dopants.

Table 3.2   Diffusion coefficients of common impurities in silicon and gallium arsenide

		Donors				Acceptors			
		$D_o^=$	$E_a^=$	D_o^-	E_a^-	D_{io}	E_{ia}	D_o^+	E_a^+
As in Si	D			12.0	4.05	0.066	3.44		
P in Si	D	44.0	4.37	4.4	4.0	3.9	3.66		
Sb in Si	D			15.0	4.08	0.21	3.65		
B in Si	A					0.037	3.46	0.41	3.46
Al in Si	A					1.39	3.41	2480	4.2
Ga in Si	A					0.37	3.39	28.5	3.92
S in GaAs	D					0.019	2.6		
Se in GaAs	D					3000	4.16		
Be in GaAs	A					7e − 6	1.2		
Ga in GaAs	I					0.1	3.2		
As in GaAs	I					0.7	5.6		

From Runyan and Bean [2] and references quoted therein. Donors are labeled with a ''D'', acceptors with an ''A'', and self-interstitials with an ''I''. All pre-exponentials are in cm²/sec and the activation energies are in electron volts.

Example 3.1

Calculate the diffusivity of arsenic in silicon at 1000°C if the concentration of arsenic is much less than the intrinsic carrier concentration and again if the arsenic concentration is 1×10^{19} cm^{-3}.

For T = 1273 K, kT = 0.110 eV. Then

$$D_i = 0.066 e^{\frac{-3.44}{0.110}} = 1.6 \times 10^{-15} \frac{\text{cm}^2}{\text{sec}}$$

According to Figure 3.4 and Table 3.1 the negative one vacancy must be considered for arsenic diffusion. Then

$$D_- = 12 e^{\frac{-4.05}{0.110}} = 1.2 \times 10^{-15} \frac{\text{cm}^2}{\text{sec}}$$

Recall from elementary semiconductor physics,

$$n = \frac{N_D}{2} + \sqrt{\left[\frac{N_D}{2}\right]^2 + n_I^2}$$

For $N_D \ll n_i$, $n = n_i$ and

$$D = D_i + D_- = 2.8 \times 10^{-15} \frac{\text{cm}^2}{\text{sec}}$$

Referring to Figure 3.3, at 1000°C, $n_i = 2 \times 10^{18}$ cm^{-3}. Then if N_D is 1×10^{19} cm^{-3}, $n = 1.04 \times 10^{19}$ cm^{-3}, and

$$D = 1.6 \times 10^{-15} + \frac{1.04 \times 10^{19}}{2 \times 10^{18}} 1.2 \times 10^{-15} = 7.8 \times 10^{-15} \frac{\text{cm}^2}{\text{sec}}$$

Figure 3-5 In interstitialcy diffusion an interstitial silicon atom displaces a substitutional impurity, driving it to an interstitial site where it diffuses some distance before it returns to a substitutional site.

A second important mechanism for diffusion in silicon relies on the presence of silicon self-interstitials. It is referred to as the interstitialcy method (Figure 3.5). In this case, an interstitial silicon atom displaces the impurity, driving it into an interstitial site. From there it moves rapidly to another lattice site, where the silicon atom is removed and becomes an interstitial. Interstitialcy is not believed to occur unless vacancy diffusion does as well. Boron and phosphorus are two impurities that tend to diffuse by both mechanisms. Either may be dominant depending on the process conditions. To find the effective diffusivity for these impurities one must add the contributions from both methods.

Figure 3-6   The kick-out (left) and Frank-Turnbull mechanisms (right).

Some impurities tend to diffuse rapidly through interstitial spaces. There are two mechanisms by which these impurities may return to the lattice; they are summarized in Figure 3.6. In the Frank-Turnbull method the interstitial impurity is captured by a vacancy. In the kick-out mechanism the impurity replaces a lattice atom. These two mechanisms are distinct from the interstitialcy method in that they do not require the presence of self-interstitials to drive the process. Impurities that tend to diffuse through this mechanism are characterized by a low interstitial solubility to substitutional solubility ratio.

3.3 Analytic Solutions of Fick's Law

Returning to the assumption of a constant diffusivity, Fick's second law is a simple differential equation that can be solved subject to various boundary conditions. In practice the dopant profiles that are of interest are sufficiently complex, and the assumption that the coefficient of diffusion is constant is sufficiently questionable that Equation 3.5 must be solved numerically. There are two sets of boundary conditions however, for which exact solutions can be derived. These solutions can be used to develop a basic understanding of diffusion processes and as rough approximations of actual profiles.

The first type of solution of Fick's law occurs when the source is fixed at the surface for all times greater than zero. Called a predeposition diffusion, the boundary conditions are

$$C(z, 0) = 0$$
$$C(0, t) = C_s \tag{3.12}$$
$$C(\infty, t) = 0$$

The solution for these conditions is given by

$$C(z, t) = C_s \, \text{erfc} \left(\frac{z}{2\sqrt{Dt}} \right), \qquad t > 0 \tag{3.13}$$

In this equation, C_s is the fixed surface concentration and erfc is a function known as the complimentary error function. The complimentary error function is tabulated in an appendix and in many math handbooks for various values. $\sqrt{Dt}$ is a common feature in the solution of diffusion problems and is known as the characteristic diffusion length.

The dose of the predeposition diffusion varies with the time of the diffusion. To obtain the dose, the profile can be integrated as

$$Q_T(t) = \int_0^\infty C(z, t) \, dz$$
$$= \frac{2}{\sqrt{\pi}} C(0, t) \sqrt{Dt} \tag{3.14}$$

This dose is measured in units of impurities per unit area, typically per cm². Since the depth of the profile is typically less than 1 μm (10^{-4} cm), a dose of 10^{15} cm^{-2} will produce a large volume

concentration ($>10^{19}$ cm^{-3}). Since the surface concentration is fixed for a predeposition diffusion, the total dose increases as the square root of the time.

The second type of solution of Fick's law is called the drive in diffusion. In this case, an initial amount of impurity Q_T is introduced into the wafer and diffused subject to the boundary condition that Q_T is fixed. If the diffusion length is much larger than the width ˆ the initial profile, the initial profile can be approximated as a delta function. Then the boundary conditions are

$$C(z, 0) = \quad 0, \quad z \neq 0$$

$$\frac{dC(0, t)}{dz} = \quad 0$$

$$C(\infty, t) = \quad 0 \tag{3.15}$$

$$\int_0^\infty C(z, t)dz = \quad Q_T = \text{constant}$$

The solution to Fick's second law for these conditions is a Gaussian centered at $z = 0$:

$$C(z, t) = \frac{Q_T}{\sqrt{\pi Dt}} e^{-z^2/4Dt}, \quad t > 0 \tag{3.16}$$

The surface concentration C_s, decreases with time as

$$C_s = C(0, t) = \frac{Q_T}{\sqrt{\pi Dt}} \tag{3.17}$$

The reader can easily demonstrate that at $x = 0$, dC/dx is zero for all $t \neq 0$. Figure 3.7 shows a plot of the predeposition and the drive in diffusions with $\sqrt{Dt}$ as a parameter.

One classic type of problem that uses these two diffusions is a predeposition followed by a drive in. Recall that one of the boundary conditions for the drive in was that the initial impurity concentration was zero everywhere except at the surface. In practice, the drive in is a good approximation as long as

$$\sqrt{Dt}_{\text{predep}} \ll \sqrt{Dt}_{\text{drive in}} \tag{3.18}$$

Now assume that boron is diffusing in a silicon wafer that has a uniform concentration of phosphorus, C_B. Also assume that $C_s \gg C_B$. Then, a depth will exist at which the concentration of boron exactly equals that of the background concentration. Since boron is a p-type dopant and phosphorus is an n-type dopant in silicon, a p-n junction will exist at this depth, which is called the junction depth x_j. If the diffusion is a drive in diffusion, one can show from Equation 3.16:

$$x_j = \sqrt{4Dt \ln\left[\frac{Q_T}{C_{\text{sub}}\sqrt{\pi Dt}}\right]} \tag{3.19}$$

If the diffusion was a predeposition diffusion, from Equation 3.13,

$$x_j = 2\sqrt{Dt} \, \text{erfc}^{-1}\left[\frac{C_{\text{sub}}}{C_s}\right] \tag{3.20}$$

Figure 3-7   Concentration as a function of depth for (a) predeposition and (b) drive in diffusions for several values of the characteristic diffusion length.

3.4 Corrections to Simple Theory

Substitutional impurities are almost completely ionized at and above room temperature. For that reason, if an electric field exists in the substrate, the total current will have both drift and diffusion components. Recall Ohm's law:

$$J = -D\frac{dC}{dz} + \mu C \mathcal{E}_x = D\left[-\frac{dC}{dz} + C\frac{q}{kT}\mathcal{E}_x\right] \tag{3.21}$$

where μ is the mobility, $\mathcal{E}$ is the electric field, and the Einstein relationship between mobility and diffusivity has been invoked. If we assume that the carrier concentration is completely determined by the doping profile, the field can be calculated directly. In a semiconductor the electric field is given by

$$\mathcal{E} = -\eta\, kT\, \frac{1}{C}\frac{dC}{dz} \tag{3.22}$$

where η is a screening factor that varies from 0 to 1. Inserting the result into Equation 3.21,

$$J = -D(1 + \eta)\frac{dC}{dz} \tag{3.23}$$

For high concentration doping ($C_{\text{doping}} \gg n_i$, C_{sub}) the profile's own electric field will enhance the movement of the impurity. This equation is identical to Fick's first law, except that the diffusivity

is multiplied by a field enhancement term $(1 + \eta)$. Furthermore, the diffusion of an impurity may be affected by the presence of other impurities through field enhancement terms.

By comparing inert, oxidizing, and nitridizing dopant diffusion experiments, some light has been shed on the processes by which various impurities diffuse [3]. The diffusivity of impurities in a semiconductor depends on the concentration of vacancies. When a semiconductor is oxidized, a high concentration of excess interstitials are generated near the oxide/semiconductor interface [4],[5]. The excess concentration decays with depth due to vacancy interstitial recombination. Near the surface these interstitials increase the diffusivity for boron and phosphorus. It is believed, therefore, that these impurities diffuse primarily by the interstitialcy process. Arsenic diffusivity is found to decrease under oxidizing conditions. An excess interstitial concentration is expected to depress the local vacancy concentration. Arsenic is therefore believed to diffuse primarily via the vacancy mechanism, at least when in oxidizing conditions. Experiments can also be done during thermal nitridization. Although this process is rarely done in practice, it injects a high concentration of vacancies into the substrate. These results have confirmed the conclusions of the earlier results by showing the opposite trends of oxidation.

To be able to predict a dopant profile after a diffusion in an oxidizing ambient, we need to find the diffusivity under these conditions. Note that, as with high concentration diffusion, the diffusivity is now a function of position. Strictly speaking, therefore, Equation 3.5 is no longer valid. To first order, however, dD/dz is small compared to dC/dz and can be ignored. Since the concentration of excess interstitials depends on the oxidation and recombination rates, the diffusivity must be a function of the oxidation rate. It has been shown that for diffusion under oxidizing conditions,

$$D = D_i + \Delta D \tag{3.24}$$

where [6]

$$\Delta D = \alpha \left[\frac{dt_{ox}}{dt} \right]^n \tag{3.25}$$

is the diffusivity enhancement or retardation due to oxidation. The exponent n has been found experimentally to be between 0.3 and 0.6. The term α may be positive (for oxidation enhanced diffusion) or negative (for oxidation retarded diffusion).

Thus far only diffusion in one dimension has been discussed. It is assumed that the dopant concentration is uniform across the wafer and therefore no net diffusion occurs laterally. This is not true near the edge of diffused patterns such as occurs under the gate of a MOSFET when diffusing the source and drain. Diffusion will occur both vertically and laterally (i.e., under the gate). The lateral diffusion is generally assumed to proceed uniformly. This leads to an effective depletion of dopant near the edge of the feature, which reduces the junction depth. Experimental verification of lateral diffused profiles is extremely difficult since dilute concentrations must be detected in very small volumes. For that reason detailed models of lateral diffusion are not as well developed.

3.5 Diffusion Coefficients for Common Dopants

In this section the vacancy and interstitialcy models of diffusion will be applied to describe the diffusion coefficients that are appropriate for each of the commonly used impurities. The diffusivity of boron in silicon has been measured over a wide range of concentrations and temperatures [7].

Referring to Fair's vacancy model the data can be fit with the intrinsic diffusivity and only the first positive vacancy term up to concentrations of 10^{20} cm^{-3}. Above this concentration not all of the boron can be accommodated into the lattice and must reside on interstitial sites or in cluster precipitates. The diffusivity of boron in this concentration range is found to be sharply reduced [8]. Figure 3.8 shows a typical diffused high concentration boron profile.

Arsenic diffuses in silicon via neutral and single negatively charged vacancies. The diffusivity of arsenic in silicon is relatively low so that arsenic is often chosen as the n-type dopant when minimal dopant redistribution is desired. Typical examples are the source/drain diffusion of submicron NMOS transistors and the emitter of many bipolar transistors. At low and medium concentrations the diffusivity is well described by simple intrinsic diffusion.

At high concentrations field enhancement is also evident. As a result

$$D_{As} \approx 2 \frac{n}{n_i} (D_i)_{As} \qquad (3.26)$$

The effect of this enhancement is to produce a very steep profile (Figure 3.9).

Arsenic also tends to form interstitial clusters at concentrations in excess of 10^{20} atoms/cm^3 that resist electrical thermal activation. This effect also tends to flatten the top of high concentration profiles. During a high temperature anneal the cluster concentration moves toward a thermal equilibrium with substitutional arsenic. The maximum carrier concentration is given by

$$C_{max} = 1.9 \times 10^{22} \text{ cm}^{-3} e^{-0.453/kT} \qquad (3.27)$$

The relative concentrations are therefore determined by the anneal temperature. The clusters are not believed to be mobile. Instead, the arsenic atoms move individually.

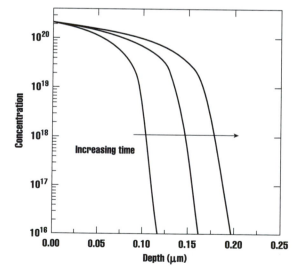

Figure 3-8 Typical profile for a high concentration boron diffusion.

Figure 3-9 Typical profile for a high concentration arsenic diffusion.

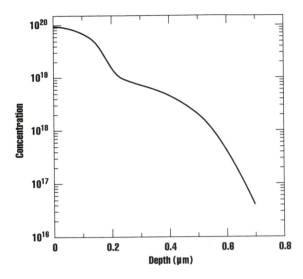

Figure 3-10 Typical profile for a high concentration phosphorus diffusion.

Phosphorus diffuses much more rapidly than arsenic. Its application to VLSI technology is limited primarily to wells (or tubs) and isolation, although its higher diffusivity is helpful in reducing the peak electric field in MOS transistors (see Chapter 16). For completeness however, high concentration phosphorus profiles will be described. Figure 3.10 shows a high concentration phosphorus diffusion. The profile consists of three regions: the high concentration region, the low concentration region, and the transition or "kink" region [9]. Near the surface the concentration is nearly constant. The diffusivity in this region is believed to have two components: D_i, the normal phosphorus atom neutral vacancy exchange; and $D_i^=$, which corresponds to positively charged phosphorus ions paired with double negatively charged vacancies to form single negatively charged pairs $(PV)^-$:

$$D_{Ph} = D_i + D_i^- \left[\frac{n}{n_i}\right]^2 \qquad (3.28)$$

Near the kink region the electron concentration falls sharply. Most of the ion vacancy pairs dissociate, and the diffusion of unpaired phosphorus ions continues into the substrate. The dissociation of the $(PV)^-$ pairs causes an excess vacancy concentration, which also increases the diffusivity in the tail region. It has been shown that the diffusivity in the tail region is given by

$$D_{tail} = D_i + D_i^- \frac{C_s^3}{n_e^2 n_i} \left[1 + \exp\left[\frac{0.3eV}{kT}\right]\right] \qquad (3.29)$$

where C_s is the surface concentration, and n_e is the electron concentration in the transition region

$$n_e = 4.65 \times 10^{21} \ cm^{-3} \ \exp\left[\frac{0.39}{kT}\right] \qquad (3.30)$$

GaAs must be heavily doped for certain optoelectronic devices and for forming ohmic contacts. Unintentional dopant redistribution in GaAs, however, tends to be a less severe problem than in silicon, since processing of GaAs normally is done at lower temperatures than silicon. Diffusion in GaAs is significantly more complicated than in silicon. It depends not only on the charge state of the vacancy and interstitial, but whether it is a gallium vacancy (V_{Ga}) or an arsenic vacancy (V_{As}). The intrinsic diffusion coefficients for several common GaAs dopants are given in Table 3.1. This section will present the diffusion mechanisms of two of the primary GaAs dopants: zinc and silicon.

Zinc is a commonly used p-type dopant in GaAs technology. Figure 3.11 shows a series of predeposition diffusions of zinc into GaAs at 600°C for various times [10]. The diffusions show a broad, flat plateau and a sharp exponential tail. The plateau is limited by the solid solubility. An early model for the diffusion of zinc developed by Weisberg [11] has two components. The first is the standard vacancy method of Fair as described earlier in this chapter. The second, which proceeds

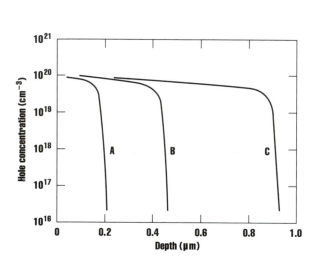

Figure 3-11 Predeposition diffusions of zinc into GaAs at 600°C for 5, 20, and 80 minutes *(after Field and Ghandi, used with permission, Electrochemical Society).*

Figure 3-12 Comparison of the multiple charge model for zinc diffusion and experimental results *(after Kahen, used with permission, Materials Research Society).*

much faster, is the Frank-Turnbull or substitutional interstitial (SI) process (Figure 3.6). In this process, zinc atoms exist in two forms: a small concentration of positively charged interstitial ions Zn^+, which diffuses rapidly, and a substitutional ion Zn^- that diffuses much more slowly by neutral vacancy exchange. While multiple positive states may participate in the interstitial diffusion, it is believed to be dominated by the singly charged ion.

In this model the Zn^+ diffuses rapidly until it encounters a V_{Ga}. It becomes trapped at this location, gives up two holes, and becomes Zn^-, which markedly reduces its diffusivity. At high concentrations this process results in a concentration dependent diffusivity of the form

$$D_{Zn} \approx AC_s^2 \tag{3.31}$$

where C_s is the concentration of substitutional zinc and A is a constant [12]. Thus the diffusion coefficient remains large until the concentration begins to drop. At this point, the diffusivity also drops, further sharpening the profile edge.

This model fails to describe a kink behavior often seen in the tail of the zinc profile. More recently, Kahen [13] modified the model to include the possibility of multiple charge states associated with the gallium vacancy. The author also included the possibility of the formation of a paired state between the substitutional negatively charged zinc ion and the interstitial positively charged zinc ion. The mechanism greatly reduces the diffusivity of the zinc. The model closely follows the kink behavior over a wide range of process conditions (see Figure 3.12).

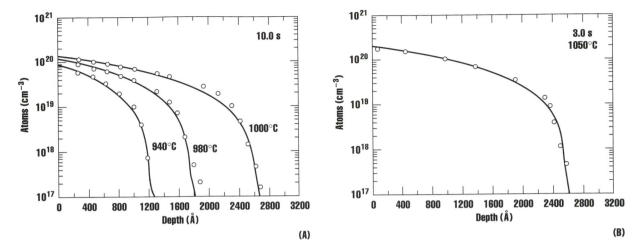

Figure 3-13 Diffusion of silicon in GaAs *(after Kahen et al., used with permission, Materials Research Society).*

One of the most common n-type dopants in GaAs is silicon (Si). Silicon is a group IV element and may be either a p-type or an n-type dopant, depending on the lattice site that it occupies. The carrier concentration then is the difference between the concentrations of silicon that reside on the two types of lattice sites. When this difference is small compared to the individual components, the semiconductor is said to be highly compensated. At high concentrations the diffusivity of silicon is concentration dependent [14],[15]. Greiner and Gibbons [14] proposed an Si_{Ga^-}–Si_{As} pair diffusion model in which a pair of impurities on adjacent (opposite type) lattice sites exchanges position with a pair of adjacent vacancies. This is believed to occur in a two step process. In this type of process the semiconductor can be highly compensated since the dopant atoms occupy both sites. The diffusivity then increases linearly with the dopant concentration.

The pair diffusion model has not been able to explain the diffusion results observed under rapid thermal annealing conditions (see Chapter 6). Furthermore, the model has been unable to reproduce the effects of substrate doping on the diffused profile. A more complete steady state model that takes into account charge effect, has recently been proposed by Yu et al. [16]. In this model, Si_{Ga^+} diffuses by exchange with an uncharged or triply negatively charged gallium vacancy. The diffusion mechanism of the Si_{As^-} atom is unknown but is assumed to be independent of the gallium atom. Coupled pairs of Si_{Ga}-Si_{As} are assumed to be immobile. Kahen et al. [17] have applied an Si_{Ga^+}/V_{Ga^-} model to fit experimental data, also with good results, although background doping was not considered. Figure 3.13 shows their results. In agreement with this model, Sudandi and Matsumoto showed that the diffusivity of silicon decreases sharply in Ga-rich LEC material where the concentration of gallium vacancies is expected to be low [18].

3.6 Analysis of Diffused Profiles

Once an impurity has diffused, it is desirable to be able to measure the impurity concentration as a function of depth and position. There are many techniques for obtaining depth profiles, but lateral profiles are quite difficult to obtain with sufficient resolution to be useful. The simplest technique for

obtaining information about the profile is to measure its sheet resistance. Instead of a concentration profile however, only a single number

$$R_s = [q \int \mu(C) C_e(z) \, dz]^{-1}$$

(3.32)

is obtained, where $C_e(z)$ is the carrier concentration and $\mu(C)$ is the concentration dependent mobility. R_s is called the sheet resistance. As discussed in Chapter 1, it is quoted in units of ohms per square ($\Omega/\square$). Sheet resistance measurements are quick and easy to perform and give the process engineer useful information, particularly if a standard or target sheet resistance is known.

The sheet resistance can be measured in a variety of ways. The simplest is the use of a four point probe (Figure 3.14a). Four point probes are available in several geometries; the most common is collinear. In this case, current is passed between the two outer probes and the voltage is measured across the inner pair. The sheet resistance is found by measuring the ratio of the voltage drop to the forced current. The result is multiplied by a geometric correction factor that depends on the probe geometry and the ratio of the probe spacing to the thickness of the diffusion [19],[20]. For collinear probes, where the probe spacing is much larger than the junction depth, the factor is 4.5325 [21]. For this method to be useful in characterizing diffused profiles in semiconductors, the underlying substrate must be insulating or must be of much higher resistivity than the layer to be measured, or the layer to be measured must form a reverse biased diode with the substrate. In the latter case, very shallow junctions can be penetrated with the probe if excessive force is used. Furthermore, the sheet resistance measurement will include the effect of the depleted region near the junction.

A second technique for measuring the sheet resistance is the Van der Pauw method [22]. The measurement is again done by contacting the edge of a sample in four places. A current is forced between one pair of adjacent contacts and the voltage is measured across the other pair (Figure 3.14b). To improve the accuracy, the probe connections are rotated 90° and the measurement repeated three times. Then the average resistance is calculated:

$$R = \frac{1}{4} \left[\frac{V_{12}}{I_{34}} + \frac{V_{23}}{I_{41}} + \frac{V_{34}}{I_{12}} + \frac{V_{41}}{I_{23}} \right]$$

(3.33)

(A)

(B)

Figure 3-14   The four point probe (a) and Van der Pauw (b) methods for determining the resistivity of a sample.

Groove wafer

After selective etch

Figure 3-15 In junction staining, a cylinder is used to groove the wafer. A doping sensitive etch then removes part of the top layer. The junction depth can be found from the known diameter of the cylinder and the measured width of the lower abraded groove.

and

$$R_s = \frac{\pi}{\ln(2)} F(Q)R \qquad (3.34)$$

where $F(Q)$ is a correction factor that depends on the probe geometry. For a square, $F(Q) = 1$. In this technique, one must be careful to correctly measure the geometry. If a square sample is assumed, the contacts must be made on the sides of the sample [23]. This can be done by breaking off a piece of the wafer in the shape of a square and making ohmic contacts; however, it is more commonly done by photolithographically patterning a Van der Pauw structure using oxide or junction isolation to restrict the diffusion geometry.

The sheet carrier concentration can also be combined with a measurement of the junction depth to provide a more complete description of the diffused profile. This is typically accomplished by beveling the wafer (see spreading resistance measurements) or by mechanically abrading a groove of known diameter in the surface of the wafer. The wafer is then immersed in a stain solution (Figure 3.15). The etch rate of the solution depends on the carrier type and concentration. P-type silicon etched in a $1:3:10$ mixture of hydrofluoric acid (HF), nitric acid (HNO_3), and acetic acid ($C_2H_4O_2$), will turn dark. A stain used for GaAs is a $1:1:10$ mixture of HF, hydrogen peroxide (H_2O_2), and water. In this case, the sample must be exposed to a bright light. The width of the stained region is measured after staining using an optical microscope with calibrated eyepieces. The junction depth can be determined from the known geometry of the bevel or groove. Limitations of accuracy and reproducibility prevent staining from being useful for junctions less than 1 μm deep. As a result, this method is much less popular than it once was.

A limitation of sheet resistance methods is that some knowledge of mobility is required to obtain even an integrated carrier concentration. The Hall effect (Figure 3.16) can be used to directly measure the integrated carrier concentration. In this measurement, a current flow in the diffused layer is also subjected to a magnetic field that is perpendicular to the flow. If one assumes that only holes are present in the diffusion, there will be a Lorentz force on each hole

$$\overline{F} = q\overline{v}x\overline{B} \qquad (3.35)$$

Holes will be deflected by this force until the component of the field perpendicular to both the current flow and the field is large enough to equal the Lorentz force:

$$\mathscr{E}_y = v_x B_z \qquad (3.36)$$

The establishment of this electric field is known as the Hall effect, and the resultant voltage is the Hall voltage,

$$V_h = v_x B_z w \qquad (3.37)$$

Figure 3-16 The Hall effect is able to simultaneously measure the carrier type, mobility, and sheet concentration.

where w is the width of the diffusion. The drift velocity of the hole can be related to the current by

$$v_x = \frac{I_x}{qwx_j\overline{C_e}}$$

(3.38)

where

$$\overline{C_e} = \frac{1}{x_j \displaystyle\int_0^{x_j} C_e\, dx}$$

(3.39)

Solving for the integrated carrier concentration,

$$\int_0^{x_j} C_e\, dx = x_j\overline{C_e} = \frac{I_x B_x}{qV_h}$$

(3.40)

This result can also be used to find the average Hall mobility of the sample if the four contacts are also used for a Van der Pauw measurement. Then the average Hall mobility is given by

$$\overline{\mu} = \frac{1}{qx_j\overline{C_e}R_s}$$

(3.41)

For a diffused profile, this average mobility is of little interest; however, the Hall mobility is an often-quoted figure of merit for the quality of an epitaxial layer with a nominally uniform concentration.

All of the preceding techniques had a serious limitation: They only provide information about the integral of the profile. Several methods can be used to measure the carrier concentration as a function of depth. The first uses the capacitance-voltage characteristic of a diode (pn junction or Schottky) or an MOS capacitor. While the MOS technique is widely used, it is more difficult to derive and requires a low interface state density at the Si/SiO$_2$ interface to be reliable. The diode method will be assumed here, but both techniques are very similar.

Assume that the structure can be described in the depletion approximation. For a one-sided step junction or a Schottky contact the depletion width is given by

$$W = \sqrt{\frac{2\varepsilon(V_{bi} + V)}{qN_{sub}}}$$

(3.42)

where ε is the dielectric constant of the semiconductor, V_{bi} is the built in voltage of the diode, N_{sub} is the substrate doping concentration and V is the externally applied voltage. The capacitance of the diode is

$$C = \frac{A\varepsilon}{W} = \sqrt{\frac{A^2 q\varepsilon N_{sub}}{2(V + V_{bi})}}$$

(3.43)

Differentiating with respect to voltage and solving for the impurity concentration,

$$N_{sub}(z) = \frac{8(V + V_{bi})^3}{A^2 q\varepsilon}\left[\frac{dC(z)}{dV}\right]^2$$

(3.44)

To measure the substrate doping then, one needs to measure the capacitance in depletion as a function of the applied voltage and find the first derivative. The doping concentration as a function of voltage for each data point can be determined using Equation 3.45, and the depth corresponding to that point may be found using Equation 3.43.

The C-V method has two significant limitations. The first is that impurity concentrations in silicon above $1 \cdot 10^{18}$ cm^{-3} cannot be measured. At these concentrations, the semiconductor becomes degenerate and acts more like a metal than a semiconductor. The second is that the depletion edges are not abrupt. Instead, they are graded over a few Debye lengths, where

$$L_D = \sqrt{\frac{\varepsilon \, kT}{q^2 C_{sub}}} \qquad (3.45)$$

Consequently, abrupt doping profiles are not well described by their carrier profiles.

The next approach to electrically profiling the carrier concentration, called *spreading resistance profilometry*, uses the dependence of the current crowding near a point contact on the local carrier concentration. Figure 3.17 shows a typical spreading resistance measurement. The sample is first beveled at a shallow angle by grinding and lapping. It is mounted in a chuck and a pair of probes are placed in contact with the surface with a predetermined force. Thin asperities of the probe are believed to penetrate the semiconductor surface to a depth of order tens of angstroms. The current is crowded into the asperity, leading to a finite resistance between the probes. If this resistance is compared to calibration standard of known concentration, methods have been developed [24] to deconvolute the resistivity to give carrier profiles. The technique can be used to measure profiles ranging from 10^{13} to 10^{21} cm^{-3}.

There are three primary limitations to spreading resistance measurements. The measurement depends critically on the reproducibility of the point contact. The probe tips must be carefully conditioned [25] and calibration standards run often. The measurement accuracy therefore depends on the level of experience of the operator. Commercial laboratories claim an accuracy of a factor of two over most ranges of doping. The second limitation of spreading resistance measurements is near surface measurement. It is difficult to accurately produce a very flat shallow angle ($<0.5°$). Furthermore, it is often a matter of judgment as to where the surface begins unless there is an insulator on the surface. Measurements below 500 Å are not generally regarded as reliable. The final limitation to spreading resistance measurements is the materials to which it can be applied. The sample is assumed to be similar (i.e., having the same dependence of mobility on doping) as the calibration standards. This is not always the case, particularly for compound semiconductors. Furthermore, GaAs undergoes significant band bending near the surface. As a result, spreading resistance is not generally done in this system.

Various approaches to electrochemical profiling have also been established [26]. These involve electrochemically etching the wafer, measuring the capacitance and/or resistivity, and then repeating the process. By measuring the change in the capacitance or the conductivity as a function of etch time, a profile of the carrier concentration can be obtained. These methods are used widely for III–V electrical depth profiling.

A variety of techniques exist to measure chemical concentrations in thin films. To be useful in evaluating diffused impurities in semiconductors, however, the technique must be sensitive to concentrations of at least 1 ppm and ideally 1 ppb. This requirement leaves only one popular approach: Secondary Ion Mass Spectroscopy (SIMS).

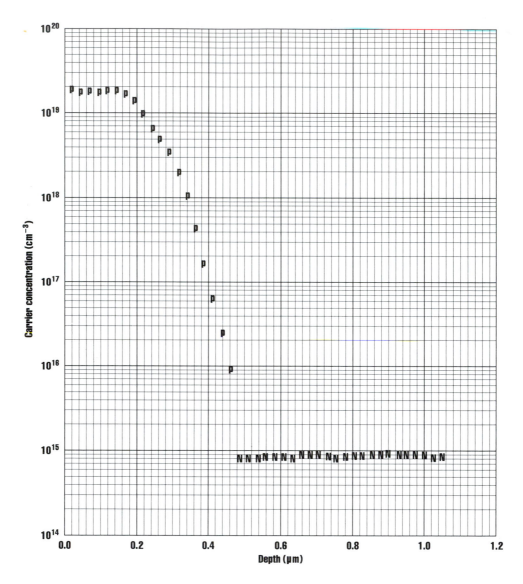

Figure 3-17   Typical spreading resistance profile showing measured carrier concentration as a function of depth *(used with permission, Solecon Labs).*

Figure 3.18 shows a typical SIMS arrangement. The sample to be tested is loaded into the instrument and the system is pumped to ultrahigh vacuum (usually about $1 \cdot 10^{-9}$ torr). The sample is then exposed to a beam of ions with an energy of between 1 and 5 keV. The energetic ions strike the surface of the sample where they destroy the crystal lattice and eject material through a process known as sputtering. (The use of sputtering as a thin film deposition process will be discussed in Chapter 12.) A fraction of the ejected material is ionized, collected, and accelerated toward a mass spectrometer. Either negative or positive ions can be collected. The instrument can be run in a

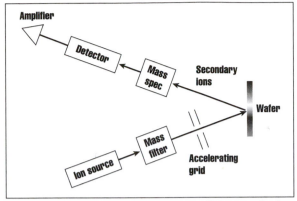

High vacuum chamber

Figure 3-18   A typical SIMS arrangement. The sample is bombarded by high energy ions. The sputtered material is mass analyzed to determine the composition of the substrate.

static mode with a slow sputter rate to survey the elements in the layer. The sputter rate can also be increased and the concentration of several impurities can be measured as a function of depth.

If the sputter rate for the substrate is known for the incident ion and ion energy, or if it can be measured after the sputter is complete, this raw data of counts versus sputter time can be transformed into counts versus depth. Frequently this is done by measuring the depth of the sputtered crater after the SIMS data have been collected. Converting the counts data into chemical species concentrations is much more difficult. The sputtering yield, the collection efficiency, the ionization efficiency, and the detector sensitivity are only approximately known and vary from day-to-day and run-to-run. The current best practice for reducing SIMS data is to run a calibration sample for each impurity immediately before and/or after the sample being measured. Implanted samples (see Chapter 5) are often used, since both the concentration and depth of the profile are well known. Even with this procedure, SIMS accuracy is at the very best about a factor of two.

SIMS sensitivities vary widely according to the instrument and the technique being used. Early SIMS systems used Ar^+ beams. These beams have low ionization efficiencies for most impurities of interest for microelectronics applications. Reactive beams such as cesium and oxygen can produce sensitivities that are several orders of magnitude better. Furthermore, conditions used to optimize positive ion production generally will not be sensitive to impurities that tend to form negative ions. To obtain a complete inventory of the impurities present, at least two runs must be made. Under optimum conditions the sensitivities that can be expected from commercial laboratories are listed in Table 3.3.

Two problems with SIMS analysis are the time and cost of such a profile. The sample must be pumped to high vacuum. Since the sample must be sputtered away during the analysis, the data

Table 3.3   Typical commercially available SIMS detection limits in single crystal semiconductor samples

Impurity	Beam	Type	Limit
Boron in silicon	O_2	+	1 to $5 \cdot 10^{14}$ cm^{-3}
Arsenic in silicon	Ce	−	$5 \cdot 10^{14}$ cm^{-3}
Phosphorus in silicon	Ce	−	$1 \cdot 10^{15}$ cm^{-3}
Oxygen in silicon	Ce	−	$1 \cdot 10^{17}$ cm^{-3}
Carbon in silicon	Ce	−	$1 \cdot 10^{16}$ cm^{-3}
Silicon in GaAs	Ce	−	$1 \cdot 10^{15}$ cm^{-3}
Magnesium in GaAs	O_2	+	$1 \cdot 10^{14}$ cm^{-3}
Beryllium in GaAs	O_2	+	$5 \cdot 10^{14}$ cm^{-3}
Oxygen in GaAs	Ce	−	$1 \cdot 10^{17}$ cm^{-3}
Carbon in GaAs	Ce	−	$5 \cdot 10^{15}$ cm^{-3}

Data courtesy Mr. Charles McGee, Charles Evans and Associates.

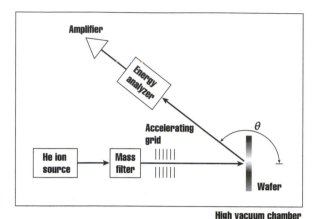

Figure 3-19   In Rutherford backscattering, the energy of the backscattered helium ions can be used to determine the depth profile of the chemical composition of the wafer.

collection is limited by the erosion rate. Samples thicker than 1 μm may take 4 to 8 hours to profile. Abrupt interfaces are also difficult to measure, particularly if they are buried. The sample erosion does not occur one atomic layer at a time; instead, the sample becomes progressively less planar. Combined with "knock on," the tendency of the incident ions to redistribute the impurities, the analysis of sharp concentration profiles can be significantly comprised. SIMS is nevertheless, a powerful and widely used tool.

Rutherford Back Scattering (RBS) is shown in Figure 3.19. In this technique, helium ions with energies in excess of 1 MeV are incident on the sample at near normal incidence. At these energies the ions penetrate deeply into the sample. Any impurities will backscatter the ions with a characteristic energy that depends on the mass of the impurity and its depth into the crystal. The backscattered helium ions are collected and the energy analyzed. The raw data from an RBS system consist of counts versus energy. The backscattered signal from an impurity atom must be larger than the signal from the host material atom in order for RBS to have reasonable sensitivity. Since the backscatter probability increases with the square of the mass of the target atom, the technique is limited to high mass impurities such as arsenic and antimony in silicon. Because the technique does not require a sputter erosion of the sample to be measured, RBS is quick and nondestructive. Its sensitivity is many orders of magnitude less than SIMS, however.

3.7 Diffusion in SiO₂

In early fabrication technologies, silicon dioxide was frequently used as a mask for ion implantation. In current technologies, silicon dioxide (SiO_2) is usually used for its insulating properties. While the presence of even moderate amounts of silicon dopant impurities in SiO_2 have little effect on the film's electrical properties, often the oxide must also serve as a barrier to impurity diffusion. One of the most dramatic examples of this application is the use of SiO_2 as the gate dielectric of a MOSFET. The polysilicon normally used as the gate electrode in this device is doped to a concentration of greater than 10^{21} cm^{-3} with phosphorus to minimize the series resistance. Directly beneath the gate oxide is the channel that is typically more lightly doped by a factor of 10^4. The gate oxide thickness in deep submicron devices is 100 Å or less. Any minute leakage of the gate electrode dopant into the channel would shift the device threshold and make the IC inoperative. This is one of the motivations for developing nitrided oxides, which have lower diffusion constants than pure SiO_2.

The diffusion mechanisms of impurities in SiO_2 has received far less study than comparable diffusion coefficients in silicon. The classic reference is Ghezzo and Brown [27], who summarized a variety of earlier references that are primarily from the 1960s. There was little agreement in the literature, particularly for values of D_o. Representative values from their summary are presented in Table 3.4.

In a more recent paper, Wada and Antoniadis [28] found that arsenic diffusivity in SiO_2 depends on the anneal ambient. In nitrogen the diffusivity was higher than in oxygen, with an activation energy of 3.7 eV, compared to 4.7 eV for O_2. In spite of the significant energy difference, the measured diffusion coefficients were nearly identical at 1200°C (4×10^{-15} cm²/sec). At 1000°C the diffusion

Table **3.4**	Diffusivity of various impurities in SiO$_2$			
Element	D_o(cm^2/sec)	E_a(eV)	C_s(cm^{-3})	Source
Boron	3×10^{-4}	3.53	$<3 \times 10^{20}$	Borosilicate
Phosphorus	0.19	4.03	8×10^{17} to 8×10^{19}	Phosphosilicate
Arsenic	250	4.90	1 to 6×10^{19}	Arsenosilicate
Antimony	1.31×10^{16}	8.75	5×10^{19}	Sb$_2$O$_5$ vapor

Data for boron and phosphorus from Ghezzo and Brown [27].

coefficients were found to be 8×10^{-17} cm^2/sec in oxygen and 1.5×10^{-17} cm^2/sec in nitrogen. At concentrations above 5×10^{20} cm^{-3}, arsenic was found to be immobile. Later work confirmed the diffusivities and the existence of the immobile high concentration phase [29], although its onset was found to be below 10^{20} cm^{-3}.

Antimony diffusion in SiO$_2$ was recently studied by van Ommen [30] using ion implantation. He found that the diffusivity of Sb in N$_2$ or dry O$_2$ ambients could be described by

$$D_{Sb} = 3.7 \times 10^{-11} \text{ (cm}^2\text{/sec)}e^{-1.32eV/kT} \qquad \textbf{(3.46)}$$

and when diffusing in a wet oxygen (O$_2$ and H$_2$O) ambient, by

$$D_{Sb} = 1.2 \times 10^{-7} \text{ (cm}^2\text{/sec)}e^{-2.25eV/kT} \qquad \textbf{(3.47)}$$

3.8 Diffusion Systems

In most modern devices feature sizes are small, often less than 1 μm. Most doping is now done by ion implantation rather than predeposition diffusion. When very heavily doped layers are required, however, diffusion tubes are sometimes used to introduce the desired dopant. These furnaces are identical to oxidation tubes, except that a controlled flow of an inert gas such as N$_2$ or Ar is used rather than O$_2$. (See Section 4.9 for a detailed description of these systems.) There are two methods used to heavily dope layers in furnace tube: solid source and liquid source doping.

Most liquid source dopants are applied through a vapor transport method. As shown in Figure 3.20, a sealed container of the liquid is immersed in a constant temperature bath to produce a known vapor pressure of the dopant above the surface of the liquid. A metered flow of an inert gas, usually nitrogen, is injected into the bubbler, and the resultant mixture of N$_2$ and dopant containing vapor is carried to the furnace. The partial pressure of dopant in the ambient of the furnace is controlled by the temperature of the bath, the pressure of the gas above the liquid, and the ratio of the flow

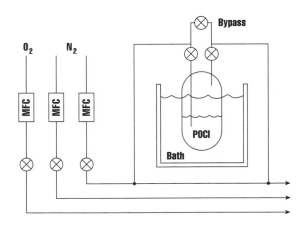

Figure 3-20   A typical bubbler arrangement for doping a silicon wafer using a POCl source. The gas flow is set using mass flow controllers (MFC).

through the bubbler to the sum of all other flows into the furnace. A bypass configuration is also typically employed to allow the flow of vapor from the bubbler to be precisely switched on and off to provide more control over the doping time. Usually an oxygen source must also be supplied to react with the incoming dopant gas.

The most common boron source is BBr_3, which has a boiling point of 90°C. It is believed that the reactions that occur in the furnace include dissociation

$$2\ BBr_3 \rightarrow 2B\ +\ 3Br_2 \tag{3.48}$$

and an oxidation

$$4B\ +\ 3O_2 \rightarrow 2B_2O_3 \tag{3.49}$$

The oxide is transported to the wafer surface where it oxidizes the surface to release free boron:

$$2B_2O_3\ +\ 3Si \rightarrow 4B\ +\ 3SiO_2 \tag{3.50}$$

The concentration of BBr_3 in the furnace must be carefully controlled, even though the surface concentration of boron will be fixed by its solid solubility in silicon at the diffusion temperature. If the BBr_3 concentration is too high with respect to the O_2 concentration, involatile silicon boride compounds will form at the surface. These compounds lead to nonuniform doping and may be difficult to remove. This can lead to contact resistance problems in the device. Liquid source phosphorus doping is normally done using a bubbler of $POCl_3$, which has a boiling point of 107°C. The reaction proceeds in a manner similar to that of BBr_3. The oxide that forms is P_2O_5. This oxide dissociates at the surface of the wafer releasing P and forming SiO_2.

Liquid sources have several disadvantages. They are often highly corrosive. Bubblers must be pressurized and have been known to explode. Because the vapor pressure above the liquid increases exponentially with temperature, the process is sensitive to small bubbler temperature changes. Furthermore, there is a significant danger of forming insoluble silicon compounds at the surface of the wafer that are invisible, but are extremely undesirable.

The dopant species can also be introduced through the use of solid sources. These discs are about the same size as the wafer and are loaded in alternate slots in furnace boats. For diffusing p-type layers in silicon boron nitride (BN), discs are commonly used. When oxidized at 750–1100°C, a thin film of B_2O_3 forms at the surface. In the presence of H_2, the volatile compound HBO_2 forms and diffuses to the surface of the wafer where a borosilicate glass is formed. This glass serves as the boron source for diffusion into the substrate. The surface concentration is typically solid solubility limited.

Both arsenic and phosphorus solid source discs can be used as n-type dopants for silicon. Phosphorus containing discs come in the form of hot pressed ammonium monophosphate ($NH_4H_2PO_4$) or ammonium diphosphate [$(NH_4)_2H_2PO_4$] in inert ceramic binders. For a more shallow profile, arsenic sources containing aluminum arsenate ($AlAsO_4$) are also available. The arsenic is transported to the wafer surface as As_2O_3, For both arsenic and phosphorus solid source diffusion, care must be taken to avoid forming an undoped thermal oxide on the surface of the wafer prior to doping. The presence of such a layer can block the diffusion of the dopants into the substrate.

3.9 SUPREM Simulations of Diffusion Profiles

It has probably become apparent to the reader that the various complications that arise in the calculation of diffused profiles, such as concentration dependent diffusivities, preclude analytic calculations for all but the simplest examples. For that reason, numerical methods have been developed for the prediction of profiles in 1-, 2-, and 3-D. Often the results can be linked to device simulators so that the effect that a change in an impurity profile would have on the device characteristics can be predicted in a straightforward manner. If the simulation results are accurate, the device designer can optimize performance and test process sensitivity with far fewer runs through the fabrication facility resulting in tremendous savings in cost and time.

While some corporations have developed proprietary software, one of the most popular packages for calculating impurity profiles is the Stanford University PRocess Engineering Module (SUPREM). SUPREM III performs detailed calculations in 1-D. SUPREM IV performs calculations in 2-D. The output of these programs are the chemical, carrier, and vacancy concentrations as functions of depth into the semiconductor. This section will use SUPREM III as a tool to investigate device impurity profiles in a more realistic manner. In the succeeding chapters the use of SUPREM III will be expanded to cover oxidation, ion implantation, and rapid thermal processing.

Before proceeding, however, a word of caution is in order. Students tend to regard the output of these programs as being infallibly correct. This is, of course, not true. The predictions of this code is only as good as the models and numerical techniques that it employs. In practice these model parameters must be rigorously checked to ensure accuracy. These programs should be regarded as calculation tools that allow the process engineer to access more complicated diffusion models. These models are also more realistic and usually, but not always, give fairly accurate results.

All diffusion process simulators are built on three basic equations. In 1-D there are the flux equation

$$J_i = -D_i \frac{dC_i}{dx} + Z_i \mu_i C_i \mathcal{E} \tag{3.51}$$

where Z_i is the charge state and μ_i is the mobility of the impurity. The continuity equation gives

$$\frac{dC_i}{dt} + \frac{dJ_i}{dx} = G_i \tag{3.52}$$

where G_i is the generation recombination rate of the impurity, and Poisson's equation

$$\frac{d}{dx}[\varepsilon\mathcal{E}] = q(p - n + N_D^+ - N_A^-) \tag{3.53}$$

where ε is the dielectric constant, n and p are the electron and hole concentrations, and N_{D+} and N_{A-} are the concentrations of the ionized donors and acceptors. These equations are solved simultaneously over a 1-D grid that the user defines.

The diffusivity used for SUPREM is based on the vacancy model of Fair. The diffusivity is calculated using Equation 3.7. The values of E_a and D_o are included in a software look-up table for boron, antimony, and arsenic in silicon. Finally, empirical models are added to take into account field aided, oxidation enhanced, and oxidation retarded diffusion.

To run SUPREM III, an input deck must be provided. This file contains a series of comments, the initialize statement, materials statements, process statements, and output statements. A description of some of the most commonly used SUPREM III commands are given in an appendix. The deck starts with the title card, which is simply a comment repeated on each page of the output. Several comment cards may follow. The user is encouraged to use these cards to document the process flow.

The next card is a control statement, the initialize card. It sets the substrate type, orientation, and doping. It also sets the thickness of the region to be simulated and it establishes a grid. The amount of grid spacing necessary depends on the doping gradients present, but typically a total of 100 to 200 grid points are used. If the default coefficients are not found to be adequate, materials cards can be used to alter the model. For example, if the user measures the diffusion coefficients of arsenic and finds that the activation energy for the neutral vacancy diffusion coefficient is different than the default, this parameter can be changed by invoking ''arsenic dix.e = (value in eV).'' Similarly, one can alter the various parameters of silicon, oxide, polysilicon, or nitride.

Once the substrate and materials are set, a series of process cards are used to call out the sequence of processing steps as they are completed. This example will begin with the diffusion card that calls out the time, temperature, diffusion ambient (oxidizing or not oxidizing), and surface concentration, if a doping process is being carried out. Multiple diffusions can be run sequentially to simulate a multistep anneal.

Finally, the chemical, active, and net impurity concentrations at each grid point can be printed or plotted. The output options vary with the particular version of SUPREM III that is installed. It is also possible to print out the materials parameters that are being used and information about each layer of the simulation. A second type of output statement involves investigating the effects of electrically biasing the top and/or bottom of the volume being simulated. These can be used to determine the sheet resistance of diffused layers and the threshold voltage of MOS structures.

Example 3.2   **Use SUPREM to simulate a predeposition diffusion.**

In this example, SUPREM III will be used to simulate a predeposition diffusion of silicon with phosphorus at 1050°C, with the surface concentration set by the solid solubility of phosphorus at that temperature. Run the following example, and then repeat the exercise using the analytic equations assuming simple intrinsic diffusion. How large is the difference? Can you explain the discrepancy?

```
Title       Example 3.2—Predeposition Diffusion of Phosphorus
Comment     Initialize the silicon substrate
Initialize  (100)Silicon Boron Concentration=1e16
+           Thickness=1.0 dX=.005 Xdx=.02 Spaces=120

Comment     Diffuse phosphorus
Diffusion   Time=20 Temperature=1050 Phosphorus Solid Sol

Print       Layers Active Concentration Phosphorus Boron Net
Plot        Active Net Cmin=1E15
Stop        End example 1
```

Example 3.3   Use SUPREM to simulate a diffused bipolar transistor.

One simple way to make an NPN bipolar transistor is to diffuse boron and arsenic from the surface into an n-type substrate. The diffused arsenic becomes the emitter, the diffused boron becomes the base, and the substrate is used as a common collector. Of critical importance to the device operation is the Gummel number, the integral of the doping profile in the base, which is inversely proportional to the current gain, β. SUPREM III will provide this number directly. Perform the following run. In the output find the net hole concentration in the second layer. This is the Gummel number in holes per square centimeter. To get a feel for the process sensitivity of the diffused NPN, rerun the process with the boron predeposition at 980°C instead of 1000°C (i.e., a 2% temperature error). By what fraction would the gain change?

```
Title       Simple NPN Diffused Bipolar Transistor
Comment     Initialize the silicon substrate
Initialize ⟨100⟩Silicon Phos Concentration=1e16
+           Thickness=2.0 dX=.010 Xdx=.030 Spaces=150

Comment     Diffuse boron
Diffusion   Time=10 Temperature=1000 Boron Solid Sol
Diffusion   Time=60 Temperature=1100 Nitrogen

Comment     Diffuse phosphorus
Diffusion   Time=7 Temperature=1050 Phos Solid Sol
Diffusion   Time=5 Temperature=1000 Nitrogen

Comment     Output statements
Print       Layers Active Concentration Phos Boron Net
Plot        Active Net Cmin=1e15
Stop        End example 1
```

3.10 Summary

This chapter reviewed the physics of diffusion and presented Fick's laws that govern diffusion. Two particular solutions were presented corresponding to drive in and predeposition diffusion. The atomistic mechanisms of diffusion were presented along with heavy doping effects. The details of diffusion for a variety of popular dopants were also discussed. At high doping concentrations the diffusion coefficient is no longer constant, but frequently depends on the local doping concentration and concentration gradient. A numerical tool, SUPREM III, was introduced that allows the student to calculate dopant profiles in the presence of these nonlinear effects.

Problems

(For all problems assume that the diffusivity can be approximated by the intrinsic diffusivity, unless specified otherwise.)

1. Assume that you have been asked to measure the diffusivity of a donor impurity in a new elemental semiconductor. What constants would you need to measure? What experiments

would you attempt? Discuss the measurement techniques that you would use to measure the chemical and carrier profiles. What problems are likely to arise?

2. Construct a semilog plot using Table 3.2 of the three contributions (D_i, D_-, and $D^=$) to the diffusivity of phosphorus as a function of temperature from 700 to 1100°C. Assume that the phosphorus concentration is 10^{19} cm^{-3}.

3. One of the high concentration effects that was not discussed was that of stress. Arsenic, for example, is a much larger atom than silicon. When high concentrations of arsenic are incorporated in the lattice it creates strain. Qualitatively discuss how this strain may affect the diffusivity.

4. Delta doping is a process used in advanced GaAs fabrication to increase the Schottky barrier height of the gate electrode. This reduces the gate electrode leakage. Delta doping is done by depositing a monolayer of a p-type dopant material directly between the gate electrode and the GaAs. Assume that the atomic surface coverage is 1.5×10^{15} cm^{-2} and that the dopant is beryllium. After the gate patterning, the source/drain is annealed at 800°C for 10 min to activate the impurity. (a) If the gate material prevents any outdiffusion from the wafer, use first order diffusion theory to calculate the junction depth if the channel is doped 1×10^{17} cm^{-3} n-type. (b) What surface concentration of Be will result? (c) Sketch the profile that you calculated using this simple theory and the profile that might actually be expected. Briefly list two reasons for the difference.

5. A silicon wafer was doped in a 1000°C predeposition diffusion with phosphorus to its solid solubility limit. The process time was 20 min. After the predeposition, the surface of the silicon was sealed and an 1100°C drive in was done. Find the drive in time necessary to obtain a junction depth of 4.0 μm. Assume a substrate concentration of 10^{17} cm^{-3}. What is the surface concentration after the drive in?

6. A 10 minute 1100°C predeposition is done using arsenic. Assume that the surface concentration reaches the solid solubility limit. Next, the wafer is annealed for 24 hours at 1000°C. Use Equation 3.27 to predict the maximum carrier concentration. How does this compare to the solid solubility limit at 1000°C? Explain the distinction between solid solubility and Equation 3.27.

7. Assume that a wafer is uniformly doped. If a Schottky contact is formed on the surface, what would the C-V curve look like?

8. Boron doped gate electrodes are desirable for certain highly scaled MOS transistors. Assume that the gate oxide thickness is 100 Å. Also assume that the concentration of boron in the oxide very close to the polysilicon gate is a constant of 10^{21} cm^{-3}. The maximum amount of boron that can be diffused into the substrate without shifting the threshold voltage outside-allowable limits is 3×10^{11} cm^{-2}. Using Table 3.4, calculate the dose of impurities that enter the semiconductor after a 4 hr anneal at 1000°C. Will this anneal produce an unacceptable threshold voltage shift?

References

1. F. J. Morin and J. P. Maita, ''Electrical Properties of Silicon Containing Arsenic and Boron,'' *Phys. Rev.* **96**:28 (1954).

2. W. R. Runyan and K. E. Bean, *Semiconductor Integrated Circuit Processing Technology*, Addison-Wesley, Reading, MA, 1990.

3. P. M. Fahey, P. B. Griffin, and J. D. Plummer, ''Point Defects and Dopant Diffusion in Silicon,'' *Rev. Mod. Phys.* **61**:289 (1989).

4. T. Y. Yan and U. Gosele, "Oxidation-Enhanced or Retarded Diffusion and the Growth or Shrinkage of Oxidation-Induced Stacking Faults in Silicon," *Appl. Phys. Lett.* **40:**616 (1982).

5. S. Mizuo and H. Higuchi, "Retardation of Sb Diffusion in Si During Thermal Oxidation," *J. Appl. Phys. Japan*, **20:**739 (1981).

6. A. M. R. Lin, D. A. Antoniadis, and R. W. Dutton, "The Oxidation Rate Dependence of Oxidation-Enhanced Diffusion of Boron and Phosphorus in Silicon," *J. Electrochem. Soc.* **128:**1131 (1981).

7. R. B. Fair, "Concentration Profiles of Diffused Dopants in Silicon," in *Impurity Doping Processes in Silicon*, F. F. Y. Wang, ed., North-Holland, New York, 1981.

8. H. Ryssel, K. Muller, K. Harberger, R. Henkelmann, and F. Jahael, "High Concentration Effects of Ion Implanted Boron in Silicon," *J. Appl. Phys.* **22:**35 (1980).

9. R. B. Fair and J. C. C. Tsai, "A Quantitative Model for the Diffusion of Phosphorus in Silicon and the Emitter Dip Effect," *J. Electrochem. Soc.* **124:**1107 (1978).

10. R. J. Field and S. K. Ghandhi, "An Open Tube Method for the Diffusion of Zinc in GaAs," *J. Electrochem. Soc.* **129:**1567 (1982).

11. L. R. Weisberg and J. Blanc, "Diffusion with Interstitial-Substitutional Equilibrium. Zinc in Gallium Arsenide," *Phys. Rev.* **131:**1548 (1963).

12. S. Reynolds, D. W. Vook, and J. F. Gibbons, "Open-Tube Zn Diffusion in GaAs Using Diethylzinc and Trimethylarsenic: Experiment and Model," *J. Appl. Phys.* **63:**1052 (1988).

13. K. B. Kahen, "Mechanism for the Diffusion of Zinc in Gallium Arsenide," in *Materials Res. Soc. Symp. Proc., vol. 163*, D. J. Wolford, J. Bernholc, and E. F. Haller, eds., MRS, Pittsburgh, 1990, p. 681.

14. M. E. Greiner and J. F. Gibbons, "Diffusion of Silicon in Gallium Arsenide Using Rapid Thermal Processing: Experiment and Model," *Appl. Phys. Lett.* **44:**740 (1984).

15. K. L. Kavanaugh, C. W. Magee, J. Sheets, and J. W. Mayer, "The Interdiffusion of Si, P, and In at Polysilicon Interfaces," *J. Appl. Phys.* **64:**1845 (1988).

16. S. Yu, U. M. Gosele, and T. Y. Tan, "An Examination of the Mechanism of Silicon Diffusion in Gallium Arsenide," in *Materials Res. Soc. Symp. Proc.*, Vol. 163, D. J. Wolford, J. Bernholc, and E. F. Haller, eds., MRS, Pittsburgh, 1990, p. 671.

17. K. B. Kahen, D. J. Lawrence, D. L. Peterson, and G. Rajeswaren, "Diffusion of Ga Vacancies and Si in GaAs," in *Materials Res. Soc. Symp. Proc.*, Vol. 163, D. J. Wolford, J. Bernholc, and E. F. Haller, eds., MRS, Pittsburgh, 1990, p. 677.

18. D. Sudandi and S. Matsumoto, "Effect of Melt Stoichiometry on Carrier Concentration Profiles of Silicon Diffusion in Undoped LEC Sl-GaAs," *J. Electrochem. Soc.* **136:**1165 (1989).

19. L. B. Valdes, "Resistivity Measurements on Germanium for Transistors," in *Proc. IRE*, **42:**420 (1954).

20. M. Yamashita and M. Agu, "Geometrical Correction Factor of Semiconductor Resistivity Measurement by Four Point Probe Method," *Japan J. Appl. Phys.* **23:**1499 (1984).

21. D. K. Schroder, *Semiconductor Material and Device Characterization*, Wiley-Interscience, New York, 1990.

22. L. J. Van der Pauw, "A Method for Measuring the Specific Resistivity and Hall Effect of Discs of Arbitrary Shape," *Phillips Res. Rep.* **13:**1 (1958).

23. D. S. Perloff, "Four-point Probe Correction Factors for Use in Measuring Large Diameter Doped Semiconductor Wafers," *J. Electrochem. Soc.* **123:**1745 (1976).

24. M. Pawlik, "Spreading Resistance: A Comparison of Sampling Volume Correction Factors in High Resolution Quantitative Spreading Resistance," in *Emerging Semiconductor Technol-*

ogy, D. C. Gupta and R. P. Langer, eds., STP 960, Am. Soc. Test. Materials, Philadelphia, 1987.

25. R. G. Mazur and G. A. Gruber, ''Dopant Profiles in Thin Layer Silicon Structures with the Spreading Resistance Profiling Technique,'' *Solid State Technol.* **24:**64 (1981).

26. P. Blood, ''Capacitance-voltage Profiling and the Characterization of III-V Semiconductors Using Electrolyte Barriers,'' *Semiconductor Sci. Techol.* **1:**7 (1986).

27. M. Ghezzo and D. M. Brown, ''Diffusivity Summary of B, Ga, P, As, and Sb in SiO_2, *J. Electrochem. Soc.* **120:**146 (1973).

28. Y. Wada and D. Antoniadis, ''Anomalous Arsenic Diffusion in Silicon Dioxide,'' *J. Electrochem. Soc.* **128:**1317 (1981).

29. R. Singh, M. Maier, H. Kräutle, D. R. Young, and P. Balk, ''Diffusion of Ion-Implanted Arsenic in Thermally Grown SiO_2 Films,'' *J. Electrochem. Soc.* **131:**2645 (1984).

30. A. H. van Ommen, ''Diffusion of Ion-Implanted Sb in SiO_2,'' *J. Appl. Phys.* **61:**993 (1987).

Chapter 4

Thermal Oxidation

One of the major reasons for the popularity of silicon ICs is the ease with which silicon forms an excellent oxide, SiO_2. This oxide is widely used as an insulator both in active devices such as MOS-FETs and in the region between the active devices, known as the field. Various methods are used for forming SiO_2. Chapter 4 will introduce thermal oxidation. This method produces oxides with the fewest defects both in the bulk and at the interface with the underlying silicon. Unfortunately, most other semiconductors do not form oxides of sufficient quality to be used in device fabrication. For that reason, this chapter will focus on silicon processing exclusively.

4.1 The Deal-Grove Model of Oxidation

The oxidation of silicon in molecular oxygen proceeds according to the overall reaction

$$Si(solid) + O_2(gas) \rightarrow SiO_2 \tag{4.1}$$

This process is called *dry oxidation* since it uses molecular oxygen rather than water vapor as an oxidant. The Deal-Grove model works well for predicting oxide thicknesses for thermal oxides larger than about 300 Å [1]. The growth of thin oxides will be discussed later in this chapter. High temperature is not necessary to grow an oxide. Silicon will oxidize in air at room temperature. Once an oxide forms, however, silicon atoms must travel through the oxide layer to react with the oxygen present at the surface of the wafer, or else oxygen molecules must travel through the oxide to reach the silicon surface where a reaction can occur. The process that drives this motion is the one discussed in the last chapter: diffusion. The diffusivity of Si in SiO_2 is several orders of magnitude smaller than the diffusivity of O_2. As a result, the chemical reaction occurs at the Si/SiO_2 interface. This has a very important effect. The interface produced by thermal oxidation has not seen the atmosphere. As a result, it is relatively free of impurities. The amount of silicon consumed due to the chemical reaction given by Equation 4.1 is about 44% of the thickness of the final oxide.

At room temperature, neither the silicon nor the oxygen molecules are sufficiently mobile to diffuse through the native oxide. After a while, the reaction effectively stops and the oxide will not get much thicker than 25 Å. For a sustained reaction to occur, the silicon wafer must be heated in the presence of an oxidizing ambient [2]. For now, assume that the ambient is O_2. Consider the diagram

Figure 4-1   Schematic diagram of the oxidant flows during oxidation.

of a growing oxide (Figure 4.1). The four oxygen concentrations in this figure are: C_g, the concentration of oxygen in the gas stream far from the wafer, C_s, the concentration of oxygen in the gas at the surface of the wafer, C_o, the concentration of oxygen in the oxide at the surface of the wafer, and C_i, the concentration of oxygen at the Si/SiO$_2$ interface. Define J as an oxygen flux, which is the number of oxygen molecules that crosses a plane of a certain area in a certain time. Now, one can then define three oxygen fluxes of interest. The first is the amount of oxygen that moves from the bulk of the gas to the surface of the growing oxide film. Recall that Chapter 2 briefly mentioned stagnant layer formation in the molten silicon of a Czochralski crucible. This layer arises due to the finite viscosity of the melt. Similarly, if a gas stream of oxygen is made to flow past the surface of the wafer, a boundary layer will exist near the surface. The gas velocity in this boundary layer ranges from zero at the surface of the wafer to the bulk gas velocity at the opposite side of the boundary layer. To a first approximation the oxygen molecules cannot be transported across this region by the gas flow: There is none. Instead, they must diffuse in a manner described by Fick's first law:

$$J_1 \approx D_{O_2} \frac{C_g - C_s}{t_{sl}} \tag{4.2}$$

where t_{sl} is the thickness of the stagnant layer. C_g can be calculated using the ideal gas law:

$$C_g = \frac{n}{V} = \frac{P_g}{kT} \tag{4.3}$$

where k is Boltzmann's constant and P_g is the partial pressure of the oxygen in the furnace. While this formulation is workable, it tends to underestimate the flux. It is more common to explicitly take into account the fact that some flow remains through much of what is called the stagnant layer by writing

$$J_1 = J_{gas} = h_g(C_g - C_s) \tag{4.4}$$

where h_g is called the *mass transport coefficient*.

The second oxygen flux is the diffusion of molecular oxygen across the growing oxide film. The concentration gradient needed to drive diffusion arises because the gas ambient acts as an oxygen source while the reacting surface acts as a sink. Then, assuming no sources or sinks of oxygen in the growing oxide the concentration varies linearly and

$$J_2 \approx D_{O_2} \frac{C_o - C_i}{t_{ox}} \tag{4.5}$$

where the diffusivity is now that of oxygen in SiO$_2$. The third flux is the flow of oxygen reacting with the silicon to form SiO$_2$. This rate is determined by chemical reaction kinetics. Since there

is an abundant supply of silicon at the reacting surface, the reaction rate and the flux are proportional to the oxygen concentration:

$$J_3 = k_s C_i \tag{4.6}$$

where the constant of proportionality, k_s is the chemical rate constant for the overall reaction described in Equation 4.1. In equilibrium these three flows must balance:

$$J_1 = J_2 = J_3 \tag{4.7}$$

Combining Equations 4.4–4.6 leaves two equations and three unknown concentrations: C_s, C_o, and C_i. Solving to find a growth rate requires one more equation. That equation is Henry's law, which says that the concentration of an adsorbed species at the surface of a solid is proportional to the partial pressure of that species in the gas just above the solid:

$$C_o = HP_g = HkTC_g \tag{4.8}$$

where H is Henry's gas constant, and the ideal gas law has been used to replace P_g. Now there are three equations and three unknowns. After some algebra one can show that

$$C_i = \frac{HP_g}{1 + \dfrac{k_s}{h_g} + \dfrac{k_s t_{ox}}{D}} \tag{4.9}$$

Finally, to obtain the growth rate, just divide the flux at the interface by the number of molecules of oxygen per unit volume of SiO_2, which is commonly labeled N_1. For oxidation by molecular oxygen, N_1 is half the number density of oxygen atoms in SiO_2, or 2.2×10^{22} cm^{-3}. The result is

$$R = \frac{J}{N_1} = \frac{dt_{ox}}{dt} = \frac{Hk_s P_g}{N_1 \left[1 + \dfrac{k_s}{h_g} + \dfrac{k_s t_{ox}}{D} \right]} \tag{4.10}$$

Assume that at time 0 the oxide thickness is t_0. The solution of this differential equation is of the form

$$t_{ox}^2 + At_{ox} = B(t + \tau) \tag{4.11}$$

where

$$A = 2D\left(\frac{1}{k_s} + \frac{1}{h} \right)$$

$$B = \frac{2DHP_g}{N_1} \tag{4.12}$$

$$\tau = \frac{t_o^2 + At_o}{B}$$

The parameters A and B are well known for a variety of process conditions. The more fundamental parameters such as diffusivity that go into A and B are rarely quoted. Furthermore, most silicon oxidation is done in atmospheric pressure. As a result, $k_s \ll h_g$, and the growth rate is nearly independent of the gas phase mass transport and, therefore, of the reactor geometry. When the oxidizing species is H_2O rather than O_2 (see below), the same equations apply, but with different values for the diffusivity, mass transport properties, reactivity, pressure of the gas, and number of molecules per unit volume.

Since A and B are proportional to diffusivity both parameters will follow an Arrhenius function. The activation energies of A and B can be calculated from the activation energy of the diffusivities and reaction rates as shown by Equation 4.12. There is a reasonably good agreement (about 10% discrepancy) between the activation energies of the diffusivities of oxygen and water in fused silica and the activation energy of B. Furthermore, since the ratio B/A eliminates the diffusivity, its activation energy should depend primarily on k_s. As expected, the activation energy of B/A is in reasonable agreement with the Si-Si bond strength.

Finally, it is worth pointing out the significance of τ, since this is a common source of confusion. It arises due to the boundary condition for the differential equation at zero time. Notice that for sufficiently thick oxides, the rate of oxidation varies with the oxide thickness. If an oxidation is begun with an initial oxide thickness of t_0, it is not accurate therefore to calculate the thickness grown in the process and simply add it to t_0. Instead, the initial thickness must be used to determine τ, and it must be added to t to obtain a total effective oxidation time. It is as though the oxidation process started at a time $-\tau$. Then at $t = 0$ the thickness of the oxide is exactly t_0.

4.2 The Linear and Parabolic Rate Coefficients

Equation 4.11 has two important limiting forms. For sufficiently thin oxides one can neglect the quadratic term. In that case

$$t_{ox} \approx \frac{B}{A}(t + \tau) \qquad (4.13)$$

On the other hand, if the oxide is sufficiently thick the term that is linear in the thickness can be ignored and

$$t_{ox}^2 \approx B(t + \tau) \qquad (4.14)$$

Because of these two limiting forms, B/A is called the *linear rate coefficient* and B is called the *parabolic rate coefficient*. These are the numbers that are usually quoted for oxidation. Figure 4.2 shows the coefficients for oxidations in pure oxygen. This type of process is called a *dry oxidation*. The plots are summarized at typical process conditions in Table 4.1.

Oxygen is not the only oxidant that can be used in this process. Another very popular one is a mixture of O_2 and H_2O. Called wet oxidation, these processes have the advantage of oxidizing at a much higher rate than dry oxidations. The fundamental reason for the higher rate is the relative size of the two molecules. The smaller water molecule diffuses more rapidly than molecular oxygen. A disadvantage is that oxides grown wet are less dense. Thus wet oxidation is typically used when a thick oxide is required that will not be subjected to any significant electrical stress. Figure 4.3 shows the linear and parabolic rate coefficients for wet oxidation. Section 4.4 will discuss the structure of the oxide in more detail.

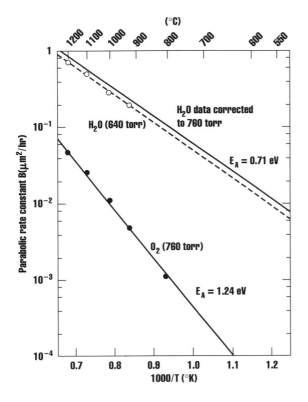

Figure 4-2   Arrhenius plot of the *B* oxidation coefficient. The wet parameters depend on the H₂O concentration and therefore on the gas flows and pyrolysis conditions *(after Deal and Grove).*

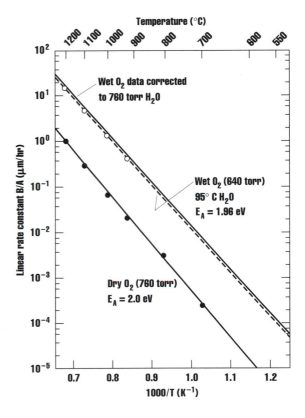

Figure 4-3   Arrhenius plot of the ratio (B/A) of the oxidation parameters *(after Deal and Grove).*

Another gas ambient used in thermal oxidation is dry molecular oxygen with a small (1 to 3%) halogen concentration. The most commonly used halogen is chlorine [3]. There are several reasons for using this mixture. Most heavy metal atoms react with Cl to form volatile (i.e., gaseous) metal chlorides. Metallic contaminants are believed to originate from the heating elements and insulation around the fused silica flow tube in which the oxidation is done. The impurities diffuse through the

Table 4.1   Dry oxidation coefficients for silicon

Temp (°C)	Dry				Wet (640 torr)	
	A (μm)	B (μm²/hr)	$\tau(h)$		A (μm)	B (μm²/hr)
800	0.370	0.0011	9		—	—
920	0.235	0.0049	1.4		0.50	0.203
1000	0.165	0.0117	0.37		0.226	0.287
1100	0.090	0.027	0.076		0.11	0.510
1200	0.040	0.045	0.027		0.05	0.720

The τ parameter is used to compensate for the rapid growth regime for thin oxides. *(After Deal and Grove.)*

wall of the furnace and may be incorporated into the growing oxide. Chlorine has the effect of constantly cleaning the gas ambient of these impurities. Oxides grown in a chlorine ambient are found to not only have fewer impurities, but also to have a better interface with the underlying silicon. The oxidation rate in O_2/Cl mixtures is higher than in pure O_2. For a 3% concentration of HCl in O_2 the linear rate coefficient doubles [4] (Figure 4.4).

HCl has been the most commonly used Cl source. Trichloroethylene (TCE) and trichloroethane (TCA) are sometimes used since they are much less corrosive than HCl. TCE has the disadvantage that it is carcinogenic. TCA has the potential of forming phosgene ($COCl_2$) at high temperature. Phosgene is a highly toxic substance also known as mustard gas. Stringent safety precautions must therefore be employed in a TCA furnace to prevent the conditions under which phosgene may form.

The Deal-Grove model accurately predicts that the oxidation rate slows as the oxide thickens. Oxides thicker than 1 µm require long exposures to very high temperatures. This will lead to dopant diffusion in the wafer that is often undesired. In this regime the oxidation rate depends on the parabolic rate coefficient, which in turn depends on C_g, the equilibrium concentration of oxygen in the gas. As shown in the derivation of the model, C_g depends on the partial pressure of the oxygen in the gas phase. Increasing the oxygen pressure in the furnace will increase the parabolic rate coefficient B and can therefore decrease the time or the temperature required for growing thick oxides. Production equipment is available for growing oxides at high pressure. They are most commonly used in bipolar technologies for isolation oxide growth and are almost always done in a wet ambient. In this appli-

Figure 4-4   The effect of chlorine on the oxidation rate *(after Hess and Deal, reprinted by permission, The Electrochemical Society).*

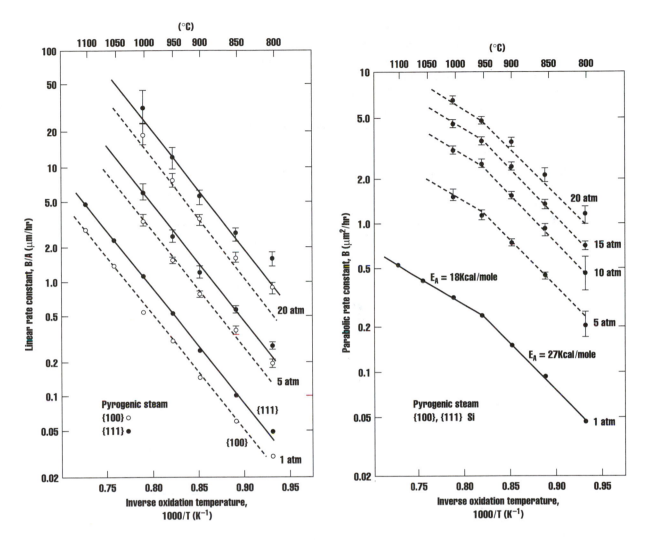

Figure 4-5 High pressure studies of the parabolic and linear rate coefficients in steam *(after Razouk et al., reprinted by permission, The Electrochemical Society).*

cation the lower oxidation temperature is required to avoid excessive diffusion of the heavily doped buried collector up into the device structure. This will be covered in more detail in Chapter 18. Figure 4.5 shows the dependence of the parabolic rate coefficient on temperature with pressure as a parameter. As expected it is proportional to the oxygen pressure [5].

Example 4.1   **A simple oxidation**

Calculate the amount of silicon dioxide grown during a 120 min 920°C steam oxidation. Assume that the wafer initially had 1000 Å of oxide.

According to Table 4.2, at 920°C, $A = 0.50$ μm and $B = 0.203$ μm²/hr. Inserting these values into Equation 4.12,

$$\tau = \frac{0.1 \text{ μm} * 0.1 \text{ μm} + 0.5 \text{ μm} * 0.1 \text{ μm}}{0.203 \text{ μm}^2/\text{hr}} = 0.295 \text{ hr}$$

Using Equation 4.11 and the quadratic formula,

$$t_{ox} = \frac{-A + \sqrt{A^2 + 4B(t + \tau)}}{2} = 0.48 \text{ μm}$$

In this case $A \approx t_{ox}$, so that neither linear nor quadratic approximations would be valid.

4.3 The Initial Oxidation Regime

The Deal-Grove model fits the oxidation rate over a broad range of parameters, if the linear and parabolic rate coefficients are known. By plotting the experimental thickness versus time for dry oxidation, however, one finds that the curve does not go through zero thickness or even through the native oxide thickness at zero time. Instead, it will typically intersect the zero time axis at a thickness of a few hundred angstroms. Figure 4.6 shows the experimentally measured oxidation rate [6] as a function of oxide thickness for a variety of temperatures. According to the Deal-Grove model the oxidation rate should approach a constant value

$$\lim_{t \to 0} \frac{dt_{ox}}{dt} = \frac{B}{A} \qquad \text{(4.15)}$$

Instead, the oxidation rate increases by a factor of 4 or more. As a result, the Deal-Grove model severely underestimates the thickness of thin oxides. Interestingly, these thin oxides are widely used in one of the most critical applications: the gate dielectric of MOS transistors. The values of τ given in Table 4.1 are used to correct the Deal-Grove model for dry oxidation to compensate for the excess growth that occurs in this initial growth regime. They should be used for dry oxidations done on bare silicon. With this correction the model accurately predicts thicknesses when $t_{ox} > 300$ Å, but predicts oxides that are thicker than reality for $t_{ox} < 300$ Å.

Several models have been put forward to explain the thin oxidation results. The models are quite controversial. Typically, they postulate one of two mechanisms for the enhanced oxidation rate. The first involves some enhancement of the arrival rate of the oxidant at the interface. Deal and Grove suggested that an electric field exists across the oxide that enhances the diffusion during the early stages of oxidation. In this model, electrons from the substrate charge surface states creates a

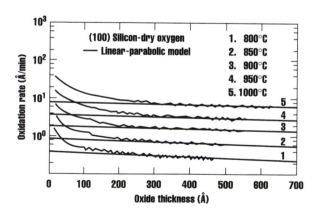

Figure 4-6   Oxide thickness dependence of the growth rate for very thin oxides oxidized in dry O_2. The substrates are lightly doped (100) silicon *(from Massoud et al. reprinted by permission, The Electrochemical Society).*

field across the oxide. The diffusing species must be O_2^-. Other authors have proposed variations on this basic mechanism [7],[8]. The problem with these models is that the oxidant must be ionized. More recent studies have found that the dominant diffusing species is neutral molecular oxygen [9]. Revesz and Evans [10] proposed that thin worm holes or microchannels with diameters of order 50 Å exist in the oxide. These holes aid in the movement of oxygen to the surface of the silicon. A difficulty with this model is that it cannot account for the increase in oxide thickness away from the point at which the microchannel intersects the silicon surface. Finally, it has been suggested that the mismatch in the thermal expansion coefficients of the oxide and silicon causes stress in the oxide and this stress may enhance the diffusivity of the oxidizing species [11]. All of these models share a fundamental flaw. The oxidation rate of silicon in the thin oxide regime is limited by the chemical reaction rate at the surface, not the arrival rate of the molecular oxygen. It is easy to show that the linear rate constant is independent of the diffusivity of the oxygen. Recently it has been suggested that the reaction rate of ionized molecules may be larger than that of neutral molecules [8] and that both ionized and neutral species may participate in the process [12].

A second class of models attempts to explain the thin oxide growth regime by appealing to an increase in the oxygen solubility in the oxide. Henry's law is only valid in the absence of dissociation or recombination of the adsorbed gas in the solid. This is not true when the oxide is very thin. Unfortunately this model has had very limited success as well. A third class of model, which is currently popular, suggests that the oxidation reaction occurs over some finite thickness. That is, the interface is not atomically abrupt [13]. The physical origin of the finite thickness is unknown, but it is suggested that some interfacial stress is relieved through an increase in defects near the silicon surface. Oxygen atoms or molecules may diffuse into the silicon slightly and react over these finite widths. The oxidation rate then follows the function

$$\frac{dt_{ox}}{dt} = \frac{B}{2t_{ox} + A} + C_1 e^{-t_{ox}/L_1} + C_2 e^{-t_{ox}/L_2} \qquad \textbf{(4.16)}$$

where L_1 and L_2 are characteristic distances over which the reaction occurs and C_1 and C_2 are constants of proportionality. L_1 is weakly temperature dependent and is typically of order 10 Å. L_2 is temperature independent with typical values of about 70 Å. This model has been shown to accurately fit experimental data even for very low temperature oxidations [14].

Recently it has been shown that the excess oxidation shown in Figure 4.6 may actually be occurring at the surface of the oxide, not at the interface [15],[16]. This is believed to be due to the diffusion of defects such as oxygen vacancies from the Si/SiO₂ interface to the surface where they react. Delarios et al. [17] have developed a detailed chemical model of the reaction, including the production of atomic oxygen at the interface by the chemical reaction of O_2. This model explains all of the experimental thin oxide data without resorting to finite reaction lengths.

4.4 The Structure of SiO₂

The form of SiO₂ that is important for microelectronic fabrication is called *fused silica*. It is amorphous and is thermodynamically unstable below 1710°C. In the temperature range of interest to most microelectronic processing, the rate of crystallization is so slow that it can be neglected. Although fused silica does not have long range order, it does exhibit a short range structure (Figure 4.7). The structure can be thought of as four oxygen atoms sitting at the corners of a triangular polyhedron. In the center of the polyhedron is a single silicon atom. Thus, each of the four oxygen atoms is approximately

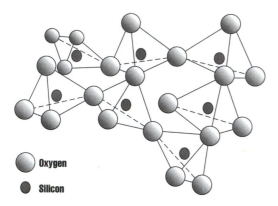

Figure 4-7   The physical structure of SiO_2 consists of silicon atoms sitting at the center of oxygen polyhedra.

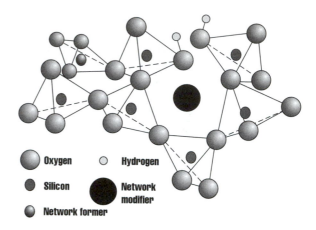

Figure 4-8   Schematic of impurities and imperfections in SiO_2.

covalently bonded to the silicon atom, satisfying the silicon valence shell. If each oxygen atom is part of two polyhedra, the valence of the oxygen would also be satisfied, and a regular crystal structure called quartz would be the result. In fused silica some oxygen atoms, called *bridging oxygen sites*, are bonded to two silicon atoms. Some oxygen atoms are nonbridging, bonded to only one silicon. Thermally grown SiO_2 then can be considered to consist primarily of a randomly oriented network of polyhedra. The larger the fraction of bridging to nonbridging sites, the most cohesive and less prone to damage the oxide is. Dry oxides have a much larger ratio of bridging to nonbridging sites compared to wet oxides.

A variety of impurities (Figure 4.8) can also exist in thermal oxides. Some of the most common are water related complexes. If H_2O is present during the growth, one reaction that can occur is the reduction of a bridging oxygen site into two hydroxyls:

$$Si : O : Si \rightarrow Si : O : H + H : O : Si \qquad \textbf{(4.17)}$$

These hydrogen atoms are only weakly bonded and can be removed under electrical stress or ionizing radiation, leaving a trap or potential charged state in the oxide. Other impurities are intentionally incorporated into thermally deposited SiO_2 to change its physical properties. Substitutional impurities that replace the silicon atom are called *network formers*. The most common network formers are boron and phosphorus. These impurities tend to reduce the bridging to nonbridging ratio, which allows the glass to flow at lower temperature. These impurities however, are normally used in deposited oxides rather than in thermal oxides.

4.5 Oxide Characterization

The oxide thickness is an important parameter of the oxidation process and, as such, many ways have been developed to measure it. This section will describe several methods for estimating the thickness of an oxide. Each has its inherent advantages and disadvantages. Most make some assumption about the oxide that may be valid only under certain circumstances. Aside from the techniques described here there are a number of thin film measurement techniques that can be used.

The first class of measurements involves a physical determination of the oxide thickness. To do this requires the production of a step in the oxide. Typically this is done with a mask followed by an etch. This type of step was introduced in Chapter 1 and will be reviewed in more detail in later chapters. Hydrofluoric acid (HF) etches oxide at a much higher rate than it etches silicon. Therefore, if a mask is applied to the wafer, the wafer immersed in HF, and then the mask removed, a step nearly equal to the oxide thickness will be left. This step can be measured using a scanning electron microscope (SEM) if it is larger than 1000 Å, or with a transmission electron microscope (TEM) if it is not. An easier approach is to use a surface profilometer. These instruments measure the surface topology by mechanically scanning the wafer while it is in contact with a needle stylus. The deflection of the needle is measured, amplified, and displayed as a function of position. Resolution of these instruments down to 2 Å is claimed by the manufacturers. This technique has the advantage that it makes no assumptions regarding the oxide other than the relative etch rates of the oxide and the silicon. Since part of the oxide must be etched to determine the thickness, this test is destructive and generally requires the use of a dedicated test wafer.

There are also several optical techniques that can be used to measure oxide thickness. The simplest is to partially immerse the unmasked wafer in dilute HF until the oxide on the submerged portion of the wafer is completely removed. Near the line between the etched and unetched oxide a slow grading of the thickness will be found. If one examines this edge under a microscope a variety of colors will be seen starting from tan (Table 4.2). These colors are due to interference between the incident and reflected light. By following the colors up to the top of the oxide an approximate thickness can be found.

There are at least two other, more sophisticated optical techniques for measuring the oxide thickness. The first is *ellipsometry*. In this technique, a polarized coherent beam of light is reflected off the oxide surface at some angle. He-Ne lasers are commonly used as a source. The reflected light

Table 4.2   Color chart for thermal dioxide (refractive index of 1.48) and silicon nitride (refractive index of 1.97). Note that multiple orders exist. An SiO_2 film that appears red may be 730 to 970 Å, 2400 to 2500 Å, or 4000 to 4400 Å.

Color	SiO_2 Thickness (Å)	Si_3O_4 Thickness (Å)
Silver	<270	<200
Brown	<530	<400
Yellow-brown	<730	<550
Red	<970	<730
Deep blue	<1000	<770
Blue	<1200	<930
Pale blue	<1300	<1000
Very pale blue	<1500	<1100
Silver	<1600	<1200
Light yellow	<1700	<1300
Yellow	<2000	<1500
Orange-red	<2400	<1800
Red	<2500	<1900
Dark red	<2800	<2100
Blue	<3100	<2300
Blue-green	<3300	<2500
Light green	<3700	<2800
Orange-yellow	<4000	<3000
Red	<4400	<3300

intensity is measured as a function of the polarization angle. By comparing the incident and reflected intensity and the change in the polarization angle, the film thickness and the index of refraction can be found. To do this definitively requires measurement at more than one incidence angle, since there is more than one thickness that will produce the same change in the light at any given angle. Ellipsometry has the advantage of being nondestructive, although it often requires that the oxide be grown on bare silicon. Since the ellipsometer's beam is quite large, it is also normally done on unpatterned wafers.

A second optical technique for measuring oxide thickness is *interference*. Commercially marked under the name of Nanospec®, this system uses light at a nearly normal incidence to the film. The reflected light intensity is measured as a function of the wavelength. When the wavelength of the light is such that the incoming and outgoing waves constructively interfere, an optical maximum will result. If the waves interfere destructively, a minimum will be seen. By measuring the difference in wavelength $\Delta\lambda$ between maxima or minima, the thickness can be determined as

$$t_{ox} = \frac{\Delta\lambda}{2n_{ox}} \tag{4.18}$$

where n_{ox} is the real part of the index of refraction of the oxide; n_{ox} is assumed to be wavelength independent in this model. The technique can measure the thickness of transparent films down to several hundred angstroms. The upper limit for thickness measurement is determined by optical losses in the film and by the ability to resolve high order peaks.

Electrical techniques are the most useful ways to characterize oxide layers. The simplest electrical measurement is the breakdown voltage. In this test, a large number of metal electrodes are formed on top of the oxide. The voltage on the capacitor is increased while the electrical current through the oxide film is measured. The leakage current is too small to measure until a high electric field is reached. Eventually, for thin oxides, a current will be detected that will rise exponentially with voltage. Within a small voltage range the current increases discontinuously, signaling an irreversible rupture of the oxide. The dielectric field strength of thermal oxide is about 12 MV/cm. Thus, by knowing the breakdown voltage, an estimate of the oxide thickness can be determined. If, as is usually the case, the thickness is known, the breakdown field can be measured. A breakdown histogram is often presented as a first order indication of oxide quality and defect density (Figure 4.9). Three groups of breakdown regions are evident. The low voltage group is known as extrinsic breakdown. These are killing defects such as pinholes in the growth process. The high voltage group are intrinsic breakdown. They typically cluster near the ultimate breakdown field of the oxide. The intermediate group is usually associated with weak spots in the oxide. The larger the fraction of breakdown events contained in the intrinsic breakdown group, the better the oxide quality. To be significant, the area of the capacitor tested should be similar to the active area on a chip.

Figure 4-9   Typical breakdown histogram for an oxide.

A popular variant of the simple breakdown histograms is the charge to breakdown test. In this technique, the oxide is stressed to a point just below the breakdown field. This may be done in constant current, constant voltage, or ramped modes. This test is also known as the Time Dependent Dielectric Breakdown (TDDB) test. The current in a TDDB test typically is of the Fowler-Nordheim type and follows the equation

$$J_{FN} = AE_c^2 e^{-B/E_c} \tag{4.19}$$

where E_c is the electric field at the point of tunneling into the oxide and A and B are constants that depend on the height of the barrier and the electron (or hole) effective mass [18]. A typical plot of the resultant curve for a constant voltage test is shown in Figure 4.10. The current through the oxide decreases due to trapping of electrons in the bulk of the oxide. The breakdown is believed to result from an accumulation of trapped positive charge near the interface. The charge to breakdown is measured by integrating the current from the start of the test until just before breakdown. It has been proposed [19] that the oxide is made up of robust and weak areas. The weak areas may be physically thinner than the rest of the oxide or perhaps have defects that easily trap charge.

An interesting aspect of this work is the application of TDDB tests to very thin oxides. A constant current TDDB test for a 62 Å oxide is shown in Figure 4.11. There is very little evidence of bulk electron trapping. This is believed to be due to the ability of the electrons to tunnel from trapped states in the oxide to either the substrate or the gate electrode [20] in a process called *tunnel annealing*. These thin oxides also show very large charge to breakdown values. This is believed to result from the small voltages required for stress conditions. For example, a 50 Å oxide requires only 5 V to develop a 10 MV/cm field. At these voltages, electrons in the oxide cannot gain sufficient energy in the oxide to suffer impact ionization and therefore, to create the holes that eventually trap at the interface. Other less efficient hole generation mechanisms must be used.

More sensitive methods of evaluating oxides use capacitance-voltage measurements. Again, a metal film must be used as an upper electrode. The wafer serves as a lower electrode. Assume for the moment that the substrate is doped p-type. If a negative voltage is applied to the gate, additional holes

Figure 4-10   Constant voltage stress measurements of a 200 Å oxide film. The sharp increase in current near the end of the trace indicates that irreversible breakdown has occurred.

Figure 4-11   Similar plot to Figure 4-10 for a 55 Å oxide. Note the lack of charge trapping and the low voltage required for stressing.

are drawn to the Si-SiO$_2$ interface. This condition is known as accumulation. Now assume that a small ac signal is added to the dc bias and the ac current is measured. The out of phase magnitude of this current is proportional to the capacitance. If the diameter of the capacitor is large compared to the oxide thickness, one can consider the device to be a parallel plate capacitor. Then

$$C_{ox} = \frac{\varepsilon_{ox} \text{Area}}{t_{ox}}$$ **(4.20)**

Again, this is a valuable method of measuring the oxide thickness. For very thin oxides however, one must also take into account the finite width of the accumulation layer in the semiconductor [14].

If the bias voltage is ramped from the negative set point to positive values, the measured capacitance will decrease. This occurs because the field changes sign, repeling the charge directly below the gate. This effectively increases the width of the dielectric, reducing the capacitance. Another way of viewing the situation is that a second capacitor, which extends from the Si/SiO$_2$ interface to the edge of the depleted region of the substrate, has been formed in series with the oxide capacitor. The voltage at which this depletion layer forms depends on the doping concentration in the substrate and the oxide thickness. If these are known, an ideal C-V curve can be calculated and compared to the predicted value.

All SiO$_2$ layers have a thin layer of positive charge near the SiO$_2$-Si interface. This is known as the "fixed" or "built in" oxide charge. It is believed that the fixed charge is related to a thin (about 30 Å) transition region near the interface that has excess silicon ions. This excess silicon broke away from the lattice during the oxidation process, but had not yet fully reacted with an oxygen molecule. The fixed charge density decreases if an inert anneal is done after the oxidation. A classic demonstration of this effect is the Deal triangle shown in Figure 4.12. Although the data shown is for ⟨111⟩ silicon, the same effect is seen in ⟨100⟩ material. The fixed charge density is dramatically decreased if the oxidation is followed by a high temperature inert ambient anneal [21]. For that reason, most oxidation processes include a short nitrogen or argon anneal before pulling the wafers from the furnace. More recently, it has been found that the roughness of the Si-SiO$_2$ interface depends on the oxidation temperature. Higher temperatures produce more abrupt interfaces [22]. This effect has been attributed to the reduction in SiO$_2$ viscosity at high temperature [23].

The effect of the fixed charge is to shift the C-V curve laterally. This will in turn change the threshold voltage at which the MOS transistor "turns on" (see Section 16.1). If the shift is reproducible and is a small fraction of the threshold voltage, this does not represent a serious problem. The fixed charge density can be calculated from this voltage shift using

$$\Delta V_g = -\frac{Q_f}{C_{ox}}$$ **(4.21)**

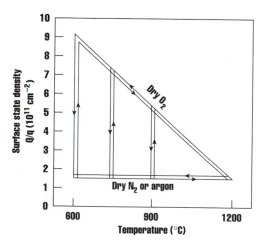

Figure 4-12 The Deal triangle showing the effects of a high temperature inert (nitrogen or argon) post oxidation anneal on interface states and fixed charge density *(after Deal et al., reprinted by permission, The Electrochemical Society).*

Often the capacitance is normalized to the area of the device. Then the fixed charge density is quoted in units of C/cm^2. This effect can be difficult to measure however, since several other effects also serve to shift the C-V trace along the voltage axis.

One of these other sources of lateral shift of the C-V trace

is mobile ionic charge, most typically sodium ions. A trace contaminant, sodium has a large mobility in silicon dioxide. These charged ions have the same effect as the oxide fixed charge, except that one must also consider the charge distribution through the oxide. Then the lateral voltage shift is given by

$$\Delta V_g = -\frac{1}{\varepsilon_{SiO_2}} \int_0^{t_{ox}} \rho(x)\, x\, dx \qquad (4.22)$$

where $\rho(x)$ is the mobile ion density. In the limit that the charge is a delta function at the interface, it is easy to show that this reverts to the formula given for the fixed charge.

The presence of mobile ionic charge represents a serious reliability problem for field effect devices. At 100°C, the diffusion coefficient for Na in SiO_2 is about $3 * 10^{-15}$ cm²/sec. If the device is left at temperature for an hour, the characteristic diffusion length is about 3000 Å. Since gate oxides of most MOSFETs are less than a few hundred angstroms, the mobile ionic charge will continuously redistribute itself in the oxide in response to changes in the gate bias. This changes the C-V curve and consequently, the threshold voltage of the device. A circuit may work initially, but fail after sufficient ionic drift.

Mobile ionic charges are easily detected by temperature bias stressing [24]. An MOS capacitor is first prepared and a C-V trace is taken. The sample is then heated above 100°C and a positive field of 2 to 5 MV/cm is applied to the gate. This bias and temperature are held for 10 to 20 min. The voltage is removed, and the capacitor is cooled to room temperature. Next the C-V trace is retaken. Finally the process is repeated, this time with a negative bias on the gate. Figure 4.13 shows typical results for a sample with positive ionic charge. Initially the ions are randomly distributed through the oxide. Under the influence of a positive gate potential, the ions drift to the interface, where they strongly shift the C-V trace. After a negative drift is applied, the ions move to the gate-SiO_2 interface where they do not affect the C-V trace. Then the mobile ionic sheet charge can be calculated from the voltage shift using Equation 4.16.

The ionic contamination can come from a variety of sources, including the furnaces themselves, wet chemicals used for processing and precleaning, and certain photoresists. Manufacturers have largely solved this problem, and VLSI grade chemicals are widely available with very low ionic concentrations. The processes that have the most stringent requirements, such as gate oxidation, may use additional precautions such as a purged, double walled furnace tube. A tube preclean consisting of 1 to 3% of HCl in an inert ambient has also been found to lower contamination levels by reacting the ions to form compounds more readily swept from the tube. Halogen containing oxidations (see Section 4.3) also reduce mobile ionic charge.

Consider the diagram of the Si-SiO_2 interface shown in Figure 4.14 [25]. In the ideal picture the interface is atomically abrupt with perfect purity; how-

Figure 4-13 Typical C-V traces from temperature bias stress measurements of an oxide contaminated with a positive ion impurity.

ever, a more realistic figure also includes oxide fixed charge and mobile ionic impurities. Another common imperfection is interface states. Due to the change from a perfectly ordered crystal to an amorphous solid, not all of the bonds at the interface will be satisfied. Since the surface atoms have incomplete valences and unsatisfied bonds, electrons and holes can readily be captured. These sites are therefore traps, whose energy depends on the local bond configuration. Since this is not fixed, there will be a distribution of trap energies throughout the band gap. Because these traps are low in energy and are located right at the interface, it is also easy for carriers to detrap. As a result, the traps can fill and empty every time the gate is biased. To understand the magnitude of the effect, consider the fact that the unreconstructed (100) silicon surface has $6.8 * 10^{14}$ atoms/cm^2. Assume that the gate oxide is 150 Å thick. Finally, assume that the allowable voltage shift is 0.1 V and that the effect of a charged interface state is the same as that of an oxide fixed charge. Then the sum of the interface trap density and oxide fixed charge density must be less than $2.3 * 10^{-8}$ C/cm^2, or less than $1.4 * 10^{11}$ defect/cm^2, or about 1 in 5000 sites. The interface therefore, must be highly perfect for typical MOS transistors.

Since the charge state of the interface traps depends on the voltage, their effects on the C-V trace are more complicated than those of fixed or mobile ion charge. Figure 4.15 shows the effect of interface states and bulk charge on the C-V trace. After stressing an ultrathin (~50 Å) oxide by passing a small current through it, both fixed charge and interface states are visible on the high frequency C-V trace. The fixed charge moves the trace to the left while the interface states reduce the slope of the C-V trace in depletion. The fixed charge in these particular devices decreases spontaneously through *tunnel annealing*. The most popular techniques of measuring the interface state density all involve taking C-V measurements. In the Terman [26] method, the measured high frequency C-V curve is compared to the ideal curve calculated from the oxide thickness, the metal semiconductor work function, the fixed charge density, and semiconductor doping concentration. Alternatively, one can measure the difference between the high and low frequency C-V traces [27] or the difference between the low frequency and theoretical low frequency curve [28]. None of these calculations are trivial. To extract the interface state density the doping concentration must also be assumed constant, which is generally not true. The reader is referred to Nicollian and Brews [29] for further information.

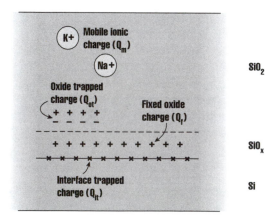

Figure 4-14 Silicon/silicon dioxide structure with mobile, fixed charge, and interface states *(©1980, IEEE, after Deal)*.

Figure 4-15 High frequency C-V traces showing the effects of interface states and fixed charge.

4.6 The Effects of Dopants During Oxidation and Polysilicon Oxidation

In nearly all oxidation processes the substrate has some doping concentration. This concentration will change during oxidation. The profile depends on the rate of diffusion of the impurity in the oxide and the segregation coefficient m, where

$$m = \frac{\text{concentration of impurity in Si}}{\text{concentration of impurity in SiO}_2} \tag{4.23}$$

Redistribution during oxidation occurs in a manner analogous to that during crystal growth.

Grove et al. [30] developed a widely adopted pictorial representation of four classes of impurities shown graphically in Figure 4.16. If $m > 1$, the oxide rejects the impurity. As a result, it accumulates in the silicon under the growing film, reaching a maximum at the interface. This effect can be offset however, if the impurity diffuses rapidly in the SiO₂. In this case, the dopant is rapidly removed from the interface. The impurity concentration in the substrate at the interface, while larger than in the oxide, is still less than the concentration in the bulk. If $m < 1$, the oxide is said to take up the dopant. Here the impurity concentration in the substrate decreases near the interface.

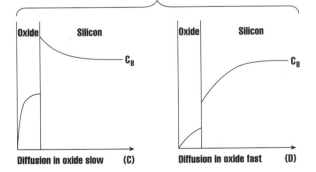

Figure 4-16   The effect of thermal oxidation on the impurity distribution in silicon and silicon dioxide. (a) Slow diffusion in oxide, $m < 1$ (boron in neutral or oxidizing ambient); (b) fast diffusion in oxide, $m < 1$ (boron in hydrogen ambient); (c) slow diffuser in oxide $m > 1$ (phosphorus, arsenic); (d) fast diffuser in oxide, $m > 1$ (gallium) *(after Grove et al.).*

Figure 4-17   Temperature dependence of boron segregation coefficient for various types of oxidations *(reprinted by permission, McGraw-Hill, after Katz).*

Figure 4.17 shows the segregation coefficient of B in Si for a variety of oxidizing conditions [31]. In these experiments, near dry oxidations means an oxidizing furnace that uses O_2 as the feed gas; however, no special provisions have been made to remove water from the ambient. Obviously, the segregation coefficient can vary, but is well described by an Arrhenius function. Phosphorus, arsenic, and antimony all have segregation coefficients of about ten [32].

The common silicon dopants all tend to enhance the oxidation rate of silicon when present in the substrate in high concentrations. Boron, which segregates into the oxide, enters and weakens the glassy structure, reducing its viscosity. When the boron surface concentration exceeds 10^{20} cm^{-3}, this also has the effect of increasing the diffusivity of molecular oxygen. As a result the parabolic rate coefficient increases [33] when the surface is heavily doped with boron (Figure 4.18).

With heavy phosphorus doping the parabolic rate coefficient shows only modest increases, but the linear rate coefficient increases rapidly for surface doping levels greater than 10^{20} cm^{-3} (Figure 4.19) [34],[35]. This is believed to occur because the segregation coefficient causes the phosphorus to accumulate at the surface. A model to explain the increased reactivity of the surface suggests that this high concentration of phosphorus shifts the Fermi energy and thereby increases the surface vacancy concentration [36]. These vacancies provide additional oxidation sites and so increase the reactivity.

The oxidation of polysilicon is of considerable interest for such applications as the production of thin oxides between poly layers in EEPROMs, the oxidation of polysilicon plugs for DRAMs, and the oxidation of poly gates that occurs during reoxidation steps. It has been found that an enhanced oxidation rate occurs due to stress at the grain boundaries [37]. Figure 4.20 shows the oxide thickness for wet and dry undoped poly oxidations for various temperatures [38]. For oxidation temperatures less than 1000°C, the oxidation rates for undoped polysilicon are larger than either (100) or (111) silicon at short times. For longer oxidations the oxide thickness falls to a level intermediate between (100) and (111) silicon. When the polysilicon is heavily doped with phosphorus, the poly oxidation rate is less than that of either (100) or (111) single crystal silicon.

Figure 4-18   Silicon dioxide thickness versus wet oxidation time for three different surface concentrations of boron *(after Deal et al., reprinted by permission, The Electrochemical Society)*.

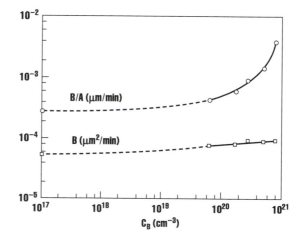

Figure 4-19   Oxidation rate coefficients for dry oxygen at 900°C as functions of the surface concentration of phosphorus *(after Ho et al., reprinted by permission, The Electrochemical Society)*.

Figure 4-20   The oxidation of undoped polysilicon in (a) wet and (b) dry ambients *(after Wang et al., reprinted by permission, The Electrochemical Society)*.

4.7 Oxidation Induced Stacking Faults

As mentioned in the previous chapter, thermal oxidation has the side effect of releasing a high concentration of silicon self-interstitials. This is the cause of the oxidation enhanced diffusion. These excess interstitial atoms constitute point defects. As mentioned in the second chapter, the free energy of a group of point defects can be reduced if they agglomerate into a larger defect. In oxidation processes, the extended defects that are usually created are Oxidation Induced Stacking Faults (OSF). Stacking faults are 2-D defects that are analogous to dislocations. They are an extra plane of atoms, of finite extent, inserted into the lattice. The OSF usually lies in the {111} plane and is bounded by partial dislocations. The defects often occur close to the Si/SiO$_2$ interface where the excess interstitial concentration is large. OSFs are very readily induced at preexisting defects such as those cause by ion implantation (see Chapter 5).

These defects can have several undesirable effects. In bipolar transistors, the dislocations that terminate the stacking fault can be a rapid diffusion path for the emitter dopant. The results are shorts across the base layer called *pipes*. In MOS technologies the primary concern is that these defects can getter heavy metal impurities that will reside in the depletion region of source/drain junctions. The impurities will act as recombination centers that will cause excess leakage in the device. Figure 4.21 shows optical micrographs of samples after a preferential etching step used to delineate the defects

Figure 4-21   Oxidation induced stacking faults in ⟨100⟩ wafers after preferential etching. As the preferential etching continues, more of the defect is exposed *(after Ravi)*.

[39]. Stacking faults in the diamond lattice lie along the {111} planes. For ⟨100⟩ wafers these planes intercept the surface along the ⟨110⟩ directions [40].

It has been shown that the length of the stacking fault is almost linearly proportional to the oxidation time [41]. Much more interesting however, is the dependence of the length on temperature [42]. As shown in Figure 4.22, there are two distinct temperature regimes. At high temperature the formation of OSFs is not favored. Stacking faults initially grow in this temperature regime but shrink with increasing oxidation time. Any existing faults will also tend to shrink. Since dopant redistribution considerations virtually preclude these high temperature oxidations however, Figure 4.22 suggests that oxidations should be done at reduced temperature. An advantage of high pressure oxidation is that, for a given growth time, the temperature can be lowered and the formation of OSF minimized. It has also been shown that the use of HCl in the ambient completely suppresses the formation of stacking faults.

4.8 Alternative Thermal Dielectrics[+]

The search for alternative dielectric systems for the MOS gate oxide has a long, and not particularly fruitful, history. Besides the thickness, only minor modifications exist between the way oxides were

Figure 4-22 Typical stacking fault length as a function of temperature for (a) dry and (b) wet oxidation. Stacking faults are not favored at either high or low temperature *(after Hu)*.

grown for 5 μm MOSFETs and the way they are being grown for many 0.5 μm MOSFETs. The impetus for an alternative dielectric is usually either a higher dielectric constant for increased current drive at a given dielectric thickness or an increase in the ability of the gate dielectric to withstand electrical stress or impurity diffusion from the gate. The penalty that is often paid is an increase in the interface state density that leads to a reduced mobility due to coulomb scattering and a poorly controlled threshold voltage. Deposited films for VLSI gate dielectrics tend to have high defect densities for the extremely thin films typically used in this application. While partially deposited films have been used commercially, only thermal films will be considered in this chapter. Other applications such as the charge storage elements used in DRAMs are able to select the materials for the electrodes, so some deposited films are more practical for that application. These films will be covered in a later chapter.

In the search for alternate thermal dielectrics, it is natural to consider the use of silicon nitride [43], which has a dielectric constant about twice that of oxide. Deposited nitride films are already used in most silicon technologies. In addition to the higher dielectric constant, nitride films are also excellent diffusion barriers that in MOS transistors can prevent impurities in the heavily doped polysilicon gate from reaching the channel. There are three difficulties with thermal nitrides. The first is that the dominant diffuser during nitridization is Si, not N_2 or NH_3. This means that the nitride film forms at the top surface, not at the Si/Si_3N_4 interface. It is not surprising then that thermal nitrides tend to have a high interface state density. Second, the thermal expansion coefficient for silicon nitride

is about twice that of silicon. Since thermal nitridization requires very high temperatures (due to its properties as a diffusion barrier), during heat up and cool down this difference leads to strain that may further degrade the interface. The third difficulty associated with thermal nitrides also arises because the films are good diffusion barriers. Restricting the furnace temperature to 1200°C, it is difficult to grow nitride layers thicker than 40 Å because the silicon diffusivity is so small [44]. Higher temperatures are generally not tolerated due to dopant diffusion in the substrate.

Ito first suggested the use of nitrided oxides as an attractive alternative to SiO_2 [45]. By exposing a thermal oxide to a high temperature NH_3 anneal, the oxide reacts to form an oxynitride: SiO_xN_y, where the values of x and y depended on the process conditions. Ammonia anneals produce a layer containing as much as 50% nitrogen at the surface of the oxide that tails off in the bulk of the film [46]. Other groups have also found a buildup of nitrogen at the SiO_2/Si interface [47],[48]. The thickness of the films increases slightly during the process and the dielectric constant increases by as much as 20%.

Ito also found that these nitrided films were more resistant to breakdown [49]. Other groups have confirmed these results and have also found that nitrided films are more resistant to radiation damage as well [50]. It has been speculated that the durability of these films is related to their properties as diffusion barriers. Hydrogen, which passivates the dangling bonds at the interface and in the oxide bulk, is believed to be released upon stressing. Oxynitrides may trap the hydrogen, preventing it from escaping the dielectric. A larger fraction of it therefore, eventually diffuses back to a trap site and repassivates the damage.

The difficulty with nitrided oxides is that these films have a high concentration of bulk electron traps under low field operation [51]. These traps appear to be associated with an increase in the concentration of OH bonds that form in the dielectric during the high temperature NH_3 anneal. It has been found that a reoxidation of a nitrided film (RNO) can reduce this liability significantly [52],[53]. These films, particularly when grown under rapid thermal processing conditions, are currently generating a great deal of interest for gate dielectrics in ULSI devices [54], where the oxide thickness is insufficient to prevent outdiffusion from the gate electrode. Perhaps more importantly, when heavily nitrided, these films suppress interface state generation during electrical stressing. As a result, NMOS transistors can be two orders of magnitude more reliable than non-nitrided oxides.

Figure 4-23   Horizontal tube oxidation/diffusion system includes computer controller, load station, and four tubes *(photo courtesy of ASM International).*

4.9 Oxidation Systems

The horizontal diffusion furnace has been the mainstay of the semiconductor industry for decades. While many innovations have been introduced, the basic idea remains unchanged. As shown in Figure 4.23, the furnace consists of four components: the gas source cabinets; the furnace cabinet, including power supplies, tubes, thermocouples, and elements; the load station; and the computer controller. As the name implies, the

gas cabinet contains the source gasses, typically in high pressure cylinders, and means for producing a controlled flow of these gasses into the furnace. Typically, sets of pressure regulators, pneumatic valves, and gas filters are used in this application. The gasses are plumbed to a shelf at the end of the furnace tube. This shelf contains additional valves and mass flow controllers (see Chapter 10) that control the flow of gasses into the furnace tube.

The furnace cabinet houses long tubes, typically four, that are stacked vertically. Furnace heating elements are wound on high purity ceramic forms large enough to accept a 150 to 300 mm diameter fused silica tube. The windings are often separated into three heating zones and fed by three phase power through a high current power controller. The furnace temperature is monitored in at least three locations with thermocouples. The thermocouple voltages are read, compared to the desired temperatures, and the error signal is amplified and fed back to the power controllers. It is not unusual for the flat zone in the center of the furnace to exhibit 0.5°C uniformity over temperature ranges of 400 to 1200°C. If wet oxidations are to be done, additional equipment is also required to ensure that the hydrogen gas completely combusts upon injection.

A. Cool down stage
B. Process stage
C. Wafer transfer stage
D. 12 cassette carousel
E. Wafer handling section
F. Processing section

Figure 4-24 Plan view of vertical furnace. Cassettes with 25 wafers are loaded into the carousel and transferred to the process stage where they are raised into the furnace *(photo courtesy of ASM International)*.

Figure 4-25 Vertical oxidation system showing carousel, robot arm, and wafers about to be raised into the furnace *(photo courtesy of ASM International).*

The wafers are loaded into fused silica holders called *boats* in the load station. The boats typically hold 25 wafers in a vertical rack. The boats are loaded onto carriers that can hold up to 200 wafers. Early carriers were actually fused silica carriages, with skids or fused silica wheels that were slowly rolled into the furnace. Concerns with particulation however, caused these carriers to be replaced by cantilevered load systems in which the boats are supported on two long rods. In these systems, the wafers never touch the furnace walls. Due to difficulty maintaining the rod linearity, many of these systems have now been replaced by soft landing systems. In these stations, the loading system carries the boats into the furnace, deposits them, and then withdraws minimizing the time at temperature for the boat support system. In each system care must be taken to avoid jostling the wafers during this process to avoid defect formation.

A microcomputer controls all of the furnace operations. Specific recipes are programmed for each process step. Disk drives are provided to store these recipes for future recall. For example, a gate oxidation step might include a preoxidation tube clean in a mixture of O_2 and HCl at 1100°C for 60 min, followed by a nitrogen purge and cool down to 800°C, followed by a slow push, or loading of the wafers in a mixture of O_2 and N_2, followed by a controlled furnace ramp to 1000°C, followed by an oxidation in O_2 and HCl, followed by a short N_2 anneal to 1050°C to reduce the oxide fixed charge, followed by a controlled cool down to 800°C, followed by a slow pull, or removal of the wafers. The microcomputer allows this to be a "one button" operation for highly reproducible results. It can also be programmed to anticipate thermal loads such as the paddle insertion by ramping the temperature just before loading to minimize temperature fluctuations. The temperature control systems are now completely digital and interface well with the furnace controller. The furnace controller also can be interfaced to a facility management computer. With the use of robotic load stations, this allows the diffusion furnaces to be completely automated.

More recently, vertical furnaces are starting to gain acceptance. As shown in Figures 4.24 and 4.25, these furnaces are similar to a horizontal furnace turned on end. Wafers are pushed from below the furnace up into the tube. These systems have four main advantages. The vertical orientation means that the wafers remain horizontal. As a result these furnaces are easier to automate. Robot wafer handlers can easily load and unload the tubes. Furthermore, the uniform spacing improves process uniformity across the wafer, particularly for large diameter substrates. Also, there is no need to deposit the wafers in the tube. There is no net force on the cantilevers except down, so they do not tend to warp with usage and can remain in the furnace. Finally vertical furnaces tend to have a smaller clean room footprint than horizontal furnaces.

4.10 SUPREM Oxidations[+]

The last chapter introduced the numerical calculation tool SUPREM to simulate the diffusion profiles of impurities in silicon. The model can also be used to perform oxidations based on the Deal-Grove model. The tool has incorporated Arrhenius functions to describe the linear and parabolic rate coefficients for wet and dry silicon coefficients, as well as a rudimentary model for chlorinated oxidation.

Oxidation processes are accessed by the same command as diffusion processes: DIFFUSION. For oxidation to occur, simply add the parameter DRYO2 or WETO2 for dry and wet oxidations, respectively. The parameters for these processes can be adjusted before invoking the DIFFUSION statement, with the DRYO2 and WETO2 commands. High pressure oxidations can be accomplished by also adding the parameter PRESSURE=x, where x is the oxygen or steam pressure in atmospheres. The pressure can be ramped by using the parameter, P.RATE=x, where x is the time rate of change of the pressure in atmospheres per minute. For chlorinated oxidations the parameter is HCL=x, where x is the percentage of HCl in the ambient. For diluted flows one can also use the commands, F.H2=x, F.H2O=x, F.HCl=x, F.N2=x, and F.O2=x, where the x values correspond to the gas flows in standard liters per minute of the corresponding gasses.

In the thin oxide regime, SUPREM III uses an empirical oxidation model [55]

$$\frac{dt_{ox}}{dt} = \frac{B}{2t_{ox} + A} + Ce^{-t_{ox}/L} \tag{4.24}$$

where B and A are the oxidation rate coefficients from the Deal-Grove model, and C and L are empirical constants. Notice that this equation closely follows the model of Massoud for $t_{ox} > L_1$, which is about 10 Å. Impurity segregation coefficients and oxidation enhanced diffusion models are also incorporated in the code so that impurity redistribution during oxidation can be modeled.

Example 4.2   **SUPREM oxidation example—gate oxidation**

```
Title      Oxidation Example—Thin Gate Oxidation in HCl
Comment    Initialize the silicon substrate
Initialize ⟨100⟩ Silicon Boron Concentration=1e17
+          Thickness=0.5 dX=.0025 Xdx=.01 Spaces=120

Comment    Push the wafers for 30 min at 800°C in N₂
Diffusion  Time=30 Temperature=27 Nitrogen T.rate=25.7667

Comment    Ramp the furnace to 1000°C over 10 min in a flow of
           10% O₂ and 90% N₂
Diffusion  Time=10 Temperature=800 F.n2=1.8 F.o2=0.2 T.rate=20
Print Layers

Comment    Oxidize the wafers for 30 min at 1000°C in a
           mixture of O₂ with 3% HCl
Diffusion  Time=30 Temperature=1000 DryO2 Hcl=3.0

Comment    Anneal the wafers at 1000°C for 10 min in a flow
           of N₂
```

```
Diffusion   Time=10 Temperature=1000 Nitrogen

Comment     Ramp the furnace to 800°C over 10 min in a flow of N₂
Diffusion   Time=10 Temperature=1000 Nitrogen T.rate=-20

Comment     Pull the wafers for 30 min at 800°C in N₂
Diffusion   Time=30 Temperature=27 Nitrogen T.rate=-25.7667

Print       Layers Chemical Concentration Boron
Plot        Active Net Cmin=1e14
Stop        End oxidation example
```

After the run is complete, find the oxide thickness. How much of the oxide was grown during the ramp? In this process it was assumed that there was no initial oxide. How would you incorporate an initial oxide (such as a thin native oxide) into your simulation? From the impurity profiles determine the boron segregation coefficient that SUPREM assumes. Which of the four categories in Figure 4.16 does boron fall into under these conditions? Taking into consideration the starting thickness and the change in the oxide thickness for the HCl process, by what percentage has the oxidation rate been enhanced by using HCl? Calculate the answer directly using the Deal-Grove model and check it using SUPREM by eliminating the HCl in the simulation and rerunning SUPREM.

4.11 Summary

This chapter introduced the topic of the thermal oxidation of silicon, presenting the Deal-Grove model. This model accurately predicts the oxide thickness of a wide range of oxidation parameters. Enhanced growth rates are seen for thin oxides. While its origin is uncertain, several models that attempt to explain the results were presented. Oxidation is also known to induce defects in the bulk silicon because of the high concentration of self-interstitials that the process produces. Techniques for avoiding these defects are described. Finally, typical oxidation systems are described and the application of SUPREM to oxidation is presented.

Problems

1. A 1000 Å gate oxide is required for some technology. It has been decided that the oxidation will be carried out at 1000°C, in dry oxygen. If there is no initial oxide, for how long should the oxidation be done? Is the oxidation in the linear regime, the parabolic regime, or between the two?

2. Repeat Problem 1 if the oxidation is done in a wet O_2 ambient.

3. It has been decided to grow the oxide in Problem 1 in two steps. A 500 Å oxide will be grown first, then the wafers will be reoxidized to a total thickness of 1000 Å. If the oxidations are carried out at 1000°C in dry O_2, calculate the time required for each of the oxidations.

4. It is necessary to grow a 1 μm field oxide to isolate the transistors in a certain bipolar technology. Due to concerns with dopant diffusion and stacking fault formation, the oxidation must be carried at 1050°C. If the process is carried out in a wet ambient at atmospheric pressure, calculate the required oxidation time. Assume that the parabolic rate coefficient is proportional to the oxidation pressure. Calculate the oxidation time required at 5 atm and at 20 atm.

5. Assume that negatively charged O_2^- has exactly twice the diffusion coefficient in SiO_2 as neutral O_2 (due to the field aided term), but 10 times the reactivity at the surface, with exactly the same activation energy for the reaction rate coefficient. Repeat Problem 1 in a source of 1 atm of O_2^-.

6. For submicron MOSFETs, it is often necessary to grow gate oxides of order 100 Å. Although process control is very difficult due to the short oxidation times involved, it is preferable to grow these oxides at high temperatures. Explain why.

7. One technique that has been used with some success to solve the problem mentioned in Problem 6 is to grow the oxides in a dilute mixture of oxygen and an inert species such as argon. Assume that we have a mixture containing 10% O_2 and 90% Ar. If we ignore the rapid oxidation regime normally associated with thin oxides and assume, as in Problem 4, that the parabolic rate coefficient B depends on the pressure of the oxidant, calculate the growth time for a 1000°C oxidation of 100 Å. Assume no initial oxide.

8. If $t_{ox} \gg 10$ Å, we can ignore the C_1 term in Equation 4.15. If we also restrict our attention to relatively thin oxides, we can make the approximation $t_{ox} \ll A$. (In the Deal-Grove model this is the linear approximation.) For that case, integrate Equation 4.15 and find an analytic expression for $t_{ox}(t)$. Take the oxide thickness to be zero at time τ.

9. Assume that L_2 is 70 Å, and C_2 is 50 Å/min. If the 100 Å oxide described in Problem 7 is grown at 1000°C in a dry oxygen ambient, and there is no initial oxide, determine the oxidation time both with and without the thin oxide rate enhancement term from the equation derived in Problem 8. Are both of the limits required for the derivation in Problem 8 strictly satisfied? If not, qualitatively describe the effect.

10. Repeat Problem 9 for a growth done in an ambient of 10% oxygen and 90% argon.

11. A thermal oxide thickness is measured both by use of a Nanospec and by measuring the accumulated capacitance. The results are found to differ by 20%, even though the same wafer was used for both measurements. Give three possible errors that might account for the discrepancy.

12. A 250 Å gate oxide is found to have a 15 mV temperature bias stress shift. Calculate the number of mobile ions per unit area in the oxide.

13. Modify the example SUPREM deck to simulate 30 min wet oxidations at 920, 1000, and 1100°C. Compare the calculated oxide thickness with the Deal-Grove predictions. Calculate the segregation coefficient from your results and compare it to the graph in Figure 4.17. Change the substrate concentration to $1.5 * 10^{20}$ cm^{-3} of boron and rerun the 920°C simulation. Compare the results to those of Figure 4.18. What does this tell you about the model used by SUPREM?

14. Use SUPREM to grow several dry oxides at 1000°C for 1, 2, 5, and 10 min. Calculate the oxide thickness as a function of time. Does your version of SUPREM attempt to model the initial oxidation regime? If not, how can a program user increase the accuracy of moderate thickness (200 Å $< t_{ox} <$ 1000 Å) dry oxidation runs?

References

1. B. E. Deal and A. S. Grove, ''General Relationship for the Thermal Oxidation of Silicon,'' *J. Appl. Phys.* **36**:3770 (1965).

2. M. M. Aptyalia, in *Properties of Elemental and Compound Semiconductors*, H. Gates, ed., Interscience, New York, 1960, p. 163.

3. R. S. Ronen and P. H. Robinson, ''Hydrogen Chloride and Chlorine Gettering: An Effective

Technique for Improving Performance of Silicon Devices," *J. Electrochem. Soc.* **119:**747 (1972).

4. D. W. Hess and B. E. Deal, "Kinetics of Thermal Oxidation of Silicon in O_2/HCl Mixtures," *J. Electrochem. Soc.* **124:**735 (1977).

5. R. R. Razouk, L. N. Lie, and B. E. Deal, "Kinetics of High Pressure Oxidation of Silicon in Pyrogenic Steam," *J. Electrochem. Soc.* **128:**2214 (1981).

6. H. Z. Massoud, J. D. Plummer, and E. A. Irene, "Thermal Oxidation of Silicon in Dry Oxygen: Growth Rate Enhancement in the Thin Regime," *J. Electrochem. Soc.* **132:**2685 (1985).

7. M. Hamasaki, "Effect of Oxidation-Induced Oxide Charges on the Kinetics of Silicon Oxidation," *Solid State Electron.* **25:**479 (1982).

8. S. M. Hu, "Thermal Oxidation of Silicon," *J. Appl. Phys.* **55:**4095 (1984).

9. D. N. Modlin, Ph.D. Dissertation, Stanford University, Stanford, CA, 1983.

10. A. G. Revesz and R. J. Evans, "Kinetics and Mechanism of Thermal Oxidation of Silicon with Special Emphasis on Impurity Effects," *J. Phys. Chem. Solids* **30:**551 (1969).

11. A. Fargeix, G. Ghibaudo, and G. Kamarinos, "A Revised Analysis of Dry Oxidation of Silicon," *J. Appl. Phys.* **54:**2878 (1983).

12. R. B. Beck and B. Majkusiak, "The Initial Growth Rate of Thermal Silicon Dioxide," *Phys. Stat. Sol.* **A116:**313 (1989).

13. H. Z. Massoud, J. D. Plummer, and E. A. Irene, "Thermal Oxidation of Silicon in Dry Oxygen Growth-rate Enhancement in the Thin Oxide Regime—II. Physical Mechanisms," *J. Electrochem. Soc.* **132:**2693 (1985).

14. K. H. Lee, W. H. Liu, and S. A. Campbell, "Growth Kinetics and Electrical Characteristics of Thermal Silicon Dioxide Grown at Low Temperature," *J. Electrochem. Soc.* **140:**501 (1993).

15. C. J. Han and C. R. Helms, *J. Electrochem. Soc.* **134:**1299 (1987).

16. F. Rochet, B. Agius, and S. Rigo, "An ^{18}O Study of the Oxidation Mechanism of Silicon in Dry Oxygen," *J. Electrochem. Soc.* **131:**914 (1984).

17. J. M. Delarious, C. R. Helms, D. B. Kao, and B. E. Deal, "Parallel Oxidation Model for Si Including Both Molecular and Atomic Oxygen Concentrations," *Appl. Surf. Sci.* **39:**89 (1989).

18. R. H. Fowler and L. W. Nordheim, "Electron Emission in Intense Electric Fields," *Proc. R. Soc.* **A119:**173 (1928).

19. C. Hu, S. C. Tam, F.-C. Hsu, P.-K. Ko, T.-Y. Chan, and K. W. Terrill, "Hot-Electron-Induced MOSFET Degradation—Model, Monitor, and Improvement," *IEEE Trans. Electron. Dev.* **ED-32:**375 (1985).

20. K. H. Lee and S. A. Campbell, "The Kinetics of Charge Trapping and Oxide Trap Recombination in Ultrathin Silicon Dioxide," *J. Appl. Phys.* **73:**4434 (1993).

21. B. E. Deal, M Sklar, A. S. Grove, and E. H. Snow, "Characteristics of the Surface State Charge of Thermally Oxidized Silicon," *J. Electrochem. Soc.* **114:**266 (1967).

22. M. T. Tang, K. W. Evans-Lutterodt, M. L. Green, D. Brasen, K. Krisch, L. Manchanda, G. S. Higashi, and T. Boone, "Growth Temperature Dependence of the Si(001)/SiO_2 Interface Width," *Appl. Phys. Lett.* **64:**748 (1994).

23. E. P. EerNisse, "Stress in Thermal SiO_2 During Growth," *Appl. Phys. Lett.* **35:**8 (1979).

24. E. H. Snow, A. S. Grove, B. E. Deal, and C. T. Sah, "Ion Transport Phenomena in Insulating Films," *J. Appl. Phys.* **36:**1664 (1965).

25. B. E. Deal, "Standardized Terminology for Oxide Charges Associated with Thermally Oxidized Silicon," *IEEE Trans. Electron. Dev.* **ED-27:**606 (1980).

26. L. M. Terman, "An Investigation of Surface States at a Silicon/Silicon Dioxide Interface Employing Metal-Oxide-Silicon Diodes," *Solid State Electron.* **5:**285 (1962).

27. R. Castagne and A. Vapaille, "Description of the SiO_2-Si Interface Properties By Means of Very Low Frequency MOS Capacitance Measurements," *Surface Sci.* **28:**157 (1971).

28. C. N. Berglund, "Surface States at Steam-Grown Silicon-Silicon Dioxide Interfaces," *IEEE Trans. Electron. Dev.* **ED-31:**701 (1966).

29. E. H. Nicollian and J. R. Brews, *Metal Oxide Semiconductor Physics and Technology*, John Wiley and Sons, New York, 1982.

30. A. S. Grove, O. Leistiko, and C. T. Sah, "Redistribution of Acceptor and Donor Impurities During Thermal Oxidation of Silicon," *J. Appl. Phys.* **35:**2695 (1964).

31. R. B. Fair and J.C.C. Tsai, "Theory and Direct Measurement of Boron Segregation in SiO_2 in Dry, Near Dry, and Wet O_2 Oxidation," *J. Electrochem. Soc.* **125:**2050 (1978).

32. A. S. Grove, *Physics and Technology of Semiconductor Devices*, John Wiley and Sons, New York, 1967.

33. B. E. Deal and M. Sklar, "Thermal Oxidation of Heavily Doped Silicon," *J. Electrochem. Soc.* **112:**430 (1965).

34. L. E. Katz, "Oxidation," in *VLSI Technology*, S. M Sze, ed., McGraw-Hill, New York, 1988.

35. C. P. Ho, J. D. Plummer, J. D. Meindl, and B. E. Deal, "Thermal Oxidation of Heavily Phosphorus Doped Silicon," *J. Electrochem. Soc.*, **125:**665 (1978).

36. C. P. Ho and J. D. Plummer, "Si-SiO_2 Interface Oxidation Kinetics: A Physical Model for the Influence of High Substrate Doping Levels. I. Theory," *J. Electrochem. Soc.* **126:**1516 (1979).

37. J. C. Bravman and R. Sinclair, "Transmission Electron Microscopy Studies of the Polycrystalline Silicon-SiO_2 Interface," *Thin Solid Films* **104:**153 (1983).

38. Y. Wang, J. Tao, S. Tong, T. Sun, A. Zhang, S. Feng, "The Oxidation Kinetics of Thin Polycrystalline Silicon Films," *J. Electrochem. Soc.* **138:**214 (1991).

39. C. L. Claeys, R. F. De Keersmaecker, and G. J. Declerck, "The Oxidation Process and Si-SiO_2 System Properties," in *The Si-SiO_2 System*, P. Balk, ed., Elsevier, Amsterdam, 1988.

40. K. V. Ravi, *Imperfections and Impurities in Semiconductor Silicon*, Wiley-Interscience, New York, 1981.

41. C. L. Clays, E. E. Laes, G. J. Declerck, and R. J. Overstraeten, in *Semiconductor Silicon 77*, H. R. Huff and E. Sirtl, eds., The Electrochem. Soc., Princeton, 1977.

42. S. M. Hu, "Anomalous Temperature Effect of Oxidation Induced Stacking Faults in Silicon," *Appl. Phys. Lett.* **27:**165 (1977).

43. S. M. Hu, "Properties of Amorphous Silicon Nitride Films," *J. Electrochem. Soc.* **113:**693 (1966).

44. T. K. Higman, R. T. Fayfield, M. S. Hagedorn, K. H. Lee, and S. A. Campbell, "Low Temperature Nitridization of Silicon by Direct Ammonia Nitridization in a Molecular Beam Epitaxy Reactor," in *Proc. North Am. Conf. MBE*, Ottawa, 1992.

45. T. Ito, T. Nozaki, H. Ishikawa, "Direct Thermal Nitridization of Silicon Dioxide Films in Anhydrous Ammonia Gas," *J. Electrochem. Soc.* **127:**2053 (1980).

46. T. Ito, T. Nakamura, H. Ishikawa, "Advantages of Thermal Nitride and Nitroxide Gate Films in VLSI Processes," *IEEE Trans. Electron. Dev.* **ED-29:**498 (1982).

47. M. L. Naiman, F. L. Terry, J. A. Burns, J. I. Raffel, and R. J. Aucoin, "Properties of Thin Oxynitride Gate Dielectrics Produced by Thermal Nitridation of Silicon Dioxide," in *IEDM Tech. Dig.*, 1980, p. 562.

48. F.H.P.M. Habraken, A.E.T. Kuiper, Y. Tammingas, and J. B. Theeten, "Thermal Nitridation of SiO_2 Films," *J. Appl. Phys.* **53:**6996 (1982).

49. T. Ito, H. Arakawa, T. Nozaki, H. Ishikawa, "Retardation of Destructive Breakdown of SiO_2 Films Annealed in Ammonia Gas," *J. Electrochem. Soc.* **127:**2248 (1980).

50. F. L. Terry, R. J. Aucoin, M. L. Naiman and S. D. Senturia, "Radiation Effects in Nitrided Oxides," *IEEE Electron. Dev. Lett.* **EDL-4:**191 (1983).

51. S. K. Lai, D. W. Dong, and A. Hartenstein, "Effects of Ammonia Anneal on Electron Trapping in Silicon Dioxide," *J. Electrochem. Soc.* **129:**2042 (1982).

52. S. K. Lai, J. Lee, and V. K. Dham, "Electrical Properties of Nitrided-oxide Systems for Use in Gate Dielectrics and EEPROM," in *IEDM Tech. Dig.*, 1983, p. 190.

53. T. Hori and H. Iwasaki, "Ultra-thin Re-Oxidized Nitrided-oxides Prepared by Rapid Thermal Processing," in *IEDM Tech. Dig.*, 1987, p. 570.

54. T. Hori, H. Iwasaki, and K. Tsuji, "Electrical and Thermal Properties of Ultrathin Reoxidized Nitrided Samples Prepared by Rapid Thermal Processing," *IEEE Trans. Electron. Dev.* **36:**340 (1989).

55. H. Z. Massoud, C. P. Ho, and J. D. Plummer, *Computer Aided Design of Integrated Circuit Fabrication Processes for VLSI Devices*, J. D. Plummer, ed., *Stanford Univ. Tech. Rep.*, 1982.

Chapter 5

Ion Implantation

The introduction of dopant impurities by predeposition diffusion was described in Chapter 3. In this process, dopant was diffused into the semiconductor from an infinite source at the surface of the wafer. The surface concentration was limited by the solid solubility, and the depth of the profile was determined by the time and diffusivity of the dopant. In principle, it seems that a more lightly doped profile could be achieved if the supply of dopant at the surface of the wafer is appropriately limited. For example, a very dilute mixture of dopant in an inert carrier gas can reduce the surface concentration. This process was used in early microelectronic technologies; however it was found very difficult to control. It was also found that the lightly doped profiles were often the most critical. The base of a bipolar transistor and the channel of a MOSFET are two examples of lightly doped profiles that must be very well controlled, since they determine the gain and threshold voltage, respectively.

In ion implantation ionized impurity atoms accelerated through an electrostatic field strike the surface of the wafer. The dose can be tightly controlled by measuring the ion current. Doses for the process range from 10^{11} cm^{-2} for very light implants to 10^{16} cm^{-2} for low resistance regions such as source/drain contacts, emitters, and buried collectors. Some specialized applications require doses of more than 10^{18} cm^{-2}. By controlling the electrostatic field, the penetration depth of the impurity ions can also be controlled. Ion implantation therefore, provides the capability to tailor the dopant profile in the substrate to some extent. Typical ion energies range from 10 keV to 200 keV. Certain special applications including forming deep structures such as retrograde wells, (see Chapter 16), require energies up to several MeV.

After considerable research demonstrations during most of the 1960s, the first commercial implanters were introduced in 1973. Despite initial reluctance, the new method of dopant introduction soon became ubiquitous. By 1980, most processes were fully implanted. Although now widely used, ion implantation also has several drawbacks. The incident ions damage the semiconductor lattice. This damage must be repaired and in some cases complete repair cannot be done. Very shallow and very deep profiles are difficult or impossible. The throughput of ion implanters is limited for high dose implants, particularly compared to diffusion processes in which 200 wafers can often be run simultaneously. Finally, ion implantation equipment is expensive. A state of the art system is about $2,000,000.

The process that this chapter describes provides a blanket dose that is essentially uniform across the wafer. To selectively dope regions of the wafer, an implant mask must be used. A variation of standard ion implantation is to focus the ion beam to a small spot and use this spot to provide a

localized process capability. These processes are called *ion beam techniques*. For example, the ions can be used directly to provide lateral variations in the doping profile across a device. This method is too expensive and too slow to find widespread application in manufacturing. Ion beams are used however, to repair mask defects and to selectively remove layers for diagnostic work.

5.1 Idealized Ion Implantation Systems

Ion implantation systems can be divided into three components (Figure 5.1): the ion source, the acceleration tube, and the end station [1]. The ion source (Figure 5.2) starts with a feed gas that contains the desired implant species. Common feed gasses for use in silicon technologies are BF_3, AsH_3, and PH_3. For GaAs technologies common gasses are SiH_4 and H_2. Most implanters set up for gaseous sources will allow any of several gasses to be selected by opening the proper valve. The flow of gas can be controlled with a variable orifice. If the desired implant species is not available in gaseous form, as shown in Figure 5.2, a solid charge can be heated and the resultant vapor used as the source. The material is heated in the oven and the vapor flows past the filament. For gaseous precursors, the oven is replaced with a simple gas feed. At high pressures, the electron current is often sufficient to maintain a glow discharge.

Ion source

Ion acceleration

Mass separation

Beam sweeping

Multiple heated
target changer
for batch
production

Figure 5-1   Schematic of an ion implanter *(after Mayer et al., reprinted by permission, Academic Press).*

Insulator   **Thermocouple**   **Corona shield**   **Oven heater**   **Heat shields**   **Filament**

Heat shields   **Gas line**   **Arc chamber**   **Extraction slit**

Figure 5-2   Schematic for a Freeman ion source. Solids can be vaporized in the oven heater while gaseous sources may be injected directly into the arc chamber *(reprinted by permission, Elsevier Science, after Freeman)*.

The gas flows into an arc chamber. This chamber has two purposes: to break up the feed gas into a variety of atomic and molecular species and to ionize some of these species. In the simplest such systems the feed gas flows through an orifice into the low pressure source chamber where it passes between a hot filament and a metal plate. The filament is maintained at a large negative potential with respect to the plate. Electrons boil off the filament and are accelerated toward the plate. As they do so they collide with feed gas molecules, transferring some of their energy. If the transferred energy is large enough, molecular dissociation can occur. BF_3, for example, breaks up into B, B^+, BF_2, BF_2^+, F^+, and a variety of other species in varying quantities. Negative ions may also be produced, but are less abundant. To improve the ionization efficiency, a magnetic field is often imposed in the region of the electron current. This produces a spiral path for the electrons, dramatically increasing the ionization probability. The positive ions are attracted to the exit side of the source chamber, which is biased at a large negative potential with respect to the filament. The positive ions then exit the source chamber through a slit. The resulting ion beam is often a few millimeters by one to two centimeters. The gas pressure in this part of the source is typically 10^{-5} to 10^{-7} torr, resulting in a stable arc between the filament and the anode. Maximum ion currents are typically a few milliamperes.

The beam now consists of a variety of species, most of them ionized. The next task is to select

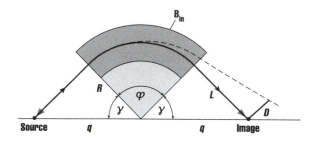

Figure 5-3   Mass separation stage of an ion implanter showing perpendicular magnetic field and ion trajectory. *D* corresponds to the displacement for an ion of M+δM.

the desired implant species. In the previous example, we may want to select only the B^+ from the beam and prevent the other species from continuing down the implanter. This is normally done with an analyzing magnet (Figure 5.3). The beam enters a large chamber that is also maintained at low pressure. A magnetic field perpendicular to the beam velocity exists in this chamber. From simple electrostatics,

$$\frac{mv^2}{r} = qvB \tag{5.1}$$

where v is the magnitude of the ion velocity, q is the charge on the ion, m is the ion mass, B is the magnetic field intensity, and r is the radius of curvature. If the ion obeys classical mechanics,

$$v = \sqrt{\frac{2E}{m}} = \sqrt{\frac{2qV_{\text{ext}}}{m}} \tag{5.2}$$

where V_{ext} is the extraction potential. The derivation of Equation 5.2 ignores the energy imparted to the ions by collisions with electrons in the source. The spread of ion energies due to this effect is of order 10 eV. Since the extraction potential is typically three orders of magnitude larger than this, Equation 5.2 provides a good approximation of the true energy.

Another slit can be made in the mass analysis chamber such that only one mass (or more accurately one charge to mass ratio) will have exactly the correct radius of curvature to exit the source (Figure 5.4). The primary limitation to the resolution of such a system is the fact that the beam has some small divergence. This arises due to the finite slit size and small variations in the ion energy. Such a system can however, readily distinguish between the isotopes of boron, B_{11}, and B_{10}.

Combining Equations 5.1 and 5.2,

$$r = \frac{Mv}{qB} = \frac{1}{B}\sqrt{2\frac{M}{q}V_{\text{ext}}} \tag{5.3}$$

Normally, the analyzing field is everywhere perpendicular to the ion velocity and the inlet and outlets are symmetric. Now assume that the field has been tuned in such a way as to allow an ion of mass M to exactly follow a circle of radius r. If an ion of mass $M + \delta M$ enters the filter the beam will be displaced by a distance [2]:

$$D = \frac{1}{2}r\frac{\delta M}{M}\left[1 - \cos\phi + \frac{L}{r}\sin\phi\right] \tag{5.4}$$

Two masses are said to be resolved when D is larger than the width of the beam plus the width of the exit slit. The best resolution occurs when r is large and M is small. (Due to beam divergence, L has less effect than this simple analysis would suggest as long as it is of order a meter or more.) Most mass filters used for IC fabrication have filters with a radius of a meter or less.

Example 5.1   **Mass Resolution**

If an analyzing magnet bends the ion beam through 45° and $L = r = 50$ cm, find the displacement D that would be seen if B_{10} is sent through the system when it is tuned for B_{11}. If the extraction potential is 20 kV, find the required field.

Since the difference in mass is 1 amu,

$$\frac{\delta M}{M} = \frac{1}{10} = 0.1$$

and

$$D = \frac{1}{2} \, 50 \text{ cm } \frac{1}{10} \, [1 - \cos 45° + \sin 45°] = 2.5 \text{ cm}$$

Since slit widths are typically a few millimeters, this analyzer could easily resolve these two masses. Obviously, heavier mass species will be more difficult to resolve and may require a larger magnet. According to Equation 5.3,

$$B = \frac{1}{0.50\text{m}} \sqrt{2 \, \frac{10 \times 1.67 \times 10^{-27}\text{kg}}{1.6 \times 10^{-19} \text{ C}} \, 2 \times 10^4 \text{V}} = 0.13\text{T} = 1.3\text{kG}$$

where the units of field are tesla (T) and kilogauss (kG).

Acceleration of the ions can be done either before or after the mass analysis. Accelerating first reduces the likelihood that ions will lose their charge before reaching the surface of the wafer, but requires a much larger magnet. The following discussion will assume analysis before acceleration. This tube is typically several meters long and must be maintained at a relatively high vacuum ($<10^{-6}$ torr). This is necessary to avoid collisions during acceleration. The beam is first focused into either a spot or a ribbon using a set of electrostatic lenses. It then enters a linear electrostatic accelerator. The accelerator consists of a set of rings attached to a voltage divider network. Adjusting the power supply that feeds the divider network adjusts the ion energy. In the event that the desired energy is less than the extraction potential, an opposing or bucking potential can be applied. This reduces the stability and spatial coherence of the beam however, and so is not often used.

The beam at this point is primarily composed of ions. Some neutrals may have reappeared. Commonly, the neutrals are ions that combined with thermal electrons:

$$B_{11}^+ + e^- \rightarrow B_{11} \tag{5.5}$$

They may also be ions that collided with other ions in the beam undergoing a charge exchange. Neutrals are highly undesirable since they will not be deflected in electrostatic end station scanning mechanisms. They will travel in a beam to the wafer where they will continuously implant near the center of the wafer. To avoid this problem, most ion implantation systems are equipped with a bend. The beam passes between the parallel plates of an electrostatic deflection system. The neutrals are

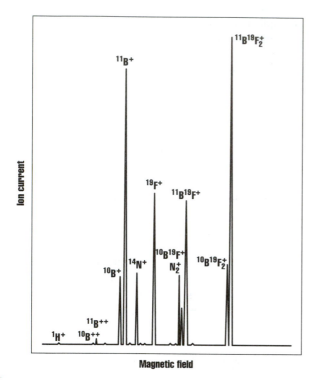

Figure 5-4   Typical mass spectrum for a BF_3 source gas *(after Ryssel and Ruge, reprinted by permission, Wiley).*

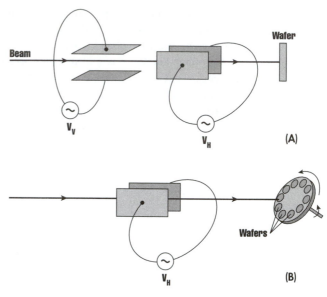

Figure 5-5   Typical scanning systems for ion implanters. (a) Electrostatic rastering commonly used in medium current machines. (b) Semielectrostatic scanning used on some high current machines.

not deflected and so do not follow the bend, but instead strike a beam stop. The ions are sufficiently deflected by the plates to continue to travel down the tube.

Near the end of the tube are additional sets of deflection plates (Figure 5.5). In many implanters both horizontal and vertical pairs of plates are used in a manner analogous to a television set. The beam is rastered back and forth and up and down, writing uniformly across the wafer. In systems designed for high throughput, the beam is only rastered in one direction. Horizontal rastering is done by mechanically moving the wafer past the beam. This is done by placing a number of wafers on the perimeter of a spinning disc. The wafers may be clamped, or as is now common, held by centrifugal force. There are a number of advantages to such a system. The wafer does not have as large a thermal load, since a batch of them are implanted at the same time. The angle of the beam with respect to the wafer remains unchanged in the mechanically scanned direction. The total pump down time is reduced by loading a batch of wafers simultaneously. In many such systems, two wheels are mounted in a side-by-side configuration. This allows one wheel to load and pump down while the other is being implanted.

In some modern high current machines, the wafers are also scanned mechanically in the radial direction as well. Electrostatic scanning requires the formation of an intense narrow beam. Because the beam current in high current machines is large however, the increased electron density will cause the beam to expand or bloom. These space-charge effects lead to poorly controlled electrostatic scanning.

Finally, one needs to control the dose of the implant. This is done in the end station by placing the wafer in a Faraday cup. The cup is simply a cage that captures all of the charge that enters it. The ion current into the wafer is measured directly by connecting an ammeter between the Faraday cup and ground. The dose is just the time integral of the current divided by the wafer area. Electrical current is easy to measure over a broad range of values. For a 6 in. wafer, typical values for implant current range from 1 μA to tens of milliamperes. To accurately measure the dose one must guard against errors due to secondary electron ejection. This process will be described in more detail in Chapter 10. It involves the creation of large numbers of electrons, many which have sufficient energy to escape the wafer, when a high energy ion strikes the surface of the wafer. To prevent secondary electron dose errors, the wafer is biased with a small positive voltage. This bias, typically tens of volts, is sufficient to attract all of the secondary electrons back to the surface of the wafer where they are reabsorbed. The ion current may represent a considerable source of energy to the wafer.

$$\text{energy} = \int \text{power} \, dt = \int IV \, dt = V \int I \, dt = VQ \qquad \textbf{(5.6)}$$

As an example, consider a typical source/drain implantation for an FET. This would involve 5×10^{15} atoms/cm^2 at an energy of 60 keV. Photoresist is commonly used as the implant mask. The energy deposited in a 6 in. wafer is about 15 kJ. This is enough to raise the temperature of the wafer by over 500°C. Furthermore, the energy is not deposited uniformly through the wafer, but rather in the top 1000 Å or so. Since photoresist is a poor thermal conductor, it can get quite hot during an implant. The photoresist may flow or be baked to such an extent that it is difficult to remove after the implant. This is particularly true at the edge of the wafer, where a thicker bead of photoresist typically forms. To avoid these problems the end stations often come with provisions to cool the wafer and thereby control the temperature. The other problem often seen in implanting through a resist mask is outgassing. The ions striking the surface break apart the organic molecules in the resist leading to the formation of gaseous H_2 that evolves from the surface, leaving behind involatile carbon [3]. Heavily implanted resist layers often have a hardened carbonized layer near the surface that is difficult to remove later. The outgassing can sufficiently raise the pressure in the end station as to cause neutralization of the beam through impact with the H_2 molecules, resulting in significant dose rate errors [4].

5.2 Coulomb Scattering°

Coulomb scattering is usually discussed while studying classical mechanics. This topic is very relevant to any discussion of ion implantation. The typical scattering experiment is shown for the laboratory frame of reference in Figure 5.6. One usually thinks of the atoms in the wafer as having a neutral charge. The incident ions however, are traveling at energies that allow them to penetrate the electron cloud around the nucleus of the atoms in the wafer. As a result, the target ions are more appropriately treated as charged ions with a screening cloud of electrons. The target ion is initially at rest in the laboratory frame of reference. An ion approaches with an incident velocity v and an impact parameter b (the distance of closest approach between the centers of the two atoms if no scattering were to occur). The target ion is assumed to be free. Once again, this is a good approximation for the high incident ion energies commonly used in microelectronic fabrication.

The objective is to calculate the amount of energy transferred to the target atom during the collision. The solution of this problem requires the conservation of energy, momentum, and angular momentum. The calculation is more involved than is warranted for our purposes and normally involves a translation into center of mass coordinates. It is covered in many undergraduate classical mechanics

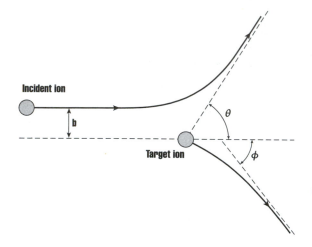

Incident ion

b

Target ion

θ

ϕ

Figure 5-6   Typical scattering problem. Figure insert shows electrostatic potential as a function of distance between the nuclei. The impact parameter is labeled b.

texts. Let us simplify the problem by considering the two balls to be hard spheres that scatter elastically. If p_i and p_t are the final momenta of the incident and target spheres and p_o is the initial momentum of the incident sphere,

$$\overline{p}_i + \overline{p}_t = \overline{p}_o \tag{5.7}$$

Conserving angular momentum,

$$L_i + L_t = L_o = p_o b \tag{5.8}$$

Conserving energy,

$$\frac{p_i^2}{2m_i} + \frac{p_t^2}{2m_t} = \frac{p_o^2}{2m_o} \tag{5.9}$$

Then one can show that the energy lost by the incident sphere is

$$\Delta E = E_o \left[1 - \frac{\sin^2 \phi}{\cos \theta \sin \phi + \cos \phi \sin \theta} \right] \tag{5.10}$$

Thus, the energy loss is proportional to the incident energy and depends on the scattering angles. These angles depend on the ion masses and the impact parameter.

5.3 Vertical Projected Range

This section will use the scattering results to discuss what happens to the energetic ions in an implanter once they reach the surface of the wafer. Some models that have been developed to describe this process will be discussed qualitatively, and the difficulties of ion implantation in modern devices will be pointed out. When an energetic ion enters a solid, it will begin to lose energy. The distance that the ion travels in the semiconductor is its *range* (Figure 5.7). As discussed in the previous section, the energy loss depends on the impact parameter. Since the ion is entering a solid, the ions in the beam will have a range of impact parameters. As a result, the nature of the energy loss mechanism can, to a good approximation, be considered probabilistic. For a given flux of ions a range of distributions will result. For a uniform beam the quantity of interest is not the total distance traveled, but rather the average depth. This quantity is the projected range R_p. The energy loss in the target material is the result of two mechanisms [5]. The first is ion electron interactions. This involves not only the valence electrons, but also the core electrons of the host material. Since a great deal of the space in the crystal is made up of

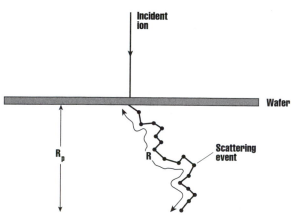

Incident ion

Wafer

Scattering event

R_p

R

Figure 5-7   The total distance that an ion travels in the solid is the range. The projection of this distance along the depth axis is the projected range, R_p.

the electron clouds from the atoms, many of these interactions will occur. Even if the electron is not in the path of the ion, energy may be transferred through coulomb interaction. For a typical semiconductor implant hundreds of thousands of these interactions occur. Furthermore, the mass ratio between the ion and an electron is of order 10^5. Any single electron ion interaction will not dramatically alter the momentum of the incident ion.

Due to the large numbers and small individual effects involved, these discrete interactions can be approximated by a continuum mechanism. That is, the effect of the electrons on the ion is very much like a particle moving through a fluid. A viscosity due to the electrons can be assigned to the crystal media. In that model, there is a viscous drag force assumed to be proportional to the velocity for energies of interest to typical microelectronic applications:

$$F_D \propto v \propto \sqrt{E} \tag{5.11}$$

Since the ion must do work to move through the media, this force can also be expressed as an energy gradient. The energy loss per unit length due to electronic stopping is given the symbol S_e and

$$S_e = \frac{dE}{dx}\Big|_e = k_e\sqrt{E} \tag{5.12}$$

where k_e is a constant of proportionality that depends on the ion and target species as [6]:

$$k_e \propto \sqrt{\frac{Z_i Z_T}{M_i^3 M_T} \frac{(M_i + M_T)^{3/2}}{(Z_i^{2/3} + Z_T^{2/3})}} \tag{5.13}$$

In Equation 5.13, Z is the charge number (number of protons) for the incident and target ions. At very high energies the viscous fluid model can no longer be invoked. Instead, S_e peaks and decreases as the incident ion energy is raised further.

The ion interaction with the lattice ions is very different from the interaction with the electrons. Consider for the time being an amorphous solid. It is known experimentally that ions penetrate semiconductor crystals to depths of thousands of angstroms. Since the atomic spacing is of the order of angstroms, about a thousand interactions can occur. Furthermore, the incident and target ions have masses of the same order of magnitude. It is possible for the incident ion to be scattered at a large angle relative to its incident velocity. As a result, nuclear interactions cannot be treated as a continuum. Instead, they must be treated as a series of discrete events. As described in the previous section, the angle at which the ion is scattered will depend on the impact parameter and on the masses and relative positions of the two ions. This means that the result of any interaction depends on all of the interactions that occurred previously, back to the first atomic layer of the solid. Since the ions are uniformly distributed over the surface of the wafer as they enter, a statistical distribution of depths will result.

To first order, Gaussian distributions can be used to model the range of depths that an ion might reach. Thus, the impurity concentration as a function of depth in an amorphous solid will be given by

$$N(x) = \frac{\phi}{\sqrt{2\pi}\Delta R_p} e^{-(x-R_P)^2/2\Delta R_p^2} \tag{5.14}$$

where R_p is the projected range, ΔR_p is the standard deviation of the projected range, and ϕ is the dose. The normalizing factors are chosen such that the integral of the profile from $x = 0$ to ∞ gives

the dose. Due to the statistical nature of the ion implantation process, the ions will also be scattered laterally, penetrating past the edges of the mask. The profile must therefore be considered in two dimensions, with both lateral and vertical standard deviations.

To find R_p one must determine the S_n, the energy loss of the incident ion per unit length of travel due to nuclear stopping. The theory of nuclear stopping is involved. The reader, if interested, is referred to [2] for an excellent treatment. It can be qualitatively described by the charged sphere model described in Section 5.2. The energy transfer during any collision will be a sensitive function of the impact parameter. The lower the impact parameter, the greater the energy loss. The average energy loss will also be a function of the ratio of the mass of the ion to that of the target atom. The smaller this ratio, the larger the average energy loss per collision will be. Finally, the average energy loss will be a function of the energy itself. The solid atom is chemically bonded. It is sitting in an approximately parabolic potential minimum. At ion low energies the average collision does not transfer enough energy to break this bond. The result is an elastic collision. The ion may change direction, but it will not lose much energy. Thus, S_n is expected to increase with ion energy at low energy. The momentum transfer is given by

$$\Delta p = \int F \, dt \tag{5.15}$$

At high velocities the collision time becomes so short that the energy loss decreases. Thus, S_n has a maximum at some energy. The maximum value of S_n can be approximated by the relation

Figure 5-8 Nuclear and electronic components of S(E) for several common silicon dopants as a function of energy *(after Smith as redrawn by Seidel, ''Ion Implantation,'' reproduced by permission, McGraw-Hill, 1983).*

$$S_n^o \approx 2.8 \times 10^{-15} \text{ eV-cm}^2 \, \frac{Z_i Z_t}{Z^{1/3}} \frac{M_i}{M_i + M_t} \tag{5.16}$$

where

$$Z = [Z_i^{2/3} + Z_t^{2/3}]^{3/2} \tag{5.17}$$

S_n has a weak energy dependence. Figure 5.8 shows the nuclear and electronic components of $S(E)$ for several common silicon dopants as a function of energy [7],[8].

Once $S_n(E)$ and $S_o(E)$ are known, one can perform the integral

$$R_p = \int_0^{R_p} dx = \int_{E_o}^0 \frac{dE}{dE/dx} = \int_{E_o}^0 \frac{dE}{S_n + S_e} \tag{5.18}$$

to get the projected range. Once known, the projected range can be used to estimate [9] ΔR_p

$$\Delta R_p \cong \frac{2}{3} R_p \left[\frac{\sqrt{M_i M_\tau}}{M_i + M_\tau} \right] \tag{5.19}$$

The projected range and its standard deviation are often obtained from LSS (Lindhard, Scharff, and Schiøtt) tables [10] or Monte Carlo simulations. These tables represent calculated values that assume a particular model for the electron density distributions of the ion and the solid atoms. The range values predicted by this model agree well with experimental results for concentrations close to the maximum. Figure 5.9 shows the projected range and implant straggle for several commonly implanted impurities [11].

Additional moments can be added to the profile to help describe the behavior at lower concentrations (Figure 5.10). The *i*th moment of a distribution is defined by

$$m_i = \int_0^\infty (x - R_p)^i N(x)\, dx \qquad \textbf{(5.20)}$$

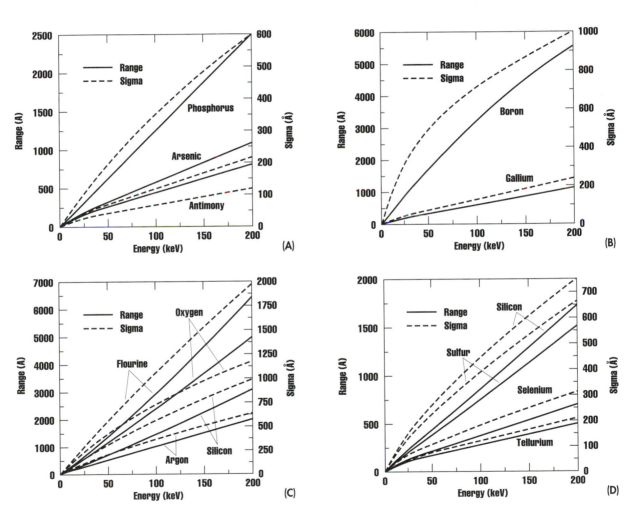

Figure 5-9   Projected range (left axis) and standard deviation (right axis) for (a) n-type, (b) p-type, and (c) other species into a silicon substrate, and (d) n-type and (e) p-type dopants into a GaAs substrate, and several implants into (f) SiO₂ and (g) AZ111 photoresist *(data from Gibbons et al.)*.

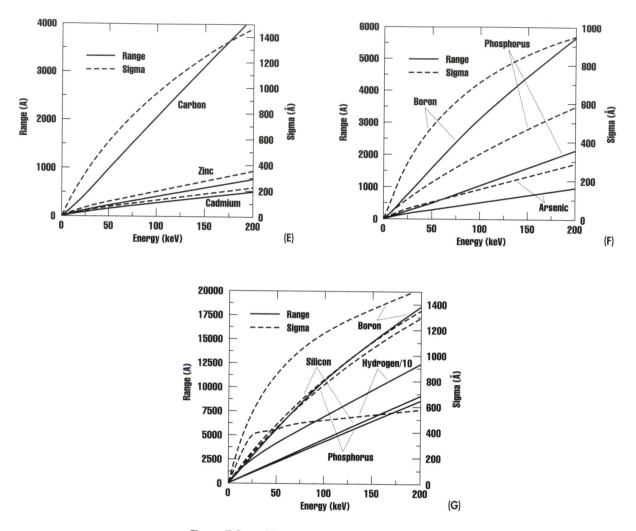

Figure 5-8 cont'd. For legend see opposite page.

The first moment is just the normalized dose. The second moment is the product of the dose and ΔR_p^2. The third moment is related to the asymmetry of the distribution. This asymmetry is usually expressed in terms of the skewness, γ where

$$\gamma = \frac{m_3}{\Delta R_p^3} \tag{5.21}$$

Negative values for the skewness represent an increased concentration on the surface side of the distribution, that is, for $x < R_p$. As an example, when implanting boron into silicon, higher concen-

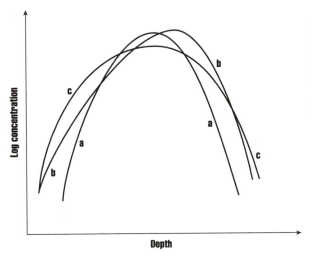

Figure 5-10 Curve (a) shows a standard Gaussian. (b) A negative skewness, which moves the peak slightly deeper and adds a tail extending toward the surface. (c) A large kurtosis flattens the peak of the distribution.

trations are found near the surface than would be predicted by Equation 5.14. This occurs because boron is much lighter than silicon and so will suffer significant backscattering. As a result, boron has large negative values of skewness, particularly at high energy where the effect is most pronounced. The fourth moment of the distribution is related to a distortion of the Gaussian peak. The distortion is expressed by the kurtosis β, where

$$\beta = \frac{m_4}{\Delta R_p^4} \tag{5.22}$$

The larger the kurtosis the flatter the top of the Gaussian. Normal Gaussians have a kurtosis of three. The values of γ and β can be found by performing Monte Carlo simulations [12], or more directly, by measuring actual profiles and fitting the results.

Boron in silicon is a rather special case. This distribution is most often described not by a modified Gaussian, but instead by a Pearson type IV distribution. Expressing this distribution in terms of the first four moments of the Gaussian

$$n(x) = n(R_p)\exp\frac{\ln[b_0 + b_1(x - R_p) + b_2(x - R_p)^2]}{2b_2}$$
$$-\frac{b_1 + 2b_1b_2}{\sqrt{4b_0b_2^3 - b_1^2b_2^2}}\tan^{-1}\left[\frac{2b_2(x - R_p) + b_1}{\sqrt{4b_0b_2 - b_1^2}}\right] \tag{5.23}$$

where

$$b_0 = -\frac{\Delta R_p^2(4\beta - 3\gamma^2)}{10\beta - 12\gamma^2 - 18} \tag{5.24}$$

$$b_1 = -\gamma\Delta R_p\frac{\beta + 3}{10\beta - 12\gamma^2 - 18} \tag{5.25}$$

and

$$b_2 = -\frac{2\beta - 3\gamma^2 - 6}{10\beta - 12\gamma^2 - 18} \tag{5.26}$$

5.4 Channeling and Lateral Projected Range

When implanting into single crystal material another complication can arise. Channeling can occur when the ion velocity is parallel to a major crystal orientation. In this situation, some ions may travel considerable distances with little energy loss (Figure 5.11), since nuclear stopping is not very effective

(A)

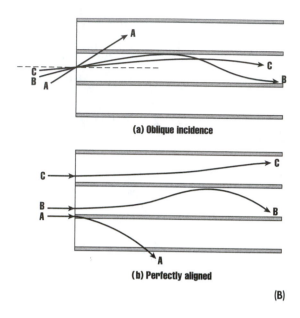

(a) Oblique incidence

(b) Perfectly aligned

(B)

Figure 5-11   (a) View of the diamond structure along a major crystal axis $\langle 110 \rangle$ and along a random direction. (b) Schematic of channeling *(reprinted by permission, Academic Press, after Mayer et al.).*

and the electron density in a channel is low. Once in a channel, the ion will continue in that direction, making many glancing internal collisions that are nearly elastic until it comes to rest or finally de-channels. The latter may be the result of a crystal defect or impurity. Channeling is characterized by a critical angle Ψ (Figure 5.12)

$$\Psi = 9.73° \sqrt{\frac{Z_i Z_t}{E d}}$$ **(5.27)**

where E is the incident energy in keV and d is the atomic spacing along the ion direction in Å. For ions with velocity vectors much larger than Ψ away from a major crystallographic orientation little channeling will occur [13]. The channeling direction need not be close to the initial ion velocity. A scattering event inside the target can redirect the incident ion along a crystallographic orientation, but the probability of this occurring is small, and so this effect is unlikely to produce a substantial distortion near the peak of the implant profile.

Figure 5-12   Critical angle at which channeling begins for common impurities in silicon *(after Ryssel and Ruge, reprinted by permission, Wiley).*

Channeling can produce a significant tail on the implant distributions. The effect is particularly pronounced when implanting light atoms on axis into a heavy matrix since the ion's atomic radius is much less than the crystal spacing. Figure 5.13 shows an extreme example of channeling: He implantation into ⟨100⟩ silicon [14]. The energy and distribution of the backscattered helium can be used to determine such crystal irregularities as impurities and defects. This analysis technique is called Rutherford Backscattering Spectroscopy (RBS). The contours of the figure correspond to helium concentrations in the wafer. The dark lines along the vertical and horizontal are the {001} directions. The very dark central line corresponds to ⟨011⟩. Similarly, one can see the {013}, the {113}, and many other crystallographic orientations. The He atoms have collected along these directions because they have channelled.

To avoid this tail, most IC implantation is done off axis. A typical tilt angle is 7°. To reduce the probability of an inadvertent line up of the crystal planes, a twist angle of about 30° is also commonly used [15]. Some ions will still scatter along the crystal axis, and channeling effects will still occur. Another way to minimize channeling is to destroy the lattice before implantation. Preamorphization of silicon can be done with high doses of Si, F, or Ar before the dopant implantation. This will be covered in more detail in Section 5.6. Some channeling reduction has also been reported by implanting through a thin screen oxide to randomize the ion velocities before entering the crystal. This has the disadvantage of unintentionally implanting oxygen due to recoil or knock-on effects.

5.5 Implantation Damage

One component of energy transfer when a high energy ion enters a wafer is collision with lattice nuclei. Many of these atoms are ejected from the lattice during the process. Some displaced substrate atoms have sufficient energy to collide with other substrate atoms to produce additional displaced atoms. As a result, the implantation process produces considerable substrate damage that must be repaired during subsequent processing. Furthermore, if the implanted species is intended to act as a dopant, it must occupy lattice sites. This process of moving a large fraction of the implanted impurities onto lattice sites is known as impurity activation. Both damage repair and implant activation are normally done by heating the wafer (annealing) after implant. Often the anneal accomplishes both tasks simultaneously. For that reason, these two processes will be treated simultaneously. The activation of implants into GaAs will not be treated in this chapter since this is now most commonly done by rapid thermal annealing, the subject of the next chapter.

Since the energy loss per collision by nuclear energy transfer is typically much larger than the binding energy of the atom in the lattice, the crystal is damaged when it is implanted. There exists a threshold dose ϕ_{th} above which the damage is complete [16]. That is, after the implant no evidence

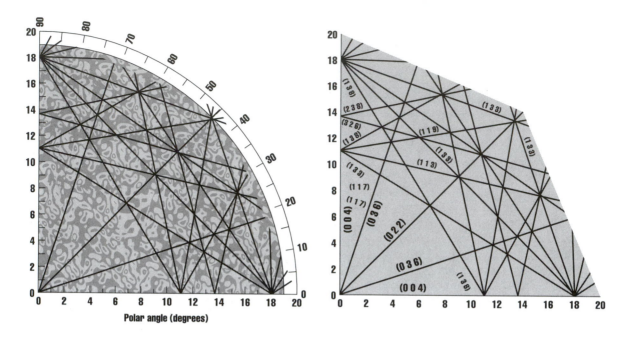

Figure 5-13   One MeV He implant intentionally done near the ⟨100⟩ axis to promote channeling. The contours represent 2% intervals of normalized backscattering yield. The darkened lines lie along the projections of the major crystal orientations in the [100] plane *(reprinted by permission, AIP, after Ziegler).*

of long range order exists, and the surface of the substrate is rendered amorphous. The critical dose depends on implant energy, implant species, target material, and substrate temperature during implantation. Figure 5.14 shows the critical dose for several impurities in silicon as functions of the wafer temperature. At high temperatures, the substrate self-anneals, and the threshold dose becomes very large. Threshold doses for light ions are also large since a greater fraction of the energy loss is electronic.

As an ion passes through the crystal, point defects consisting of interstitials and vacancies are created by direct interaction or through collisions with recoiled target atoms. The defects created by the implantation process are called *primary defects* [17]. Figure 5.15 shows RBS spectra for an unimplanted wafer and a wafer implanted with 2×10^{15} cm^{-2} boron at 200 keV. The dashed line represents the contribution of the random component of the beam to the spectra. The shaded area represents the signal from the displaced silicon atoms. From the area of the shaded region, it is estimated that the interstitial silicon concentration is 7×10^{16} cm^{-2}, roughly 35 times the implant dose.

Secondary defects occur when an implanted wafer is annealed. As previously mentioned, point defects have a high energy in the crystal. This energy can be reduced by recombination or by agglomeration into extended defects [18]. Typically, these defects take the form of small point defect clusters such as divacancies or condense into higher dimensional defects like dislocation loops. For boron implanted into silicon, it appears that secondary defects occur when the Si interstitial concentration produced by the implant is $\geq 2 \times 10^{16}$ cm^{-2}. Ions such as P or Si that have masses comparable to that of silicon have larger critical interstitial concentrations, typically about 5×10^{16} cm^{-2}. It has

Figure 5-14 Critical implant dose required to amorphize a silicon substrate as a function of substrate temperature for several common silicon dopants *(after Morehead and Crowder).*

Figure 5-15 RBS spectra showing the backscattered He intensity as a function of energy. Curve (b) is an unimplanted wafer. The curve labeled (a) is a wafer after a 2×10^{15} B implant at 200 keV. The dashed line represents the component of the spectrum corresponding to the remaining crystalline silicon. The shaded area represents the contribution of displaced atoms *(reprinted by permission of Elsevier Science, after Schreutelkamp et al.).*

been suggested that implant light atoms like B form isolated defects while heavier ions form more extended defects [19]. The interstitials created by moderate mass ions are bound in these defect clusters, and few are free to agglomerate in larger extended defects. Schreutelkamp et al. have shown that secondary defects do not form for heavy ions ($Z > 69$) except for MeV implants. Even then, the critical interstitial concentration for a heavy ion like Sb exceeds 10^{17} cm^{-2}. Heavy ions, therefore, tend to amorphize the substrate before secondary damage is generated.

Annealing processes that minimize secondary defects are of considerable technological interest. In these applications, high temperatures are required to ensure that all of the dopant is activated and few residual extended defects remain. Isochronal annealing curves display the active carrier concentration normalized to the dose as a function of the annealing temperature for a fixed annealing time. Typical times are 30 or 60 min. Unless otherwise stated, the anneals are done in nitrogen. Figure 5.16 shows the result of an isochronal anneal experiment for boron [20]. At low temperature the carrier concentration is dominated by point defects. As the anneal temperature is raised, point defect repair begins with some vacancies capturing nearby interstitials. This reduces the net trap concentration in the substrate, raising the free carrier concentration. In the 500 to 600°C range the diffusivity of the defects is raised sufficiently to cause agglomeration and the formation of extended defects. This is particularly true at higher boron fluences where the damage is large. Finally, at high temperature, these extended defects are annealed out and the active carrier concentration approaches the implant dose. This requires temperatures of 850°C to over 1000°C, with high fluence implants requiring the highest anneal temperatures.

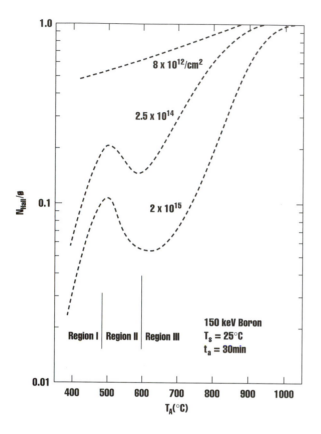

Figure 5-16 Fraction of implanted boron activated in silicon for several isochronal anneals *(after Seidel and MacRae, reprinted by permission, Elsevier Science).*

When the substrate has been rendered amorphous, either due to the implant itself or because of a preamorphization step (Section 5.6), the crystallinity is repaired by solid phase epitaxy (SPE). In theory, the crystal reforms using the underlying undamaged substrate as a template. In so doing, most of the impurities are incorporated into the growing lattice on nearly equal footing with the displaced substrate material. In discussing the annealing behavior of these amorphous layers, one must first recognize that amorphous layers need not extend to the surface. Figure 5.17 shows the damage induced in the substrate versus depth for a variety of arsenic implants [21]. Obviously, the degree of damage near the surface is much less than that further into the substrate. Buried amorphous layers are therefore possible by using high energy implants. Defects must penetrate the crystalline region before reaching the surface and being annihilated. Furthermore, SPE regrowth fronts will begin at both sides of the amorphous region. The plane where these fronts meet contains defects that can degrade device performance [22].

Generally, the annealing process for an amorphous layer involves an SPE regrowth of the layer. This regrowth can be done at temperatures as low as 600°C, since the regrowth velocity of silicon at this temperature is greater than 300 Å/min for ⟨100⟩ and about one order of magnitude less for ⟨111⟩ [23]. (For high dose implants the SPE rate depends on the implanted species.) Thus, a 30 min anneal at 600°C will regrow almost a μm of material, which is well in excess of the amorphization depth of any reasonable implant. The activation of impurities occurs at much lower temperatures because of SPE. Figure 5.18 shows isochronal curves for the activation of phosphorus in single crystal and amorphous materials. A large percentage of the implant is activated even at 550°C.

The major concern with SPE regrowth of amorphous layers is residual defects. These layers include not only simple 1-D defects like dislocation loops, but also 2-D and 3-D defects like twins and stacking faults. The defects may originate from micro islands of single crystal material that are slightly displaced from their original position or become displaced during the SPE. These islands also serve as nucleation centers for regrowth. Defects will be formed when these disparate growth fronts meet. To reduce the concentration of the defects to an acceptable level, high temperature anneals are usually required (of order 1000°C) [24]; however, even these high temperature anneals may not be sufficient to remove all of the damage. Figure 5.19 shows the annealing characteristics of high dose arsenic implants. Even after 1000°C anneals, residual damage remains. The damage is concentrated in the regions near the edge of the implant. High temperature anneals also allow the activation of impurities in the implant tail regions, outside the amorphous layer. To avoid excessive diffusion during the high temperature anneal, the point defect concentration is first lowered by a low temperature SPE step.

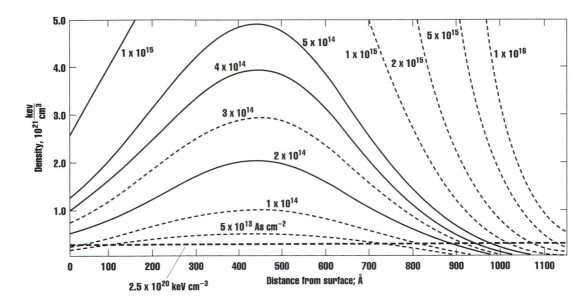

Figure 5-17 Damage density distribution for 100 keV arsenic implanted into silicon with dose as a parameter *(reprinted by permission, AIP, after Prussin et al.).*

Figure 5-18 Isochronal annealing of phosphorus in silicon, with dose as a parameter. The solid lines correspond to implants that amorphized the substrate *(reprinted by permission, AIP, after Crowder and Morehead).*

Figure 5-19   Cross-sectional TEMs image of implant damage in submicron areas for a high dose arsenic implant (5×10^{15} cm^{-2} at 25 keV) immediately after implant, after an 800°C anneal, and after a 1000°C anneal. Notice that even at the highest temperature residual damage remains at the edges of the implanted regions *(reprinted by permission of Elsevier Science, after Gyulai).*

5.6 Shallow Junction Formation $^+$

In the production of very small devices, it is necessary to reduce the junction depths. Most of the work in shallow junction formation has concentrated on the formation of very shallow source/drain junctions for submicron CMOS. Since boron has a much larger diffusivity than arsenic, shallow P+/n junctions have proven to be the more difficult to form. This section will therefore concentrate on shallow P+ junction formation, but many of the techniques have been applied to n-type diffusions as well. One of the most popular techniques for shallow junction formation, rapid thermal processing, is explicitly excluded since it will be covered in detail in the next chapter.

The formation of shallow boron junctions is limited by three effects. The boron atom is very light. As a result, it has a deep implant projected range. Lowering the implant energy to less than 10 keV by bucking the extraction potential is often undesirable due to concerns with beam stability. One

way to solve this problem is to use molecular implants. The molecule is believed to dissociate upon impact. BF_2 has been a popular molecular implant species. The kinetic energy associated with the boron atom is given by

$$E_B = E_{BF_2} \frac{M_B}{M_{BF_2}} = E_{BF_2} \frac{11}{49} \qquad (5.28)$$

assuming that the mass 11 isotope of boron is used. Due to problems associated with fluorine electrical activity and defect cluster formation [25] and the increasing availability of very low energy implantation equipment, the use of BF_2 for shallow P+ junctions is declining.

The second problem associated with shallow junction formation is channeling. Even in randomly oriented wafers channeling occurs [26] due to boron's high probability for deflection into a major crystallographic direction. This effect can be minimized by preamorphizing the wafer with a high dose implant of a heavier atom before boron implantation. One popular choice is germanium. Ozturk and Wortman [27] have demonstrated about a 20% reduction in the junction depth (assuming $N_{sub} = 10^{16}$ cm^{-3}) when the wafers were preamorphized with a low energy Ge implant at a dosage of 3×10^{14} cm^{-2} before BF_2 implantation. No difference in diode characteristics were seen [27]. Junction depth reductions in excess of 40% have been reported for atomic B implants that are preamorphization with silicon implantation [28]. One disadvantage of these techniques is that the amorphous region must be recrystallized, and this recrystallization may leave residual defects.

The third problem associated with shallow junction formation is the anomalous diffusion that occurs during the high temperature anneal [29]. The effect is most pronounced in the tail region [30]. This effect is sometimes called the *transient annealing effect*. It has also been observed for phosphorus implants [31]. The cause is still uncertain. One model suggests that since only about 20% of the boron is substitutional as implanted [32], the anomalous diffusion occurs interstitially [33]. This model fails to explain the anomalous diffusion effects when the implant is done on axis. This latter result has reinforced theories that the transient diffusion effect is associated with damage and, in particular, with extended defects formed during annealing [34],[35]. More recently, Schreutelkamp has proven that the transient diffusion mechanism for B in silicon is the addition of the kickout-diffusion mechanism (see Chapter 3). Due to implant damage at the beginning of the anneal, a high concentration of self-interstitials is present. These interstitials kick out the substitutional boron to interstitial sites where it rapidly diffuses. As the damage is annealed, the excess self-interstitial concentration drops and so diffusivity decreases until late in the anneal when extended defects are annihilated and give up excess interstitials that again drive the transient diffusion process.

5.7 Buried Dielectrics[+]

Thus far only one application of implantation, the introduction of dopants into the semiconductor substrate, has been discussed. A second application of the process is in device isolation. In GaAs technologies, this is a widely applied technique used to provide a surface isolation between adjacent devices fabricated in a conducting layer on an insulating substrate. Protons (H^+) are used most commonly, although other materials have been investigated. From an implantation process standpoint, there is little difference between this application and doping. We will therefore defer any further discussion of this isolation technique until Chapter 17. This section will cover the use of implantation to form silicon on insulator (SOI). This is done by performing a high energy implant of N^+ to form Si_3N_4 in a process called SIMNI [36] or by performing a high energy implant of O^+ to form SiO_2 in a process called SIMOX. The latter is much more popular.

Separation by IMplanted OXygen (SIMOX) was first reported by Izumi [37]. The advantages of forming a buried insulator include increased radiation hardness, increased circuit speed, and in-

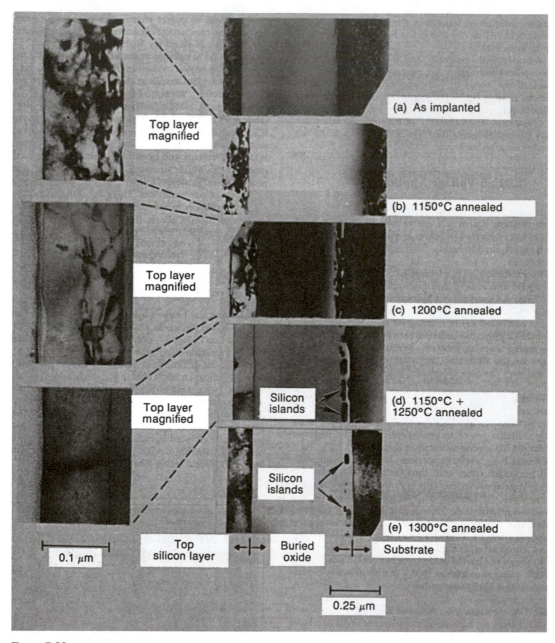

Figure 5-20 Cross-sectional TEMs of SIMOX layers for different anneal conditions. Samples were implanted at 150 keV with an oxygen dose of 2.25 × 10^18 cm^−2 *(after Lam, reprinted by permission, © 1987 IEEE).*

creased packing density. The typical SIMOX process involves implanting the wafer with 150 to 300 keV O^+ at doses of about 2×10^{18} cm^{-2}. For most implanters, this involves a very long implantation time. Often the implant is done on axis to intentionally channel and minimize the near surface damage. The implant step produces a nearly amorphous layer with about twice as many oxygen atoms as silicon near the projected range. To minimize surface damage the wafer must be held at temperatures of at least 400°C during the implant [38]. This is very important because the amorphous oxide layer

that will later be formed under the silicon prevents a recrystallization from the undamaged portions of the substrate. The implanted oxygen will also tend to diffuse from the projected range to form a broad flat concentration plateau if the wafer is held at about 500°C during the implant. The oxygen concentration in this region is the atomic concentration required to form SiO_2 [39].

Following the implant a very high temperature step must be done to form the oxide. Typically the anneal is done under a deposited oxide cap. The anneal conditions are at least 1300°C for several hours. Some authors recommend annealing at temperatures over 1400°C [40]. Figure 5.20 shows a set of cross-sectional transmission electron micrographs of SIMOX wafers after various thermal anneals. Obviously, the crystal quality of the surface layer improves dramatically with anneal temperature. Residual oxygen atoms left in the implant tail also serve as thermal donors that can affect the carrier concentration in the active device regions. Fortunately, thermal anneals of 1300°C also reduce the thermal donor density to less than 10^{15} cm^{-3} [41].

Potential problems with SIMOX include heavy metal impurities that lead to enhanced junction leakage, pinhole density, material quality, and thickness uniformity. Metallic contamination is present in all implants; however, typical SIMOX implant doses are nearly a thousand times those of conventional source/drain implants. The implanter components must be carefully designed for SIMOX use to avoid this problem. For example, silicon apertures and beam stops may be used rather than stainless steel or tungsten. Pinholes in the buried oxide occur when a particle is on the surface of the wafer during the SIMOX implant. The particle screens or partially screens the O_2 implant. The result is a singularity in the dielectric as it rises to the surface. Careful wafer cleaning and attention to particulate control can minimize this problem. Another concern is the dislocation density in the surface silicon. While the material quality has improved dramatically, it still has of order 10^3 dislocations/cm^2. CMOS ICs can be fabricated quite well in SIMOX material, and 256k SIMOX SRAMs have been reported. Recently, even bipolar for BiCMOS applications have been fabricated in this technology. Minority carrier lifetimes of about 10^{-7} sec have been observed.

One of the primary concerns for SIMOX is still cost, particularly in commodity components. Recently, implanters have been specifically developed for the SIMOX process to address these concerns. Beam currents in these systems typically exceed 100 mA, allowing higher throughput and therefore reducing cost. Using these implanters, metallic contamination has been held to less than 10^{11} cm^{-2} and pinhole density less than 0.2 cm^2. Thickness uniformity of the single crystal layer is approaching 50 Å over 6 in. wafers.

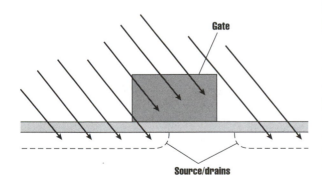

Figure 5-21   Simple shadowing example for deeply scaled MOSFET. As a result of the tilt angle one of the source/drain diffusions does not extend to the channel leading to poor I-V characteristics.

5.8 Ion Implantation Systems— Problems and Concerns

Modern implanters have several technical problems that the idealized system described at the beginning of the chapter did not address. This section will review some problems besides damage annealing, that make the life of implant engineers interesting. Figure 5.21 shows a common problem for implantation of wafers with topology: shadowing. Any nonplanar masking layer will cast an implant shadow because of the tilt and twist angles of the implant. Within the last five years the decrease in the allowable thermal cycles and resultant diffusion, along with the increasing reliance on self aligned structures to minimize parasitic capacitance, has increased the importance of shadowing effects. In a typ-

Figure 5-22   Variability of the angle between the incident beam and the surface normal in simple electrostatically scanned systems.

ical example, the I-V characteristics of an MOS transistor will be asymmetric with respect to the choice of source and drain contacts. The newest generation of implanters allows both the implant angle and the twist to be varied while the wafers are under high vacuum. This allows multiple implant angles to be run efficiently. Often four angles are run in a "quad" arrangement.

The problem of shadowing can be further exaggerated by using electrostatic scanning as shown in Figure 5.22. The scanning plate deflection can cause the implantation angle to vary from 5° to 9° across the wafer. Another problem with the tilt angle in electrostatically scanned implantation systems is that the dosage of the wafer will vary from one side to the other. The problem becomes more pronounced with large wafers where the electrostatic angle can be 2° or more. High current implanters, in which the wafers are mechanically scanned in both directions, do not suffer from this problem; however, the wafer in these systems heats during the implant unless carefully cooled. Because of this heating and the edge clamping process used, the wafers frequently flex, causing further implant angle uncertainty.

The increasing use of large diameter silicon wafers makes the task of producing highly uniform implants more difficult. Current and Keenan have also demonstrated that production implanters can have considerable variability from machine to machine, particularly at low dosages [42]. Besides the problems discussed in the previous section, dose rate monitoring errors can be present due to several effects. Any neutral species that strike the wafer will not produce an electric current and so will not be counted. Neutrals can be produced in the beam by recombination with residual gas atoms or thermal electrons [43]. Dosage errors can also be the result of incomplete secondary electron recapture. To prevent this problem, the wafer is held in a *Faraday cup* and the stage is typically held at a small positive bias. Electron flood guns are also commonly used to maintain charge neutrality and compensate for secondary electron loss.

Ion implantation is often assumed to be an extremely clean process. It is done in high vacuum, and the implant species is selected to correspond to the desired isotope of the desired species. Implanters have four types of commonly seen contamination. As with all processing equipment, one type of contamination is particles. If the particles fall on the wafer before implantation, they will screen the wafer from the implant. A later clean may remove the particle, but its unseen shadow remains behind as a virtually undetectable killing defect. Most particles are caused by improper wafer handling, improper pump down procedures, clamping procedures, the use of unfiltered gas for venting, and in high current systems, the spinning disk [44]. The use of clampless wafer mounting in particular, has been found to have a significant effect on particle generation in ion implanters.

The other three types of contamination are related to much smaller impurities that may fall on the wafer during implantation. To guard against them, implants are often done through a thin screen oxide that is removed after implantation. Hydrocarbons are produced in high current machines when vapors from diffusion pumps backstream into the beamline and are polymerized during implantation. These compounds are extremely resistant to subsequent chemical cleans. This problem can be avoided by carefully baffling the diffusion pumps, or as is often done, by using cryo or turbo pumps. Heavy

metal impurities such as Fe, Cr, and Ni are sometimes seen when stainless steel apertures are used [45]. New machines commonly use graphite or copper free aluminum for these components to avoid this effect. Finally, it is common to see some cross contamination, particularly on low dose implants that follow a high dose implant of a different species. It is therefore important that lifetime killing impurities never be implanted with semiconductor implanters. Many larger fabrication facilities will often dedicate an implanter to a particular impurity to avoid any cross contamination concerns.

5.9 Implanted Profiles Using SUPREM$^+$

The process simulator SUPREM, which was introduced in Chapter 3 to estimate diffusion profiles, and later used in Chapter 4 for oxidation, may also be used to calculate implant profiles. This feature is very useful, since in practice most impurities are implanted. Simulated impurities can be implanted, activated, and diffused to compare to real profiles. SUPREM contains data for the implant parameters for most common dopants. If an unusual material is to be used, SUPREM will use range and standard deviation information that the user must provide at that step to simulate the implant. SUPREM also handles the implantation of films through multiple layers. Typically, the program will predict the profile based on simple Gaussian, two-sided Gaussian, Pearson type IV, or dual Pearson type IV distributions. Some versions of the program are also able to predict the profile based on Boltzmann transport and Monte Carlo methods.

A typical SUPREM run that would be used to simulate an N+ source/drain region is given below.

Example 5.2   **Predeposition source/drain implant**

```
Comment      Initialize the silicon substrate
Initialize ⟨100⟩ Silicon Boron Concentration=1e16
+            Thickness=1.0 dX=.005 Xdx=.02 Spaces=120

Comment      Grow the Gate Oxide
Diffusion    Time=20 Temperature=950 DryO2 HCl=3%
Diffusion    Time=10 Temperature=950
Print        Layers

Implant      Arsenic Energy=60 Dose=5e+15 Tilt=7
Diffusion    Time=30 Temperature=550
Diffusion    Time=10 Temperature=1000
Print        Layers Active Concentration Boron Arsenic Net
Plot         Active Net Cmin=1E15
Stop         End Example 5.2
```

5.10 Summary

This chapter introduced the technology of ion implantation. The components of a modern ion implanter were described and some limitations of ion implanting impurities were presented. Most implant profiles can be described by a Gaussian distribution. Additional moments of the Gaussian including skewness and kurtosis are sometimes used to better approximate experimentally observed profiles. After implantation the impurities must be annealed. The anneal step activates the impurities and repairs the implant damage. Different annealing recipes are called for depending on the amount of damage

in the substrate. In silicon technologies, implantation can also be used to form buried insulators through high dose oxygen implantation. Finally the use of SUPREM to simulate implantation was covered.

Problems

1. A 30 keV implant of B^{11} is done into bare silicon. The dose is 10^{12} cm^{-2}. (a) What is the depth of the peak of the implanted profile? (b) What is the concentration at this depth? (c) What is the concentration at a depth of 3000 Å (0.3 μm)? (d) The measured concentration is found to be an order of magnitude larger than the value predicted in part (c), although the profile agrees with answers (a) and (b). Give a possible explanation, assuming that the measured value is correct.

2. A MOSFET threshold voltage adjust implant is done through a gate oxide of 150 Å. The implant species is boron at 30 keV. Estimate the fraction of boron implanted in the oxide. (You may have to use the approximation that $x \ll R_p$ for the Gaussian.)

3. A mass spectrometer as described in the chapter is used for element extraction in an implanter. Calculate the magnetic field necessary to extract silicon (mass 28) if the extraction potential is 20 keV and the radius of curvature for the analyzer is 30 cm. Explain why one might also see some N_2 in the implanted profile if the source cabinet has a small vacuum leak.

4. A typical high current implanter operates with an ion beam of 2 mA. How long would it take to implant a 150 mm diameter wafer with O^+ to a dose of 1×10^{18} cm^{-2}?

5. We can approximate the mass of an atom as twice the atomic charge number. Based on this calculate k_e as a function of Z_i assuming that the target is silicon. Repeat it for a germanium substrate and plot both curves. Discuss the significance.

6. The depth of the junction of the source/drain region in an MOSFET must be reduced as the gate length is scaled. It is highly desirable to produce low resistivity junctions thinner than 0.1 μm. Is this a significant problem for ion implantation? Justify your answer for both N^+/p and P^+/n junctions. What are the major problems in forming these structures?

7. Use SUPREM to implant boron into silicon with a dose of 10^{13} cm^{-2} and an energy of 100 keV. Record the chemical concentration (i.e., Print Boron Chemical). Plot this profile and a Gaussian distribution using the implant parameters found in Figure 5.9. Discuss the differences.

8. In an earlier chapter you used SUPREM to model a double diffused NPN bipolar transistor. Repeat the exercise now using ion implantation. Assume that the wafer is uniformly doped n-type at 10^{16} cm^{-3} and the wafer serves as the collector. The desired device profile includes an emitter doping concentration of 10^{20} cm^{-3}, a peak base doping of 10^{18} cm^{-3}, a base width of 0.2 μm, and a Gummel number (the total amount of net dopant in the base) of about 10^{13} cm^{-2}. Which impurities would you use for the base and the emitter? Use SUPREM to develop an implant and activation cycle that would produce this profile. Plot the net active concentration to demonstrate your solution. Why is such a profile more controlled than the corresponding double diffused profile?

References

1. J. W. Mayer, L. Erickson, and J. A. Davies, *Ion Implantation in Semiconductors, Silicon and Germanium,* Academic Press, New York (1970).

2. G. Dearnaley, J. H. Freeman, R. S. Nelson, and J. Stephen, *Ion Implantation*, New Holland Amsterdam, 1973.

3. T. C. Smith, ''Wafer Cooling and Photoresist Masking Problems in Ion Implantation,'' *Ion*

Implantation Equipment and Techniques, H. Ryssel and H. Glawischnig, eds., vol. 11, Springer Series in Electrophysics, Springer-Verlag, New York, 1983, p. 196.

4. P. Burggraaf, "Resist Implant Problems: Some Solved, Others Not," *Semiconductor Int.* **15:**66 (1992).

5. J. F. Gibbons, "Ion Implantation in Semiconductors—Part I, Range Distribution Theory and Experiments," *Proc. IEEE* **56:**295 (1968).

6. Y. Xia and C. Tan, "Four-parameter Formulae for the Electronic Stopping Cross-section of Low Energy Ions in Solids," *Nucl. Inst., Meth.* **B13:**100 (1986).

7. B. Smith, "Ion Implantation Range Data for Silicon and Germanium Device Technologies," *Res. Studies*, Forest Grove, OR, 1977.

8. T. E. Seidel, "Ion Implantation," in *VLSI Technology*, S. M. Sze, ed., McGraw-Hill, New York, 1983

9. J. Lindhard and M. Scharff, *Phys. Rev.* **124:**128 (1961).

10. J. Lindhard, M. Scharff, and H. Schiøtt, "Range Concepts and Heavy Ion Ranges," *Mat. Fys. Med. Dan. Vidensk. Selsk* **33:**14 (1963).

11. J. F. Gibbons, W. S. Johnson, and S. W. Mylroie, *Projected Range Statistics*, Dowden, Hutchinson, and Ross, Stroudsburg, Pa, 1975.

12. W. P. Petersen, W. Fitchner, and E. H. Grosse, "Vectorized Monte Carlo Calculations for the Transport of Ions in Amorphous Targets," *IEEE Trans. Elect. Dev.* **30:**1011 (1983).

13. D. S. Gemmell, "Channeling and Related Effects in the Motion of Charged Particles Through Crystals," *Rev. Mod. Phys.* **46:**129 (1974).

14. J. F. Ziegler, "Channeling of Ions Near the Silicon ⟨100⟩ Axis," *Appl. Phys. Lett.* **46:**358 (1985).

15. N. L. Turner, "Effects of Planar Channeling Using Modern Ion Implant Equipment," *Solid State Technol.*, Feb. 1985, p. 163.

16. F. F. Morehead and B. L. Crowder, "A Model for the Formation of Amorphous Silicon by Ion Implantation," in *First International Conference on Ion Implantation*, F. Eisen and L. Chadderton, eds., Gordon and Breach, New York, 1971.

17. R. J. Schreutelkamp, J. S. Custer, J. R. Liefting, W. X. Lu, and F. W. Saris, "Preamorphization Damage in Ion-Implanted Silicon," *Mat. Sci. Rep.* **6:**275 (1991).

18. T. Y. Tan, "Dislocation Nucleation Models from Point Defect Condensations in Silicon and Germanium," *Materials Res. Soc. Symp. Proc.* **2:**163 (1981).

19. B. L. Crowder and R. S. Title, "The Distribution of Damage Produced by Ion Implantation of Silicon at Room Temperature," *Radiation Effects*, **6:**63 (1970).

20. T. E. Seidel and A. U. MacRae, "The Isothermal Annealing of Boron Implanted Silicon," in *First International Conference on Ion Implantation*, F. Eisen and L. Chadderton, eds., Gordon and Breach, New York, 1971.

21. S. Wolf and R. N. Tauber, *Silicon Processing for the VLSI Era*, vol. 1, Lattice Press, Sunset Beach, CA, 1986.

22. M. I. Current and D. K. Sadana, "Materials Characterization for Ion Implantation," in *VLSI Electronics—Microstructure Science* **6,** N. G. Einspruch, ed., Academic Press, New York, 1983.

23. L. Csepergi, E. F. Kennedy, J. W. Mayer, and T. W. Sigmon, "Substrate Orientation Dependence of the Epitaxial Growth Rate for Si-implanted Amorphous Silicon," *J. Appl. Phys.* **49:**3906 (1978).

24. J. A. Pals, S. D. Brotherton, A. H. van Ommen, and H. J. Ligthart, "Recent Developments in Ion Implantation in Silicon," *Mat. Sci., Eng.* **B4:**87 (1989).

25. T. O. Sedgwick, *Nucl Instr. Meth.* **B37/38:**760 (1989).

26. D. R. Myers and R. G. Wilson, "Alignment Effects on Implantation Profiles in Silicon," *Radiation Effects* **47:**91 (1980).

27. M. C. Ozturk and J. J. Wortman, "Electrical Properties of Shallow P+-n Junctions Formed by BF_2 Ion Implantation in Germanium Preamorphized Silicon," *Appl. Phys. Lett.* **52:**281 (1988).

28. H. Ishiwara and S. Horita, "Formation of Shallow P+n Junctions by B-Implantation in Si Substrates with Amorphous Layers," *Jap. J. Appl. Phys.* **24:**568 (1985).

29. T. E. Seidel, D. J. Linscher, C. S. Pai, R. V. Knoell, D. M. Mather, and D. C. Johnson, "A Review of Rapid Thermal Annealing (RTA) of B, BF_2 and As Implanted into Silicon," *Nucl. Instrum. Methods B* **7/8:**251 (1985).

30. T. O. Sedgwick, A. E. Michael, V. R. Deline, and S. A. Cohen, "Transient Boron Diffusion in Ion-implanted Crystalline and Amorphous Silicon," *J. Appl. Phys.* **63:**1452 (1988).

31. G. S. Oehrlein, S. A. Cohen, and T. O. Sedgwick, "Diffusion of Phosphorus During Rapid Thermal Annealing of Ion Implanted Silicon," *Appl. Phys. Lett.* **45:**417 (1984).

32. H. Metzner, G. Suzler, W. Seelinger, B. Ittermann, H.-P. Frank, B. Fischer, K.-H. Ergezinger, R. Dippel, E. Diehl, H.-J. Stöckmann, and H. Ackermann, "Bulk-Doping-Controlled Implant Site of Boron in Silicon," *Phys. Rev. B.* **42:**11419 (1990).

33. L. C. Hopkins, T. E. Seidel, J. S. Williams, and J. C. Bean, "Enhanced Diffusion in Boron Implanted Silicon," *J. Electrochem. Soc.* **132:**2035 (1985).

34. R. B. Fair, J. J. Wortman, and J. Liu, "Modeling Rapid Thermal Diffusion of Arsenic and Boron in Silicon," *J. Electrochem. Soc.* **131:**2387 (1984).

35. A. E. Michael, W. Rausch, P. A. Ronsheim, and R. H. Kastl, "Rapid Annealing and the Anomalous Diffusion of Ion Implanted Boron," *Appl. Phys. Lett.* **50:**416 (1987).

36. K. J. Reeson, "Fabrication of Buried Layers of SiO_2 and Si_3N_4 Using Ion Beam Synthesis," *Nucl Instrum. Methods* **B19–20:**269 (1987).

37. K. Izumi, M. Doken, and H. Ariyoshi, "CMOS Devices Fabricated on Buried SiO_2 Layers Formed by Oxygen Implantation in Silicon," *Electron. Lett.* **14:**593 (1978).

38. H. W. Lam, "SIMOX SOI for Integrated Circuit Fabrication," *IEEE Circuits, Devices* **3:**6 (1987).

39. P.L.F. Hemment, E. Maydell-Ondrusz, K. G. Stevens, J. A. Kilner, and J. Butcher, "Oxygen Distributions in Synthesized SiO_2 Layers Formed by High Dose O^+ Implantation into Silicon," *Vacuum* **34:**203 (1984).

40. G. F. Celler, P.L.F. Hemment, K. W. West, and J. M. Gibson, "Improved SOI Films by High Dose Oxygen Implantation and Lamp Annealing," in *Semiconductor-on-Insulator and Thin Film Transistor Technology*, A. Chiang, M. W. Geis, and L. Pfeiffer, eds., *Mat. Res. Soc. Symp. Proc.* **53**, Boston, 1986.

41. S. Cristoloveanu, S. Gardner, C. Jaussaud, J. Margail, A.-J. Auberton-Hervé, and M. Bruel, "Silicon on Insulator Material Formed by Oxygen Ion Implantation and High Temperature Annealing: Carrier Transport, Oxygen Activation, and Interface Properties," *J. Appl. Phys.* **62:**2793 (1987).

42. M. I. Current and W. A. Keenan, "A Performance Survey of Production Ion Implanters," *Solid State Technol.*, p. 139, Feb. 1985.

43. H. Glawischnig and K. Noack, "Ion Implantation System Concepts," in *Ion Implantation Science and Technology*, J. F. Ziegler, ed., Academic Press, Orlando, 1984.

44. P. Burggraaf, "Equipment Generated Particles: Ion Implantation," *Semiconduct. Int.* **14:**10, p. 78 (1991).

45. E. W. Haas, H. Glawischnig, G. Lichti, and A. Bleicher, "Activation Analytical Investigation of Contamination and Cross Contamination in Ion Implantation," *J. Electronic Mat.* **7:**525 (1978).

Chapter 6

Rapid Thermal Processing

The last few chapters have discussed the redistribution of impurities at high temperature. For small devices this redistribution is often very undesirable and so there has been a great deal of emphasis in recent years on low temperature processes that minimize diffusion. Some processes, such as implant annealing however, are not as effective at low temperature. Certain types of implant damage cannot be annealed out unless high temperatures are achieved. Furthermore, some dopants require anneal temperatures of at least 1000°C to be completely activated. The other avenue to reducing diffusion is to reduce the time at temperature. Standard furnace annealing is ill suited to short time anneals. The wafers in such a system heat from the edges inward. To avoid excessive temperature gradients that might lead to warpage, the wafers must be heated and cooled very slowly [1]. As a result, even though the annealing time could be short, the long temperature ramps result in significant diffusion. At the same time, there has been an increasing emphasis on processes that operate on one wafer at a time rather than large batches. These single wafer processes provide the best uniformity and reproducibility, particularly for large wafer sizes. Rapid thermal processing (RTP) describes a family of single wafer hot processes that have been developed to minimize the thermal budget of a process by reducing the time at temperature in addition to, or instead of, reducing the temperature.

RTP was originally developed for implant annealing. While this application is still common, the use of rapid thermal approaches to processing has spread to oxidation, chemical vapor deposition, and epitaxial growth. Central to all rapid thermal processes is a set of common problems: heating and cooling the wafer uniformly, being able to maintain a uniform temperature during the process, and measuring the wafer temperature. The next few sections will describe the nature of these problems and the approaches that rapid thermal systems have adopted to address them. The chapter will then review some applications of the technology starting from the traditional implant annealing and the more recently developed oxidation and nitridization and concluding with the formation of silicides. The use of RTP for chemical vapor deposition (RTCVD) and epitaxial growth will be delayed until later in the book, since these topics also require an understanding of general chemical vapor deposition techniques.

Rapid thermal processes can be divided into three broad classes by the type of heating that is carried out: adiabatic, thermal flux, and isothermal [2]. The first demonstration of RTP used an adiabatic heat source [3]. In this approach, fast pulses of light in a broad beam heat only the front surface (of order a few microns) of the wafer as long as the pulse length is short compared to the thermal time constant of the substrate. Adiabatic systems are often powered by broad beam coherent

sources such as excimer lasers. While this type of annealing system allows the shortest time at temperature, it has several important drawbacks. These include poorly controlled temperatures, poorly controlled anneal times, large vertical temperature gradients, and large capital equipment costs. Thermal flux systems use an intense spot source such as an electron beam or a focused laser that is scanned across the wafer. The scan period must be short compared to the thermal time constant or large lateral thermal gradients will result. While this type of system has been used for research, the defects caused by lateral thermal nonuniformity are usually large enough to prevent their use for IC fabrication. Isothermal heating uses a broad beam of radiation to heat the wafer for many seconds. These systems may have minimal temperature gradients across and through the wafer. Typically they are powered by incoherent sources such as an array of tungsten-halogen lamps. The wafer rests on quartz pins in isothermal systems. Quartz is selected because of its chemical stability and its low thermal conductivity. This arrangement is sometimes called *thermal isolation*. This chapter will concentrate on isothermal RTP systems since almost all systems use this design.

6.1 Gray Body Radiation, Heat Exchange, and Optical Absorption°

There are four types of heat transfer that may be of interest to semiconductor processing: conduction, convection, forced flow, and radiation. Thermal conduction is the diffusion of heat through a solid or gas. The heat flow through a cross-sectional area A of a solid or an immobile gas or liquid is given by

$$\dot{q}(T) = k_{th}(T)A\nabla T \tag{6.1}$$

where k_{th} is the thermal conductivity of the material. Dividing both sides by the area produces Fick's first law in terms of heat transfer. Since most of the optical energy in rapid thermal processing is absorbed in the first few μm of the wafer, thermal conduction through the wafer plays an important role in the final temperature distribution. When considering thermal conduction in a gas however, one must also take into account gas flow since it can alter the rate of heat transfer. If the flow is caused by an externally applied pressure gradient it is said to be forced flow. Examples include flows caused by gas injection or pumping. Flows that are in response to temperature gradients in an otherwise closed system are called *natural flows*. The movement of water in a heated pot is a natural flow. An effective heat transfer can be defined as

$$\dot{q} = h(T - T_\infty) \tag{6.2}$$

where T_∞ is the temperature of the gas far from the wafer and h is an effective heat transfer coefficient that depends on both free and forced flow. For most geometries h is a function of temperature and the position on the wafer.

The amount of power that can be delivered by gas flow is rather limited. As a result, most rapid thermal systems use radiative heat transfer as the primary method of heat exchange. One of the basic parameters of radiative heat transfer is the spectral radiant exitance $M_\lambda(\lambda,T)$, the amount of power radiated by a body into a perfectly absorbing environment (black box) per unit surface area of the emitting object and per unit wavelength of the radiation. Planck's radiation law gives the spectral radiant exitance as

$$M_\lambda(T) = \varepsilon(\lambda)\,\frac{c_1}{\lambda^5(e^{c_2/\lambda T} - 1)} \tag{6.3}$$

where $\varepsilon(\lambda)$ is the wavelength dependent emissivity of the emitting body and c_1 and c_2 are the first and second radiation constants given by 3.7142×10^{-16} W-m^2 and 1.4388×10^{-2} m-K, respectively. When $\varepsilon = 1$, the emitting source is said to be a black body.

When $M_\lambda(\lambda, T)$ is integrated over all wavelengths from 0 to ∞, and it is assumed that the emissivity is wavelength independent, the result is the total exitance $M(T)$, which is given by the Stefan-Boltzmann equation

$$M(T) = \varepsilon \sigma T^4 \tag{6.4}$$

where σ is the Stefan-Boltzmann constant, 5.6697×10^{-8} W/m^2-K^4. Comparing Equations 6.2 and 6.4, the amount of power radiated by an object is proportional to the fourth power of the temperature, while the power conducted through thermal conduction is proportional to the temperature difference between the object and the background. As a result, radiation is the dominant heat transfer mechanism at high temperature while thermal conduction is more important at low temperature. Since most rapid thermal systems operate in the high temperature regime, radiation exchange is the dominant heat exchange mechanism.

Differentiating Equation 6.1 and setting the result equal to zero, one can define λ_{max} as the wavelength at which the emitted power is maximized. Then

$$\lambda_{max} = \frac{0.2898 \text{ cm-K}}{T} \tag{6.5}$$

This relationship can be used to translate the temperature of an emitting body into a corresponding color temperature. Many lamps are specified this way since it would be difficult to measure their temperature directly.

When radiation is incident on the surface of the wafer it may be reflected, absorbed, or transmitted. $\rho(\lambda, T)$ is defined as the fraction of reflected radiation and $\tau(\lambda, T)$ as the fraction of the transmitted radiation. According to Kirchhoff's law

$$\varepsilon(\lambda, T) = 1 - \rho(\lambda, T) - \tau(\lambda, T) \tag{6.6}$$

For opaque materials $\tau(\lambda, T) = 0$, and

$$\varepsilon(\lambda, T) = 1 - \rho(\lambda, T) \tag{6.7}$$

Once the emissivities of two bodies are known, the net power transfer between them can be calculated. We might take these two bodies to be the wafer and the lamp array, for example. Let ε_1 and ε_2 be the average emissivities for all wavelengths being considered for the two bodies. Then the net power transfer from body 1 to body 2 is

$$s = \dot{q}_{1 \rightarrow 2} - \dot{q}_{2 \rightarrow 1} = \sigma(\varepsilon_1 T_1^4 - \varepsilon_2 T_2^4)A_1 F_{A1 \rightarrow A2} \tag{6.8}$$

where A_1 is the area of body 1, and F_{A2-A1} is a geometric constant called the *view factor* or

$$F_{A1 \rightarrow A2} = \frac{1}{A_1} \int_{A_2} \int_{A_1} \frac{\cos \beta_1 \cos \beta_2}{\pi r^2} dA_1 dA_2 \tag{6.9}$$

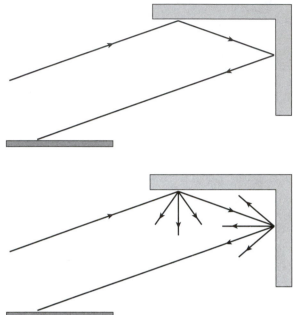

Figure 6-1 Geometry for calculating the view factors between two surfaces, A_1 and A_2.

Figure 6-2 Possible optical paths among three surfaces including only single reflections. This diagram assumes diffuse reflections.

configuration factor. $F_{A1 \rightarrow A2}$ is the fraction of the total solid that the area A_2 subtends from the surface A_1. The view factor can be calculated directly using the angles β_1 and β_2 as being the angles from the surface normals as shown in Figure 6.1 [4]. Alternatively, the view factors for many simple geometries have been tabulated [5].

Consider what happens if a third surface like a reflector is added (Figure 6.2). One might first consider three pair-wise interactions, but that would only include the power radiated by the heated surfaces. The reflector in particular may be water cooled, keeping the surface temperature low. Yet the reflector surface may transmit a great deal of power because it increases the effective view factor between the wafer and the lamp. The reflector also allows the wafer to radiate to itself. If all possible reflections are included, the complexity of pair-wise scheme increases substantially. Alternatively, one can treat all of the surfaces simultaneously, including reflections, using matrix methods [6],[7].

6.2 High Intensity Optical Sources and the Reflecting Cavity

Most isothermal systems use either W-halogen lamps or long arc noble gas discharge lamps as power sources (Figure 6.3). The W-halogen lamp consists of a tightly wound spiral tungsten filament encased in a fused silica envelope. The spiral winding increases the surface area of the lamp, thereby increasing the radiation efficiency. The spiral filament may be linear with a connector on both sides of the bulb or the filament may be shaped for single ended termination. The quartz envelope of these lamps

contains a halogenated gas. One common component is PNBr$_2$ [8]. Tungsten evaporates from the heated filament and deposits on the walls of the envelope. As the walls heat the halide gas reacts with the tungsten to form volatile tungsten-halides that diffuse to the much hotter filament where they decompose and redeposit the tungsten. This process has an intrinsic feedback mechanism. As more tungsten deposits on the envelope, the increased quartz heating increases the etching reaction rate. This feedback prevents excessive buildup of tungsten.

W-halogen lamps emit as blackbodies. Virtually all of the radiation is in the 0.8 to 4.0 μm range. Assume that it is necessary to be able to raise the temperature of a 150 mm wafer to 1100°C. Since the emissivity of silicon is approximately 0.7, the total exitance at this temperature (Equation 6.2) is about 2.5 kW. Each tungsten lamp consumes approximately 100 W of electrical energy per centimeter of lighted length. About 40% of that energy is converted into light in a typical filament lamp; however, that number varies significantly from lamp to lamp. Since only a fraction of the emitted optical power is collected by the wafer, it would be necessary to use an array of fifteen to thirty 8 in. lamps to heat the wafer.

Long arc noble gas discharge lamps contain two refractory metal electrodes sealed in a fused silica envelope with a noble gas such as krypton or xenon. The electrical properties of the discharge are determined by the electrode spacing, the inner diameter of the envelope, and the gas composition and pressure. When ignited with a high voltage pulse (roughly 2 kV per centimeter of lighted length) the gas is ionized and a dc path is established. These lamps emit strongly in the visible part of the

Figure 6-3   Tungsten-halogen lamps (left) and noble gas arc lamps (right) for use in rapid thermal systems.

Figure 6-4   Various chamber designs include: (a) the reflecting cavity and (b) a windowed system using an intense source and a shaped reflector.

spectrum. Their spectra includes a very high temperature electron plasma that looks like a gray body with $\lambda_{max} \approx 200$ nm, along with discrete line spectra corresponding to the electronic transitions of the fill gas.

Discharge lamps consume up to 700 W/cm of lighted length. Optical conversion efficiencies are about 45%. Discharge lamps must be water cooled to be able to withstand these high power densities. Particular care must be taken to ensure that the quartz to metal seals and the metal electrodes remain cool. Using the same criteria of heating a 150 mm wafer to 1100°C, one would only need two to four 8 in. lamps; however, this advantage in power density is partially offset by the need to water cool and electrically isolate the lamps for igniting them. Furthermore the lamps must be run by a more costly regulated dc source, while the filament lamps may operate directly from a simple line (i.e., 60 Hz) source.

Various chamber geometries have been used to optimize the power collection efficiency (Figure 6.4). At the same time, one must also try to design the chamber so that the wafer can achieve and maintain a uniform temperature. Many current generation RTP system designs use the reflecting cavity approach. In this design the wafer resides in a quartz flow tube. Linear W-halogen lamps are above and below the flow tube. An N_2 purge may be used to cool the lamps and the outer surface of the quartz flow tube. The entire assembly is encased in a gold coated box. The surface of the gold is roughened to ensure diffuse reflection. The purpose of this arrangement is to randomize the optical path so as to distribute the radiation uniformly across the wafer.

The use of reflecting cavities has been only partially successful in achieving a uniform temperature across the wafer. The wafer edge tends to be more cool than the center primarily due to three effects (Figure 6.5). For a finite sized array of lamps, the view factor from the outside of the wafer is less than that of the center. The edge of the wafer is almost completely blocked from direct radiation exchange with the lamps. Finally, for common geometries, the edge of the wafer is more effectively cooled than the center. Collectively, these edge effects may lead to temperature gradients of several tens of degrees (Figure 6.6). These temperature gradients lead to process nonuniformity, and if severe enough, lead to slip and wafer warpage.

Various approaches have been used to minimize thermal nonuniformity in RTP systems. In order to compensate for the increased edge losses, radiant power to the edge of the wafer must be increased. Current generation systems do this by shaping the reflectors or lamp spacings, or changing the transmissive properties of the fused silica flow tube. These approaches suffer from the fact that the amount of excess radiation required depends on the process temperature. Thus, the power distribution can only be optimized for one temperature and then only if the lamps do not age. Furthermore, temperature ramps require nearly uniform radiant distributions. As a result, the wafer edges overheat during tem-

Figure 6-5   Causes of thermal nonuniformity includes: (a) a reduced view factor to the lamp array for large *r*, (b) very small view factors along the wafer edge, and (c) nonuniform gas phase heat transfer *(after Campbell, 1994)*.

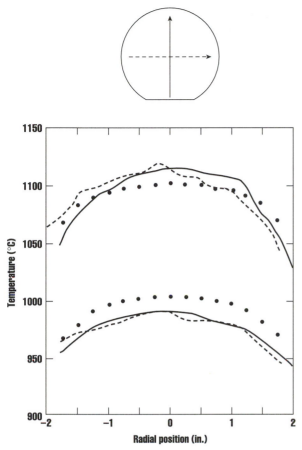

Figure 6-6   Typical wafer temperature distributions across the wafer in an early generation rapid thermal system *(after Lord, ©1988 IEEE).*

perature ramp up [9]. The most promising approach to solving this problem is to divide the lamps into heating zones that can be independently controlled. For example, one set of lamps are arranged geometrically in such a way that they heat the outermost part of the wafer more than the center. A second set might heat the center of the wafer more effectively than the edges. A third might provide roughly uniform illumination. By choosing the proper combination of power settings, a nearly uniform temperature can be obtained for a broad range of process conditions. This type of arrangement is called a *multizone heater*. As will be discussed in more detail later in the chapter, it can be used most effectively if the temperature distribution across the wafer is either known or can be predicted.

6.3 Temperature Measurement

One of the most difficult tasks associated with rapid thermal processing is accurate and reproducible temperature measurement. The wafer temperature is used in a feedback loop to control the lamp power output. In most pieces of process equipment the wafer temperature is measured using a thermocouple embedded in the wafer holder or susceptor. This is not possible in RTP since there is no susceptor. The thermocouple can be placed as a point contact on the surface of the wafer, but the thermal impedance associated with such a contact will result in a substantial temperature difference between the wafer and the thermocouple. Furthermore, heat loss through the lead will cool the wafer locally, leading to stress and process nonuniformity in the vicinity of the contact. Finally, direct contact thermocouple measurements may allow chemical reactions with the silicon wafer. These reactions contaminate the wafer and shift the characteristics of the thermocouple.

For these reasons all temperature measurements in RTP systems are done indirectly. The most popular techniques are pyrometry (optical) or thermoelectric detectors. The most common thermoelectric detector for RTP is the thermopile. This device operates by the Seebeck effect. A bimetal junction is suspended on a thin membrane. When the junction is heated it develops a small voltage. A second junction shielded from the radiation serves as a reference. The magnitude of the voltage difference between the two junctions varies linearly with the temperature difference between the junctions [10].

The second widely used method is pyrometry. Most pyrometers operate by measuring the radiant energy received in some band of energies, assuming the source to be a gray body of known emissivity, and converting the power to a source temperature using the Stefan-Boltzmann relationship. Care must be taken in selecting the operating wavelength to ensure that any lamp output at that wavelength is

completely filtered before entering the chamber. Special purpose glasses or other optical filters can be used for this purpose. Most commercial systems monitor some band in the mid infrared (3–6 μm). Arc lamps produce insignificant radiation in this band. If W-halogen lamps are used, the flow tube itself can be used to filter out most of the radiation from the lamps in this band. In that case, a hole or thinned window in the flow tube must be provided to allow the pyrometer a line of sight to the wafer. The window thickness must be well known or temperature errors as large as 100°C can result [11]. Another concern with pyrometry is that some process gasses can absorb energy in the IR and so reduce the apparent wafer temperature [12]. Depending on the gas composition, pressure, and temperature, strong absorption bands can exist anywhere from 2 to 10 μm [13].

The major caveat in optical pyrometry is that the effective emissivity must be accurately determined. The effective emissivity includes both intrinsic and extrinsic contributions. The intrinsic emissivity is a function of the material, the surface finish, the temperature, and the wavelength at which it is measured. Fortunately, the intrinsic emissivity of bare silicon wafers is reasonably well characterized [14] as a function of temperature (Figure 6.7). At low temperature the emissivity depends on the doping concentration for energies less than the bandgap. As the temperature increases above 600°C, the semiconductor becomes intrinsic. Free carriers are present in sufficient numbers to produce a nearly wavelength independent emissivity. The extrinsic emissivity relates to the amount of radiant energy from other sources that is reflected back onto the spot being measured and so increase the apparent temperature. As such, it must be measured in situ and may change as the chamber reflectivity changes. Furthermore, the presence of layers such as polysilicon and silicon dioxide can dramatically alter the apparent emissivity (Figure 6.8) at the measurement wavelength due to interference effects [15].

Figure 6-7 The emissivity of silicon as a function of wavelength with temperature as a parameter. Above 600°C the wafer is intrinsic *(after Sato, reprinted by permission,* Japanese J. Appl. Phys.*).*

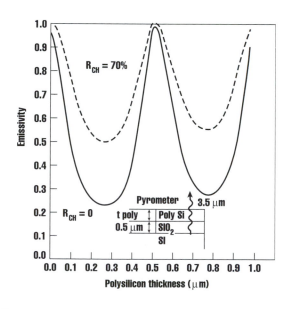

Figure 6-8 The effect of polysilicon thickness on effective emissivity for a polysilicon/SiO₂/silicon structure in nonreflective and 70% reflective cavities *(after Hill and Boys, reprinted by permission, Plenum Publishing).*

Example 6.1

The major source of uncertainty in pyrometry is an uncertainty in the emissivity. If the wafer temperature is 1000°C, what wavelength is most desirable to minimize the effect of this uncertainty?

According to Equation 6.3,

$$M_\lambda T = \varepsilon(\lambda) \frac{c_1}{\lambda^5(e^{c_2/\lambda T} - 1)}$$

Differentiating and assuming that M and λ are fixed and that $e^{c_2/\lambda T} \gg 1$,

$$\frac{dT}{T} = - \frac{\lambda T}{c_2} \frac{d\varepsilon}{\varepsilon}$$

If the pyrometer is operating at 5 µm, a 5% uncertainty in the emissivity produces a 22°C error in the temperature. On the other hand, a pyrometer operating in the 0.94–0.96 µm range produces only a 4°C temperature error for the same 5% error in emissivity.

Commercial RTP systems can control the wafer center temperature for a bare wafer process such as an implant anneal to about 10°C; however, the absolute temperature is generally less well known due to emissivity uncertainties. Two color pyrometers measure the emitted power at two wavelengths to attempt to correct for emissivity changes. Implicit in their operation is the assumption that the ratio of the emissivity at these two wavelengths is fixed. This is not always a good assumption, particularly when the process involves a film growth or deposition. Of course, the emissivity will change from that of the substrate material to the emissivity of the material deposited during the process. More problematic however, is the situation that occurs when the film is at least partially transmissive. When the thickness of the film becomes an integral multiple of one fourth of the measurement wavelength, destructive interference takes place that dramatically changes the apparent emissivity. For these processes temperature reproducibility is severely degraded. It is possible to measure the emissivity directly by measuring the reflectivity of the wafer and using Kirchoff's law (Equation 6.7) [16]. In order for this scheme to work the measurement must be done in visibility where the wafer is opaque [17]. The pyrometry should be done at wavelengths as close as possible to that of the reflection measurement; however, in practice this is difficult to achieve.

Various nonpyrometric techniques have been investigated to avoid the problems associated with assuming an emissivity. Several authors have demonstrated the use of thermal expansion as a direct measure of the temperature. Diffraction gratings patterned on the surface of the wafer can be used with projection moiré interferometry [18] or in a diffraction order measurement [19]. In either case, the movement of the grating via thermal expansion can be measured and used to infer temperature changes. Of course, these methods require an appropriate pattern on the surface of the wafer that must be properly aligned with the optical probe beam and detector. Furthermore, noise can be a problem at high temperature due to strong gas flow conditions caused by the large thermal gradients in the chamber. Alternatively, the wafer diameter itself can be measured optically [20]. The demonstrated repeatability of this technique is 1%.

Another interesting approach to measuring temperature is the use of acoustic waves. It has been

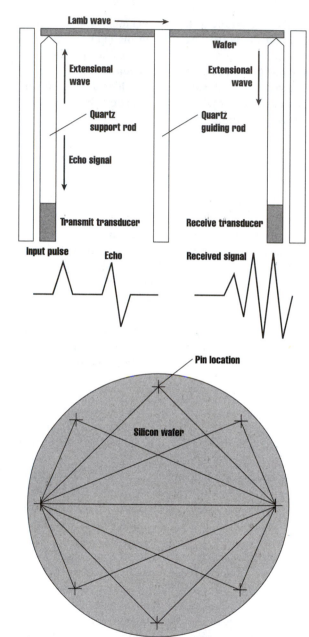

Figure 6-9   Acoustic temperature measurement side and top views *(after Degertekin et al., used by permission, Materials Research Society).*

found that the velocity of sound in silicon is a linear function of the substrate temperature [21]. Acoustic waves can be launched through one of the quartz support pins and the wave detected using another pin (Figure 6.9). Since the transducers can act as both acoustic sources and sensors, there are $\frac{1}{2}(n^2-n)$ possible paths. Because the measured velocity is averaged over the path, the temperature profile across the wafer can be extracted by taking multiple measurements along different paths [22]. This information is extremely valuable for multizone heating arrangements.

6.4 Thermoplastic Stress°

The existence of the thermal gradients in the wafer gives rise to thermoplastic stress. If the stress becomes too large, defects such as dislocations and slip can result. Assuming that the substrates are isotropic, that dislocations do not occur, and that the temperature gradients vertically through the wafer can be neglected, one can estimate the thermal stresses on the wafer. Due to radial symmetry in $T(r)$, the shear stress will be zero. The radial and angular stress components are given by

$$\sigma_r(r) = \alpha E\left[\frac{1}{R^2}\int_0^R T(r')r'dr' - \frac{1}{r^2}\int_0^r T(r)r'dr'\right] \quad \textbf{(6.10)}$$

and

$$\sigma_s(r) = \alpha E\left[\frac{1}{R^2}\int_0^R T(r')r'dr' - \frac{1}{r^2}\int_0^r T(r)r'dr' - T(r)\right] \quad \textbf{(6.11)}$$

where α is the linear thermal expansion coefficient, E is Young's modulus, and R is the radius of the wafer.

Figure 6.10 shows a typical plot of the radial dependence of these stresses for a simple process consisting of a linear ramp to temperature, an anneal at temperature, and a linear ramp down. Under steady state conditions the wafer edges are slightly cooler than the center. Then the radial stress has a maximum near the center of the wafer and goes to zero at the edge. The tangential stress rises from zero at the center and is typically much larger than the radial stress component. Since this stress is so large and since defects will nucleate more readily at the wafer edge, most rapid thermal processing induced slip

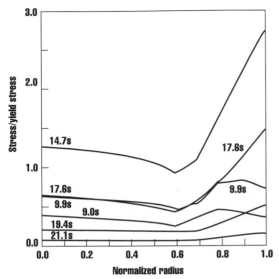

Figure 6-10   Normalized stress versus position on a wafer during a heating transient (*after Lord, ©1988, IEEE*).

is seen at the edge of the wafer. The diamond structure tends to yield along the ⟨110⟩ directions in the (111) planes and so for p-type (100) wafers, the slip lines are parallel and perpendicular to the major flat orientation.

The yield strength is the limit at which a material will plastically deform. For silicon the yield strength is described by the Haassan formula

$$\sigma_{\text{yield}} = A\left[\frac{\dot{e}}{\dot{e}_o}\right]^{1/n} e^{E_a/kT} \tag{6.12}$$

where $\dot{e}$ is the strain rate, $\dot{e}_o$ is a reference strain rate normally taken to be 10^{-3} sec^{-1} [23].

The exact value of σ_{yield} depends on the oxygen and dopant concentrations in the wafer and the previous processing. Typical values include $A = 3630$ Pa, $E_a = 1.073$ eV, and $n = 2.45$ [24]. At high strain rates the yield strength is limited to 60 MPa, while at low strain rates σ_{yield} does not decrease below about 3.5 MPa [25]. Thus, the wafer can tolerate higher strains during transients, but the strain generated during transients is generally much larger than that generated in the steady state since most RTP systems are optimized for steady state uniformity.

6.5 Rapid Thermal Activation of Impurities

The last chapter introduced ion implantation. Because of its ability to produce well controlled impurity doses whose concentration may in fact exceed the solid solubility, the technique has become almost universally accepted. As device sizes have been reduced, the concentration gradients have increased, and the maximum allowable dopant redistribution has decreased. The implant damage in these wafers must still be removed by annealing. Depending on the implant species energy and dose, this may require temperatures as high as 1100°C [26]. The basic reason that rapid thermal processing was developed was to access these high temperatures while minimizing the thermal budget by reducing the time at temperature.

One of the most attractive features of RTP is that the wafer may not reach thermal equilibrium. This means that the electrically active doping profiles can actually exceed the solid solubility. Arsenic, in particular, is found to require only very short anneals to achieve a high level of activation [27]. The arsenic atoms have insufficient time to form clusters and condense out into inactive defects [28]. If the activation is too incomplete however, the excess arsenic atoms seem to contribute a deep level [29]. These levels can be efficient generation/recombination centers leading to electrical leakage if they are close to p-n junctions.

It has generally been observed that low temperature and reduced time anneals of implanted species produce chemical junctions that are deeper than those predicted by simple diffusion theory [30],[31]. Diffusivity enhancements for boron in silicon have been reported to be as much as a factor of 1000×. The origin of this enhancement is believed to be residual implant damage. The wafer is believed to have a high concentration of vacancies and self-interstitials after an implant [32]. This effect is sometimes called transient effects or transient enhanced diffusion. It has been shown that the

Table 6.1   The activation energies for steady state intrinsic diffusion and transient diffusion in silicon. All energies are in eV.

	Steady-State	Transient
Boron	3.5	1.8
Arsenic	3.4	1.8
Phosphorus	3.6	2.2

Data taken from Fair [33].

increase in the junction depth of low dose implants is proportional to the square root of the implant energy. Furnace anneal processes often attempt to annihilate excess point defects through an extended treatment at temperatures of 500 to 650°C before heating to the activation temperature. RTP anneals may also include a brief low temperature step for the same reason.

The activation energies for transient enhanced diffusion for the three most common silicon dopants are shown in Table 6.1. Although initially somewhat controversial, it is now generally agreed that transient effects during the diffusion of arsenic are observed, but are much less pronounced than for boron. These transient effects decay with some characteristic time constant that is related to the rate of defect annihilation in the substrate.

It has also been shown that for boron and BF_2 not all of the chemical impurities are activated in RTP. Furnace anneals always activate the low concentration implant tails, but may not fully activate the region near the peak concentrations due to thermodynamic considerations such as the solid solubility. It has been observed, however, that after RTP anneals the low concentration boron tails may not be fully activated, giving rise to an electrical junction that is more shallow than the chemical one [34],[35]. Figure 6.11 shows the difference between the chemical profile (solid line) and electrically active profile (open points) after a 30 sec anneal at 1000°C for both boron and BF_2 implants. Both peak and tail concentrations are not fully activated. The nonactivated impurities are believed to be due to the formation of inactive boron interstitial pairs [36].

Most implant activation in GaAs requires the use of a capping layer, usually Si_3N_4 or SiO_xN_y, to prevent the outdiffusion of arsenic from the wafer [37]. The cap layer is typically deposited by plasma enhanced chemical vapor deposition. As with silicon, the wafer is first annealed at close to 650°C for about 60 sec to reduce the point defect density. Activating the n-type impurity silicon in GaAs requires temperatures of 800 to 1000°C. Capping layers may be difficult to remove and due to the thermal expansion mismatch, can lead to slip formation in the wafer. It has also been shown that GaAs implant activation can be done without capping layers using a specially designed graphite susceptor (Figure 6.12) [38]. The wafer is placed inside a graphite cavity that is two or three times thicker than a typical wafer [39]. It has been shown that this type of annealing leads to an improved surface quality and a reduced particle count [40] compared to standard capped RTA. Finally capless anneals can also be done

Figure 6-11   Chemical and active boron profiles after incomplete activation in RTP (*after Kinoshita et al., used by permission, Materials Research Society*).

Figure 6-12   Photograph and schematic for capless arrangement for annealing GaAs *(after Kazior et al., ©1991 IEEE)*.

by laying another GaAs wafer on top of the wafer to be annealed in the RTP. Care must be taken to ensure that the surface oxides have been removed from the upper wafer.

6.6 Rapid Thermal Processing of Dielectrics

Deep submicron devices require the production of very thin oxides of silicon. One technique for growing these oxides is to slow the oxidation rate by reducing the oxidation temperature. The difficulty with this approach is that the fixed charge and interface state density both tend to increase as the growth temperature is reduced. As a result, rapid thermal oxidation (RTO) appears to be an attractive alternative, allowing short time oxidations at suitably high temperatures. This indeed has become one of RTP's areas of application. Alternatively, it is possible to slow the oxidation rate by diluting the oxygen with argon or some other inert gas. While this approach is capable of growing high quality oxides, the diffusion of impurities in the substrate is still a concern.

Virtually all RTO is done with dry oxygen. Early results showed that oxide thickness increased linearly with time, with an oxidation rate of approximately 3 Å/s at 1150°C [41]. The oxides were electrically robust with good breakdown characteristics [42]. Rapid thermal oxidation uniformity is affected by the thermoplastic stress in the wafer caused by nonuniform temperatures and can result in a growth rate enhancement near the wafer edge where the stress is a maximum [43]. This effect is most pronounced for low temperature and short time oxidations.

Numerous papers have been published regarding the oxidation rate of silicon during RTP. A representative set of data [44] is shown in Figure 6.13. A cohesive model of the oxidation rate during RTO has been difficult to develop for several reasons. The first is that RTO normally occurs completely within the initial oxidation regime. As described in Chapter 4, the origin of this effect is still in dispute. Furthermore, a wide variability in the experimentally measured oxidation rates have been reported in the literature. This can be related to the difficulty in measuring the wafer temperature

Figure 6-13   Typical data for oxide thickness as a function of time for a rapid thermal oxidation process *(after Moslehi et al., 1985).*

during growth. It is generally believed that as much as a 50°C temperature variability can be deduced from the literature [45]. As a result most RTO models are empirical.

As mentioned in an earlier chapter, very thin gate oxides have reliability concerns. It has been found that nitridizing an oxide can increase the ability of impurities to diffuse through the oxide and can reduce carrier trapping under electrical stress [46]. Heavy nitridization reduces the mobility of the inversion layer by 20 to 50% [47] and leads to an increase in electron trapping. This is believed to be due to an increase in hydrogen incorporation in the oxide [48]. Using RTP, a light nitridization can be done [49],[50] that avoids most of this mobility reduction. It has also been shown that using RTP to introduce small amounts (parts per million) of fluorine in an oxide produces excellent breakdown characteristics. This fluorine can be incorporated through the use of NF_3 in a rapid thermal processor [51].

One significant potential advantage of RTP is its potential to be used as a hot wafer/cold wall process if the chamber walls are properly cooled. The importance of this observation is that multiple processes can be run in a single chamber since the walls do not contaminate sequential process steps. The wafer temperature can be used to start and stop these reactions. The gas flowing in the chamber can be changed while the wafer is cool, preventing any chemical reaction. The advantage of this technique, of course, is the minimization of wafer handling and resultant contamination. Multiple layers can be deposited in this way with well controlled interfaces. This capability was exploited very early in the development of RTP [52] by thermally growing a thin oxide and then depositing a layer of polycrystalline silicon using chemical vapor deposition to fabricate MOS capacitors. The capacitors were found to have excellent interfacial properties. Later work by the same group demonstrated that the lower silicon electrode could be grown selectively in silicon dioxide windows followed by oxidation of the epitaxial layer and finally polysilicon deposition so that both interfaces of the dielectric were formed in the RTP reactor [53].

6.7 Silicidation and Contact Formation

As will be described in some detail in Chapters 15, 16, and 18, for some device applications the resistivity of even heavily doped silicon is too large. In those cases, it is common to form metal silicides on top of the exposed silicon to reduce the resistivity. Metal silicides are often formed by depositing a thin metal layer on top of the wafer and then heating it to drive a silicidation reaction. Common metals for silicide formation include Ti, Pt, W, Ta, Mo, and Co. RTP allows the careful control of the silicidation temperature and ambient to minimize impurities and to promote the growth of the most desirable stoichiometry and phase of the silicide. In some cases the wafer has a partial coverage of SiO_2 before metal deposition. The metal does not react with the oxide and so can be easily removed wet chemically after the silicide reaction.

$TiSi_2$ is one of the most desirable films for many applications due to its low resistivity (13 $\mu\Omega$-cm). During the silicide formation anneal however, silicon can diffuse along the grain boundaries of the $TiSi_2$, leading to an overgrowth of the silicide on top of the edges of the oxide (Figure 6.14) [54].

Figure 6-14   Results of a 900°C 30 sec formation of TiSi₂ in an oxide window. The ragged edges indicate silicide formed over the edges of the window *(after Brat et al., used by permission, The Electrochemical Society).*

This overgrowth can be minimized by first annealing at a low temperature to form TiSi. The unreacted metal is then stripped and a high temperature RTA is done to form the desired phase (C54) of the silicide. This phase of TiSi₂ requires an anneal of at least 750°C [55],[56], with some reports indicating that as high as 900°C is required [57]. When the initial silicidation is carried out in a pure N_2 ambient, lateral growth of the silicide is further impeded [58]. The titanium silicidation reaction is sensitive to ppm levels of H_2O or O_2 in the ambient since oxidation and silicidation will compete at the interface. Oxygen will also dramatically increase the sheet resistance of the silicide. Silicides formed in N_2 containing 10 ppm of O_2, for example, have nearly forty times the resistivity of silicides formed in 0.1 ppm O_2 [59]. RTP has a significant advantage over standard furnace anneals with regard to oxygen contamination, because the smaller RTP flow tubes reduce backstreaming from the air [60].

Another application of RTP for silicon technologies is the formation of barrier metals. These conducting films are used to prevent the interdiffusion of the silicon substrate and the aluminum based alloys used for interconnecting devices. One common diffusion barrier is TiN. The film can be formed by reacting titanium with N_2 or NH_3 during a rapid thermal anneal, but this method produces high resistance contacts when small geometries are used [61]. It is preferred instead to deposit the TiN with close to perfect stoichiometry and then anneal the films in an RTA with a nitridizing ambient. This is sometimes called *grain boundary stuffing* since the residual nitrogen is believed to occupy surface sites along the grain boundaries of the polycrystalline TiN. This nitrogen passivates the grain boundary and reduces the grain boundary diffusivity.

Rapid thermal processing has also been used for contact formation in GaAs technologies. As will be discussed in more detail in Chapter 15, low resistance ohmic contacts to n-type GaAs are formed by depositing and thermally annealing a layer containing a mixture of gold and germanium. While early technologies used conventional ovens for this annealing, it has recently been shown that very desirable contacts can be formed reproducibly using a short anneal at about 450°C [62]. The interfacial layer between the metal and semiconductor is considerably broader in conventional anneals [63], and the surface of the contact is more smooth, which improves circuit yield. Since temperature is a critical parameter in forming these contacts, a backside metallization on the GaAs wafer is sometimes used to ensure a uniform temperature across the wafer [64].

6.8 Advanced Rapid Thermal Processing Systems

As discussed earlier in the chapter, one of the principal problems with RTP is thermal uniformity. New RTP systems have recently been developed that use nontraditional chamber and/or heating designs to improve the thermal uniformity in RTP (Figure 6.15). As of this writing many of these systems are not yet commercially available, but this is likely to change in the next few years.

Perhaps the most radical approach to solving thermal uniformity in RTP is the resistively heated

or quartz lamp heated bell jar (Figure 6.15a). The wall is either made from SiC [65] or quartz [66]. The chamber walls are typically kept several hundred degrees above the maximum wafer temperature. The lower part of the chamber is kept cool. The wafer is rapidly heated in the chamber. The heating rate is controlled by the velocity of the wafer elevator and is typically limited to about 100°C/sec [67]. The reactor is capable of processing very large wafers without any evidence of wafer warpage or slip.

The second approach (Figure 6.15b) also suspends the wafer in a cylindrical tube. Unlike the hot wall approach described above, this method relies on several hexagonal rings of lamps above and below the wafer [68]. The lamps that are coplanar or nearly coplanar with the wafer are used to heat the wafer edges. Lamp rings further from the plane of the wafer are more effective at heating the wafer center. As mentioned previously, this type of heating arrangement is called *multizone*. Each lamp ring or pair of lamp rings (the chamber is symmetric above and below the wafer plane) is controlled by a separate power supply. It has been shown that thermal uniformity of 1.5°C can be achieved across a 150 mm wafer [69] in this type of system.

Another promising method for obtaining uniform temperatures in a rapid thermal system is the flat ring approach to adaptive multizone heating [70],[71]. In this system, the lamp assembly is divided into three or four concentric rings (Figure 6.15c). The reflector is shaped to partially focus radiation on the wafer. Single ended tungsten-halogen lamps are the heat source. They are mounted through holes that penetrate the reflecting rings. Just as for the hexagonal chamber each lamp ring preferentially heats different parts of the wafer. By either using multiple point temperature sensing or a simple model of temperature profiles as a function of process conditions as input to the control algorithm;

Figure 6-15 New designs for high uniformity RTP include: (a) the hot wall system (after Roozeboom and Parekh), (b) the multizone hexagonal system (*after Cho et al.*), and (c) the multizone cylindrical system (*after Moslehi et al., 1991, used with permission of Materials Research Society*).

Figure 6-15 cont'd.   For legend see p. 141.

the output of each ring can be independently controlled so as to maintain the most uniform temperature distribution.

6.9 Summary

Rapid thermal processing was developed to enable short time high temperature implant annealing. Generally, the wafer rests on quartz pins in a flow tube and is heated using a bank of high intensity filament lamps (Figure 6.16). Problems with RTP include temperature measurement and thermal

Figure 6-16   Photograph of a commercial rapid thermal processor.

uniformity of the wafer. Excessive temperature gradients across the wafer cause thermoplastic stress that may lead to wafer warpage and/or slip. Rapid thermal processing has been extended to include silicide and barrier metal formation, thermal oxidation, chemical vapor deposition, and epitaxial growth.

Problems

1. Undoped silicon is nearly transparent for photons with energies close to or larger than the bandgap (1.1 eV). Assume that a wafer is transparent for $\lambda \geq 1$ μm. If the wafer is heated with a tungsten-halogen lamp operating at 2000 K, what fraction of the incident energy is transmitted through the wafer? Would GaAs absorb more or less of the energy? Why?

2. The primary mechanism for absorbing this infrared radiation is by free carriers (electrons and holes). As the wafer heats up, the intrinsic carrier concentration increases (see Chapter 3). What effect will this have on RT processes? Roughly sketch a plot of temperature versus time for an undoped silicon wafer heated with a constant power tungsten-halogen lamp.

3. Large diameter wafers are difficult for RTP at high temperature without inducing thermoplastic induced slip at the edge of the wafer. Describe the reasons for this slip. If the slip is more pronounced after high temperature ramps, what does this tell you about the radiation pattern at the surface of the wafer?

4. What problems might you expect if you try to use SUPREM to model rapid thermal annealing?

5. Why has little work been done on rapid thermal oxidation in a wet ambient? (Hint: What are RT oxides used for? Are wet oxides suitable for this application?)

6. Explain the reasons for multizone heating in RTP.

References

1. S. M. Hu, "Temperature Distribution and Stresses in Circular Wafers in a Row During Radiative Cooling," *J. Appl. Phys.* **40**:4413 (1969).

2. C. Hill, in *Laser and Electron Beam Solid Interactions and Materials Processing*, J. F. Gibbons, L. D. Hess, and T. W. Sigmon, eds., Elsevier, North Holland, New York, 1981.

3. T. O. Sedgewick, "Short Time Annealing," *J. Electrochem. Soc.* **130**:484 (1983).

4. S. A. Campbell, "Rapid Thermal Processing," in *Computational Modeling in Semiconductor Processing*, M. Meyyappan, ed., Artech, New York, 1994.

5. J. R. Howell, *A Catalog of Radiation Configuration Factors*, McGraw-Hill, New York, 1982.

6. K. Knutson, S. A. Campbell, and F. Dunn, "Modeling of Three-Dimensional Effects on Temperature Uniformity in Rapid Thermal Processing of Eight Inch Wafers," *IEEE Trans. Semicon. Man.* **7**:68 (1994).

7. K. L. Knutson, *Theoretical and Experimental Investigations of Thermal Uniformity in Rapid Thermal Processing and Reaction Kinetics in Chemical Vapor Deposition for a Dichlorosilane/Hydrogen System*, PhD Dissertation, University of Minnesota (1993).

8. J. R. Coaton and J. R. Fitzpatrick, "Tungsten-Halogen Lamps and Regenerative Mechanisms," *IEE Proc.* **127A**:142 (1980).

9. S. A. Campbell and K. L. Knutson, "Transient Effects in Rapid Thermal Processing," *IEEE Trans. Semicon. Man.* **5**:302 (1992).

10. F. Roozeboom, "Temperature Control and System Design Aspects in Rapid Thermal Processing," in *Materials Res. Soc. Symp. Proc.* **224**:9 (1991).

11. B. Brown, Proc. 9th European RTP Users Group Meeting, Harlow, U.K., 1992.

12. A. J. LaRocca, in *The Infrared Handbook*, W. L. Wolfe and G. J. Zissis, eds., Environmental Research Institute of Michigan, Ann Arbor, 1989.

13. F. Roozeboom, "Rapid Thermal Processing Status, Problems and Options After the First 25 Years," *Mat. Res. Soc. Symp. Proc.* **303**:149 (1993).

14. T. Sato, "Spectral Emissivity of Silicon," *Jap. J. Appl. Phys.* **6**:339 (1967).

15. C. S. Hill and D. Boys, "Rapid Thermal Annealing Theory and Practice," in *Reduced Thermal Processing for ULSI*, R. A. Levy, ed., Plenum, New York, 1989.

16. A. T. Fiory, C. Schietinger, B. Adams, and F. G. Tinsley, "Optical Fiber Pyrometry with In-Situ Detection of Wafer Radiance and Emittance—Accufiber's Ripple Method," *Mat. Res. Soc. Symp. Proc.* **303**:139 (1993).

17. J.-M. Dihlac, C. Ganibal, and N. Nolhier, "In-Situ Wafer Emissivity Variation Measurement in a Rapid Thermal Processor," *Materials Res. Soc. Symp. Proc.* **224**:3 (1991).

18. S. H. Zaidi, S.R.J. Brueck, and J. R. McNeil, "Non Contact 1°C Resolution Temperature Measurement by Projection Moiré Interferometry," *J. Vac. Sci. Technol.* **B10**:166 (1992).

19. S.R.J. Brueck, S. H. Zaidi, and M. K. Lang, "Temperature Measurement for RTP," *Mat. Res. Soc. Symps. Proc.* **303**:117 (1993).

20. B. Peuse and A. Rosekrans, "In-Situ Temperature Control for RTP Via Thermal Expansion Measurement," *Mat. Res. Soc. Symp. Proc.* **303**:125 (1993).

21. B. A Auld, *Acoustic Fields and Waves in Solids*, vol. 1, Wiley, New York, 1973.

22. F. L. Degertekin, J. Pei, Y. J. Iee, B. T. Khuri-Yakub, and K. C. Saraswat, "In-Situ Temperature Monitoring in RTP by Acoustical Techniques," *Mat. Res. Soc. Symp. Proc.* **303:**133 (1993).

23. A. E. Widmer and W. Rehwald, "Thermoplastic Deformation of Silicon Wafers," *J. Electrochem. Soc.* **133:**2405 (1986).

24. J. R. Patel and A. R. Chaudhuri, "Macroscopic Plastic Properties of Dislocation Free Germanium and Other Semiconductor Crystals. I. Yield Behavior," *J. Appl. Phys.* **34:**2788 (1963).

25. H. A. Lord, "Thermal and Stress Analysis of Semiconductor Wafers in a Rapid Thermal Processing Oven," *IEEE Trans. Sem. Man.* **1:**105 (1988).

26. K. S. Jones and G. A. Rozgonyi, "Extended Defects from Ion Implantation and Annealing," in *Rapid Thermal Processing Science and Technology*, R. B. Fair, ed., Academic Press, Boston, 1993.

27. A. Leitoila, J. F. Gibbons, T. J. McGee, J. Peng, and J. D. Hong, "Solid Solubility of As in Si Measured by CW Laser Annealing," *Appl. Phys. Lett.* **35:**532 (1979).

28. J. L. Altrip, A.G.R. Evans, J. R. Logan, and C. Jeynes, "High Temperature Millisecond Annealing of Arsenic Implanted Silicon," *Solid State Electron.* **33:**659 (1990).

29. J. L. Altrip, A. G. Evans, N. D. Young, and J. R. Logan, "The Nature of Electrically Inactive Implanted Arsenic in Silicon After Rapid Thermal Processing," *Mat. Res. Soc. Symp. Proc.* **224:**49 (1991).

30. J. R. Marchiando, P. Roitman, and J. Albers, "Boron Diffusion in Silicon," *IEEE Trans. Electron. Dev.* **TED-32:**2322 (1985).

31. R. B. Fair, "Low-Thermal-Budget Process Modeling with the PREDICT™ Computer Program," *IEEE Trans. Electron. Dev.* **TED-35:**285 (1988).

32. Y. Kim, H. Z. Massoud, and R. B. Fair, "The Effect of Ion Implantation Damage on Dopant Diffusion in Silicon During Shallow Junction Formation," *J. Electron. Mater.* **18:**143 (1989).

33. R. B. Fair, "Junction Formation in Silicon by Rapid Thermal Annealing," in *Rapid Thermal Processing Science and Technology*, R. B. Fair, ed., Academic Press, Boston, 1993.

34. H. Kinoshita, T. H. Huang, and D. L. Kwong, "Modeling of Boron Diffusion and Activation for Non-Equilibrium Rapid Thermal Annealing Application," *Mat. Res. Soc. Symp. Proc.* **303:**259 (1993).

35. S. R. Weinzierl, J. M. Heddleson, R. J. Hillard, P. Rai-Choudhury, R. G. Mazur, C. M. Ozburn, and P. Potyraj, "Ultrashallow Dopant Profiling via Spreading Resistance Measurements with Integrated Modeling," *Solid State Technol.* (Jan. 1993) p. 31.

36. I. W. Wu and L. J. Chen, "Characterization of Microstructural Defects in BF_2^+ Implanted Silicon," *J. Appl. Phys.* **58:**3032 (1985).

37. S. S. Gill and B. J. Sealy, "Review of Rapid Thermal Annealing in Ion Implanted GaAs," *J. Electrochem. Soc.* **133:**2590 (1986).

38. S. J. Pearton and R. Caruso, "Rapid Thermal Annealing of GaAs in a Graphite Susceptor—Comparison with Proximity Annealing," *J. Appl. Phys.* **66:**2482 (1989).

39. T. E. Kazior, S. K. Brierley, and F. J. Piekarski, "Capless Rapid Thermal Annealing of GaAs Using a Graphite Susceptor," *IEEE Trans. Sem. Man.* **4:**21 (1991).

40. S. K. Brierley, T. E. Kazior, and F. J. Piekarski, "Optimization of Capless Rapid Thermal Annealing for GaAs MESFETs," *Mat. Res. Soc. Symp. Proc.* **224:**451 (1991).

41. J. Nulman, J. P. Krusius, and A. Gat, "Rapid Thermal Processing of Thin Gate Dielectrics—Oxidation of Silicon," *IEEE Electron Dev. Lett.* **EDL-6:**205 (1985).

42. M. M. Moslehi, K. C. Saraswat, and S. C. Shatas, *Proc. SPIE* **623:**92 (1986).

43. R. Deaton and H. Z. Massoud, "Effects of Thermally Induced Stresses on Rapid Thermal Oxidation of Silicon," *J. Appl. Phys.* **70**:3588 (1991).

44. M. M. Moslehi, S. C. Shatas, and K. C. Saraswat, "Thin SiO_2 Insulators Grown by Rapid Thermal Oxidation of Silicon," *Appl. Phys. Lett.* **47**:1353 (1985).

45. J-M Dilhac, *Mat. Res. Soc. Symp. Proc.* **146**:333–338 (1989).

46. M. H. Moslehi and K. C. Saraswat, "Thermal Nitridation of Si and SiO_2 for VLSI," *IEEE Trans. Electron. Dev.* **ED-32**:106 (1985).

47. M. A. Schmidt, F. L. Terry, Jr., B. P. Mather, and S. D. Senturia, "Inversion Layer Mobility of MOSFETs with Nitrided Oxide Gate Dielectrics," *IEEE Trans. Electron. Dev.* **35**:1627 (1988).

48. T. Hori, H. Iwasaki, and K. Tsuji, "Electrical and Physical Properties of Ultrathin Renitrided Oxides Prepared by Rapid Thermal Processing," *IEEE Trans. Electron. Dev.* **36**:340 (1989).

49. T. Hori and H. Iwasaki, "Improved Transconductance Under High Normal Field in Nitrided Oxide MOSFETs," *Electron. Dev. Lett.* **10**:195 (1989).

50. T. Hori, "Inversion Layer Mobility Under High Normal Field in Nitrided-Oxide MOSFETs," *IEEE Trans. Electron. Dev.* **37**:2058 (1990).

51. J. Kuehne, G. Q. Lo, and D. L. Kwong, "Chemical and Electrical Properties of Fluorinated Oxides Prepared by Rapid Thermal Oxidation of Si in O_2 with Diluted NF_3," *Materials Res. Soc. Symp. Proc.* **224**:367 (1991).

52. J. C. Sturm, C. M. Gronet, and J. F. Gibbons, "Limited Reaction Processing: *In-Situ* Metal-Oxide-Semiconductor Capacitors," *IEEE Electron. Dev. Lett.* **EDL-7**:282 (1986).

53. J. C. Sturm, C. M. Gronet, C. A. King, S. D. Wilson, and J. F. Gibbons, "*In-Situ* Epitaxial Silicon-Oxide-Doped Polysilicon Structures for MOS Field-Effect Transistors," *IEEE Electron. Dev. Lett.* **EDL-7**:577 (1986).

54. T. Brat, C. M. Osburn, T. Finsted, J. Liu, and B. Ellington, "Self Aligned Ti Silicide Formation by Rapid Thermal Annealing Effects," *J. Electrochem Soc* **133**:1451 (1986).

55. R. W Mann, C. A. Racine, and R. S. Bass, "Nucleation, Transformation, and Agglomeration of C54 Phase Titanium Disilicide," *Materials Res. Soc. Symps. Proc.* **224**:115 (1991).

56. R. Beyers and R. Sinclair, "Metastable Phase Transformation in Titanium-Silicon Thin Films," *J. Appl. Phys.* **57**:5240 (1985).

57. M. Bariatto, A. Fontes, J. Q. Quacchia, R. Furlan, and J. J. Santiago-Aviles, "Rapid Titanium Silicidation: A Comparative Study of Two RTA Reactors," *Materials Res. Soc. Symp. Proc.* **303**:95 (1993).

58. S. S. Iyer, C.-Y. Ting, and P. M. Fryer, "Ambient Gas Effects on the Reaction of Titanium with Silicon," *J. Electrochem. Soc.* **302**:2240 (1985).

59. C. M. Osborn, H. Berger, R. P. Donovan, and G. W. Jones, "The Effects of Contamination on Semiconductor Manufacturing Yield," *J. Environ. Sci.* **31**:45 (1988).

60. C. M. Osborn, "Silicides," in *Rapid Thermal Processing, Science and Technology,* R. B. Fair, ed., Academic Press, Boston, 1993.

61. G. D. Yao, Y. C. Lu, S. Prassad, W. Hata, F. S. Chen, and H. Zhang, "Electrical and Physical Characterization of TiN Diffusion Barrier for Sub-Micron Contact Structure," *Materials Res. Soc. Symp. Proc.* **303**:103 (1993).

62. M. A. Crouch, S. S. Gill, J. Woodward, S. J. Courtney, G. M. Williams, and A. G. Cullis, "Structural and Electrical Properties of Ge/Au Ohmic Contacts to n-type GaAs Formed by Rapid Thermal Annealing," *Solid State Electron.* **33**:1437 (1990).

63. J.B.B. Oliveira, C. A. Olivieri, J. C. Galzerani, A. A. Pasa, L. P. Cardoso, and F. C. dePrince, "Characterization of AuGeNi Ohmic Contacts on n-GaAs Using Electrical Measurements, Auger Spectroscopy, and X-ray Diffractometry," *Vacuum* **41**:807 (1990).

64. "Investigation of the Uniformity of Ohmic Contacts to N-type GaAs Formed by Rapid Thermal Processing," *Solid State Electron.* **36:**295 (1993).
65. C. Lee and G. Chizinsky, "Rapid Thermal Processing Using a Continuous Heat Source," *Solid State Technol.* **32**(1):43 (1989).
66. C. Lee, U.S. Patent 4857689 (1989).
67. W. DeHart, *Microelectronic Man. Technol.* **14**(7):44 (1991).
68. F. Roozeboom and N. Parekh, "Rapid Thermal Processing Systems: A Review with Emphasis on Temperature Control," *J. Vac. Sci. Technol.* **B8:**1249 (1990).
69. B.-Y. Cho, P. Vandenanbeele, and K. Maex, "Evaluation of a Vertical Tube Concept for RTP," *Materials Res. Soc. Symp. Proc.* **303:**165 (1993).
70. P. P. Apte, S. Wood, L. Booth, K. C. Saraswat, and M. M. Moslehi, *Materials Sci. Symp. Proc.* **224:**209 (1991).
71. M. M. Moslehi, J. Kuehne, R. Yeakley, L. Velo, H. Najm, B. Dostalik, D. Yin, and C. J. Davis, "*In-situ* Fabrication and Process Control Techniques in Rapid Thermal Processing," *Materials Sci. Symp. Proc.* **224:**143 (1991).

Part III

Unit Processes 2: Pattern Transfer

The preceding part of the book presented processes required to introduce, activate, and diffuse dopants, and to grow oxides in silicon. A limitation of these processes is that they are done to the entire wafer. The essence of IC fabrication is the ability to transfer information from the IC designer's workstation to the semiconductor wafer. This third part of the book will deal with the processes involved in this transformation: lithography and etching. These are extremely important, since the lateral dimension of many layers and the proximity in which structures can be placed often determine the speed of the circuit. Pattern transfer is so important that a technology is often referred to in terms of its feature size (e.g., half-micron CMOS).

Where the telescope ends, the microscope begins. Which of the two has the grander view?[1]

As discussed in Chapter 1, most lithography is done in a two-step process. First, the design is broken down into layers of information. Each layer is a map for the location of one film on the finished IC. Usually these layers are printed as photomasks. Once a photomask has been made, it can be used to pattern wafers by shining a light through it. The light falls onto a photosensitive material that is then developed to try to reconstruct the original layer image. The production of photomasks will not be explicitly discussed here. The process is very similar to that of lithography on wafers. Both optical lithography (Chapter 7) and electron beam lithography (Chapter 9) are used in photomask generation.

As the desired feature size continues to shrink, optical lithography is becoming much more difficult. As a result, new lithographic processes have been developed. Some, like excimer laser steppers and phase contrast masks are techniques that extend optical lithography to smaller features. Others use very short wavelength nonoptical radiation to reproduce fine lines. These processes have not yet replaced optical lithography, but may become important in the next decade. Chapter 9 will discuss two of the most popular new processes: electron beam and x-ray lithography.

---

[1]Victor Hugo, French poet/novelist/dramaticist (1802–1885)

Once the photosensitive material has been patterned, it is often used as a stencil for reproducing the image in the underlying film. This is usually done subtractively by depositing the film everywhere and removing the unwanted material. The photosensitive material then must serve as a mask for this etching. Chapter 11 will cover both standard etching and an additive process called *liftoff* that is sometimes used for GaAs fabrication.

Contents

diffusions are done at high concentrations at which the simplifying assumptions used in the solution derivation are simply not valid. In the area of lithography even the simplest solutions of the Fresnel equations are beyond the scope of the book. In this text a widely used, powerful simulation program called SUPREM III* has been used to provide more meaningful examples of the sort of real-world dopant redistribution problems that the microelectronic fabrication engineer might face. This software is intended to augment, not replace, learning the fundamental equations that describe microelectronic processing. Typical installations include VAX-, SUN-, Apollo-, and DOS-based microcomputers. The book also enriches the basic material with additional sections and chapters on process integration for various technologies and on more advanced processes. This additional material is in sections marked with a ''+''. If time does not permit covering these sections, they may be omitted without loss of the basic content of the course.

Finally, one has to acknowledge that, no matter how many times this material is reviewed, it cannot be guaranteed to be free of all of the (hopefully) minor errors. In the past, publishers have provided errata when errors were sufficiently numerous or egregious. Even when an errata is published, it is very difficult to get it in the hands of people who have already bought the book. This means that the average reader is often unaware of most of the corrections until a new revision of the book is released. Typically, this is five years or more. This book will have an errata file that anyone can access at any time. This file will also include minor additions to the book that were not available at press time. It will be updated regularly so that you can be assured of the highest degree of accuracy and timeliness in the book. The file will be a simple ASCII text file that can be down-loaded if you have access to a system on the Internet with ftp file-transfer capability. Once in the home directory of your computer, the procedure is as follows (the anticipated machine prompts are italicized):

ftp oz.ee.umn.edu
login: anonymous
password: (please put your internet id/address here)
ftp> cd /users/campbell
ftp> get microfab.err
ftp> quit

The file microfab.err will be transferred to your system. It can be read on UNIX systems with any editor that you might have (eg. vi or ed) or with the ''more'' command. It can also be loaded into any word processing program. I encourage you to download the file after you buy the book and make the indicated changes. You may want to check the contents of this directory for other files that might be of interest. You can do this by using the ls command. If you find something that you feel needs correction or clarification, I invite you to notify me at my e-mail address, Campbell@ee.umn.edu. Please be sure to include your justification, citing published references.

Minneapolis   **S.A.C.**
September 1995

---

*SUPREM III is a trademark of the Board of Trustees of the Leland Stanford Junior University.

Preface

Magic. That is the word that I often hear used to describe the process of fabricating integrated circuits. Society regards the electronics industry, and microelectronics in particular, as a pinnacle of "high-tech." Even engineers, sometimes the very ones who design the circuits being built, consider the fabrication process to be something mysterious. There are a variety of reasons for this phenomenon. The rate of technology evolution is so rapid that narrow specialization is required in order to stay current. The fabrication line further requires a number of subspecialties, each of which is rooted in some combination of physics, chemistry, statistics, chemical engineering, electrical engineering, and mechanical engineering. Jargon develops among specialists that reduces information flow outside the discipline.

The intent of this book is to introduce microelectronic processing to a wide audience. It is primarily intended as a textbook for seniors and/or first-year graduate students, but may also be used as a reference for practicing professionals. Both silicon and GaAs are covered, although the emphasis is on silicon-based technologies. The book assumes one year of physics, one year of mathematics (through simple differential equations), and one course in chemistry. Most students with electrical engineering backgrounds will also have had at least one course in semiconductor physics and devices including *pn* junctions and MOS transistors. This material is extremely useful for the last four chapters and is reviewed in the first sections of Chapters 16, 17, and 18 for students who find they are a bit rusty. One course in basic statistics is also encouraged, but is not required for this course.

Microelectronics textbooks necessarily divide the fabrication sequence into a number of unit processes that are repeated to form the integrated circuit. The effect is to give the book a survey flavor: a number of loosely related topics each with its own background material. Most students have difficulty recalling all of the background material. They have seen it once, two or three years and many final exams ago. It is important that this fundamental material be reestablished before studying new material. Distributed through each chapter of this book are reviews of the science that underlies the engineering. These sections, marked with an "°", also help make the distinction between the immutable scientific laws and the applications of those laws, with all of the attendant approximations and caveats, to the technology at hand. Optical lithography, for instance, may have a limited life, but diffraction will always be with us.

A second problem that arises in teaching this type of course is that the solution of the equations describing the process often cannot be done analytically. Consider diffusion as an example. Fick's laws have analytic solutions, but they are only valid in a very restricted parameter space. Predeposition

To my most important collaborator, Suellen

Oxford University Press

Oxford New York
Athens   Auckland   Bangkok   Bombay
Calcutta   Cape Town   Dar es Salaam   Delhi
Florence   Hong Kong   Istanbul   Karachi
Kuala Lumpur   Madras   Madrid   Melbourne
Mexico City   Nairobi   Paris   Singapore
Taipei   Tokyo   Toronto

and associated companies in
Berlin   Ibadan

Copyright © 1996 by Oxford University Press, Inc.

Published by Oxford University Press, Inc.,
198 Madison Avenue, New York, New York, 10016-4314
http://www.oup-usa.org/

Library of Congress Cataloging-in-Publication Data

Campbell, Stephen A., 1954–
    The science and engineering of microelectronic fabrication / Stephen
A. Campbell.
        p.   cm.—(Oxford series in electrical engineering)
    Includes bibliographical references and index.
    ISBN 0-19-510508-7 (cl)
    1. Semiconductors—Design and construction.   I. Title.
II. Series.
TK7871.85.C25   1996
621.3815'2—dc20   95-35577
                                                                          CIP

Printing (last digit): 9 8 7 6 5 4 3 2 1

Printed in the United States of America
on acid-free paper

The Science and Engineering of Microelectronic Fabrication

Stephen A. Campbell

University of Minnesota

New York   Oxford

Oxford University Press

1996

The Science and Engineering of Microelectronic Fabrication

The Oxford Series in Electrical Engineering

M. E. Van Valkenburg, Senior Consulting Editor
Adel S. Sedra, Series Editor, Electrical Engineering
Michael R. Lightner, Series Editor, Computer Engineering

Chapter 7

Optical Lithography

7.1 Lithography Overview

Figure 7.1 shows a flow chart by which most ICs are designed. It begins with the identification of the chip function. If it is a complicated function, it may be broken down into several levels of subfunctions. The subfunctions are laid out on a floor plan that allocates space on a chip or a chip set for the future design. At this point, the designer may construct a high level model of the chip to test functionality and to get an estimate of its performance. The designer then assembles the chip from predesigned pieces of circuitry called *cells*, custom designing whatever is necessary. The layout of these cells is predicated on a set of design or layout rules. (See Figure 7.2 for an example of a few rules for a technology.) These rules are a contract between the fabrication facility and the designer. For each level, the layout rules will govern the smallest feature allowed, the smallest spacing allowed, the minimum overlap for features on this level to those on another level, the minimum spacing to underlying topology, and so on. If these rules are followed, the fabrication facility is obligated to ensure that the chip functions as designed. After the design is complete, it will be checked against the layout rules to ensure compliance. Finally, additional simulations may be run from the actual layout. If they do not meet specifications, the design is modified until they do. Depending on the sophistication of the computer automated design (CAD) tools, some or all of the above processes are automated. In the most advanced CAD systems, the tool is fed a high level description of the function to be implemented along with a file that contains the design rules. The program will generate a chip layout along with performance estimates. The designer then oversees the process and makes manual adjustments as necessary to improve performance.

The interface between the designers and the fabrication facility occurs through the production of photomasks. Each photomask contains an image of one layer of the process. Depending on the aligner to be used, the photomask may be the same size as the finished chip or an integral factor of that size. Figure 7.3 shows some typical photomasks. The masks may be the same size as the final image on the wafer (1×) or the image on the mask may be reduced during the exposure. Common reductions are 5× and 10×. Photomasks are fabricated on various types of fused silica. The most important properties for the mask include a high degree of optical transparency, a small thermal expansion coefficient, and a flat, highly polished surface that reduces light scattering. On one surface of this glass is a patterned opaque layer. In most masks the opaque layer is chromium. After the masks

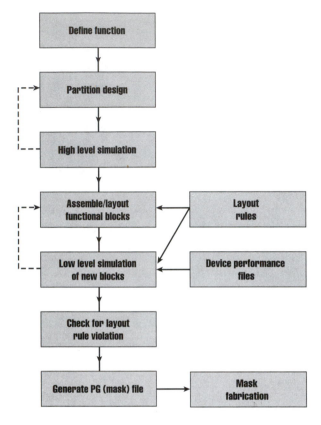

Figure 7-1   Simplified IC design process flow diagram.

7.1	Contact size (fixed)	1.0μm x 1.0μm
7.2	Minimum contact to contact space	1.2μm
7.3	Minimum active area overlap of contact	1.2μm
7.4	Minmum contact to gate spacing	0.8μm
	Maximum contact to gate spacing	
	For standard device performance file	1.5μm

etc.

Figure 7-2   Excerpt of typical design rule set. This portion deals with first metal rules for a particular technology.

have been patterned, the pattern may be verified by checking it against the database. Any unwanted chrome can be removed by laser ablation. Any pinholes in the chrome can be repaired with an additional deposition. This is a critical step, since the photomask may be used to pattern tens of thousands of wafers. Any defect in the mask large enough to be reproduced, will be on every wafer made from the mask.

Lithography is the most complicated, expensive, and critical process in mainstream microelectronic fabrication. (Several good references exist on lithography for students who are interested in specializing in this area. Early classic books include Stevens [1] and Bowden et al. [2]. Elliott's book deals primarily with the topic of the next chapter but is also a reasonable reference [3]. A recent book by Moreau [4] is one of the most comprehensive.) Using DRAMs as an example, Table 7.1 shows some of the lithographic requirements expected over several generations of ICs [5]. Lithography accounts for nearly one third of the total fabrication cost, a percentage that is rising. A typical silicon technology will involve 15–20 different masks. For some BiCMOS processes, as many as 28 may be used. Although traditional GaAs technologies have required far fewer masking levels, the number is now increasing as well. Furthermore, the technology performance is often predicated on the ability of producing very fine lines. Even the evaluation of the performance of a lithographic process is difficult. A memory manufacturer might require very tight control on a critical feature size over a billion transistors per wafer and over tens of thousands of wafers. A device researcher, on the other hand, might be perfectly satisfied if 50% of the features fall into some acceptable range on one wafer. Thus, it is all too easy to wind up comparing apples to oranges when evaluating photolithographic performance.

Figure 7.4 shows a cross schematic of a simple system for optically exposing a wafer. An optical source on the top is used to shine light through the mask. The image is projected onto the wafer surface, which is coated with a thin layer of photosensitive material known as *photoresist*.

Figure 7-3 Typical photomasks including (from left) a 1x plate for contact or projection printing, a 10x plate for a reduction stepper, and a 10x plate with pellicles.

Optical photolithography can therefore be divided into two parts. The design and operation of the exposure tool that makes the image of the photomask at the surface of the wafer is mainly a problem in optical system design. The other half of the equation is the chemical processes that occur once the radiation of the image is absorbed in the photoresist and the pattern is developed. The dividing line is the pattern of radiation that strikes the surface of the wafer, also called the *areal image* of the mask. This chapter will review the basic physics of optical exposure. The tools discussed are all optical. They use either visible or ultraviolet (UV) light as the exposing radiation. They are called *aligners* because they have a dual purpose. Not only must they reproduce the image of a particular layer, they must also align that layer to the previous ones.

There are three primary measures of performance for an aligner (Table 7.2). The first is the resolution, the minimum feature size that can be exposed. Resolution is not a fixed number for a given

Table 7.1 Expected lithographic requirements for projected DRAM technology generations (from [1]). The minimum feature size represents the size at which the DRAMs are expected to be ultimately manufactured.

DRAM Size	Chip Size (mm²)	Chips in Exposure Field (# in mm × mm)	Minimum Feature Size (μm)	Overlay Accuracy (μm)
4 Mb	80–90	2 in 15 × 15	0.8–0.7	0.35
16 Mb	120–140	2 in 18 × 18	0.6–0.5	0.25
64 Mb	170–200	2 in 20 × 20	0.4–0.3	0.15
256 Mb	250–300	1 in 22 × 22	0.25	0.10
1 Gb	350–400	1 in 25 × 25	0.18	0.07

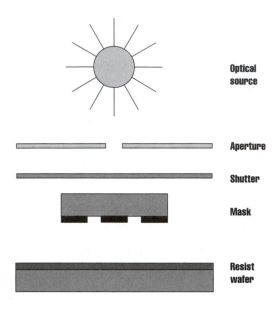

Optical source

Aperture

Shutter

Mask

Resist wafer

Figure 7-4 Schematic of a simple lithographic exposure system.

aligner, although equipment manufacturers sometimes seem to lead one to believe otherwise. Resolution depends on the ability of the photoresist to reconstruct the pattern from the areal image. As already mentioned, very small features may be resolved with a particular optical tool and resist system, but the dimensional control is so poor that they cannot be reliably used. To be useful, resolution figures are often quoted as a minimum feature size that can be resolved and still maintain a certain feature tolerance. A typical number might be a 3σ distribution of linewidths with no more than 10% variation.

The second performance metric for the aligner is registration, which is a measure of the overlay accuracy from layer to layer. Again, good measurements of registration are statistical in nature. If registration errors are completely random the mean error is zero. The width of the registration error distribution as measured by a figure like 3σ is a good indicator of overlay performance. This number depends on the experience of the operator and the state of the wafer. Other factors that affect registration include film structure and feature topology. The third primary performance measure is throughput. Electron beam systems have excellent resolution, and the registration can be quite good. The throughput for a typical IC pattern that contains 10^6 transistors can be less than one wafer per hour, making it unacceptable in many applications.

With solid measures of these metrics, one must still bear in mind the intended technology. For most ULSI technologies the 3σ registration must be about one third the minimum feature size for the lithographic tool to be fully useful. Throughput and process uniformity are also extremely important. Other technologies, such as GaAs MESFET, may have more lax requirements for registration and throughput, but may require excellent resolution. Even within a technology, different layers may have different requirements. Thus, some levels may be exposed on one type of tool and other levels on another. Using different types of processes for different layers is called *mix and match lithography*.

In discussing the optics of lithography, one must distinguish between those problems in which

Table 7.2 The effects of some of the resist parameters on process outcomes. XX indicates a strong effect, while – indicates little effect.

	Resolution	Registration	Wafer to Wafer Control	Batch to Batch Control	Throughput
Exposure system	XX	XX	X	XX	XX
Substrate	X	X	XX	X	X
Mask	X	X	–	X	X
Photoresist	XX	X	XX	XX	XX
Developer	X	–	XX	XX	X
Wetting agent	–	–	XX	X	–
Process	X	X	XX	XX	XX
Operator*	X	XX	XX	X	XX

*Operator row entries are for manual aligners. Many advanced tools use automated alignment and exposure systems that dramatically reduce the dependence of the results on the skill of the operator.

all dimensions are large compared to the wavelength of light and those in which this criterion is not satisfied. For example, in discussing an optical system that includes a light source, a reflector, and a lens, all of the geometries are of order one centimeter or larger. In such a system, the light can be treated as a particle traveling in straight lines between the components. The analysis used in this situation is called *ray tracing*. On the other hand, once the light passes through a mask where the feature sizes on the mask approach the optical wavelength, one must consider properties such as diffraction and interference. These phenomena require a description of light as an electromagnetic wave.

7.2 Diffraction°

This section will develop a basic understanding of the physics of lithographic exposure. This discussion can be found in any introductory optics text, in much more detail than can be presented here [6]. The integrals derived will be considerable simplifications, and even these can only be solved numerically. Nevertheless, without this background material, it is difficult to understand the lithographic limitations and the reason that one system is superior to another.

One starts with the understanding that light propogates as an electromagnetic wave. These waves can be described as

$$\mathcal{E}(\bar{r}, v) = E_o(\bar{r})e^{j\phi(\bar{r}, v)} \tag{7.1}$$

where E_o is the electric field intensity, j is the imaginary number, ϕ is the phase of the wave, $\bar{r}$ is the position, and v is the frequency of the wave. Huygen's principle says that any local disturbance in an optical system, such as a mirror or a photomask, can be considered to generate a large number of spherical wavelets that propagate outward from the point of disturbance. To find the disturbed electromagnetic wave, one must sum over all of the wavelets.

Figure 7.5 shows the application of Huygen's principle to a system of interest. The upper figure shows the formation of point sources on the surface of the mask. The lower figure shows a point source of light being used to illuminate a mask. On the mask is a single feature: a long narrow aperture of width $2W$ and length $2L$. Here the approximation will be made that the transparent region of the mask does not affect the incoming wavefront. The aperture can be divided into a large number of differential rectangular elements of width dx and length dy. The points P_w and P_m are on the wafer and mask, respectively. Each rectangular element generates a wavelet. The sum of the contributions to the exposure at a point on the surface of the wafer can be found using the integral

Figure 7-5   Huygen's principle applied to the optical system shown in Figure 7-4. A point source is used to expose an aperture in a dark field mask.

$$\mathcal{E}(R') = j\frac{A}{\lambda}\int\int_\Sigma \frac{e^{-jk(R,R')}}{RR'}\,d\sigma \tag{7.2}$$

where A is the amplitude of the sine wave at a unit distance from the source, and Σ is the solid angle

Figure 7-6   Typical near field (Fresnel) diffraction pattern.

subtended by the aperture in the mask. Once one finds the electric field, multiplying it by its complex conjugate can be used to calculate the electromagnetic intensity at the surface of the wafer. If one lets W and $L \rightarrow \infty$, Equation 7.2 represents the summation of an infinite series of spherical waves that exactly reproduces the undisturbed wave. For such a wave the intensity

$$I = \mathcal{E}\,\mathcal{E}^* = E_o e^{j\phi} E_o e^{-j\phi} = E_o^2 \qquad \textbf{(7.3)}$$

If, on the other hand, the aperture is of finite extent, the electric field is the superposition of plane waves with different phases. In the simplest case, where the aperture has been divided into only two elements,

$$\begin{aligned} I &= [E_1 e^{j\phi_1} + E_2 e^{j\phi_2}][E_1 e^{j\phi_1} + E_2 e^{j\phi_2}] \\ &= E_1^2 + E_2^2 + 2E_1 E_2 \cos(\phi_1 - \phi_2) \end{aligned} \qquad \textbf{(7.4)}$$

The interesting part of the equation of course, is the cross term due to interference between the wavelets. It gives rise to the oscillations that are a characteristic part of the diffracted image.

In practice, the solution of Equation 7.2 is quite complicated, even for this very simple geometry. In photolithography one is only interested in two limiting cases. If Equation 7.2 is solved subject to the simplifying assumption that

$$W^2 \gg \lambda \sqrt{g^2 + r^2} \qquad \textbf{(7.5)}$$

where r is the radial distance between the center of the diffraction pattern and the observation point, the result is near field or Fresnel diffraction.[2] The image of this type of diffraction is shown in Figure 7.6. The edges of the image rise gradually from zero, and the intensity of the image oscillates about the expected intensity. The oscillations decay as one approaches the center of the image. The oscillations are due to constructive and destructive interference of Huygen's wavelets from the aperture in the mask. The amplitude and period of these oscillations depend on the size of the aperture.

When W is small enough that the inequality in Equation 7.5 is no longer valid, the oscillations are large. When W is very large however, the oscillations rapidly die out, and one approaches simple ray tracing. Then, by geometric arguments the width of the image at the surface of the wafer is increased by an amount ΔW given by

$$\Delta W = W \frac{g}{D} \qquad \textbf{(7.6)}$$

The other diffraction extreme occurs when

$$W^2 \ll \lambda \sqrt{g^2 + r^2} \qquad \textbf{(7.7)}$$

---

[2]Actually, as $W \rightarrow \lambda$, even this approximation is no longer valid. Vector diffraction theory must be used, giving rise to polarization effects.

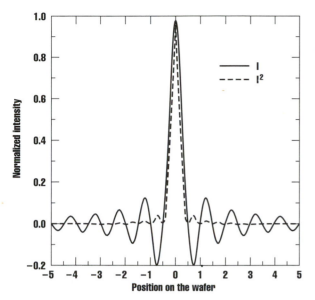

Figure 7-7 Typical far field (Fraunhofer) image.

This is called far field or Fraunhofer diffraction. Equation 7.7 is called the *Fraunhofer criterion*. Then Equation 7.2 can be simplified considerably. The intensity as a function of position on the surface of the wafer is given by

$$I(x,y) = I_e(0)\left[\frac{(2W)(2L)}{\lambda g}\right]^2 I_x^2 I_y^2 \qquad (7.8)$$

where $I_e(0)$ is the flux density (typically expressed in W/cm^2) in the incident beam and

$$I_x = \frac{\sin\left[\dfrac{2\pi xW}{\lambda g}\right]}{\dfrac{2\pi xW}{\lambda g}} \qquad (7.9)$$

$$I_y = \frac{\sin\left[\dfrac{2\pi yL}{\lambda g}\right]}{\dfrac{2\pi yL}{\lambda g}} \qquad (7.10)$$

where L is the length of the lines and spaces. In Equations 7.9 and 7.10, x and y are the coordinates of the observation point on the surface of the wafer. A plot of this function and its square in 1-D is shown in Figure 7.7. The function has a sharp maximum at $x = 0$ and goes through 0 at integer multiples of one half.

Real systems are much more complicated than these simple expansions, however. The light source is not a point, but a finite volume. It may also emit a number of wavelengths. The light is collected through a lens/mirror assembly. Each of the optical components will have some imperfections such as local distortions and aberrations. The mask itself will reflect, absorb, and phase shift the incident radiation. Reflections on the surface of the wafer further complicate matters. As a result, the image produced on the surface of the wafer (the areal image) can only be approximated numerically. Even then sophisticated software is required. Several commercial programs are available to do these calculations. In the following sections, some simple approximations will be presented that are frequently used to determine resolution in place of these more exact calculations of aerial intensity.

7.3 The Modulation Transfer Function and Optical Exposures

When discussing the resolution of a system it is customary to discuss a series of lines and spaces called a *diffraction grating* rather than a single aperture. If the Fraunhofer criterion is met, one can roughly approximate the areal image by the superposition of the individual intensities. (Actually, the interference between peaks must be taken into account, but if the peaks are well separated this is a modest effect.) Figure 7.8 shows a construct of the normalized intensity from such a grating as a function of position on the wafer. The intensity no longer reaches 1, and the minimum intensity is no longer 0. Instead, define I_{max} as the maximum intensity of the radiant pattern and I_{min} as the minimum

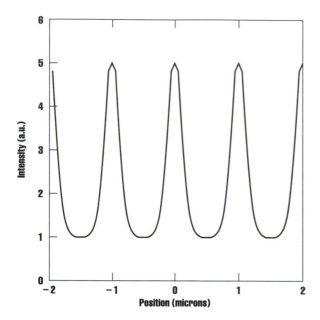

Figure 7-8   Far field image for a diffraction grating.

intensity. In Figure 7.8, they are about 0.52 and 0.17, respectively.

The modulation transfer function (MTF) of an image can be defined as

$$\text{MTF} = \left(\frac{I_{max} - I_{min}}{I_{max} + I_{min}} \right) \qquad \textbf{(7.11)}$$

The MTF is a strong function of the period of the diffraction grating. As the period of the grating decreases, MTF decreases. Physically, one can think of the MTF as a measure of the contrast in the areal image. The higher the MTF, the better the contrast. For the grating in Figure 7.8, the MTF is about 0.5.

Figure 7.9 shows a plot of areal intensity for a grating mask, based on simple Fresnel diffraction. Superimposed on this intensity plot are two simplified resist response indicators. In Figure 7.9a, the resist responds ideally: A single line exists at an exposure energy density D_{cr}. All regions of the photoresist that receive exposures greater than D_{cr} will completely dissolve during the develop process. All regions of the wafer that receive exposures below D_{cr} will not be attacked during the develop process. As the widths of the lines and spaces are decreased, diffraction only makes small changes in the widths of the lines and spaces until I_{min} multiplied by the exposure time $> D_{cr}$ or I_{max} multiplied by the exposure time $< D_{cr}$. Figure 7.9b shows a more realistic resist response model. The resist now has two critical exposure energy densities. For $D < D_0$, the resist will not dissolve in the developer. For $D > D_{100}$, the resist will completely dissolve in the developer. For the intermediate shaded regions ($D_0 < D < D_{100}$), the image will partially develop. As W

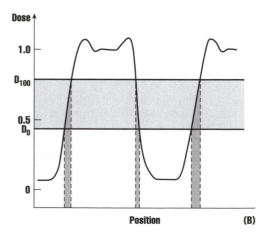

Figure 7-9   Plot of dose versus position on the wafer. Dose is given by the intensity of the light in the aerial image multiplied by the exposure time. Typical units are mJ/cm².

decreases the MTF goes down, and the areal intensity quickly enters a regime where the diffraction grating can no longer be perfectly reproduced on the wafer. The point at which this occurs depends on the values of D_0 and D_{100} and therefore, on the resist being used. In common resist systems when the MTF is less than about 0.5, the image can no longer be reproduced.

7.4 Source Systems and Spatial Coherence

The next few sections will use these concepts to discuss actual aligners. We will start with the component common to all aligners: the source system. This consists of the light source itself and any reflecting and/or refracting optics used to collect, collimate, filter, and focus the source. From the previous discussion, one should have a sense that the wavelength of the exposing radiation is a critical parameter of the lithography process. All things being equal, the shorter the wavelength, the smaller the feature size that can be exposed. Of course, the exposure requires that a certain amount of energy be deposited. Furthermore, this energy must be deposited uniformly across the wafer. To maintain reasonable exposure times, one requires an intense source at these short wavelengths. This section will review some of the most popular optical sources.

The most common type of optical source for photolithography is the high pressure arc lamp. Arc lamps are the brightest incoherent sources available. They are also used to pump lasers. Long gap arc lamps were mentioned in Chapter 6 since they are sometimes used to heat wafers in rapid thermal processors. Arc lamps for lithography have much shorter gaps as shown in Figure 7.10. The lamp consists of two conducting electrodes sealed in a fused silica envelope. The electrodes, one pointed and the other round, end in a gap of about 5 mm. Most lamps contain a mercury vapor at a pressure close to 1 atm when the lamp is cold. To ignite the lamp, a high voltage spike is applied across the gap. The spike voltage must be sufficient to ionize the gas. Several kilovolts are common strike potentials. The mixture of partially ionized gas, electrons, and energetic neutral species inside the tube is called a *plasma* or a *glow discharge*. This topic will be discussed in Section 10.4. For now, appreciate that the plasma will conduct current. Most arc lamp power supplies also include a boost supply, typically a capacitor charged to several hundred volts, that ensures the plasma stability in the moments after the strike. The ionized gas in the lamp is very hot and the pressure in the bulb during operation may reach 40 atm. Typical electrical power dissipation by a photolithography arc lamp bulb is 500–1000 W. The emitted optical power is a little less than half of that.

During operation the lamp contains two optical sources. The high temperature electrons in the arc act as a very hot gray body source, radiating power as (Equation 6.3)

$$M_\lambda(T) = \frac{\epsilon(\lambda, T) \, C_1}{\lambda^5 \, (e^{c_2/\lambda T} - 1)} \tag{7.12}$$

Figure 7-10 Photograph of a typical high pressure short arc mercury lamp (*courtesy Osram Sylvannia*).

where $M_\lambda(T)$ is the energy density distribution, and C_1 and C_2 are the first and second optical constants as defined in Chapter 6. The electrons in the arc lamp plasma typically have temperatures of order 40,000 K. This corresponds to a peak emission at a wavelength of 75 nm, which is very deep in the ultraviolet. Since this energy is

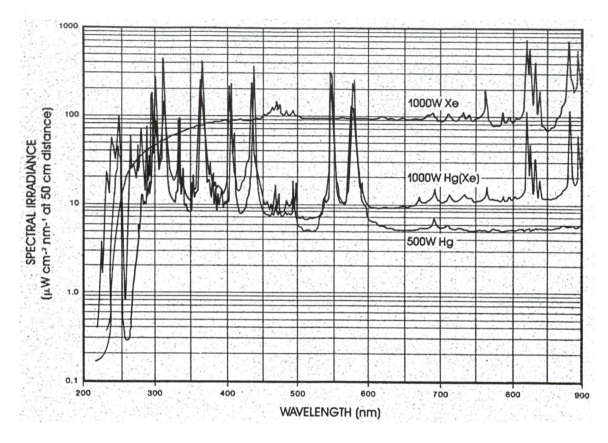

Figure 7-11   Line spectra for mercury and xenon arc lamps *(courtesy of Oriel Corporation).*

above the bandgap of the fused silica envelope, much of it will be absorbed before it leaves the lamp housing. Lamp manufacturers sometimes add impurities to the fused silica to enhance this absorption, since this highly energetic emission results in ozone production in the lamp assembly.

The second optical source in the lamp is the mercury atoms themselves. Collisions with the energetic electrons push the electrons of the mercury atoms into bound high energy states. When they decay into lower energy states, they emit optically at the wavelength corresponding to the energy transition. These line spectra are so sharp that they can be used to identify the primary species in the plasma. Figure 7.11 shows the line spectra of a typical mercury lamp [7]. The lines have been named according to their energy. Aligners often filter out all but a single line. At this writing, the older optical exposure equipment is g-line (436 nm), but i-line (365 nm) usage is increasing rapidly. To extend the use of arc lamps deeper into the UV, xenon can be used as the fill gas. Xenon has a strong line at 290 nm, with much weaker lines at 280, 265, and 248 nm. Obtaining suitable optical components for use at 290 nm, however, is still a concern.

During operation, energetic mercury ions bombard the negative electrode. As they do so, they eject small amounts of the electrode material through a physical process known as *sputtering*. (See Chapter 12.) Some of the sputtered electrode coats the inside wall of the fused silica housing. Furthermore, the high temperature seen by the inside surface of the lamp may also cause a slow devitrification of the fused silica, leaving a cloudy white appearance. The combination of sputtering and devitrification reduces the output intensity. The additional absorbed energy also causes the lamp en-

(1)	**Super high pressure mercury lamp**
(2)	**Ellipsoidal concave mirror**
(3)	**First mirror**
(4)	**Collimator**
(5)	**ND filter**
(6)	**L-39 filter**
(7)	**Fly's eye lens**

(8)	**Second mirror**
(9)	**Condenser lens**
(10)	**Mask plane**

Figure 7-12   Schematic of a typical source assembly for a contact/proximity printer *(after Jain).*

velope to get hotter. Ultimately the lamp will fail, often by exploding, severely damaging the aligner's optics. For that reason, aligner lamps are usually cooled with fans during operation and are replaced after a set number of hours of use.

The simple optical design of the source in Figure 7.4 is of course, not very practical. Only a very small fraction of the radiation from the lamp will reach the wafer. Unless the wafer is spherical (not usually a desired condition) the power density at the surface will be nonuniform. There are, therefore, four primary objectives for the design of the optical source system. The first is to collect as much of the radiation as possible. Without such collection, exposure times are impractically long. The second objective is to make the radiated intensity uniform over the field of exposure, preventing some parts of the wafer from being overexposed while others are underexposed. The third objective is to collimate and shape the radiation to the extent needed. Normally perfect collimation is not desired; instead a few degrees of divergence is often used. Finally, the source must select the exposure wavelength(s). Figure 7.12 shows a schematic of a source assembly for a simple but practical aligner.

Collection is most often done by the use of parabolic reflectors. By placing the arc at the focal point of the reflector, all of the radiation so captured is collimated. Arc lamps do not radiate uniformly. The plasma acts as a diffuse, semitransparent source. The electrodes and bulb shape the plasma so as to maximize the optical intensity perpendicular to the arc direction. For a properly designed reflector, little radiation is lost from the bottom of the lamp. The dark spot at the top of the lamp allows one to have a hole in the reflector to bring in the power leads and exhaust cooling air. The finite extent of the arc prevents such an arrangement from being perfectly collimated. If less collimation is desired, the lamp can be moved away from the focal point.

To increase the uniformity of the optical source, some type of optical integrator must be used. One common component is a fly's eye lens. This is a large fused silica lens containing many small lenslets. The lenslets decollimate the incident light and a second objective recollects the light and reshapes it to the desired dimensions. A second approach that addresses both collection and integration is the use of fiber optic bundles. One end of the fibers either surrounds the arc source or is used immediately after the parabolic reflector. The fibers are mixed so as to provide a uniform source at the other end of the bundle. Optical fibers are sometimes used with arc lamps but are commonly used for excimer laser sources.

The wavelength selection is done through a set of filters. It is common to use a cold mirror to absorb the infrared radiation from the lamp. This prevents unintentional exposure and heating of the components further downstream. The wavelength selection can be done through the use of optical notch filters or a series combination of high pass and low pass filters. A mechanical shutter completes the source assembly.

Since most arc lamps are efficient emitters in the near UV and visible wavelengths, they are inefficient in the deep UV. Excimer lasers are the brightest optical sources in this part of the spectrum.

The word excimer is a concatenation of the words excited and dimer. True excimers therefore, contain excited dimers (molecules with two atoms of the same element like F_2). A molecule that has one or more electrons in excited energy levels will be designated using an asterisk, as F_2^*. Excimer lasers are more properly called *exciplex lasers* since most modern excimers contain high pressure mixtures of two or more elements. These elements do not react when they are in the ground state, but if one or both is excited, a chemical reaction will proceed. In most excimers, one of the precursors is a halogen or halogen containing compound such as NF_3 and the other is a noble gas. A common example is XeCl where the reaction that leads to lasing is

$$Xe^* + Cl_2 \rightarrow XeCl^* + Cl \qquad \textbf{(7.13)}$$

The excited molecule emits in the deep UV, returning it to the ground state where it immediately dissociates. If enough energy is supplied to maintain a large population of the noble species in the excited state, lasing will continue to occur. The energy is usually supplied by 10–20 kV arcs across two flat plate electrodes spaced from 1–2 cm apart. The arc can be strobed at rates up to several hundred Hertz.

Some common excimer laser sources are given in Table 7.3 [8]. Of primary interest is the lasing wavelength and the power. A typical resist exposure dose is several hundred mJ/cm^2. A laser must be able to produce about 1 J at the surface of the wafer so that a field several centimeters on a side can be exposed in no more than 1 sec. To achieve this, the laser should put out close to 10 W. Although XeCl puts out considerable power, it is not much further into the deep UV than arc lamps. The combination of high power and deep UV lines make ArF and KrF attractive sources for advanced optical lithography. F_2 excimers have also been used at 157 nm for contact lithography [9], but the low output power makes them impractical for production use.

Excimers emit strongly in a multimode fashion with a relatively poor spatial coherence. That is, excimer laser beams are more divergent than beams from argon ion lasers. This is a strong drawback to many laser applications, but it is actually an advantage for lithography. Lasers with high spatial coherence induce speckle, which occurs when phase variations are introduced into wavefronts by surface reflections [10]. For comparison an Ar-ion laser has a bandwidth of <0.0001 nm, while a free running excimer has a bandwidth of about 1 nm. In lithography applications, excimer sources are often line narrowed to a few picometers, but are still much broader than an argon ion laser. Figure 7.13 shows a typical speckle pattern for a narrow line laser exposure.

The containment of a high pressure halogen, along with high voltage sometimes used to pump the laser, makes safety a concern. Early excimer systems were considered exotic, dangerous, and unreliable. Furthermore, the very high energy density pulses (10 kW/cm^2) tended to degrade the

Table 7.3 Practical excimer sources that are potentially useful in semiconductor photolithography

Material	Wavelength (nm)	Max Output (mJ/pulse)	Frequency (pulses/sec)	Pulse Length (ns)
F_2	157	6	10	20
ArF	193	175–300	400	15
KrCl	222	125	150	15
KrF	248	300–1500	500	25
XeCl	308	200–1500	500	20

Data taken from Jain.

Figure 7-13 Speckle pattern obtained from exposing a pattern using a narrow linewidth laser as the optical source *(after Jain).*

Figure 7-14 Optical train for an excimer laser stepper *(after Jain).*

optical quality of the lenses of early excimer systems by devitrifying the silica. This image of excimer lithography has improved considerably, and currently exposure tools based on excimer sources are used in development lines at a number of IC manufacturers. Figure 7.14 shows the optical train of an excimer based 10:1 reduction stepper (Section 7.6). Excimers must still be serviced after about 10^7 pulses and replaced after about 5×10^7 pulses, but advances in laser design are expected to push these numbers to 2×10^8 and 10^9, respectively (about a month between laser servicing, assuming 100 Hz operation, and 50% duty cycle) [11],[12]. One of the primary remaining problems for deep UV excimer lithography is the development of suitable commercial photoresists. Typical optical resists are almost opaque at these wavelengths. All of the excimer energy is deposited in the top layers of the resist [13], resulting in poor image formation. This topic will be discussed further in the next chapter.

7.5 Contact/Proximity Printers

The simplest type of aligner is a contact printer. In contact printing, the mask is pressed against the resist coated wafer during exposure. The primary advantage of contact printing is that these small features can be made using comparatively inexpensive equipment. Figure 7.15 shows a typical contact

Figure 7-15 Typical contact exposure system *(courtesy of Karl Suss).*

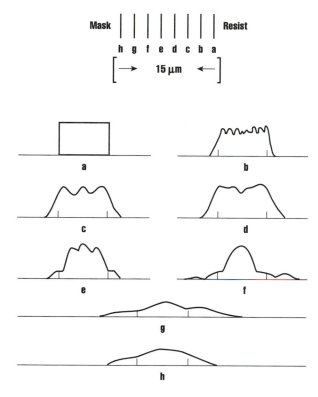

Figure 7-16   Intensity as a function of position on the wafer for a proximity printing system where the gap increases linearly from $g = 0$ to $g = 15$ μm *(after Geikas and Ables).*

exposure system. The mask is held chrome side down in a frame just below the microscope objectives. Vernier screws are then used to move the wafer with respect to the mask. Once the wafer is aligned to the mask the two are clamped together, the microscope objectives are retracted, and the wafer/mask assembly is wheeled into the exposure station. Here radiation from a high intensity lamp (housed in the black box, Figure 7.15, upper right) is used to expose the wafer. After exposure, the carriage is returned to the inspection station for unloading. The figure insert shows the wafer ready for unloading.

Ideally, the entire wafer is in contact with the mask. Because of this contact the gap between the wafer and the optical disturbance (photomask) goes to zero and diffraction effects are minimized. Thus the MTF should be ≈ 1.0 for essentially any feature size. Actually, due to the finite resist thickness the gap cannot be zero. Furthermore in real contact printers, the mask contact varies across the wafer surface. This occurs since neither the wafer nor the mask is perfectly flat. Pressures ranging from 0.05–0.3 atm are used to push the mask into more intimate contact with the wafer. This is called the *hard contact mode of exposure.* In the most extreme cases, thin film masks are sometimes used to help promote the contact. Then the resolution is limited primarily by light scattering in the resist. Features as small as 10 Å [14],[15] have been produced using this method, with extremely thin resists as demonstration vehicles. In more useful resists, features as small as 0.1 μm have been demonstrated using contact printing and deep submicron sources [16],[17]. It is easier to use these sources in contact printing since the optics are so simple. Resolution using more common sources is about 0.5 μm.

The major disadvantage of hard contact lithography is defect generation due to the contact between the resist coated wafer and the photomask. Defects are generated both on the wafer and on the mask on every contact cycle. For this reason, contact printers are typically limited to device research or other applications that can tolerate high defect levels. Proximity printing was developed to avoid defect generation. In this type of exposure tool, the mask floats off the surface of the wafer, typically on a cushion of nitrogen gas. The gap between the wafer and mask is controlled by the flow of nitrogen into this space. Separations of 10–50 μm are typical. Since there is no longer any (intentional) contact between the wafer and the mask, defect generation is sharply reduced.

The problem with proximity printing is a reduction in the resolution. Consider a mask consisting of a single aperture of width W in a dark field. Assume that this mask is exposed using a monochromatic, nondivergent light source, such as a broad beam laser. Figure 7.16 shows the areal intensity of such an image as a function of the gap [18]. When g is small such that

$$\lambda < g < \frac{W^2}{\lambda}$$

(7.14)

Table 7.4   Maximum allowable proximity gap for near and deep UV sources as a function of the feature size normalized to the gap required for 2.5 μm resolution with a deep UV source

Feature Size (μm)	Maximum Gap for Near UV Source	Maximum Gap for Deep UV Source
2.5	0.63	1.0
2.0	0.37	0.61
1.0	0.08	0.24
0.5	0.05	0.07

Data taken from Lin.

the system is in the near field region of Fresnel diffraction. The areal image produced by such an aperture is a well known function of λ, g, and W, and closely approximates the ideal image. The small intensity oscillation near the edges of the pattern is called *optical ringing*. In a real optical system in fact, the beam will not have perfect spatial coherence and so even the modest ringing pictured here will not exist. As the gap is increased however, eventually

$$g \geq \frac{W^2}{\lambda} \qquad (7.15)$$

Then the image approaches that of far field Fraunhofer diffraction, as shown in Figure 7.8.

When the gap is very large, the image is severely degraded. As a rule of thumb, one can say that features less than

$$W_{\min} \approx \sqrt{k\lambda g} \qquad (7.16)$$

cannot be resolved in this type of optical printing, where k is a constant that depends on the resist process. Typical values of k are close to 1. For a gap of 20 μm and an exposure wavelength of 436 nm (g-line), the minimum feature size for proximity printing is about 3.0 μm. This however, would require an excellent resist process. A minimum feature size with some process latitude would be about 50% larger than this. To further improve the resolution, it is necessary to reduce the wavelength or the gap. Reducing the gap increases the risk of contact. A second problem with small gaps is that the variations in the gap that may be caused by wafer or mask nonplanarity, dirt particles, resist beads, and unintentional tilt, lead to linewidth variations across the wafer. A more attractive alternative is to reduce the wavelength. Systems used to expose fine features therefore, typically operate deep in the UV. This process is limited primarily by high brightness sources and compatible optical materials for refractive optics. Table 7.4 shows the maximum allowable gap as a function of feature size for near and deep UV sources [19] for a 10% image distortion.

7.6 Projection Printers

Projection printers were developed to obtain the high resolution of contact printing without the defects. Although some GaAs ICs are built with contact printing, projection printers have become by far the most widely used exposure tool for the manufacture of ICs. This section will introduce the basic

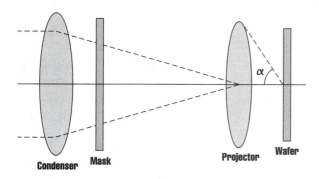

Figure 7-17   Schematic for the optical train of a simple projection printer.

equations and then will discuss both common types of systems: scanners and steppers. Figure 7.17 shows a simple example of a Kohler projection lithography system. The mask is held between the condenser and a second set of lenses called the projector or objective. The purpose of the projector is to refocus the light onto the wafer. In some cases, the light from the condenser is not collimated but instead is focused on the plane of the projector.

Part of the light from the mask in Figure 7.17 has been diffracted to a large angle. To try to reimage the pattern onto the wafer one must, at minimum, collect that diffracted light. The numerical aperture (NA) of the system is defined as

$$NA = n \sin(\alpha) \tag{7.17}$$

where α is one half the angle of acceptance of the objective lens, and n is the refractive index of the media between the objective and the wafer. In the case of optical aligners, the exposure is typically done in air where $n = 1.0$. Typical values for NA range from 0.16–0.5.

The resolution of a projection system can be limited by imperfections in the optical train. These imperfections may take the form of lens irregularities such as aberrations, inclusions, or distortion, or the separation between the mask and the objective may be incorrect. In most aligners used in IC manufacturing however, the optics are sufficiently well made such that the resolution is limited by the ability of the optical train to collect and reimage the light. This limit is referred to as Rayleigh's criteria and is given by

$$W_{min} \approx k \frac{\lambda}{NA} \tag{7.18}$$

where k is a constant that again depends on the ability of the resist to distinguish between small changes in intensity (typically k is of order 0.75). This means that an aligner with an NA of 0.4, together with a 436 nm source, can be used to image lines as small as 0.8 μm. Since the mask does not contact the wafer during exposure, defects are not created by projection aligners.

One route to finer lines is to develop higher NA lenses. Progress has been made in this area, but this has a technical price. The depth of focus can be described as the distance along the optical train that the wafer can be moved and still keep the image in focus. For a projection system this is given by

$$\sigma = \frac{\lambda}{NA^2} \tag{7.19}$$

Increasing the numerical aperture increases the resolution linearly, but decreases the depth of focus quadratically. Again taking $\lambda = 436$ nm with a NA of 0.2, $\sigma = 10.9$ μm. Increasing the NA to 0.4 reduces the depth of focus to 2.7 μm. Maintaining the depth of focus across a 200 mm wafer is difficult since the wafer topology alone may be as much as 2 μm unless some planarization is done.

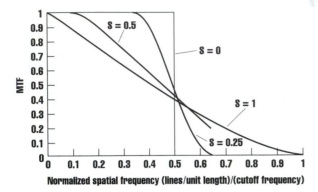

Figure 7-18   Modulation transfer function as a function of the normalized spatial frequency for a projection lithography system with spatial coherence as a parameter.

Added to this are the wafer bow and flatness at that stage of the technology. Some compromise must therefore be achieved between resolution and depth of focus.

The other avenue to reducing the minimum feature size is using shorter wavelength sources. Most lithography tools now in use have moved from g-line to i-line. One of the difficulties in this transition is optics. Virtually all projection printers use diffractive rather than reflecting optics. The photon energy at i-line is rather large (3.3 eV). It becomes difficult to manufacture large diameter lenses with a sufficiently high degree of perfection that are completely transparent at this wavelength. Small inclusions of water, for example, may cause dark spots in the mid and deep UV. This problem is greatly exaggerated for excimer sources. The photon associated with the ArF excimer has an energy of 6.3 eV. Perfect fused silica has a band edge of about 9 eV, but since it is an amorphous material, the band edge is not sharp. Making optical components of sufficient quality and size for these wavelengths is extremely challenging.

Finally, the image resolution is also a function of the spatial coherence of the source. As a rough measure of the spatial coherence, first consider a circular source of some known diameter. The light from this source passes through an aperture known as the pupil, where a condenser lens attempts to collimate the beam. Then the spatial coherence S is approximately given by

$$S \approx \frac{\text{source image diameter}}{\text{pupil diameter}} \tag{7.20}$$

Figure 7.18 shows the modulation transfer function of a diffraction grating period on the mask [20]. Such a mask consists of equally sized lines and spaces, each of width W. The ordinate of the plot is the spatial frequency, given by

$$\nu_{\text{ap}} = \frac{1}{\Gamma} = \frac{1}{2W} \tag{7.21}$$

The spatial frequency has been normalized to the Rayleigh criterion,

$$\nu_o = \frac{1}{W_o} = \frac{\text{NA}}{0.61\lambda} \tag{7.22}$$

For a source with perfect spatial coherence ($S = 0$), the MTF drops abruptly at the Rayleigh criterion. It is not surprising that for a given grating frequency less than the highest resolvable, the MTF decreases as S increases. Interestingly however, in grating periods less than that proscribed by the Rayleigh limit, the MTF increases with decreasing spatial coherence. Depending on the critical modulation transfer function of the resist, it may be preferable to use a radiation source that does not have perfect spatial coherence.

Example 7.1   Resolution limits for optical projection lithography

It has been determined that a resist is capable of resolving images with an MTF of 0.4. If the exposure tool has a NA of 0.35, an exposure wavelength of 436 nm, and a S of 0.5, what is the minimum feature size that this tool can resolve? What is the depth of focus? If this source were replaced with an i-line source (365 nm), how would these numbers change?

According to Figure 7.18, an MTF of 0.4 with $S = 0.5$ corresponds to a normalized spatial frequency of $0.52\nu_0$. Evaluating Equation 7.22, $\nu_0 = 1.32 \ \mu m^{-1}$. Then, the resolution is 0.684 line pairs per micron. This corresponds to a pitch (the sum of the minimum line and space widths) of 1.46 μm or a linewidth of 0.73 μm. If an i-line source is used instead and nothing else changes, the resolution improves to 0.61 μm. Evaluating Equation 7.19, the depth of focus is 3.56 μm for g-line and 2.98 μm for i-line.

1:1 scanning projection aligners revolutionized the microelectronics industry during the 1970s. Prominent among these are the scanning mirror projection aligners first developed by Perkin-Elmer

Figure 7-19   Schematic for the operation of a scanning mirror projection lithography system *(courtesy of Canon U.S.A.).*

[21],[22]. In these systems the wafer and mask are clamped into a cartridge that scans a narrow arc of radiation across the mask and the wafer. Figure 7.19 shows a schematic of the scanning system. During the scan, an arc of light that shines through the photomask reflects off two spherical mirrors. The scanning feature ensures that the light stays near the center of the optical system. A major advantage of this type of system is that the exposure optics can be completely reflective. Aside from the source itself, no large expensive quartz lenses are required. The NA of a typical mirror projection system however, is only about 0.16. This type of system can be used to geometries of about 2.0 μm at very high throughput. Figure 7.20 shows front and side views of a commercial scanning mirror projection system. The full optical path is superimposed on the side view. Because of their high throughput, ICs are still fabricated using this type of equipment, but they are becoming less competitive. Excimer sources have also been applied to scanning systems using a cylindrical lens and mirror arrangement to transform the rectangular beam into the required crescent shape [23]. Using these sources one can obtain 1.0 μm features.

1:1 scanning systems are being replaced by step and repeat projection aligners. These systems

(A)

Figure 7-20   Photographs of a 1x mirror projection system: front view (a) and side view (b) with optical path schematic superimposed *(photos courtesy of Canon U.S.A.)*.

Fig.7-1 **Right-side View And Projection And Illumination System**

Scanning direction

Exposure direction

(B)

❶ 2 kW super high pressure mercury lamp	❹ Arc monitor	❾ Mirror	⓮ Filter shutter	⓳ Carriage
❷ Ellipsoidal mirror	❺ Mirror	❿ Arcuate slit	⓯ Mirror	⓴ Trapezoidal miror
❸ Mirror	❻ Fly's eye lens	⓫ Spherical mirror 1	⓰ Spherical mirror 2	㉑ Concave mirror
	❼ Mirror	⓬ Mirror	⓱ Eyepiece	㉒ Convex mirror
	❽ Condenser lens	⓭ Detector	⓲ Mask	㉓ Wafer

Figure 7-20 cont'd.

are refractive, frequently with a reduction built in. Early steppers used 10:1 reduction lens; 5:1 lens are now more common. This means that only a small region of the wafer, called a *field* (typically 0.5–3 cm^2), is exposed at a time [24]. This allows systems to be built with very high NAs and therefore, high resolution. Between exposures the wafer must be mechanically moved to the next field, hence the common name *steppers*. While features as small as 0.1 μm have been demonstrated [25], they involved unacceptable constraints such as very small fields of view and very minimal focus/exposure latitude. State of the art i-line steppers can repeatably pattern features less than 0.5 μm over fields of more than 2.5 cm on a side [26]. Excimer based steppers are widely believed to be capable of production at the 0.35 μm level [27] and may be capable of 0.25 μm IC production.

In addition to using shorter wavelength radiation, high resolution in a stepper can also be achieved through the use of a high numerical aperture lens. The technical challenge is to produce a very large lens with little aberration that is highly transmissive in the deep UV. This has been an area of continual progress. State of the art lenses have evolved from NAs of 0.28 in 1978, to 0.38 in 1985, to nearly 0.5 today. At the same time, the image field size has increased from 5 mm on a side to nearly 14 mm on a side. Next generation steppers promise to have fields of about 25 mm on a side, with numerical apertures in excess of 0.5.

Figure 7-21   Photograph of a step and repeat lithography aligner. The optical column is enclosed in the large environmental chamber to minimize temperature fluctuations *(photo courtesy of Canon U.S.A.).*

The first steppers used global alignment and focusing. In such a system, the focus and alignment is set at one time for the entire wafer. Newer systems are capable of sight by sight control of these variables. The system can automatically adjust the alignment and focus at every field on the wafer. Thus they are able to avoid many of the depth of focus problems normally associated with high NA systems, greatly reducing the effects of wafer warpage and distortion. This also makes the use of large diameter wafers much more feasible. To accommodate very large numerical apertures combined with short wavelengths in new i-line systems, the newest generation of machines are also beginning to use field by field leveling systems [28] that automatically adjust the height of the wafer for each field to keep the image in focus.

The primary disadvantage of steppers is throughput. While scanning projection printers can achieve throughput of nearly 100 wafers/hr, steppers typically operate at 20–50 wafers/hr. The throughput of the system is determined by

$$T = \frac{1}{O + n \cdot [E + M + S + A + F]} \tag{7.23}$$

where n is the number of die per wafer, E is the exposure time, M is the stage movement time per exposure, S is the stage settling time, A is the site by site alignment time (if used), F is the autofocus time (if used), and O is the overhead associated with loading/unloading the wafer, prealigning it, then moving the wafer under the column and preforming the global alignment. Some of this can be done concurrent with the previous wafer exposure to reduce or eliminate the O term. Since it is common for n to be 50–100, the total of the times inside the bracket is critical to the commercial success of steppers. It must be 2 sec or less in order for the tool to be practically useful. Figure 7.21 shows a commercial step and repeat lithography tool. The exposure tool itself is kept in the large controlled ambient box (left). The rack (right) contains the controller and alignment station.

7.7 Advanced Mask Concepts$^+$

As the quest for producing finer features at ever higher densities continues to challenge lithographers, new methods have emerged to improve the defect density and resolution by the method of mask fabrication. Several proposed methods such as predistorted reticles and the use of optical proximity correction are so difficult that they are unlikely to be implemented on a large scale in the near future. This section will discuss three current or potential mask improvements. Pellicles are already in widespread use. Antireflective coatings and phase contrast masks are in limited manufacturing applications at this writing.

The widespread acceptance of reduction stepper systems have made mask making much easier. Instead of making a mask with 0.5 μm lines, a 5$\times$ stepper only requires a mask with 2.5 μm minimum features. Furthermore, small defects such as particles probably would not be imaged on the surface of the wafer. If the particles are large enough to be printed however, unlike a scanning photolithography system the stepper will repeat the defect in every field across the wafer. In some cases, the die size is large enough that only one die can be printed per exposure area. A defect on such a mask will make every die on the wafer nonfunctional. Very careful attention to potential defects are therefore essential in stepper plates.

The simplest technique for defect detection on photomasks is optical inspection. This is an extremely tedious process that is unlikely to be effective. The types of defects to be detected are clear defects such as pinholes, notches, and missing geometries, and opaque defects such as bridges and particles. Other mask defects include scratches and chips from improper handling and runout/runin and magnification errors. Visual inspection may also introduce defects due to the handling and the long proximity of the mask with relatively particle prone humans.

Two types of automated mask inspection systems have been developed. The easier of the two to implement is the die to die comparison technique. In this system, the mask is inspected at two nominally identical locations by illuminating the bottom while maintaining two inspection objectives on the top that are set at a fixed interval. The system then logs the locations of all positions for which the intensities measured at the two objectives are not sufficiently similar. Of course, such a system cannot work with stepper plates that have only one die on the mask. In that situation, the mask geometries must be directly compared to the database from which the mask was made. This method tends to generate many more false errors than a direct compare, but is still a good screen for a subsequent visual inspection.

The previous steps are often used to ensure that the stepper plate is free of defects as it leaves the mask shop. To minimize the impact of particles in the fab, stepper plates are often pellicalized. In this process, a thin coating of transparent material similar to Mylar is stretched over a cylindrical frame on either side of the mask. The frame stands off the membrane at a distance of about 1 cm from the surface of the mask. The purpose of the pellicle is to ensure that any particles that fall on

Figure 7-22   Basic concept of phase shift masks as described by Levenson.

the mask are kept outside the focal plane of the optical system [29]. Damage to pellicles when using high energy sources such as excimers is still a problem at this writing.

In any optical system, some fraction of the light will not follow the desired optical train, but instead will be lost. In enclosed systems, such as those used for photolithography, light rays may reflect off of multiple surfaces and strike the surface of the wafer, degrading the areal image. One problem with pellicles is the loss of transmission and increased light scattering from the surface of the Mylar. For example, the light scatter in a Perkin Elmer III scanning projection printer has been shown to degrade 3σ linewidth tolerances for a 2 μm line from ±0.31 μm to ±0.42 μm [30]. Even the light scatter from the chromium lines in the mask can be a significant source of linewidth variation, particularly in deep submicron patterns [31]. One new mask technique to reduce this problem is to use a 10% antireflective coating on the lens side of the mask.

Perhaps the most dramatic improvement in resolution that masks can provide is through the use of phase shifting. The concept of phase shifting the optical image, which was first suggested by Levenson [32], is illustrated in Figure 7.22. A mask containing a diffraction grating is overcoated with a phase shifting material at twice the period of the grating in such a manner that every other aperture is covered by the material. The thickness and refractive index of the material is sufficient to exactly shift the phase of the light by 180° with respect to the light that does not see the phase shift material. Ideally the material would not attenuate, reflect, or scatter the incoming light. The result of the phase shifting is that the tail of the diffracted distributions from adjacent features would destructively rather than constructively interfere. This dramatically improves the modulation transfer function at the surface of the wafer and therefore, the resolution. While phase shifted masks (PSM) remains largely a development tool, they have recently received an enormous amount of interest. Fujitsu Ltd., Toshiba Corp., and Matsushita Electric have all recently announced fully functional 64 Mb DRAMs fabricated using PSM [33]. It is likely the use of i-line lithography and PSM will extend optical steppers to the 0.35 μm level. Coupled with an excimer source, optical steppers will manufacture 0.25 μm technologies. They will be used for the 256 Mb DRAM and perhaps even produce the 0.18 μm features required for the 1 Gb DRAM.

Since Levenson's introduction of the concept, a myriad of phase shifting techniques have been suggested. Some, such as Levenson's original proposal, add films to the surface. Others accomplish the phase shifting by etching the quartz mask itself to a sufficient depth [34]. While these techniques work very well on diffrac-

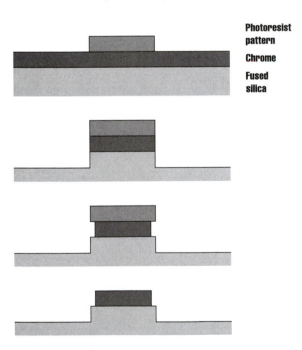

Figure 7-23   Self aligned method for phase shifting the radiation at the edges of a pattern.

tion gratings, application to actual patterns is rather difficult. As an alternative, one can use self aligned phase shifters. In this type of technology, only the region near the edge of the mask feature is exposed to the phase shifted radiation [35]. This has the effect of sharpening the areal image, but in a straightforward manner that is easy to implement. Figure 7.23 shows a typical self aligned method [36], sometimes called *rim phase shifting*. The chromium pattern is used to expose a layer of resist spun on top of the reticle. After development, a layer of resist remains which is perfectly aligned with the chromium pattern. The resist layer is used as a mask to etch the quartz plate. Finally, the resist is used to undercut the chromium a controlled distance, and the resist is stripped. What remains is a quartz ledge around each chromium line. This ledge acts as a local phase shifter that dramatically improves contrast and the repeatability of the linewidth of small features.

7.8 Surface Reflections and Standing Waves

Until now, this chapter has considered only the optical train up to the wafer. In order to get an exposure through the thickness of the photoresist, it must be partially transparent. The light that penetrates the resist may reflect back off the surface of the wafer and change the optical energy deposited in the resist. The required exposure time in a photolithography process therefore, depends on the films on the wafer. A specular metal layer on the surface will require a shorter exposure than exposures over less reflective films.

One problem with surface reflections is image control. This is particularly a problem for late process steps where significant topology may have been developed. The problem is shown schematically in Figure 7.24. In this case, the metal line is running into an oversized contact. The contact reflects the light and exposes the pattern, resulting in poor definition of the line in the contact and possibly a hole in the metal. Figure 7.25 shows a real pattern with this problem. Resist lines running over patterned reflective sublayers have poorly controlled linewidths. An expanded view of one part of the pattern is shown at right. The resist line that enters the contact hole is almost completely gone.

Several solutions have been developed to avoid this problem. It is possible to control the reflectivity of a layer by changing the deposition parameters, but this often requires some compromise of the deposited film characteristics. A better solution is to avoid topology and planarize the layers. This avoids not only unintentional side exposures, but a host of other problems, both for lithography and for metal deposition. The planarizing layer may be temporary, such as a thick absorbing resist layer applied under a thinner imaging resist. This type of process will be covered in the next chapter. The second solution is to planarize the layers through etching processes. This will be discussed in Chapter 15.

One solution to surface topology effects that has been applied with considerable success is the use of antireflective polymers under the photoresist [37]. Figure 7.26 shows resist lines patterned on specular metal films with the same topology as in Figure 7.25. In the figure at right, a 2700 Å antireflective layer was spun on before the photoresist. After resist develop, a short etch in an oxygen plasma was used to remove the exposed antireflective material. Side reflections are completely eliminated and dimensional control is excellent. (For comparison, a micrograph of a standard resist line

Figure 7-24 Light from the exposed regions can be reflected by wafer topology and be absorbed in the resist in nominally unexposed regions.

Figure 7-25   Resist lines for patterning the metal running diagonally from lower left. On the right is an expanded view of the geometry described in Fig. 7.25 *(after Listvan et al.).*

Figure 7-26   Micrographs of resist lines on reflective metal without and with a 2700 Å antireflecting layer under the resist *(after Listvan et al.).*

on a similar substrate is included as well.) The drawback of such a process is its complexity. The layer must be applied and baked before the resist is applied. It must not interact with the resist, and it must be noncontaminating. The layer must be etched after resist develop, and it must be stripped after resist removal.

An examination of micrographs of most resist lines will reveal the presence of ridges along the resist edges. These ridges are caused by standing waves in the aerial image. Standing waves are due to constructive and destructive interference between the incident and reflected light. Figure 7.27 illustrates the effect. As a result of this interference, the intensity of the light in the resist varies with

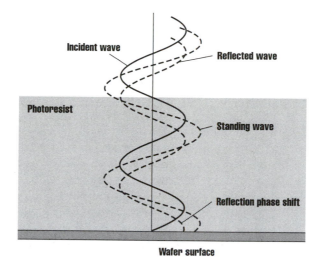

Figure 7-27   The formation of standing waves in the resist.

thickness. With the change in the exposure comes a change in the resist dissolution rate (see Chapter 8). The use of an antireflecting coating completely eliminates standing wave patterns. For good resist processes however, the lateral amplitude of the wave on the edge of the resist is much less than 500 Å, and so standing waves are not a major problem.

7.9 Alignment

It was suggested at the outset of this chapter that photolithographic processes should be judged by three metrics: resolution, registration, and throughput. One should not underestimate the importance of registration in microelectronic fabrication. It limits packing density and therefore, circuit performance. To understand why, consider the following analogy. A skyscraper is to be built floor by floor. Each floor is to be built on the ground. All plumbing and electrical are done before assembly of the building. Connections are made automatically when each floor is placed on the previous one. If one hopes to get water and power on the top floor, each level must line up very closely to those above and below it. To accomplish this feat the tradesmen decide to top each floor off with rubber connectors on each pipe and electrical contact pads on each wire. Needless to say, if 10^6 of the connections must be made at each level, the size of the pads and connectors could easily determine the amount of space left for the actual office space. If the building is 200 feet on a side and 50% of the space is allocated for these passthroughs and the spaces between them, they must be about 2 mm on a side. Assuming that the wires and pipes are 1 mm in diameter, the required building alignment for the construction is ±0.5 mm.

A reasonable rule of thumb for ULSI lithography is that registration errors should be no more than one fourth to one third of the resolution. As already mentioned, steppers have the potential for aligning each exposure field individually. In order to utilize this site by site alignment economically, an automated alignment system must be used. Typically this is done by reflecting light off the surface of the wafer and back through the mask. A box within a box arrangement is used, and the wafer is

Misalignment   **Run out**

Figure 7-28   Two typical registration errors.

moved until the widths of the intensity peaks on either side of the center box are equal. Often a single global alignment is done manually first to ensure that each die will be properly aligned to minimize the search time.

Figure 7.28 demonstrates two possible types of misalignments. Simple x, y, and θ errors are called *misregistration* and are due to poorly aligning the mask and wafer. In many steppers, the wafer is not aligned directly to the mask. Instead, both the wafer and mask are aligned to the optical train of the exposure system in two separate steps. Misregistration errors can occur if these two alignment schemes do not exactly match. A periodic process of baseline correction is done to prevent these systematic errors. Runout or runin is a net difference in the distance between die. On $1\times$ systems like contact/proximity and scanning projection printers, runout is often caused by changes in the physical size of the wafer after it has gone through a series of high temperature steps. Steppers reduce this problem in a global alignment mode and almost completely eliminate it using a good site by site scheme. This type of error can easily be caused by small changes in the temperature of the mask. Fused silica is typically the material of choice for the mask, due in part to its low thermal expansion coefficient ($5 \times 10^{-7}°C^{-1}$). Preventing an expansion of more than 0.1 μm across a 8 in. plate requires temperature control of about 0.75°C. This is not straightforward to obtain when large amounts of optical energy must be transmitted through the mask. Advanced exposure systems are used in submicron steppers to control the temperature of the mask to minimize this problem. Die rotation and misalignment due to alignment marker distortion are also possible. To find the total overlay tolerance one must sum the squares of the individual components:

$$T = \left[\sum T_i^2 \right]^{1/2}$$

(7.24)

7.10 Summary

This chapter concentrated on the production of the areal image, the optical intensity as a function of position on the surface of the wafer. For the small features of interest in IC production, diffraction effects are extremely important. Simple contact printers can be used for pattern structures to less than 1 μm, but these systems are highly defect prone. To avoid this problem the mask can be floated above the wafer in a process known as proximity printing, but at a cost of degraded resolution. Projection lithography systems capable of submicron resolution were introduced. To achieve increased resolution in either type of optical system, it is desirable to use shorter wavelengths of exposing radiation. While mercury arc lamps are currently the most widely used source, excimer lasers are beginning to make contributions and may dominate in the next five years. Finally, methods to increase resolution through mask making were introduced, primarily the use of phase shifted masks. If deep UV excimer sources and phase shifted masks are combined, it is likely that optical lithography can be extended to and perhaps below 0.25 μm production.

Problems

1.   Some arc lamps produce a significant amount of energy in the deep UV because of the high energy electrons in the plasma. Ozone creation is therefore, a significant concern. (a) Calculate

the plasma temperature required for the blackbody component of the radiation to be maximum at 200 nm. (b) If the volume of the lamp is 0.1 l and the gas is at 1 atm at room temperature, use the approximation

$$E \approx \frac{3}{2} kT$$

where k is the Boltzmann constant, to determine the internal energy of the gas. This is a rough approximation that assumes the ionization of the gas to be negligible and that all species are at the same temperature. Note that one mole of gas occupies 22.8 l at room temperature and 1 atm.

2. Show that for $x \ll \frac{\lambda g}{w}$ and w small enough to satisfy the Fraunhofer criteria that I(x,y) is independent of position in the x direction. Explain the significance of this result.
3. Use Equations 7.8, 7.9, and 7.10 in a 3-D graphing program to plot the areal intensity versus x and y for a 3×1 μm aperture. Assume that $\lambda = 436$ nm and $g = 25$ μm.
4. Plot resolution and depth of field as a function of exposure wavelength for a projection aligner with 100 nm $\leq \lambda \leq$ 500 nm. Use k = 0.75 and NA = 0.26. On the same plots, recalculate these functions for NA = 0.41. Discuss the implications of these plots for the technologist that must manufacture transistors with 0.5 μm features.
5. A proximity aligner is used to expose 1 μm apertures. The gap is 25 μm. The separation between the mask and the g-line source is 0.5 m. Is the Fresnel criterion (Equation 7.5) satisfied? At what feature size would the inequality no longer be valid? How does this compare with the simple feature size prediction of Equation 7.16?
6. Repeat problem 5, if the source is replaced with an i-line lamp and with an ArF laser.
7. A particular resist process is able to resolve features whose MTF is $\geq$ 0.3. Using Figure 7.18, calculate the minimum feature size for an i-line aligner with a NA = 0.4 and S = 0.5.

References

1. G. Stevens, *Microphotography*, Wiley, New York, 1967.
2. M. Bowden, L. Thompson, C. Wilson, eds., *Introduction to Microlithography*, American Chemical Soc., 1983.
3. D. Elliott, *Integrated Circuit Fabrication Technology*, McGraw-Hill, New York, 1982.
4. W. M. Moreau, *Semiconductor Lithography, Principles, Practices, and Materials*, Plenum, New York, 1988.
5. P. Burggraaf, "Lithography's Leading Edge, Part 2: I-line and Beyond," *Semicond. Int.* **15(3)**:52 (1992).
6. For example, M. V. Klein, *Optics*, Wiley, New York, 1970.
7. M. Bowden and L. Thompson, *Introduction to Microlithography*, M. Bowden, L. Thompson, and M. Lacombat, eds., American Chemical Soc., 1983.
8. K. Jain, *Excimer Laser Lithography*, SPIE Optical Engineering Press, Bellingham, WA, 1990.
9. H. Craighead, J. C. White, R. E. Howard, L. D. Jackel, R. E. Behringer, J. E. Sweeney, and R. W. Epworth, "Contact Lithography at 157 nm with an F_2 Excimer Laser," *J. Vac. Sci. Tech.* **B1**:1186 (1983).
10. P. Concidine, "Effects of Coherence on Imaging Systems," *J. Opt. Soc. Am.* **56**:1001 (1966).

11. P. Oesterlin, P. Lokai, B. Burghart, W. Muckenheim, H.-J. Kahlert, and D. Basting, ''Advanced Lithography Excimer System: Absolute Wavelength Adjustment, 5 Watt Average Power,'' *Proc. SPIE* **1088**:404 (1989).

12. R. K. Brimacombe, T. J. McKee, E. D. Mortimer, B. Norris, J. Reid, and T. A. Znotins, ''Performance Characteristics of a Narrow Band Industrial Excimer Laser,'' *Proc. SPIE* **1088**:416 (1989).

13. M. S. Hibbs, ''Optical Lithography at 248 nm,'' *J. Electrochem. Soc.* **138**:199 (1991).

14. A. Voschenkov and H. Herrman, ''Submicron Resolution Deep uv Photolithography,'' *Electron. Lett.* **17**:61 (1980).

15. A. Yoshikawa, S. Hirota, O. Ochi, A. Takeda, and Y. Mizushima, ''Angstroms Resolution in Se-Ge Inorganic Resists,'' *Jap. J. Appl. Phys.* **20**:L81 (1981).

16. H. Smith, ''Fabrication Techniques for Surface-Acoustic-Wave and Thin-Film Optical Devices,'' *Proc. IEEE* **62**:1361 (1974).

17. B. Lin, ''Deep UV Lithography,'' *J. Vac. Sci. Technol.* **12**:1317 (1975).

18. G. Geikas and B. Ables, Kodak Photoresist Seminar, p. 22 (1968).

19. B. Lin, in *Fine Line Lithography*, R. Newman, ed., North-Holland, Amsterdam, 1980, p. 141.

20. J. E. Roussel, ''Submicron Optical Lithography?'' *SPIE Proc.* **275**, Semicond. Microlithography, p. 9 (1981).

21. D. A. Markle, Solid State Technol., p. 50, (June 1979).

22. M. C. King, ''New Generation of Optical 1:1 Projection Aligners,'' in *Developments in Semiconductor Microlithography IV, Proc. SPIE* **174**:70 (1979).

23. R. T. Kerth, K. Jain, and M. R. Latta, ''Excimer Laser Projection Lithography on a Full-Field Scanning Projection System,'' *IEEE Elec. Dev. Lett.* EDL-7, p. 299 (1986).

24. P. Burggraaf, ''Wafer Steppers and Lens Options,'' *Semicond. Int.*, p. 56, (Mar. 1986).

25. K. Hennings and H. Schuetze, *SCP Solid State Technol.*, p. 31, (July 1966).

26. M. A. van den Brink, B. A. Katz, and S. Wittekoek, ''A New 0.54 Aperture i-line Wafer Stepper with Field by Field Leveling Combined with Global Alignment,'' in *Optical/Laser Microlithography IV*, V. Pol, ed., *Proc. SPIE* **1463**:709 (1991).

27. R. Unger, C. Sparkes, P. DiSessa, and D. J. Elliott, ''Design and Performance of a Production-Worthy Excimer-Laser-Based Stepper,'' in *Optical/Laser Microlithography IV*, V. Pol, ed., *Proc. SPIE* **1674**:708 (1992).

28. R. Unger and P. DiSessa, ''New i-line and Deep-UV Optical Wafer Steppers,'' in *Optical/ Laser Microlithography IV*, V. Pol, ed., *Proc. SPIE* **1463**:709 (1991).

29. R. Herschel, ''Pellicle Protection of Integrated Circuit Masks,'' in *Semiconductor Microlithography VI, Proc. SPIE* **275**:23 (1981).

30. P. Frasch and K. Saremski, ''Feature Size Control in IC Manufacturing,'' *IBM J. Res. Dev.* **26**:561 (1982).

31. B. J. Lin, ''Phase-Shifting and Other Challenges in Optical Mask Technology,'' 10th Ann. Symp. Microlithography, *Spie* **1496**:54 (1990).

32. M. D. Levenson, N. S. Visnwathan, and R. A. Simpson, ''Improving Resolution in Photolithography with a Phase Shifting Mask,'' *IEEE Trans. Elec. Dev.* ED 26, p. 1828 (1982).

33. G. E. Flores and B. Kirkpatrick, ''Optical Lithography Stalls X-rays,'' *IEEE Spectrum* **28(10)**:24 (1991).

34. A. K. Pfau, W. G. Oldham, and A. R. Neureuther, ''Exploration of Fabrication Techniques for Phase-Shifting Masks,'' in *Optical/Laser Microlithography IV*, V. Pol, ed., *Proc. SPIE* **1463**:124 (1991).

35. A. Nitayama, T. Sato, K. Hashimoto, F. Shigemitsu, M. Nakase, ''New Phase-Shifting Mask with Self-Aligned Phase-Shifters for a Quarter-Micron Photolithography,'' *Tech. Dig. IEDM*, p. 3.3.1 (1989).

36. Y. Yanagishita, N. Ishiwata, Y. Tabata, K. Nakagawa, and K. Shigematsu, ''Phase-Shifting Photolithography Applicable to Real IC Patterns,'' in *Optical/Laser Microlithography IV*, V. Pol, ed., *Proc. SPIE* **1463**:124 (1991).

37. M. A. Listvan, M. Swanson, A. Wall and S. A. Campbell, ''Multiple Layer Techniques in Optical Lithography: Applications to Fine Line MOS Production,'' *Optical Microlithography III: Technology for the Next Decade,* Proc. SPIE **470:**85 (1983).

Chapter 8

Photoresists

The last chapter discussed the production of the aerial image, the pattern of radiation produced at the surface of the wafer during an optical exposure. To transfer a pattern the radiation must strike a photosensitive material, and it must change the properties of that material in such a way that a replica of the mask is left on the surface of the wafer after the photolithography process is complete. The photosensitive compound used in microelectronics is called *photoresist* or simply, *resist*. This chapter will discuss the effects of radiation on the properties of photoresist. It will concentrate on novolac based systems since they are most commonly used in IC fabrication.

8.1 Photoresist Types

One of the most basic categories into which photoresist can be divided is its polarity. Following exposure of the photoresist to the light, it is immersed in a developer solution. Positive photoresist responds to the light in such a way as to make the exposed regions dissolve more quickly during the development process. Ideally, the unexposed regions will remain unchanged. Negative photoresist responds in the opposite manner. Unexposed regions of the resist will dissolve in the developer, while exposed regions remain behind. Positive resists tend to have the best resolution and are therefore much more popular for IC fabrication. They will be covered in more detail; however, negative resists will also be briefly discussed.

The photoresists used for IC fabrication normally have three components: a resin or base material, a photoactive compound (PAC), and a solvent that controls the mechanical properties, such as the viscosity of the base, keeping it in a liquid state. In positive resists, the PAC acts as an inhibitor before exposure, slowing the rate at which the resist will dissolve when placed in a developing solution. Upon exposure to light, a chemical process occurs by which the inhibitor becomes a sensitizer, increasing the dissolution rate of the resist. Ideally, the inhibitor would completely prevent any dissolution of the photoresist, and the enhancer would produce an infinite dissolution rate. Of course, this is not achieved in practice.

The two most utilitarian metrics for the performance of a resist are sensitivity and resolution. Sensitivity refers to the amount of light energy (usually measured in mJ/cm^2) necessary to create the chemical change described above. The more sensitive a photoresist is, the faster the process will be, since for a given exposure intensity, a shorter exposure time will be necessary. Resolution refers to the smallest feature that can be reproduced in a photoresist. As explained in the last chapter, this strongly depends on the exposure tool and the photoresist process itself. But even in a fixed exposure tool, this metric in particular carries with it a considerable uncertainty.

8.2 Organic Materials and Polymers°

Many of the compounds that will be considered in this chapter are carbon based organic molecules. Since this is commonly outside the area of expertise of many of the students who might use this text, this section will introduce these materials and review a few of their relevant properties. Like silicon, carbon has four electrons in its valence shell. It requires four additional electrons to complete the shell. Unlike silicon however, carbon readily bonds with itself allowing it to form complex chains and long repetitious molecules. Life on Earth is based on this ability. Carbon also readily bonds with hydrogen and materials to the right of it in the Periodic Table. The molecules discussed in this chapter are made up of carbon, oxygen, hydrogen, and nitrogen, with small amounts of additional elements. Over the last decade, methods have also been developed to combine hydrocarbons with heavy metals such as gallium to produce a class of materials known as organometallics. These saturated aliphatic (open chain) compounds will be discussed separately in another chapter. This section will concentrate on two classes of carbon compounds: aromatic rings and long chain polymers.

Aromatic rings consist of six carbon atoms arranged in a planar hexagonal structure. In the simplest such compound, benzene, a single hydrogen atom, is attached to each carbon atom (Figure 8.1). Often these hydrogens are left off of the chemical symbol for brevity. The carbon atom gains two electrons by covalently bonding to two nearest neighbor carbon atoms, it gains one electron by bonding to the hydrogen. The last unpaired electron from each atom participates in a delocalized bond that takes the shape of a doughnut around the benzene molecule. It is this highly mobile delocalized pi-electron which gives the aromatics their unique properties.

Figure 8-1   Diagram of simple benzene aromatic ring. The delocalized pi-bond electrons are in a ring that surrounds the nuclei. The symbols indicate old and current ring notation.

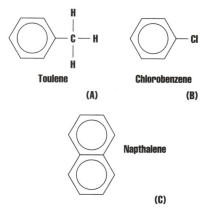

Figure 8-2   Some aromatic based compounds based on: (a) single site substitution, (b) double site substitution, and (c) aromatic condensation.

Figure 8-3 (a) Polyethylene, an example of a simple polymer. (b) Branch chain polymers, (c) Cross linking.

A wide variety of compounds can be made by making simple variations of the benzene ring. As shown in Figure 8.2, toluene, a common solvent, is formed by replacing one of the hydrogens by a methyl (CH_3) group. Similarly, single hydrogen substitutions can produce phenol (OH), aniline (NH_2), chlorobenzene (Cl), and styrene ($HC=CH_2$). (An ''$=$'' sign in the molecular formula indicates that two electrons are participating in the bond.) An organic acid called carboxylic acid can be formed by adding a carboxyl group (COOH). This compound in particular will be important in the discussion of positive photoresists. The xylene class of solvents is formed by replacing two hydrogen atoms with methyl groups. Finally, aromatic rings can attach directly to each other. The simplest such combination, napthalene, is shown in Figure 8.2. Much larger molecules can be formed in linear and 2-D arrays of aromatic rings. Graphite is an example of such a compound. Many of the suspected carcinogens in cigarette smoke are smaller condensed aromatics. Finally, not all aromatic rings have six carbon atoms. Rings can also form with five atoms, although they are less common. Often these five atom rings are attached to one or more benzene rings.

Polymers are very large molecules formed by linking together many smaller repeating units called *monomers*. A polymer can contain as few as five monomers, or it may contain thousands. Typical polymers include plastics, rubbers, and resins. Because of carbon's ease of bonding to itself, many polymers are carbon based. The simplest polymer is polyethylene (Figure 8.3). It consists of a long chain of carbon atoms, each bonded to two hydrogen atoms. The monomer for polyethylene then is CH_2. A polymer may also be branch chained as shown in Figure 8.3. These molecules are stronger and have a higher density. Finally, polymers may cross link, that is, bond with themselves or with other polymers. Again, this increases the strength, and more importantly for this application, it reduces the ability of these molecules to dissolve in typical solvents. On the other hand, if polymers are broken apart into a number of shorter chains, the molecules are more readily dissolved.

As with the benzene ring, many common compounds are simple variations of the basic polyethylene chain. If one hydrogen on every other carbon is replaced by a chlorine atom, the resultant material is polyvinyl chloride (PVC), which is commonly used to make plastic pipes for plumbing; PVC can also be softened to make simulated leather and raincoats. If a benzene ring is substituted for the hydrogen, the resultant material is acrylic, a common plastic.

8.3 Typical Reactions of DQN Positive Photoresist

The most popular positive resists are referred to as DQN, corresponding to the photoactive compound (DQ) and matrix material (N), respectively. These resists evolved from materials used to make blueprints [1]. The matrix material in DQN resists is a thick resin called *novolac*. As shown in Figure 8.4, novolac is a polymer whose monomer is an aromatic ring with two methyl groups and an OH group. Most novolac is used as an adhesive when making plywood. By itself, it dissolves easily in an aqueous

Figure 8-4 Meta-cresol novolac, a commonly used resin material in g- and i-line applications. The basic ring structure may be repeated from 5 to 200 times.

solution. Solvents are added to the resin to adjust the viscosity. This is an important parameter for the application of the resist to the wafer. Most of the solvent is evaporated from the resist before the exposure is done and so plays little part in the actual photochemistry. The solvents used in positive resists are generally combinations of aromatic compounds such as xylene and various acetates, although newer resists targeted toward deep UV applications are moving away from these compounds.

The most commonly used photoactive compounds (PACs) in these resists are the diazoquinones (DQ), such as the one shown in Figure 8.5. The part of the molecule from SO_2 down, which includes the two lower aromatic rings, is specific to particular resist manufacturers and play only a secondary role in the exposure process. We will therefore simplify the DQ molecule by representing it with a generic R. In this state, the PAC acts as an inhibitor, reducing the dissolution rate of the resist in the developer by a factor of 10 or more. This occurs by a chemical bonding of the PAC and the novolac at the surface of the resist where it is exposed to the developer [2], although the exact mechanism is still the subject of debate [3]. Recent work has also suggested that for the inhibitor mechanism to be operative, thermal cycles such as the softbake are necessary after application [4].

The nitrogen molecule (N_2) in the PAC is weakly bonded. As shown in Figure 8.6, the addition of UV light will free the nitrogen molecule from the carbon ring

Figure 8-5 Diazo-quinone (DQ), the most commonly used photoactive compound for g- and i-line applications. The right hand ring is not an aromatic but has a double bond.

Figure 8-6 Photolysis and subsequent reactions of DQ upon UV exposure.

leaving behind a highly reactive carbon site. One way to stabilize the structure is to move one of the carbons outside the ring. The oxygen atom is then covalently bonded to this external carbon atom. This process is known as a Wolff rearrangement. The resultant molecule shown in Figure 8.6 is called ketene. In the presence of water, a final rearrangement occurs in which the double bond to the external carbon atom is replaced with a single bond and an OH group. This final product is carboxylic acid.

This process works as a PAC because the starting material will not dissolve in a base solution (one with a pH > 7). If the PAC is added to the matrix in about a 1:1 mixture, the photoresist is almost insoluble in a base solution. Carboxylic acid, on the other hand, readily reacts with and dissolves in base solutions [5]. This dissolution occurs for two reasons. The resin/carboxylic acid mixture will rapidly take up water. The nitrogen released in the reaction also foams the resist, further assisting the dissolution [6]. The chemical reaction that occurs during this dissolution is the breakdown of the carboxylic acid into water soluble amines such as aniline and salts of K (or Na, depending on the developer). As already mentioned, the novolac resin itself is already water soluble and so dissolves easily. This process continues until all of the exposed resist is removed. It is not unusual for the dissolution rate of the photoresist to change by more than an order of magnitude under this process. Only light, water, and the ability to remove the nitrogen gas are required to drive the process. Typical developer solutions are KOH or NaOH diluted with water.

One of the great advantages of DQN photoresists is that the unexposed areas are essentially unchanged by the presence of the developer, since it does not penetrate the resist. Thus, a pattern of small lines on a clear field that are imaged onto a positive resist keep their linewidth and shape. Another advantage is that novolac is a long chain polymer that is fairly resistant to chemical attack. The photoresist pattern is therefore a good mask for subsequent plasma etching. Most negative resists work by cross polymerization, a process in which the large resin molecules attach to each other to become less soluble. A typical negative resist is an azide sensitized rubber such as cyclized polyisoprene. Negative photoresists have very high photospeeds and adhere well to the wafer without pretreatment. The primary disadvantage of these resists is swelling. Swelling refers to a broadening of the linewidth during the development phase, which takes place in organic solvents rather than aqueous solutions. An after develop bake will typically cause the lines to return to their original dimension, but this swelling and shrinking process often causes the lines to be distorted. Closely spaced lines may come in contact during the swelling phase. As a result, negative resists are generally not suited to features less than 2.0 μm, unless very thin imaging layers are used in a multilayer process. Pinholes are also a serious concern in such an application.

8.4 Contrast Curves

As mentioned above, resolution is a very useful metric for resist performance, but it is highly dependent on the exposure tool. A function known as contrast (γ) is used to characterize the resist more directly. The contrast of a resist is measured by first coating a wafer with a layer of photoresist. Assume for now that a positive tone resist is used. The resist thickness is measured and the resist then given a uniform exposure of light for a small period of time. The exposure dose then is just the light intensity (in mW/cm^2) multiplied by the exposure time. Next, the wafer is immersed in a developer solution for a fixed period of time. Finally, the wafer is removed from the developer, rinsed off, dried, and the remaining resist thickness is measured. If the light intensity was not too large, little of the PAC has changed from an inhibitor to an enhancer, so the thickness should be about the same as the

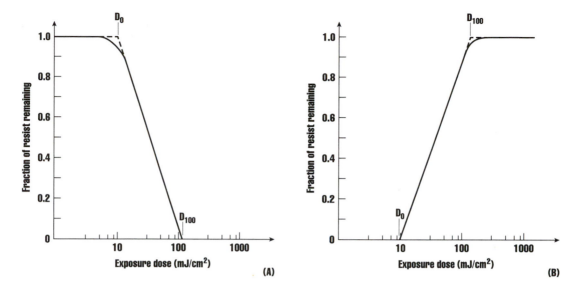

Figure 8-7   Contrast curves for idealized resists: (a) positive tone and (b) negative tone.

original thickness. The experiment is then repeated for increasingly larger doses. If one normalizes the remaining resist thickness and plots it versus the logarithm of the incident dose, a γ or contrast curve as shown in Figure 8.7 will be obtained. The curve has three regions: low exposure where almost all of the resist remains, high exposure where all of the resist is removed, and the transition region between these two extremes.

To derive a numerical value for the contrast of a photoresist, first approximate the steeply sloped portion of the curve by a straight line. The line extends from the lowest energy dose for which all of the resist is removed. This energy density is called D_{100}. The dose at which the line has a y-value of 1.0 is approximately the lowest energy needed to begin to drive the photochemistry. This energy is called D_0. Then the contrast is defined as

$$\gamma = \frac{1}{\log_{10}\dfrac{D_{100}}{D_0}}$$ (8.1)

which is simply the slope of the line.

Contrast can be thought of as a measure of the ability of a resist to distinguish between light and dark portions of the mask. For the moment, the significance of contrast may not be apparent. Consider the exposure of a diffraction grating. As discussed in Chapter 7, the radiant intensity varies smoothly near the edges of the lines and spaces. The higher the contrast of the resist, the sharper the line edge. Typical optical photoresists have a contrast of 2 to 3. This means that D_{100} is

Table 8.1   Contrasts for several commercial resists for different wavelengths. The AZ formulations are products of Shipley.

λ (nm)	AZ-1350	AZ-1450	Hunt 204
248	0.7	0.7	0.85
313	3.4	3.4	1.9
365	3.6	3.6	2
436	3.6	3.6	2.1

Data from Leers [7].

$10^{1/3}$ to $10^{1/2}$ times larger than D_0. Furthermore, contrast curves are not fixed for a given resist. They depend on the development process, the softbake and post exposure bake processes, the wavelength of the exposing radiation, the surface reflectivity of the wafer, and several other factors. One task of the lithographer is to adjust the photoresist processing to maximize the contrast while maintaining an acceptable photospeed. Table 8.1 shows typical contrasts for several resists for various wavelengths.

Example 8.1

Consider the contrast curve obtained from a commercially available resist, shown in Figure 8.8. Assume that the areal image can be approximated by two straight lines. The intensity is a constant from $x = 0$ to $x = 1.0$ μm and varies linearly from $x = 1.0$ to $x = 2.5$ μm (Figure 8.8b). Then, we can multiply this exposure energy by the time to obtain a profile of dose as a function of position (not shown). Finally, we can map the dose plot onto the contrast curve and determine the resist profile as shown in Figure 8.8c.

For typical resists low exposure dosages are less than 50 mJ/cm². At these energies, the resist profile depends primarily on the low exposure and transition regions of the contrast curve. These exposures produce shallow angle resist profiles that are less dependent on the quality of the areal image. For doses greater than about 150 mJ/cm², the exposed regions are typically well above D_{100}. Then, the resist profiles depend primarily on the optical image and light scattering and absorption in the resist and the profiles are fairly sharp. While a sharper image is usually desirable, it comes at a cost of a longer exposure time and therefore slower throughput. Most exposures are done in the moderate or high exposure regimes. Subsequent comments shall be confined to these regimes.

Once the light begins to penetrate the resist its intensity will decrease as

$$I = I_0 \, e^{-\alpha z} \tag{8.2}$$

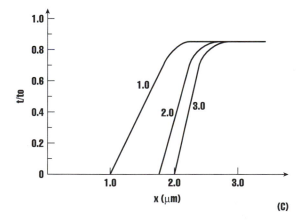

Figure 8-8 (a) Measured contrast curve for a commercial DQN resist, (b) Simple areal image, (c) Resultant profile for 1, 2, and 3 sec exposures.

where α is the optical absorption coefficient in the photoresist, having units of inverse length. In general, D_0 is independent of the resist thickness. D_{100} is inversely proportional to the absorbance A, where

$$A \equiv \frac{I(T_R)}{I_0} = e^{-\alpha T_R} \tag{8.3}$$

where T_R is the resist thickness. Then one can show that [8]

$$\gamma = \frac{1}{\beta + \alpha T_R} \tag{8.4}$$

where β is a dimensionless constant. As one might reasonably expect, γ increases as the thickness of the resist decreases. If the resist is too thin however, it may have poor step coverage as it covers the topography. It may also be unable to withstand the etch of the underlying layers. Some compromise must therefore be made between absolute resolution and more practical resist parameters.

8.5 The Critical Modulation Transfer Function

Another resist figure of merit that can be determined from the contrast is the critical modulation transfer function (CMTF). It is approximately the minimum optical modulation transfer function necessary to obtain a pattern. The critical modulation transfer function is defined by

$$\mathrm{CMTF}_{\mathrm{resist}} = \frac{D_{100} - D_0}{D_{100} + D_0} \tag{8.5}$$

It can be found using the contrast as

$$\mathrm{CMTF}_{\mathrm{resist}} = \frac{10^{1/\gamma} - 1}{10^{1/\gamma} + 1} \tag{8.6}$$

*Optional steps

Figure 8-9   Typical process flow in a photolithography step.

A typical value for the CMTF is about 0.4. The function of the CMTF is to provide a simple test for the resolution of a photoresist. If the MTF of an areal image is less than the CMTF, that image will not be resolved. If it is greater, it is possible to resolve the image. Like the contrast, it gives us a single number for the resolution.

Example 8.2

A new photoresist is being developed for use in an i-line stepper. If the stepper has NA = 0.45 and a spatial coherence of 0.5, what is the minimum feature size when $\gamma = 2.0$? If the resist has $\gamma = 3.0$, what will the minimum feature size be?

Using Equation 8.6, the CMTF of the two resists are 0.519 and 0.366. Setting these two numbers equal to the MTF in Figure 7.18, the resolutions are 0.42 and 0.52 lines per unit length normalized to the cutoff frequency. For an i-line stepper with a 0.45 NA lens, the cutoff frequency is 2.02 μm^{-1}. Then, the resolution of the exposure tool using these two resists would be 0.85 and 1.05 line pairs per micron or minimum feature sizes of 0.59 and 0.48 μm.

8.6 Applying and Developing Photoresist

The steps in a photolithographic process are listed in Figure 8.9. For positive resists, the wafer must typically be pretreated before resist application in order to obtain a smooth, uniform coverage of the photoresist with good adhesion of the resist to the wafer. The first step in the pretreatment is typically a dehydration bake. Done at 150–200°C, in either vacuum or dry nitrogen, this step is intended to drive off most of the adsorbed water on the surface of the wafer. At this temperature, about one monolayer of water remains on the surface. Dehydration bakes can be done at much higher temperatures to further remove all of the adsorbed water, but these high temperature bakes are much less common.

Immediately following the bake the wafer is often primed with hexamethyldisilazane (HMDS), which acts as an adhesion promoter. Vapor priming is done by suspending the wafer above a container of the high vapor pressure HMDS liquid and allowing the vapor to coat the surface of the wafer. Liquid HMDS can also be applied directly on the wafer by dispensing a fixed volume and spinning the wafer to spread out the liquid to a very thin uniform coating. Using either method, one monolayer of HMDS bonds readily with the surface of the wafer, even if it is partially hydroxilated. The other side of the molecule bonds readily with the resist.

The wafer is coated with photoresist after it is primed. The most common application method is spin coating. The wafer is first mounted on a vacuum chuck, which is a flat, hollow metal disc connected to a vacuum line. The chuck has a number of small holes in the surface. When a wafer is placed on its surface, the vacuum draws the wafer into intimate contact with the chuck. A predetermined amount of resist is then dispensed on the surface of the wafer. Torque is applied to the chuck to rapidly accelerate it at a controlled rate up to a maximum rotational speed, usually 2000–6000 rpm. The acceleration stage is crucial to obtaining good uniformity since the solvents begin evaporating from the resist as soon as it is dispensed. The wafer is spun at this speed for a fixed period of time, then decelerated in a controlled manner to a stop. A variation of this method, called a dynamic dispense, is to apply some or all of the resist while the wafer is spinning at low speed. This allows the resist to spread across the wafer before the high speed spin.

Resist thickness and thickness uniformity are critical parameters in developing a good photolithography process. Thickness is not a strong function of the dispense amount. Typically, less than 1% of the dispensed resist remains on the wafer after spinning. The rest flies off during the spin. To avoid a redeposition of this material spinners have splash guards around the chuck. The thickness of the resist is primarily determined by its viscosity and the spin speed. Higher viscosities and slower spin speeds will produce thicker layers of photoresist. The resist thickness is found to vary with spin speed as

$$T_R \propto \frac{1}{\sqrt{\omega}} \tag{8.7}$$

A typical process might be a 30 sec spin at 5000 rpm to produce a resist thickness of about 1.0 μm. At this point, the resist has a tacky consistency since less than one third of the solvent remains.

After spinning, the wafers must undergo a softbake or prebake. The function of this step is to

drive off most of the solvent in the resist and to establish the exposure characteristics [9]. The dissolution rate in the developer will be highly dependent on the solvent concentration in the final photoresist. Generally shorter times or lower temperature softbakes lead to an increased dissolution rate in the developer and so to a higher sensitivity, but at the cost of lower contrast. High temperature softbakes can actually begin to drive the photochemistry of the PAC, leading to resist dissolution of unexposed regions in the developer. In practice, the softbake cycle is empirically determined through trial and error by optimizing the contrast while retaining an acceptable photosensitivity. Typical softbake temperatures are 90–100°C. Times range from 30 sec on a hotplate to 30 min in an oven. The solvent concentration remaining after softbake is usually about 5% of the original concentration.

After softbake the wafer is exposed. This process was described in some detail in the previous chapter. After exposure, the wafer must be developed. Nearly all

Figure 8-10   Contrast curves for Megaposit SPR500 resist using MF CD-26 developer *(courtesy Shipley)*.

positive resists use alkaline developers such as KOH dissolved in water. During the develop cycle, the carboxylic acid reacts with the developer to form amines and metallic salts. In doing so, it consumes the KOH. Care must be taken to constantly replenish the developer if a consistent process is to be maintained. In simple immersion developing, this is done by periodically dumping and refilling the tank, for example, after a certain number of wafers have been processed. To maintain a more consistent develop process, the develop time is often increased between tank refills. A method more commonly used in manufacturing is to ensure a supply of fresh developer for each wafer by dispensing it on track equipment in a process known as puddle develop [10] or by loading a batch of wafers into a spray developer system [11].

Like many chemical reactions the develop process is very temperature sensitive. It is important therefore, to control the develop temperature carefully if accurate control of linewidths is to be maintained. Developer temperature often must be controlled to better than 1°C. In the case of spray developers, the temperature drop associated with the adiabatic expansion as the developer is forced from the nozzle must be taken into account. To compensate for this effect, heated spray nozzles are sometimes used.

The develop process can affect the resist contrast and therefore the resist profiles. Figure 8.10 shows the contrast curves for Megaposit® SPR500™ [12] photoresist developed using MF CD-26. The film thickness for this curve was 1.085 μm, and the exposure was done with an i-line source. The softbake was done for 60 sec using a 90°C hotplate, and a 60 sec 110°C hotplate bake was done after exposure. When using a 60 sec single spray puddle application of the developer, the contrast was 3.14. However, if two spray dispenses are done, each of which is 30 sec long, the contrast increases to 4.12. Obviously, the photolithography process is a multidimensional space and optimization of the process is an important and challenging task.

High temperature bakes may also be done after develop. Sometimes called a hardbake, this is used to harden the resist against further energetic processes such as ion implantation and plasma etching. Figure 8.11 shows 1.0 μm line and space and large feature resist profiles for 60 sec hotplate bakes of varying temperatures. The resist is Shipley Megaposit SPR-2FX, which is used for submicron g-line exposures. At sufficiently high temperatures, the resist profile begins to reflow. This can be used to produce a shallow profile so that during a subsequent etch the angle can be reproduced in the underlying film. The reflow temperature depends on the resist to be used. Some resists, formulated for use in energetic processes, do not flow below 200°C.

There are several types of systems available for

Hardbake Temp./Time   SPR-2FX 1.3
1.0 μm L/S   100 μm Pad

No hardbake

105°C/60 sec.

110°C/60 sec.

115°C/60 sec.

120°C/60 sec.

125°C/60 sec.

Figure 8-11   Resist profiles of a 1 μm line and space and a large feature in SPR-2FX resist for different hardbake temperatures *(courtesy Shipley)*.

Figure 8-12   Top view of an advanced photoresist processing system including cassette load and unload, resist application, bake and develop stations, and a central robot *(courtesy Silicon Valley Group).*

resist processing. The simplest, and the one most likely to be encountered in a university laboratory, is a pair of convection ovens for processing batches of wafers through hard and softbakes and a single wafer spinner. Often resist is applied using a syringe. Even this crude and inexpensive equipment can be used to fabricate submicron features, but uniformity and reproducibility leave much to be desired. At the other extreme, industrial fabrication facilities normally use automated photoresist processing systems (Figure 8.12) sometimes called tracks. In these systems, wafers move from a storage cassette to a hotplate or infrared lamp oven where a dehydration bake is done. Next, they are moved to a dispense station where HMDS is applied and spun, then photoresist is dispensed and spun. Next, they move to a second hotplate or infrared lamp for a softbake. Finally, they are sent to a second storage cassette from which they are taken for exposure. Cool plates are often used after the hotplates to ensure a repeatable temperature during dispense and to avoid contamination of the wafers by the receiving cassette.

Some tracks are linked to the exposure tool in a cluster arrangement for automatic exposure. These tools have the highest overall consistency and reproducibility. In this application, the automated resist processing equipment need not be integrated into a stand alone unit. Instead, separate bake, coat, and develop modules may be arranged in a circular fashion along with the exposure tool. At the center of the cluster, a robot arm can pick and place the wafers at the appropriate station. Such systems can schedule the wafers for resist processing in a consistent manner to make the time between coat and expose and the time between expose and develop the same for each wafer. They can also reduce the particulation associated with the belt transport mechanisms typically found in tracks [13]. In either type of automated system, photoresist thickness variations of less than 50 Å (0.5%) across a wafer and less than 100 Å wafer to wafer are commonly attained. Due to the nature of photoresist and the number of mechanical parts in these systems however, fully automated systems require frequent attention.

8.7 Second Order Exposure Effects

An important parameter when selecting a resist is its absorption spectrum. If the wafer is to be exposed using an aligner with a particular line source, one must know the α of the resist at that wavelength. If α is too large (say more than one over the thickness of the resist), only the top of the resist will effectively be exposed. Upon develop, the lower part of the resist will be left behind, and the wafers will appear underdeveloped. If α is too large, little of the light will be absorbed during the exposure and long exposure times will be required. An advantage of the diazoquinones is that they absorb strongly at both g- and i-lines of mercury but do not absorb very well in the mid UV and the visible ranges [14]. This allows the use of room light in the lithography area, as long as the deep UV components have been filtered out. Typical DQ sensitizers also do not absorb well in the deep UV.

One must also take into account the absorbance of the resin portion of the resist. Light absorbed by the resin does not reach the PAC and so does not drive the photochemistry. Pure novolac is water-white. Due to an oxidation step in the resist fabrication process, it turns an orange-brown [15]. All novolac compounds absorb strongly in the deep UV. Novolac based resists are therefore not appropriate for exposures in the deep UV, since the resin itself will absorb most of the light. Modified novolac however, can be used down to about 250 nm [16]. As will be discussed later in this chapter, it is desired to use the DQN resist as an opaque layer for planarizing a subsequent resist layer, a high temperature post bake will increase the absorptivity of the layer in the mid UV.

To further complicate matters, for most resists the absorbance changes on exposure. Typically it decreases. The actinic absorbance is defined as the difference between the unexposed and the exposed absorbance (See Figure 8.13). This effect is also known as bleaching. The advantage of bleaching is that it can provide a more uniform exposure. As the top layers of the resist are exposed, they become partially transparent, allowing a fuller exposure of the lower layers.

When exposing photoresist over topology a new set of problems arises. The last chapter described the effects of surface reflections on the areal image. A second problem that exists in the presence of topology is linewidth variations due to resist thickness variation.

Figure 8-13   Total absorbance of a layer of SPR511-A resist before and after exposure. The difference between the two curves is the actinic absorbance *(courtesy Shipley).*

Figure 8-14   Cross-sectional view of resist as it covers a vertical step.

Since the photoresist is a viscous film, it does not coat conformally. Instead, it will tend to smooth out the surface topology (Figure 8.14). On top of a step it will be thinner than the nominal thickness. Immediately next to a step it will be thicker than the nominal thickness. Since the step height is often about the same as the resist thickness, these variations can be substantial. The resist thickness changes will lead to linewidth variations: a bulging of the photoresist line just off the step and a necking at the step. These linewidth variations are caused by an effective overexposure on the step and an effective underexposure just off the step. In the case of metal interconnect, these variations can be a serious reliability concern, since the linewidth necking near the step often occurs near a poor region of step coverage in the deposition and near a point where a large gradient exists in the thermal conductivity to the substrate. All of these effects tend to make it likely that a metal line will fail in operation.

One solution to this problem is to planarize the layers if possible. This will be discussed in Chapter 15, since planarization has many more impacts than just lithography and is being aggressively pursued as part of an advanced interconnect strategy. Another solution is the use of multilayer resists. In a typical application, a thick planarizing polymer is first applied. The polymer is called locally planarizing since it can smooth steps in the topology but does not affect changes in height that occur over long distances such as the wafer bow. The lower level may also be opaque to the exposing radiation to prevent surface reflections (see Section 7.8). The upper layer of photoresist may be quite thin to improve the contrast and minimize light scattering. In some cases a third material such as SiO_2 may be deposited between the two polymers to prevent their interaction and to act as an etch mask for subsequent pattern transfer. This is particularly useful if the lower level is PMMA or a similar polymer with a low etch resistance.

After application and softbake the upper resist is exposed and developed. The photoresist is then used as a mask to etch the underlying polymer, which in turn is used as the processing mask. This allows the upper or imaging photoresist to be quite thin and therefore have high contrast. It is how very fine features are produced in contact lithography. The drawback of such approaches, of course, is process complexity. In applications where small features are required however, planarization of the substrate in at least some of the process steps is almost mandatory.

8.8 Advanced Photoresists and Photoresist Processes [+]

This section will review new resist technologies that show promise for future deep submicron lithography. The categories are somewhat arbitrary. All of the topics covered are really intended for deep UV application although some are not listed as such. Silicon containing resists can be used as a contrast enhancement layer or they can be a resist themselves.

The last chapter discussed the utility of deep UV sources such as excimer lasers and indicated that these exposure systems are now being used by a number of IC manufacturers. One of the major difficulties with this application is the lack of good deep UV resists. Novolac compounds begin to strongly absorb just below 250 nm, making KrF (248 nm) exposures marginally acceptable, but precluding the much more desirable ArF (193 nm) exposures. Polymethyl methacrylate (PMMA) is a prototype deep UV resist. This is a resist much favored by lithographers intent on displaying the ultimate in resolution, but otherwise has limited utility as an imaging layer. Since PMMA is most commonly used as an e-beam resist, its structure will be discussed in more detail in the next chapter. Briefly, however, it is a long chain polymer consisting of alternating segments of H–C–H and CH_3–C–$COOCH_3$. Often this chain is compressed or ''rolled up.'' Under deep UV exposure the long chains can be broken and one or more of the carbon atoms left with an unsatisfied bond, or either the methyl (CH_3) group or the ester ($COOCH_3$) side chain can be affected. If the main chain breaks, the resulting short molecules are more easily dissolved in a developing solution. Microfoaming by the dissociated gaseous products (CH_3, CH_3OH, and $HCOOCH_3$) also increases the dissolution rate [17].

PMMA has two primary drawbacks. The first is that the plasma etch tolerance of the resist is very low, lower in fact, than most films to be etched. Therefore, unless a very thick layer of PMMA is used to protect a very thin film, the resist will disappear during the etch before the film does. Furthermore, the dissociation of the PMMA changes the chemistry of the etch and frequently leads to polymeric deposits on the surface of the wafer. The other primary limitation of PMMA is a low sensitivity. Typical optical exposure doses are greater than 200 mJ/cm^2. Various sensitizers can be added to PMMA [18]. In some cases, these additives can also increase the etch resistance. A second approach to increase the sensitivity is to expose the wafer at an elevated temperature. This has the additional advantage of increasing the contrast. Contrasts of 7 have been reported [19] for exposures at 140°C.

While the sensitivity may be improved, the poor etch tolerance and short shelf life of PMMA based resists have forced the development of alternative deep UV resists. There are two goals: finding a suitable matrix material and finding an improved PAC. While alternate materials, such as poly(4-hydroxystyrene), are much more transparent than more commonly used novolac compounds, they are extremely difficult to inhibit to a sufficient degree [20]. The substitution of dimethyl groups has reduced the dissolution rate of this matrix material by a factor of 500, close to what is necessary to a high performance resist [21]. An alternative imaging scheme for deep UV resists is to use a completely different chemical path. One interesting idea currently being developed is to employ uncapping of an acidic functionality by acid catalyzed removal of the protecting

Table 8.2   Current or soon to be commercially available deep UV resists including tone and sensitivity [23]

Source	Name	Pos	Neg	Dose (mJ/cm^2)	Status
Hoechst	DX-59	X	X	<50	Prototype sampling
IBM and Shipley	APEX-E	X		75	Limited commercial
OCG	CAMP-6	X		100	Available
Shipley	XP-89131		X	20–30	Available—excimer
	XP-90236		X	<20	Available—broadband
	XP-2198	X		30	Prototype sampling
	SNR200		X		Available
UCB-JSR	KRF		X	20–30	Available
	KRF	X		20–40	Available

groups [22]. A few commercially available deep UV resists and their current status are given in Table 8.2.

Contrast enhancement layers (CELs) allow the use of an optical lithography exposure tool for smaller features than would otherwise be attainable. The basic process involves a material spun onto a resist coated wafer after the softbake step [24]. This material must be nominally opaque to the exposing wavelength, but undergoes a bleaching reaction on exposure that renders it transparent. Effectively, the use of a CEL transfers the mask to this upper layer, which is in hard contact with the resist. After exposure, the contrast enhancement layer is stripped before develop. CEL is particularly important for deep UV resists since the optical sources may be less intense and the matrix materials may tend to absorb the radiation.

A technical figure of merit that is sometimes quoted for deep UV resists is its quantum efficiency. This is essentially the ratio of the number of photoevents in the resist to the incident photon flux density. Ideally, one would want this ratio to be as large as possible. In recent years, a very interesting development has been the work on chemically amplified resists (CARs). In these resists, an additional photoactive compound is added to the matrix and photosensitizer. On exposure to light, the chemical amplification agent acts in such a way as to greatly increase the primary photochemical process. For example, an early CAR contained onium salts. On exposure to light, these salts dissociate to produce an acid that catalyzes bond breaking/bond forming mechanisms dramatically increasing the quantum efficiency [25] and therefore the system throughput. Exposure dosages as low as 10 mJ/cm^2 have been reported. This resist system has been used in the manufacture of 1 Mb DRAMs [26] and in 16 Mb DRAM pilot line operations [27]. One liability of these salts is the metallic contaminants left in the resist. Other workers have demonstrated CARs using nonmetallic components [28],[29]. While the use of chemical amplification in mid UV applications can be thought of as a manufacturing enhancement, they play a much more vital function in the deep UV resists currently being developed. Here the absorbance may be low enough that the use of these resists without chemical amplification is impractical. Almost all resists currently being developed for deep UV applications use some type of chemical amplification [30]. Common problems with CARs include a deterioration of the image during the time between exposure and post exposure bake and a deterioration of the surface of the resist after coating when exposed to room air.

Inorganic resists fall into a class of resists known as *charge transfer compounds*. In these systems, insolubility is brought about by a change in the polarity of the resist molecule. The prototypical inorganic resist is

Figure 8-15 Processing sequence for Ag/Se-Ge resists *(after Yoshikawa et al., reprinted by permission, AIP).*

Figure 8-16   (a) 0.2 μm and (b) 0.15 μm lines imaged in a 300 Å poly(n-butylsilyne) resist over a novolac resin *(after Kunz et al., used by permission, SPIE).*

Ag-doped Ge-Se. In this process, a 2000 Å layer of Ge-Se is first deposited by sputtering or evaporation (Chapter 12). Next, a 1000 Å layer of Ag is plated on top of the Ge-Se layer in a $AgNO_3$ containing bath. On exposure to light from 200 to 460 nm [31], a photodoping process creates Ag_2Se. This can readily be dissolved in a potassium-iodide acid solution [32]. After the excess silver removal, the image can be dry developed in a CF_4 or other fluorine containing plasma [33]. As shown in Figure 8.15, both positive and negative tones can be achieved. Inorganic resists have the advantage of very high contrast ($\gamma \approx 7$), resulting in the ability to produce fine lines even in g-line exposure systems [34]. Due to their thin film nature, they require a thick planarizing underlayer. The deposited Ge-Se layer tends to have pinholes. Many more defects may be added during the plating process. As a result, these inorganic resists have not yet found much acceptance.

The second category of new resist materials are called *dry developable*. The most interesting class of these materials are the silicon containing resists. Organosilanes were first proposed as resist sources by Hofer et al. [35]. Their motivation was as follows: Since the matrix materials such as novolac tend to absorb in the deep UV, one must try to construct an imaging process that will occur in a very thin resist layer, but can be transferred into a thicker underlayer through an anisotropic plasma etch (see Chapter 11). The top layer must be highly resistant to such plasma processes. Silicon containing resists are attractive since they form SiO_2, which protects the underlying resist [36] in an oxygen plasma.

The field of silicon containing resists gained considerable momentum when it was discovered that polymeric silicon materials called polysilynes could be fabricated [37]. Upon exposure to deep UV radiation, significant bleaching was observed due to the replacement of the Si-Si network with a cross linked siloxane network [38]. The exposed resists can be developed as negative resists in non-polar solvents such as toluene or xylene, which will dissolve the unexposed areas. Wet developed contrasts greater than 7 are obtained in this system. Figure 8.16 shows 0.2 and 0.15 μm lines imaged in a 300 Å poly(n-butylsilyne) resist on top of a novolac resin. After 110 mJ/cm^2 exposure and 15 sec toluene development, the underlying novolac resist was etched in an oxygen plasma designed to etch vertically. The resists can also be developed in a selective plasma etch system. If the etch is selective for silicon to silicon dioxide, a positive image will result. If the etch is selective for silicon dioxide over silicon, a negative tone will develop. The most commonly used process, which etches silicon selectively to silicon dioxide, is an HBr plasma. At present, a contrast of 5 has been obtained in such a system. It has also been found that a separate polysilyne resist layer need not be applied. Instead, the surface of the resist can be silylated by gaseous [39] or liquid [40] exposures at elevated temperatures or by exposure to a suitable plasma environment [41]. As with the bilayer process, high contrasts and deep submicron images can be obtained.

8.9 Summary

This chapter reviewed the basic chemistry of the most popular positive tone resists called DQN. These resists have good photospeeds for mid UV applications, but the novolac based resins are not appropriate for deep UV applications. One of the most important resist figures of merit, contrast, was discussed. Several methods were explored for determining the minimum feature size once the contrast curve and the optical system are known. An introduction to the resist processing sequence and typical resist processing equipment was given. Finally, new types of resists that have high contrast and may be appropriate for deep UV applications were discussed.

Problems

1. Calculate the CMTF for AZ-1450 at the four wavelengths listed in Table 8.1. Assuming NA = 0.4, use Figure 7.18 to determine the minimum feature size for an aligner with $S = 0.5$ using this resist at the various wavelengths.
2. Discuss why the resist contrast might decrease at short wavelengths (high photon energies). Include in this discussion the effects of these photons on both the matrix and the PAC.
3. Derive Equation 8.6.
4. Explain why resist bleaching is a desirable effect.
5. Would you expect the bilayer inorganic/novolac method discussed in the chapter to work well over moderate (<5000 Å) topography? Justify your answer. If surface reflections are a problem, how could that be addressed?
6. Assume that a wafer is being exposed with a proximity printer. In the far field limit in one dimension,

$$I(x) \approx I_1(0)\left[\frac{2W}{\lambda g}\right]^2 I_x^{\ 2}$$

where $I_1(0)$ is the linear flux density in the incident beam at the center of the aperture, and I_x is given by Equation 7.9. Assume that we are using a 1.0 μm thick positive tone resist with $D_0 = 30$ mJ/cm^2 and $D_{100} = 100$ mJ/cm^2. For $D < D_0$ assume that no resist is removed in the develop process. For $\lambda = 436$ nm (0.436 μm), $g = 10$ μm, and $I_1(0) = 100$ mW-μm^2/cm^2, calculate the resist profiles (thickness versus x) for $W = 1$ μm and exposure times of 1 sec, 2 sec, 3 sec, and 4 sec. Using the distance over which the resist is 0.5 μm thick as the width of the developed space, plot the space width versus exposure time for these four points. You may wish to set up your coordinate system so that the center of the exposure is at $x = 0$. Then the spacing on the mask goes from -1 μm to $+1$ μm. The problem is symmetric about the origin, and you only have to calculate and plot half of the resist profile.

References

1. P. Hanson, "Unconventional Photographic Systems," *Photogr. Sci. Eng.* **14**:438 (1970).
2. J. Pacansky and J. Lyerla, "Photochemical Studies on a Substituted Napthalene-2,1,Diazo-oxide," *J. Electrochem. Soc.* **124**:862 (1977).
3. M. Furuta, S. Asaumi, and A. Yokota, "Mechanism of Dissolution of Novolac-diazoquinone Resist," in *Advances in Resist Technology and Processing VIII*, H. Ito, ed., in *SPIE Proc.* **1466**:477 (1991).
4. V. Rao, L. L. Kosbar, C. W. Frank, and R. F. W. Pease, "The Effect of Sensitizer Spatial Distribution on Dissolution Inhibition in Novolac/Diazonapthoquinone Resists," in *Advances in Resist Technology and Processing VIII*, H. Ito, ed., in *SPIE Proc.* **1466**:309 (1991).
5. C. G. Willson, "Organic Resist Materials—Theory and Chemistry," in *Introduction to Microlithography*, L. F. Thompson, C. G. Willson, and M. J. Bowden, eds., *Advances in Chemistry Series* **219**, Amer. Chem. Soc., Washington DC, 1983.
6. W. Hinsberg, C. Willson, and K. Kanazawa, "Use of a Quartz Crystal Microbalance Rate Monitor to Examine Photoproduct Effects on Resist Dissolution," in *Proc. SPIE Resist Technol.* **539**:6 (1985).
7. D. Leers, "Investigation of Different Resists for Deep UV-Exposure," *Solid State Technol.*, p. 91, March 1981.
8. W. M. Moreau, *Semiconductor Lithography, Principles, Practices, and Materials*, Plenum, New York, 1988, p. 31.
9. B. D. Washo, "Rheology and Modeling of the Spin Coating Process," *IBM J. Res., Dev.* **21**:190 (1977).
10. R. F. Leonard and J. A. McFarland, "Puddle Development of Positive Resist," in *SPIE Proc.* **275**, *Semiconductor Microlithography VI* (1981).
11. D. Burkman and A. Johnson, "Centrifugal On-Center, Flood Spray Development of Positive Resist," *Solid State Technol.*, p. 125, May 1983.
12. Megaposit and SPR500 are trademarks of Shipley Corporation.
13. S. Clifford, B. Hayes, and R. Brade, "Results of Photolithographic Cluster Cells in Actual Production," in *Proc. SPIE* **1463**:551 *Optical/Laser Microlithography IV*, (1991).
14. G. Willson, R. Miller, D. McKean, T. Thompkins, N. Clecak, and D. Hofer, *Amer. Chem. Soc.* **40**:54 (1983).
15. A. Knop, in *Applications of Phenolic Resins*, Springer-Verlag, Berlin, 1979.
16. E. Gipstein, A. Duano, and T. Thompkins, "Evolution of Pure Novolac Cresol-Formaldehyde Resins for Deep U.V. Lithography," *J. Electrochem. Soc.* **129**:201 (1981).
17. P. Van Pelt, "Processing of Deep-Ultraviolet (UV) Resists," in *SPIE Proc.* **275**:150 (1981).
18. U.S. Patent 4,405,708 (1983) U.S. Phillips.

19. K. Harada and S. Sugawara, "Temperature Effects on Positive Electron Resists Irradiated with Electron Beam and Deep-UV Light," *J. Appl. Polymer Sci.* **27**:1441 (1982).

20. G. Pawlowski, T. Sauer, R. Dammel, D. J. Gordon, W. Hinsberg, D. McKean, C. R. Vicari, and C. G. Willson, "Modified Polyhydroxystyrenes as Matrix Resins for Dissolution Inhibitor Type Photoresists," in *Proc. SPIE* **1262**:391 (1990).

21. K. Przybilla, H. Röschert, W. Spieß, Ch. Eckes, S. Chaterjee, D. Khanna, G. Pawlowski, and R. Dammel, "Progress in Deep UV Resists," in *Advances in Resist Technology and Processing VIII*, H. Ito, ed., in *SPIE Proc.* **1466**:144 (1991).

22. J. G. Maltabes, S. J. Holmes, J. R. Morrow, R. L. Barr, M. Hakey, G. Reynolds, W. Brunsvold, C. G. Willson, N. Clecak, S. MacDonald, and H. Ito, "1× Deep UV Lithography with Chemical Amplification for One Micron DRAM Production," in *Proc. SPIE* **1262**:2 (1990).

23. P. Burggraaf, "Lithography's Next Stepper," *Semiconductor Int.* **16**(2):38 (1993).

24. B. F. Griffing and P. R. West, "Contrast Enhancement Lithography," *Solid State Technol.*, p. 152, May 1985.

25. E. Reichmanis and C. W. Wilkins, in *Microelectronic Polymers*, M. S. Htoo, ed., Marcel-Dekker, New York, 1989.

26. J. G. Maltabes, S. J. Holmes, J. R. Morrow, R. L. Barr, M. Hakey, G. Reynolds, W. R. Brunsvold, C. G. Willson, N. J. Clecak, S. A. MacDonald, and H. Ito, *Advances in Resist Technology and Processing VII*, M. P. C. Watts, ed., in *Proc. SPIE* **1262**:2 (1990).

27. S. Holmes, R. Levy, A. Bergendahl, K. Holland, J. Maltabes, S. Knight, K. C. Korris, and D. Poley, *Optical/Laser Microlithography III*, V. Pol, ed., in *Proc. SPIE* **1264**:61 (1990).

28. C. Renner, U.S. Patent 4,371 605 (1983).

29. W. Brunsvold, W. Montgomery, and B. Hwang, "Non-metallic Acid Generators for i-line and g-line Chemically Amplified Resists," in *Advances in Resist Technology and Processing VIII*, H. Ito, ed., in *SPIE Proc.* **1466**:368 (1991).

30. P. Burggraaf, "What's Available in Deep-UV Resists," *Semicond. Int.*, p. 56, September 1994.

31. G. Benedikt, U.S. Patent 4,571,375 (1986).

32. Y. Yoshikawa, O. Ochi, H. Nagai, and Y. Mizushima, "A Novel Inorganic Photoresist Utilizing Ag Photodoping in Se-Ge Glass Films," *Appl. Phys. Lett.* **29**:677 (1977).

33. Japanese Patent 82,50430; *Chem. Abstr.* **97**:48231 (1982).

34. E. Ong and E. L. Hu, "Multilayer Resists for Fine Line Optical Lithography," *Solid State Technol.*, June 1984.

35. D. C. Hofer, R. D. Miller, and C. G. Willson, "Polysilane Bilayer UV Lithography," in *SPIE Proc.* **469**:16 (1984).

36. G. N. Taylor, M. Y. Hellman, T. M. Wolf, and J. M. Zeigler, "Lithographic, Photochemical, and O_2 RIE Properties of Three Polysilane Copolymers," in *SPIE Proc.* **920**:274 (1988).

37. P. A. Bianconi and T. W. Weidman, "Poly(n-hexylsilyne): Synthesis and Properties of the First Alkyl Silicon [RSi]$_n$ Network Polymer," *J. Amer. Chem. Soc.* **110**:2342 (1988).

38. R. R. Kunz, P. A. Bianconi, M. W. Horn, R. R. Paladugu, D. C. Shaver, D. A. Smith, and C. A. Freed, "Polysilyne Resists for 193 nm Excimer Laser Lithography," *Advances in Resist Technology and Processing VIII*, in *SPIE Proc.* **1466**:218 (1991).

39. K.-H. Baik, L. Van den Hove, A. M. Goethals, M. Op. de Beeck, and R. Borland, "Gas Phase Silylation in the Diffusion Enhanced Silylated Resist Process for Application to Sub-0.5 μm Optical Lines," *J. Vac. Sci. Technol. B* **8**:1481 (1990).

40. K.-H. Baik, L. Van den Hove, and R. Borland, "Comparative Study Between Gas- and Liquid-Phase Silylation for the Diffusion-Enhanced Silylated Resist Process," *J. Vac. Sci. Technol. B* **9**:3399 (1991).

41. K. Kato, K. Taira, T. Toshihiko, T. Takahashi, and K. Yanagihara, "Effective Parameters of DESIRE Process to Controlling Resist Performance at Sub-half to Quarter Micron Rule," in *Resist Technology and Processing IX*, H. Ito, ed., in *SPIE Proc.* **1672**:415 (1992).

Chapter 9

Nonoptical Lithographic Techniques[+]

The last two chapters discussed optical lithography as the primary tool for imaging patterns on the wafer surface. It was pointed out that resolution limits are a serious concern for this process. Predicting the location of these limits has proven to be an exercise in frustration for those who work in alternate patterning processes. Optical lithography has been extended through the use of steppers with high numerical aperture lenses, deep UV sources, and now, phase contrast masks. Even the most ardent optical lithography supporters, however, believe that the technology cannot be extended much beyond, and perhaps to 0.18 μm. If feature sizes are to be scaled to 0.1 μm or below, new lithographic techniques must be developed. These methods are collectively called *nonoptical lithography*. Recall in Chapter 7 that the Rayleigh limit of resolution is inversely proportional to the wavelength. A general feature of nonoptical lithographies then is the use of very short wavelength energy sources. The two sources considered in this chapter, x-ray and electron beam lithography, have wavelengths so small that diffraction no longer defines the lithographic resolution.

9.1 Interactions of High Energy Beams with Matter[°]

The two most likely exposure sources for a nonoptical lithography system are short wavelength photons (x-ray) and high energy electrons (e-beam). The interactions that occur in the photoresist and in the underlying layers are somewhat similar for the two sources. This section will review the physics of these interactions before going on to the process itself.

Typical x-ray sources used for lithography emit photons with energies of 0.1 to 10 keV. When these photons are incident on a solid, there are many possible interactions, but the two most likely ones are photoelectric absorption and the Compton effect. As shown in Figure 9.1, both processes involve an interaction between photons and electrons. For energies much less than 10 keV, photoelectric emission dominates [1] and the ejected electron carries with it almost all of the incident photon energy. The capture cross section for the photoelectric process depends on the mass of the target material. At higher energies, the Compton process dominates. It can be thought of as a collision between an electron initially at rest and a photon of energy hc/λ and momentum h/λ. A portion of the incident photon energy is lost to the electron during this scattering event. Since the energy needed to liberate an electron from the solid (the work function) is typically two to three orders of magnitude lower in energy than the incident photon energy, electrons in the target can be approximated as

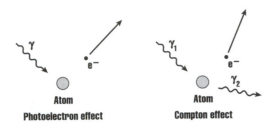

Photoelectron effect   **Compton effect**

Figure 9-1   The two dominant interaction processes for high energy photons with matter: Compton scattering and the photoelectric effect.

free and therefore the cross section for the Compton effect depends only on the electron density. By conserving both momentum and energy, one can show that

$$\lambda_2 - \lambda_1 = \lambda_c(1 - \cos \theta) \tag{9.1}$$

where λ_c is the Compton wavelength (0.0243 Å), and θ is the angle between the incident and final photon momentum. For x-ray wavelengths of 1 Å, only a small fraction of the energy is lost during a single Compton scattering event. As a result, high energy x-rays will penetrate a considerable distance into many solids.

Most x-ray lithography is done at energies well below 10 keV ($\lambda \gg 1$ Å) where photoelectric absorption dominates. As will be discussed in the next section, the majority of the incident photon energy will ultimately be dissipated by secondary electrons generated by impact ionization. In that respect, x-ray lithography and e-beam lithography use similar exposure mechanisms in the resist, once the initial photoelectron event occurs. An important difference between the two processes is that the secondary electrons generated in the resist during x-ray lithography are usually about an order of magnitude lower in energy that those from e-beam systems. As a result, the distance over which the energy is spread in x-ray lithography is much less.

Due to the finite size of the capture cross section, the incident photons are not absorbed at the surface of the resist. Instead, they will penetrate to some depth until absorption occurs. A reasonable approximation is to consider the resist to be an amorphous solid with a single capture cross section. One can define $\alpha(\lambda)$ as the wavelength dependent absorption coefficient

$$\alpha(\lambda) = \sigma(\lambda)n \tag{9.2}$$

where n is the number density of the target material. At typical exposure wavelengths, the absorption coefficient is a function of the photon energy, but is of order 1 to 10 μm^{-1} for most resists. The absorption coefficient also has discontinuities corresponding to the energies of the core levels of the atoms (Figure 9.2). Since resist layers are typically 0.1 to 1.0 μm thick, this has the advantage of providing a nearly uniform exposure through the depth of the resist, although a great deal of exposure energy is required to fully drive the exposure process. X-ray interaction with the underlying materials, in particular x-ray induced damage however, is also an important consideration for x-ray lithography because such a large fraction of the incident energy will pass through the resist and fall onto the wafer itself.

In electron beam lithography, the interactions of electrons and solids are of primary importance. When energetic electrons enter matter, they can pass through undeflected, scatter elastically, or scatter inelastically. Elastic scattering events occur primarily through interaction with the atomic nucleus and can result in deflections larger than 90°. This type of interaction can be treated classically as Rutherford scattering from a screened coulomb potential. The ratio of the scattered current to the incoming current is given by

$$R = \frac{1}{\sqrt{\theta^2 + \left[\dfrac{\lambda}{2\pi a} \right]^2}} \tag{9.3}$$

where λ is the wavelength of the electron and a is the atomic radius that can be approximated by

$$a \approx \frac{0.9 a_o}{Z^{1/4}} \tag{9.4}$$

where a_o is the Bohr radius (0.529 Å) and Z is the atomic number of the target.

There are a number of potential inelastic energy loss mechanisms including low energy (<50 eV) secondary electron generation, inner shell excitations leading to x-ray and Auger electron emission, electron-hole pair creation followed by recombination and photon emission (cathodoluminescence), and phonon excitation. Finding the path of electrons in the solid, therefore, is similar to finding implantation profiles (Chapter 5). One can do a Monte Carlo calculation where a fixed number of electrons are launched over some finite spot size. The path of each electron will be unique. In this technique, each type of possible interaction must be modeled. One can build up a distribution after simulating the passage of a large number of electrons.

Alternatively, the process can be treated statistically using the Bethe equation [2]:

$$\frac{dE}{dx} = \left[\frac{N_A e^4}{2 \pi \varepsilon_o^2} \right] \left[Z \frac{\rho}{A} \right] \left[\frac{1}{E} \ln \frac{E}{66 J} \right] \tag{9.5}$$

where x is the distance into the target, N_A is Avogadro's number, Z is the atomic number of the target, A is the atomic weight, ρ is the mass density of the target, and J is the mean ionization potential, which can be approximated by

$$J(\text{eV}) \approx 11.5 * Z. \tag{9.6}$$

Then, the projected range R_p can be found according to

$$\int_0^{R_p} dx = R_p = \int_{E_o}^0 \left[\frac{dE}{dx} \right]^{-1} dE \tag{9.7}$$

Figure 9-2   Absorption coefficients for some common materials as a function of photon energy *(after Glendenning and Cerrina, reprinted by permission, Noyes Publications).*

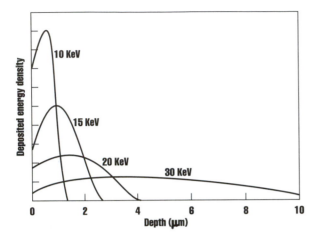

Figure 9-3 Deposited energy density for various energy electron beams incident on silicon as a function of depth.

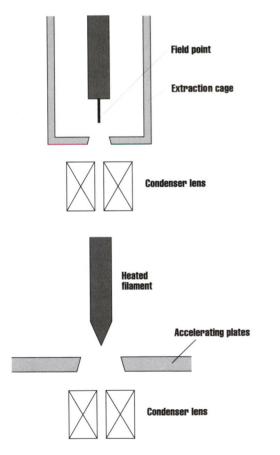

Figure 9-4 Simplified cross section schematics of field emission and thermionic emission electron guns.

Figure 9.3 shows a plot of deposited energy density for various energy electron beams incident on silicon. At typical e-beam energies (25 to 50 keV), most of the energy is deposited at depths greater than 1 μm. As a result, a great deal of the damage is in the substrate. When high energy electrons strike a surface, they may also induce chemical changes. This happens primarily by the cascade of relatively low energy secondary electrons. This will be discussed later in the chapter.

9.2 Electron Beam Lithography Systems

Electron beam lithography (EBL) systems may be used for either mask generation or to directly write patterns on the wafer. Currently, EBL is the technology of choice for mask generation due to its ability to accurately define small features. Because of this book's focus on IC fabrication processes however, we will limit the discussion to direct write EBL. Most direct write systems use a small electron beam spot that is moved with respect to the wafer to expose the pattern one pixel at a time. Several versions of projection and proximity electron beam lithography systems have also been developed [3]. The short penetration length of electrons precludes the use of a solid substrate like quartz for the mask, however. A very thin membrane mask can be used, or else a stencil mask with cutouts through which the beam can pass [4] is needed. These mask difficulties are sufficient to make projection and proximity e-beam an unattractive technology for IC manufacturing, and so we will focus on conventional EBL systems. Direct write EBL systems can be classified as raster scan or vector scan, with either fixed or variable beam geometries. Each type of system has advantages. The selection depends on the type of writing it is designed to do.

All electron beam systems have in common the need for an electron source with a high intensity (brightness), high uniformity, small spot size, good stability, and a long life. Brightness is measured in units of amperes per unit volume per steradian. Electrons can be removed from the cathode of the gun by heating the cathode (thermionic emission), applying a large electric field (field aided emission), a combination of the two (thermal field aided emission), or with light (photo emission). Figure 9.4 shows cross sections of two simple electron guns that might be used for EBL. The most common source is thermionic due to its high brightness. The filament material must be chosen to minimize its evaporation from the cathode and so produce the longest life possible. One of the primary figures of merit for the gun is the emitted electron current density given by

$$J_c = AT^2 e^{-E_w/kT} \tag{9.8}$$

where A is Richardson's constant for the material (typically 10 to 100 A/cm²-K²) and E_w is the effective metal work function. These electrons are accelerated and some fraction of them is collected. The measure of the collected electron energy is the brightness, β. While an increase in the emitted current density also generally increases β, if the increase in β decreases the collection efficiency, the percent increase in the brightness is not as large as the percent increase in the current density.

Most thermionic sources use either tungsten, thoriated tungsten, or lanthanum hexaboride (LaB₆). Tungsten filaments allow operation at pressures as high as 0.1 mtorr, but their current density is only about 0.5 A/cm². As a result, their brightness is less than $2 * 10^4$ A/cm³-sr. Thoriated tungsten

Field emission electron gun and first lens

Beam limiting aperture (200 μm)

Second lens and auxilliary blanker

Blanking plates and knife edge

32 μm deflection plates, X and Y

4 μm deflection plates, X and Y

32 μm deflection plates, X and Y

Third lens assembly

280 μm deflection coils

Fast focus correction coil

Writing surface

Figure 9-5 System schematic of an early EBES system. The basic column is similar in current generation EBL systems *(after Herriot et al., reprinted by permission, ©1975 IEEE).*

cathodes have somewhat lower brightness at the same filament current and require higher vacuum (0.01 mtorr), but their maximum current density can be as large as 3 A/cm². Lanthanum hexaboride (LaB₆) cathodes are the most popular [5], with a current density over 20 A/cm² and a brightness of nearly 10^6 A/cm³-sr. LaB₆ filaments require a vacuum of at least 10^{-6} torr however, and must be well protected against sudden vacuum loss.

A concern with all filament sources is the apparent source size or crossover diameter. Because a large volume of wire is heated, the thermionic electron sources produce broad beams. The energy distribution from such a source is also quite broad. This leads to focusing problems similar to those found in large optical partially incoherent sources. For typical LaB₆ sources, the crossover diameter is about 10 μm. To get to a 0.1 μm spot, a demagnification of 100× is required. Even though the source has a large current density, the inherent source brightness is greatly reduced in order to achieve deep submicron resolution.

Recently it has been shown [6] that tips made from Zr/W/O annealed in forming gas (90% N₂ and 10% H₂) can be used in thermal field aided emission guns to produce crossover diameters as small as 200 Å. As a result, lithography systems using such a gun [7] can produce spots of 100 Å with current densities [8] as high as 1000 A/cm². To obtain acceptable field stability, however, this type of gun requires a constant vacuum of at least 1×10^{-8} torr. In spite of this limitation, this type of gun shows considerable promise for future high resolution, high throughput EBL systems.

Once an electron stream is generated, it must be formed into a narrow beam. This is done in most practical EBL systems through two or three electrostatic lenses and a variety of apertures and knife edges. Figure 9.5 shows a typical lens arrangement. The final spot diameter on the surface of the wafer is given by

$$d^2 = d_o^2 + d_s^2 + d_c^2 \qquad\qquad (9.9)$$

where d_s is the spherical aberration, d_c is the chromatic aberration due to the nonzero energy distribution, and d_o is the perfect lens diameter, which is limited by the finite source size and space-charge effects. In commercial EBL systems, d_s and d_c can be made negligibly small and final spot sizes of less than 0.1 μm can be routinely achieved.

Most EBL systems use Gaussian beams, that is, a beam whose intensity varies approximately as a Gaussian with radius from the center of the beam. As will be described later in this chapter, the primary disadvantage of electron beam lithography is throughput. The process is too slow for eco-

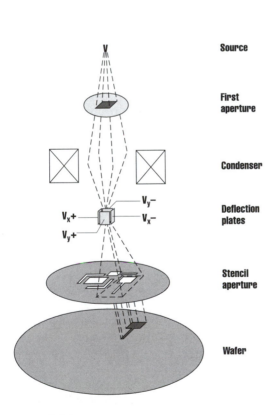

Figure 9-6   The use of stencil masks with EBL in order to improve system throughput. This type of exposure is only useful when exposing highly repetitive designs.

Figure 9-7   A variable shaped beam exposure system using mechanical beam stops for beam shaping. The broad beam exposes many pixels simultaneously, but dimensional control is not as reliable as standard EBL.

nomically manufacturing most ICs. To improve throughput, special purpose EBL systems have been produced with shaped beams that combine direct write and projection lithography [9]. In one variation, a stencil mask is placed above the wafer (Figure 9.6) with a small number of predetermined geometries to be replicated many times (as in a DRAM or a bubble memory). Also included is a blank area through which the beam can be directed for nonstandard geometries. When the beam is sent through this area, it can be rastered as in a standard EBL.

A second and much more popular method of producing a variable shaped beam is illustrated in Figure 9.7. In this case, the beam is partially intercepted by a square aperture [10]. By changing the deflection, the width and length of the line can be varied. Alternatively, x and y shaping can be done separately. Furthermore, an electrostatic rotation lens can be used to produce rectangles at 45° (or any other angle) to the primary axis. Since most images are made up of rectangles, this is an extremely efficient manner of performing EBL [11]. Hundreds or even thousands of pixels can be exposed simultaneously. Linewidth is somewhat more difficult to control in this method, but for many applications the increase in throughput is sufficient to compensate for the reduction in linewidth control. Figure 9.8 shows a matrix of exposures ranging from $0.15 \times 0.15 \ \mu m$ to $2.0 \times 2.0 \ \mu m$ done in a single pass in a variable beam system [12].

In addition to beam shaping, another important factor to direct write throughput is the beam trajectory. Raster scanning (Figure 9.9) was used in the first EBL systems including the early Bell Labs EBES [13]. It is a direct descendent of scanning electron microscopes. In a typical system, the electron beam first passes through a pair of blanking plates that can deflect the beam into a beam stop. A second pair of plates is used to scan the beam in one direction, while the stage is mechanically scanned in the direction perpendicular to the beam scan. The area to be scanned is divided into fields a few millimeters on a side. The scan trajectory is defined as the path that the beam would follow if every pixel were to be exposed. The data is divided into a bit map organized serially along the scan trajectory. To improve edge resolution the e-beam spot size is typically 1/2 to 1/5 of the minimum feature size, and several passes are taken through each exposure area.

Figure 9-8   Exposure matrix in a variable shaped beam system *(after Hohn, reprinted by permission of SPIE).*

Figure 9-9   A comparison of scanning methodologies; raster scan (left) and vector scan (right).

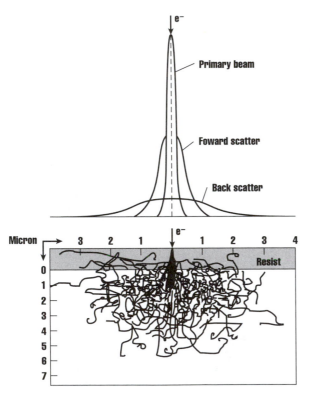

Figure 9-10   Monte Carlo simulation of electron trajectories during an EBL exposure. The upper curve indicates the forward and backscattered components of the beam *(after Hohn, reprinted by permission, SPIE).*

If one is free to choose the polarity of the image, one can take advantage of the fact that much less than 50% of the die must be exposed in a typical pattern. In a raster scan method, every pixel must be scanned serially, and so the exposure time is nearly independent of the pattern. The pattern is written by opening and closing a shutter. Vector scan EBL systems have been developed to improve the throughput by directing the electron beam to only those parts of the chip to be exposed [14]. In this method, the digital location of each area to be exposed is fed to x and y digital to analog converters (DACs). The beam is directed only to those pixels that must be exposed. Software must be used to sequence the pixels so as to minimize the time consumed by beam deflection.

Vector scan systems have a second important advantage over raster scan systems. Ignoring system aberrations and nonlinearities, the precision of the image address is determined simply by the width of the digital word. By using high speed, wide word DACs, each pixel can be placed on an extremely fine grid, without actually requiring the beam to access each pixel [15]. As an example, a 14 bit DAC provides 600 Å resolution over a 0.1 mm field. More commonly, multiple high speed 8 or 10 bit DACs are used in parallel to provide extremely high resolution over fields of order of a few millimeters. The speed of the DACs is a critical concern in determining system throughput for vector scan systems; however, this is a very active area of technology development.

One of the other major areas of concern for electron beam lithography is pattern distortion due to proximity effects. This refers to the tendency of scattered electrons to expose nearby areas that may not be intended for exposure. The effect is much more severe for electron beams than for optical exposures. The cause of these exposures can be readily seen by examining a Monte Carlo simulation of the electron trajectories from a typical e-beam exposure (Figure 9.10). We can divide the scattering types into forward and backscatter. Forward scatter occurs over a small range of angles with respect to the incident velocity and leads to a slight broadening of the image. Backscatter contributes a large area fog of exposure. To minimize the effects of this broadening most high resolution EBL is done in a thin resist layer. The image is then transferred to a thicker masking layer that can be used to obtain the desired

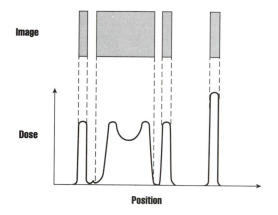

Figure 9-11 Small and large figures to be patterned with EBL requires position dependent dosage to compensate for proximity effects.

pattern. The incident beam energy has to be optimized for a given resist layer. The backscattering range, which depends on the square of the incident energy, can be quite large. For a high energy beam this dispersal may be sufficient to reduce the exposure dose enough to prevent a significant change in the resist properties over the entire area unless the resist has a very high sensitivity. Otherwise, the exposure dose must be corrected for backscatter as well.

If two pattern features are very close to each other, the overlapping trajectories, primarily from the forward scattering, lead to significant unintentional exposure [16] or proximity effect. A number of algorithms have been developed that can be used to modify the exposure time (and therefore, dose) per pixel to compensate for proximity effects. The more comprehensive of these must be done off line by taking the fractured data for a design, running it through the proximity correction algorithm, and producing a dose corrected fractured data file that can be used to run the exposure. Generally these algorithms calculate the concentration of surrounding features, suitably weighted by their distance from the area to be exposed, and subtract the amount of backscattered exposure that one expects to see [17]. A simple illustration of large and small features is shown in the upper part of Figure 9.11. The dose, corrected for proximity effect, taken along a line through the center of the features, is shown in the lower part of the figure. The center of the large pad receives a lower dose than nominal, and the isolated line at right receives a larger dose. Figure 9.12 shows a pattern consisting of lines and spaces, large pads with small rectangular holes, and isolated small rectangles. The upper part of the figure shows the desired intensity pattern, the energy distribution without proximity correction, and the corrected dose map. The lower half of the figure shows the results with and without correction when the smallest feature is 0.25 μm. Without proximity correction, the smallest features are not resolved, and the longest line shows a significant linewidth variation that depends on the local image density [18].

Proximity correction algorithms usually begin by describing the scattered beam as a double Gaussian:

$$I = I_o\left[e^{r^2/2\alpha^2} + \eta_E e^{r^2/2\beta^2}\right]$$
(9.10)

To apply the correction, one needs to know the proximity parameters: α (forward scattering), β (backscattering), and η_E (ratio of energy deposited due to backscattering to the energy deposited due to forward scattering) [19]. These parameters depend primarily on the resist used, the underlying substrate, and most importantly, the accelerating voltage. Often these parameters are found by performing Monte Carlo simulations. Once the beam spreading is known, the deposited energy at each point can be determined and the dosage map adjusted until all of the points to be exposed receive a nearly uniform dose.

Proximity effects can be completely eliminated in a new approach to EBL: scanning tunneling microscope EBL [20]. In this technique, a field emission probe is brought into extremely close proximity to the surface of the wafer and kept there using a feedback control by monitoring the field emission current from the tip. Alternatively, atomic force techniques can be used. Incident electron energies in scanning tunneling microscopy are 4 to 50 eV. The technique has been used to expose resists and to do direct surface modification down to features of 200 Å. A major concern for this

(A)

(B)

Figure 9-12   (a) Map of desired deposited energy, uniform dose deposited energy, and dose map needed to produce a uniform deposited energy, (b) electron micrograph of EBL exposures with and without proximity correction (reprinted by permission, SPIE, after Chang, et al., 1992).

Figure 9-13   Photograph of a JEOL-6000FS EBL system *(courtesy of JEOL Corporation).*

technique is throughput, since the stage and/or tip must be mechanically scanned. Throughput of these systems is much less than that of traditional e-beam lithography. If a system consisting of a large array of these tips could be fabricated and controlled however, STM lithography could become a viable technique.

9.3 Electron Beam Lithography Summary and Outlook

Unlike optical lithography, the resolution of EBL is well below any device geometries desired by the mainstream IC manufacturers. Even by the mid 1970s, electron beam lithography had demonstrated the capability to write lines and spaces less than 100 Å in width [21]. Commercially available electron beam lithography systems (Figure 9.13) are now capable of routinely writing features with resolutions of well below 0.1 μm (Table 1.1). The cost of these systems would seem to be high at about $5M per unit; however, commercial excimer stepper systems for optical lithography are now about $3M.

Table 9.1   Critical specifications for a JEOL JBX-6000FS/SFE direct write electron beam lithography system designed for 0.1 μm device production

Maximum current density	2.0 A/cm²
Beam diameter	0.005 to 0.2 μm
Cathode type	Thermionic field emission (Zr/O/W)
Accelerating voltage	25/50 kV
Writing accuracy	0.06 μm (2σ)
Minimum linewidth	0.02 μm

The major concern with EBL is throughput. EBL can be thought of as a serial process, with the pattern information transferred to the wafer one pixel at a time. Exposures using masks, on the other hand, proceed in a massively parallel manner with every pixel being exposed simultaneously. High brightness sources, vector scan systems, and low inductance deflection coils combined with large bore lenses have all been developed over the last few years to improve throughput, but the technology remains at best an order of magnitude slower than optical lithography [22]. For example, the commercial system shown in Figure 9.13, which is intended for device production, can expose a wafer containing 10,000 0.1 × 200 μm devices in about 20 min. A typical IC wafer, however, contains hundreds of millions of transistors, making exposure times extremely long. Because of the throughput limitations, it is unlikely that EBL will ever be seen as a production technology for the bulk of the IC industry as long as a viable alternative exists.

Aside from mask making, EBL has two primary applications. Variable beam systems have been used with high brightness sources to provide a fine feature prototyping or limited production capability. Using EBL, it is straightforward to quickly modify patterns and produce a limited number of test die, since no mask making step is required. Combined with its high resolution, this capability makes EBL a very attractive technique for research and advanced prototyping. Typical applications in this market include the manufacture of high speed GaAs ICs such as millimeter microwave (MMIC) devices. The small wafer size, low device count, and the small volume of wafers, combined with the need for deep submicron resolution make EBL an ideal choice. The other primary application of EBL is device research. Again, EBL is ideally suited to this task because of its excellent resolution and its ability to rapidly image small quantities of demonstration vehicles. Furthermore, when throughput is not a serious concern, as is the case in discrete transistor research, high resolution EBL systems can be made at relatively low cost by converting scanning electron microscopes.

9.4 X-ray Sources

The second type of nonoptical lithography to be considered uses x-rays as the radiation. There are three x-ray sources that can be used for x-ray lithography (XRL). In order of increasing intensity (and complexity) they are electron impact, plasma, and storage rings. As with electron beam sources, ideally the x-ray source should be as small and as bright as possible. All of the x-ray sources must be operated under vacuum. Unlike EBL however, most wafers are exposed at 1 atm in x-ray lithography. This improves system throughput by avoiding the need to pump the wafers to high vacuum. Thin beryllium windows are used on the x-ray source to extract the x-rays. A film as thin as 25 μm can withstand a 1 atm pressure differential across a small (<1 cm) diameter. Windows up to 6 cm in diameter have been developed for large area exposures [23]. Beryllium windows age with x-ray exposure and so must be replaced periodically.

The simplest type of x-ray source is the electron impact source. It uses a high energy electron beam incident on a metal target [24]. As discussed in Section 9.1, when energetic electrons strike a target, one of the primary energy loss mechanisms is through core level electron excitation. When these excited electrons fall back to the core level, x-rays are emitted. These x-rays form a discrete line spectra whose energies depend on the target material. In addition, a continuous spectrum of Bremmsstrahlung ra-

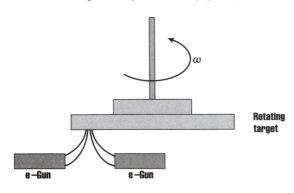

Figure 9-14   A simple rotating electron impact x-ray source uses electron beams focused on a rotating tungsten anode.

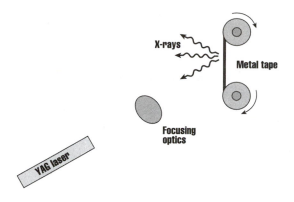

Figure 9-15   Laser plasma heated x-ray source uses a focused high intensity pulsed laser to ablate a metal film. The superheated metal atoms radiate x-rays.

diation is emitted due to the deceleration of the charged electrons [25].

One of the primary limitations of electron impact sources is power dissipation. If the target gets too hot, it will begin to evaporate. For this reason, refractory metals such as tungsten and molybdenum are often used for targets. The simplest sort of x-ray source, shown in Figure 9.14, is somewhat similar to the electron beam evaporator that will be discussed in a later chapter, except that in this case, the hearth charge is water cooled to prevent evaporation of the target [26]. To allow higher power densities, the water cooled anode may be rotated at 7000 to 8000 rpm to dissipate the heat over a larger area [27]. In these systems, an electrical power dissipation of up to 20 kW can be achieved.

The next most intense class of sources use plasmas. There are two common types of plasma x-ray sources: laser heated and electron discharge heated plasmas. Of these, laser heated are the most popular. While laser heated systems have been built with excimers [28], most use a pulsed Nd glass slab laser. The laser energy is generated in 10 ns, 20–25 J pulses of 1.053 μm radiation [29]. This energy is focused into a 200 μm diameter spot at the surface of a thin metal film (Figure 9.15). The power density (of order 10^{13} W/cm^2) in each pulse is sufficient to vaporize the film. The super heated metal vapor radiates x-rays with wavelengths from 8 to 20 Å.

The laser heated plasma is well suited for lithography: It is an intense source with a very small diameter. The exposure energy can be metered using a photodiode and the dose controlled digitally by changing the number of pulses. The metal film is frequently coated on a tape, allowing the target to be advanced as rapidly as the laser can be pulsed. Current state of the art YAG lasers can be pulsed at 2 Hz. Typical conversion energies from the laser output to x-ray energy is about 10%, or about 2 joules per pulse.

If the distance between the source and the wafer is 10 cm and no reflectors or other optics are used, the energy density at the surface of the wafer is given by

$$E_d = \frac{E_o}{2\pi r^2} \qquad (9.11)$$

which is a little less than 4 mJ/cm^2 per pulse or 8 mW/cm^2. As a comparison, rotating anodes in a similar geometry supply about 1.5 mW/cm^2, and stationary anodes supply about 0.2 mW/cm^2. If the resist requires an exposure of 40 mJ/cm^2, and power loss in the mask and exposure tool is ignored (see below), 10 pulses will be required. This will take about 5 sec per exposure which, because of the large field size, may be adequate for IC production. Such systems are commercially available for 0.5 μm feature sizes. Because of the short wavelength of the exposing radiation, these systems have excellent dimensional control. One published report listed the critical dimension control (3σ) over two wafer lots as 0.041 μm for a 0.5 μm feature [30].

Synchrotrons are the brightest x-ray sources for lithography. While the cost of the storage ring is high, the fact that a single ring can support a large number of exposure tools would make synchrotrons attractive for production in x-ray lithography. The basic electron storage ring is shown in Figure 9.16. Electrons are injected into the ring, where they are maintained at 10^6 to 10^9 eV. The energy of the electrons in the ring is given by

Electron storage ring source

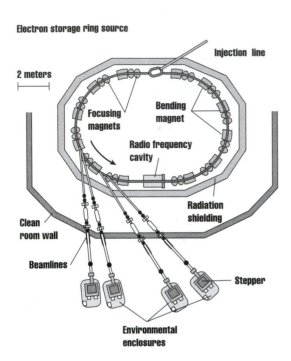

Figure 9-16   Basic schematic of an electron storage ring for XRL. Several exposure stations are indicated *(after Glendenning and Cerrina, reprinted by permission, Noyes Publications).*

$$E \approx 0.3 \, \frac{GeV}{T - m} \, B \cdot r \qquad (9.12)$$

where B is the magnetic field of the bending magnet in tesla, and r is the radius of curvature.

The energy lost in the emission process is made up in the RF cavity, where electrons are accelerated back to the storage ring energy. Electron injection must typically be repeated a few times per day, since the electron current will slowly decay due to leakage. The electrons may be injected as energies close to those of the storage ring, in which case they are first sent through a linear accelerator, or they may be injected at much lower energies and accelerated to speed in the synchrotron.

At each bending magnet location an intense x-ray beam is emitted. The simple storage ring shown in Figure 9.16 could support 12 exposure tools. When a nonrelativistic charged particle is accelerated in a circular path with angular frequency ω, it emits power as a dipole

$$\frac{dP}{d\Omega} = \frac{q^2 r^2 \omega^4}{4 \pi c^3} \frac{1}{\sin^2 \theta} \qquad (9.13)$$

where c is the speed of light and Ω is the solid angle. In synchrotron sources, the electrons are usually highly relativistic. This leads to a Doppler (blue) shift of the radiation energy and a narrowing of the angle of emission. The result is that the median wavelength of such a machine is given by

$$\lambda_c(\text{Å}) = 5.6 \, \frac{r(\text{m})}{E(\text{GeV})^3} = \frac{18.64}{B(T)E(\text{GeV})^2} \qquad (9.14)$$

Most synchrotron sources that are currently operational were not designed as x-ray lithography sources and so have median wavelengths of 20 Å. New systems designed for this application have median wavelengths of about 10 Å.

Finally, the size of the storage ring is worth noting. Many current generation storage rings are quite large, with a radius of over 10 m [31]. The limitation is one of magnetic field. Large water cooled electromagnets can generate about 20 kG (2 T) of field. For a 1 GeV beam, this will produce a turn radius of about 1 m, for a total ring diameter of perhaps 5 m. Such a ring, however, requires a great deal of power to operate. Large bore superconducting magnets, on the other hand, can produce fields of 50 kG (5 T), resulting in turning radii of about 0.5 m. With this type of technology, compact storage rings with diameters of 2 m can be built [32]. The cost of operating these magnets is primarily that of the liquid helium refrigeration. If high critical current and high T_c superconducting wires can be made in the future, these compact storage rings would be extremely attractive.

9.5 X-ray Exposure Systems

Unlike electrons and optical photons, it is very difficult to construct any type of optics for x-rays. Most current generation x-ray exposure tools therefore, are 1 to 1 proximity printers or steppers. If the exposure tool is intended for IC manufacturing, it must be a stepper, since it is not currently possible to fabricate x-ray masks larger than a few inches on a side. Although it is possible and may be desirable to construct a reduction stepper, these machines are not yet available. This section will concentrate on a simple proximity exposure tool (Figure 9.17). A reflector may be used to collect and collimate the source if a rotating anode or plasma source rather than a synchrotron is used, delivering it to a narrow aperture. But unlike optical systems, simple polished metal reflectors are extremely inefficient. A helium column or a vacuum exists between the wafer and the beryllium window to avoid absorption in the gas. The wafer must be aligned to the mask optically and the mask clamped to the wafer carrier. The carrier assembly is then positioned in the beam and the x-ray shutter is opened. If the system is a stepper, the stage is moved under the mask before the next exposure is done. After all of the exposures are complete, the carrier is rolled out, and the wafer is removed for developing.

Clearly, with 10 Å photons, the resolution limitation will not be diffraction effects. If one attempts to use the point source machine shown in Figure 9.17 to expose a pattern of lines and spaces, one can begin to appreciate the limits of x-ray lithography. Consider Figure 9.18. Let d denote the source aperture, D be the aperture to mask spacing, G is the mask to wafer gap, W_m is the linewidth on the mask, W_w is the linewidth on the wafer, and r_m and r_w are the distance from the inner edge of the line on the mask and the wafer, respectively, to the centerline of the exposure system. As a result of the geometry, one can show that a pattern distortion given by

$$r_w = r_m + G \frac{r_m}{D} \qquad (9.15)$$

Figure 9-17   Simple proximity x-ray lithography aligner. The basic system is very similar to optical proximity systems.

Figure 9-18   Geometry of the exposure system shown in Figure 9-17.

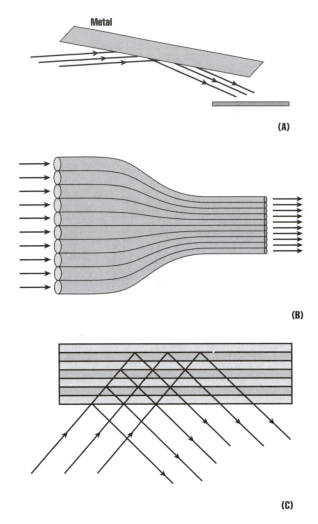

Figure 9-19   Possible choices for x-ray optics systems include glancing angle metal mirrors (a), Kumakhov lenses (b), and multilayer mirrors (c).

occurs. For a gap of 25 μm, a D of 1 m, and an r_m of 2 cm, the shift is about 0.5 μm. This effect may be tolerable if all features are exposed on the same tool or the masks made so as to compensate for the distortion, as long as the mask to wafer spacing can be well controlled.

A more serious limitation is penumbral blur. Referring to Figure 9.18, if one assumes the mask to be perfectly opaque, the finite source size will still lead to continuous shadow edges. The width of the outer shadow is considered the blur δ, given by

$$\delta = G\,\frac{d}{D} \qquad \textbf{(9.16)}$$

The size of this blur is often considered to be a very coarse measure of the resolution of an x-ray exposure tool. For the dimensions that we considered before, using a 4 mm aperture, the resolution is of order 0.1 μm.

Of course, penumbral blur is just a simple geometric effect. If one would produce a uniform wide area beam, this problem could be avoided as well. While synchrotrons naturally produce a large area beam with a small divergence, it is not very uniform [33]. Either type of source could clearly benefit from x-ray lenses. There are three basic types of lenses currently being studied. The first and simplest is the glancing incidence mirror. As shown in Figure 9.19a, this is simply a highly polished metal surface, which the beam strikes at a very low angle. Although such a system can be used in a simple application such as to raster the incident beam from a synchrotron, one cannot build up complicated optical systems with it.

The Kumakhov lens uses a large number of small diameter glass capillary tubes to collect and redirect the radiation. Often the tubes may be drawn down to focus the radiation (Figure 9.19b). X-rays that enter the tube at small angles internally reflect off of smooth interior walls until they exit on the other side of the lens [34]. The current generation of Kumakhov lenses are over a meter in length but have highly nonuniform aerial intensities.

A third strategy for x-ray optics uses a multilayer mirror approach [35]. Originally developed for x-ray astronomy, these mirrors use alternating layers of two elements with widely different electron concentrations (Figure 9.19c). One material has a high mass (the scatterer) and the other has a light mass (the spacer). A common combination is molybdenum and silicon. At each layer, only a small fraction of the incoming plane wave is reflected. If the spacer layer thickness is chosen properly, the reflected waves from the scattering layers interfere constructively. In order to have high efficiency reflection, the layer interfaces must be very smooth [36]. If the layer thicknesses are chosen properly, over half of the incident energy can be reflected for low energy x-rays [37]. Furthermore, if the mirror

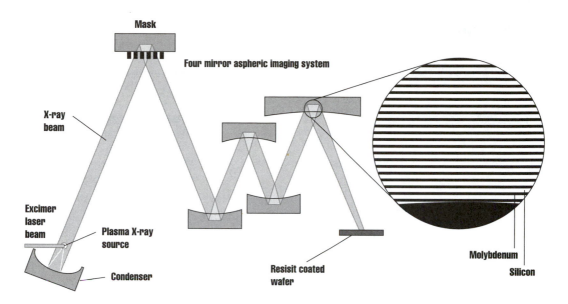

Figure 9-20   An x-ray projection lithography system using x-ray mirrors and a reflective mask *(after Zorpette, reprinted by permission, ©1992 IEEE).*

surface is properly shaped not only can collimation be achieved, but so can reduction. If these mirrors allow a reduction stepper to be made with a very high intensity source, many of the mask distortion problems would be considerably reduced.

Perhaps the most intriguing aspect of multilayer lenses is the possibility that the masks can be made on a reflecting surface rather than on a membrane. Not only would this solve most of the current generation x-ray mask problems (see below), it would allow the construction of relatively inexpensive x-ray projection printers [38]. These systems must use soft x-rays ($\lambda > 50$ Å) [39], where the reflectivity is large [40]. This type of system has already demonstrated feature sizes of 0.1 μm [41]. Figure 9.20 shows a system designed around this principle. The mask is scanned from left to right while the wafer is scanned from right to left. Projection x-ray systems have several shortcomings. Since many layers are required, the defect density per layer must be extremely small. The spectral bandwidth is only 2 to 3%, limiting the power transmission unless a highly monochromatic source is used. The stability of the mirrors and coating after long term x-ray exposures is a concern. Finally, the requirements for machining and polishing tolerances in the mirror fabrication process are currently beyond the state of the art [42].

9.6 X-ray Masks

One of the most difficult aspects of x-ray lithographic technology is mask production. Since no material is highly transmissive for x-rays, the mask substrate must be a low Z thin film [43]. The mask information is transmitted using a patterned absorber layer on this membrane. Table 9.2 compares some of the films used most often for x-ray masks. The substrate materials are listed roughly in order of increasing desirability. For comparison, two common absorbers, tungsten and gold, have been included. While diamond and SiC have the most desirable properties, they are also the most difficult to produce [44]. Due to the ease of fabrication, the most common membrane material at this writing is nonstoichiometric Si_3N_4. The film is most commonly deposited using low

Figure 9-21 X-ray mask blank fabrication process produces a membrane stretched across a mechanical support ring.

Figure 9-22 Additive process for x-ray mask fabrication.

pressure chemical vapour deposition. The temperature and gas compositions can be used to control the stress in the membrane [45].

All x-ray mask processes begin with the fabrication of a x-ray mask blank. As shown in Figure 9.21, in the typical blank fabrication process, a silicon wafer is first coated with the membrane material. In the case of silicon membranes, the surface of the wafer may be doped to be able to selectively etch the substrate [46]. The back of the wafer is patterned lithographically to protect the outer ring. The membrane material on the back side of the wafer is removed, then a long wet chemical silicon etch is done to remove most of the wafer. Once the wafer has been removed, the remaining ring is epoxy mounted to a Pyrex® ring for additional strength and mechanical stability. Once the wafer blank has been completed, the mask must be made. A typical additive sequence of x-ray mask fabrication is

Table 9.2 Mechanical properties of typical x-ray mask materials

Material	Young's Modulus	Thermal Exp Coefficient ($°C^{-1}$) $\times 10^{-6}$	Thickness for 50% Transmission (μm)
Silicon nitride	3.36	2.7	2.3
Silicon	1.3	2.6	5.5
Boron nitride	1.8	1.0	3.8
Silicon carbide	3.8	4.7	3.6
Diamond	11.2	1.0	4.6
Tungsten	*	4.5	0.8 for 10 dB
Gold	*	14.2	0.7 for 10 dB

Starting material: mask blank

Deposit: Tungsten

Pattern and RIE tungsten

Figure 9-23   Subtractive process for x-ray mask fabrication.

shown in Figure 9.22. A thin tantalum/gold-plating base is deposited on the completed blank, and a thick stencil resist is spun on and baked. Thin layers of chromium and imaging resist are deposited on the stencil resist. The top resist is exposed, developed, and used to pattern the chromium layer. The chromium is used as a hard mask in a vertical etch of the stencil resist. The chromium is then stripped and gold is plated up selectively in the resist trenches [47]. Finally, the resist is stripped, and the plating base may be removed.

A subtractive process is shown in Figure 9.23. The mask blank is coated with tungsten and a resist is applied, patterned, and used as a mask to etch the tungsten. The resist is stripped to complete the process. Each technique has advantages. The subtractive process is simpler and should produce fewer defects, since it does not require a plating step. It requires the development of a vertical etch of tungsten that can stop on the membrane. This is much more difficult than achieving etch selectivity of resist to silicon since the common etchant for both silicon and tungsten is fluorine. The lack of selectivity tends to limit the aspect ratio of tungsten absorbers. Gold can be used as an absorber in the subtractive process but it suffers from redeposition problems during the patterning step [48]. Redeposition can be avoided through the use of high energy ion milling [49], but this process can easily damage the membrane. Stress control is critical in the tungsten deposition process to avoid peeling and mask distortions. Stress is not a problem for a plating process; however, many silicon fabrication facilities are extremely reluctant to introduce gold because of possible contamination problems.

Pattern distortion is a serious problem in x-ray masks. In addition to placement errors associated with the e-beam writing of the mask, serious distortion may arise from nonuniform clamping of the mask in the system [50]. Other sources of distortion include stress changes or differential thermal expansion of the mask materials. The stress on the mask blank may change during the addition of the various layers on the mask. After exposure and patterning, the mask can relax back toward its previous condition. During exposure thermal expansion, differences among the ring, the membrane, the absorber, and the resist can also cause local distortions. For that reason, choosing materials with matching expansion coefficients is very desirable. To provide optimal stability, the membrane stress should be large compared to the additional layers, but not so large as to lead to buckling and warpage. It should also be as thin as possible to minimize exposure times. This can be overcome with high brightness x-ray sources, but the mask must also be optically transparent for most alignment techniques.

9.7 Summary and Outlook for X-ray Lithography

The attractive feature of x-ray lithography is its potential for use in volume manufacturing. Figure 9.24 shows a commercial x-ray stepper that has been evaluated as a prototype for manufacturing. The major obstacle to the near term implementation of simple proximity printing is the mask. A number of groups are working on x-ray lithography, and improvements are being made; however, many issues are far from resolved. Mask distortion and aging at 0.1 μm feature sizes are extremely serious. Many also question the ability of the mask to tolerate the handling implicit in day-to-day use in a production environment. Questions regarding mask cleaning and repair have not been resolved. Finally, there is no commercial supplier of masks, although there is some movement toward the creation of a central facility to serve as an initial x-ray mask house. If the equipment issues with soft x-ray projection systems can be resolved however, it would be the most attractive technology.

Figure 9-24   Photograph of a commercial x-ray exposure tool *(courtesy Karl Suss).*

9.8 E-beam and X-ray Resists

Image production in the resist is very different in x-ray and e-beam lithography than for traditional optical lithography. In an optical process, the energy of the absorbing photon is well defined. The photon energy is given by

$$E = h\nu = \frac{hc}{\lambda} \tag{9.17}$$

For an i-line source the photon energy is 3.4 eV. In contrast, a high concentration of secondary electrons with a wide range of energies is produced on exposure to either electron beams or x-rays. Rather than designing the resist so that a single chemical reaction is driven by the exposure, the resist must be designed so that the desired reaction occurs preferentially, but recognize that many reactions will occur. Because the energetic beams penetrate the resist well into the substrate, undesirable re-

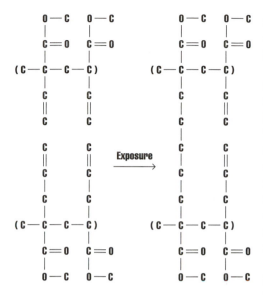

Figure 9-25   Cross linking of an e-beam resist where the basic PMMA structure has been modified through the addition of a C=C side chain to promote cross linking.

actions (damage) in the substrate must also be considered. This section will discuss intentional chemical reactions in the resist. The next section will discuss exposure damage.

In resists, two types of chemical changes are of particular interest. The photoresists that are often used in nonoptical lithography consist of long chain carbon polymers (see Section 8.2). During irradiation, atoms on adjacent chains may be displaced and the carbon atoms will bond directly. This process is known as *cross linking*. Highly cross linked molecules dissolve more slowly in a developer solution. A material in which cross linking is the dominant reaction during exposure is a negative resist. As discussed in Chapter 8, radiation can also disrupt the polymer chains, rendering them more soluble in developer. A material in which chain scission is the dominant reaction during exposure is a positive resist.

The most important resist criteria are contrast and sensitivity for the exposure type and energy, and resistance to damage during plasma etching. One of the most commonly used positive photoresists for high resolution work is polymethyl methacrylate (PMMA). The monomer for PMMA is

$$-[CH_2 - CCH_3(COO(CH_3))]- \qquad (9.18)$$

It has poor to fair sensitivity, but good contrast (typically 2 to 4). In PMMA, both cross linking and fragmenting of the polymeric chains occur (Figure 9.25 shows cross linking in a modified PMMA), but the rate of scission is much larger than that of cross linking. The fragments are about 100 Å long. Like many simple hydrocarbon polymers however, it stands up very poorly in plasma etching environments. As a result, many PMMA-like resists are limited to mask making where only a thin Cr wet etch on a flat substrate is required. When these resists are used for direct writing, PMMA is typically used as the imaging resist for a liftoff process. A variety

Table 9.3   Listing of a few common resists for e-beam and x-ray lithography. For comparison typical numbers for a common optical resist are given as well.

Resist	Tone	EBL Sens ($\mu C/cm^2$)	EBL Contrast	XRL Sens (mJ/cm^2)	XRL Contrast
PMMA	+	100	2.0	6500	2.0
PBS	+	1	2.0	170	1.3
EBR-9	+	1.2	3.0		
Ray-PF	+			125	*
COP	−	0.5	0.8	100	1.1
GMCIA	−	7.0	1.7		
DCOPA	−			14	1.0
Novolac Based	*	200–500	2–3	750–2000	~2

"*" indicates that the value is process dependent.

Chloromethyl styrene

Epoxies

Vinyl

Figure 9-26   Common groups used to promote cross linking.

of derivatives of PMMA have also been demonstrated. One worthy of note is EBR-9®, where the polymer structure has been replaced by

$$—[CH_2 - CCl(CO_2(CH_2CF_3))]— \qquad (9.19)$$

Manufactured by Toray in Japan, EBR-9® has one of the highest sensitivities for positive resists, roughly 15 times that of PMMA (Table 9.3).

A number of negative resists have also been fabricated. These resists have components on the polymer chain that enhance the likelihood of cross linking. Typical cross linking components include chloromethyl styrene, epoxies, and vinyl groups (see Figure 9.26). During exposure the polymers readily cross link at these positions, reducing the solubility of the resist in developer. The negative resists generally have sensitivities as good as or better than the best positive resists, but have low contrasts and are prone to swelling during the develop cycle.

X-ray exposures are typically done in thick resist layers and produce vertical sidewalls with well controlled feature sizes. In contrast, thin layers of imaging resists are often used in e-beam lithography to minimize scattering effects. The pattern is transferred into a hard mask by liftoff or etching. When the resist must be directly used in an etching environment, either a novolac based optical photoresist or a novolac/sulfone copolymer is used. These resists are even less sensitive than PMMA. As an example, Shipley AZ-1350, a commonly used positive tone optical resist, acts as a negative resist under e-beam exposure with a contrast of 3.2 and requires a dose about 10 times that required for PMMA.

9.9 Radiation Damage in MOS Devices

One of the consequences of the use of energetic beams such as x-rays and e-beams for lithography is that the beam is not confined to the resist layer. Either the beam itself, the secondary electrons, or high energy radiation generated by the beam will lead to unintentional and often undesirable chemical changes in the underlying layers [51]. The most sensitive of the common IC structures to these effects are oxides and oxide/semiconductor interfaces. This section will review some of the primary results of radiation exposure and discuss the repair of these effects. For a comprehensive review, the reader is referred to Ma and Dressendorfer [52].

It has been observed that irradiation causes the fixed oxide charge density, the neutral trap density, and the interface state density of MOS devices to increase (see Chapter 4 for a discussion of these defects). The fixed charge shifts the threshold voltage of the MOS transistor. The interface states shift the threshold and reduce the carrier mobility. Neutral traps in the oxide can cause the device to degrade more rapidly under electrical stress. Trapped charge changes the local field in the oxide and may ultimately lead to destruction. As shown in Figure 9.27, the process by which these three effects arise is quite complex. It may involve a direct bond breaking mechanism during the absorption event, or the damage may be caused by the absorption of carriers created in the oxide during the exposure. From Figure 9.27 it is apparent that the carriers in the oxide may fall into traps, or they may create traps through bond breaking. An example of the bond breaking mechanisms believed to occur during the radiation of MOS devices is shown in Figure 9.28. In this example, an oxygen vacancy leads to

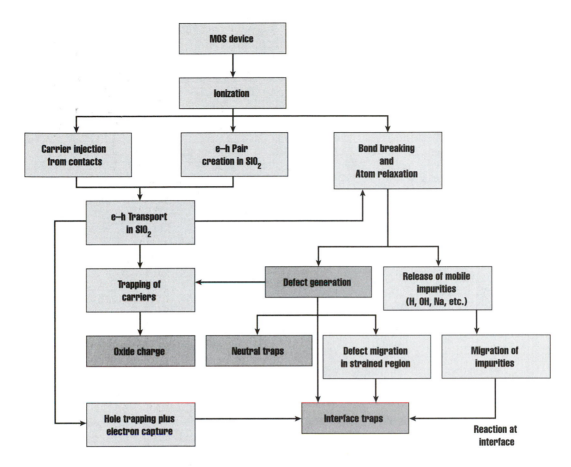

Figure 9-27 Radiation damage flow path *(after Ma and Dressendorfer, reprinted by permission, Wiley).*

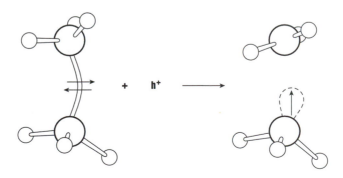

Figure 9-28 An example of a trap creation process believed to occur upon x-ray irradiation of MOS structures. The larger atoms are silicon, the smaller are oxygen. Due to an oxygen vacancy, the two silicon atoms are initially bonded together.

a strained Si-Si bond. Holes created during exposure are captured, causing the SiO_2 tetrahedra to relax back, leaving a permanently unsatisfied bond [53]. At the Si/SiO_2 interface, it is believed that hole capture releases atomic hydrogen that had originally passivated defect sites.

Since the number of ionizing events in the oxide increases with oxide thickness, the damage is most pronounced in thick oxides. This dependence can be approximated by

$$N_{fc}, N_{it} \approx kt_{ox}^n \tag{9.20}$$

where k is a constant, N_{fc} and N_{it} are the fixed charge and interface state density, respectively, and n is approximately 2 [54],[55]. At a typical e-beam exposure dose (2×10^{-5} C/cm^2 or about 10^{14} electrons/cm^2), a MOSFET with a 500 Å gate oxide will experience a fixed charge increase of about 10^{12} cm^{-2}.

Most of the increase in fixed charge density and interface state density can be repaired by a 400°C anneal in forming gas [56]. This is a significant result, since the final anneal of most silicon technologies is done at temperatures at least that high. Neutral traps, on the other hand, require anneal temperatures of at least 700°C. There are indications [57] that these traps are not completely removed until the anneal temperature approaches 1000°C. As a result, any back end energetic beam processes such as metal patterning may leave neutral traps that cannot be removed, and therefore these technologies will be more susceptible to later damage.

When dealing with deep submicron devices with very thin gate oxides, it has been shown that oxides grown with dry oxygen at low temperature are quite radiation hard [58], but some radiation effects remain. Irradiated unannealed n-channel devices with 50 Å gate oxides were found to degrade about twice as fast as unirradiated devices under electrical stressing [59]. The efficiency of neutral trap annealing of these thin oxides is found to depend on the annealing ambient [60] with pure hydrogen showing good effects even at 450°C. As a result, oxide damage is not expected to be a serious barrier to x-ray lithography of deep submicron devices.

9.10 Summary

Because of diffraction effects, optical lithography is not expected to be capable of being extended to feature sizes of much less than 0.2 μm. Two possible candidates for nanoscale lithography are electron beam (EBL) and x-ray (XRL) lithography. EBL has demonstrated the ability to produce feature sizes of about 100 Å, far exceeding device requirements for the foreseeable future. The process is too slow however, to make most IC manufacturing viable. High throughput machines using beam shaping, raster scan, and large field sizes can be made, but throughput remains nearly an order of magnitude less than optical steppers. X-ray lithography has the potential for volume production since, like optical lithography, it is a parallel rather than a serial process. The best x-ray sources include synchrotrons and laser heated plasmas. XRL masks must be fabricated on thin membranes. Distortion of the mask due to mounting, thermal effects, and exposure aging are all serious concerns. In addition the area of x-ray mask repair has not been addressed in any detail.

Problems

1. Construct a table of the wavelength and energy of: g-line, i-line, and ArF laser optical sources, 10 Å x-ray photons, and the De Broglie wavelength of 10 keV electrons. The De Broglie wavelength of an electron is given by

$$\lambda = \frac{h}{p} = \frac{hc}{\sqrt{2mc^2qV}} = \frac{12.25 \text{ Å-V}^{1/2}}{\sqrt{V}}$$

Use Equation 7.16 to predict the proximity printing diffraction limited minimum feature size for each source if the proximity gap is 10 μm and $k = 1$.

2. If one examines the Monte Carlo results for an e-beam exposure shown in Figure 9.11, some of the scattering events involve scattering angles (the angle between the incident and outgoing velocity vectors) of greater than 90°. Explain briefly why this is seen far more often in e-beam lithography than in ion implantation.

3. Estimate the parameters in Equation 9.10 for the beam broadening indicated in the upper half of Figure 9.10. Use the lateral scale on the lower part of the diagram. (I_o will be arbitrary.)

4. Refer to Figure 9.5 for an e-beam column. Where might you expect to observe x-rays from this column? Where would they be particularly intense?

5. An x-ray mask is made up of tungsten absorbing patterns on a silicon nitride membrane. If the membrane heats 10°C during exposure and the field size is 2.5 cm on a side, how much can the field be distorted due to thermal expansion? Using approximation that the maximum allowable distortion is about one fourth the minimum feature size, what is the maximum temperature rise allowable on the mask if the minimum feature size is 0.1 μm?

6. How thin must a silicon nitride membrane be for 90% transmission at the wavelength used in Table 9.2?

7. According to Equation 9.16, the penumbral blur can be minimized if the aperture or the gap is reduced. Explain the trade offs involved for each of the two parameters.

References

1. F. B. McLean, H. E. Boesch Jr., and T. R. Oldham, "Electron-Hole Generation, Transport, and Trapping in SiO$_2$," in *Ionizing Radiation Effects in MOS Devices and Circuits*, T. P. Ma and P. V. Dessendorfer, eds., Wiley-Interscience, New York, 1989.

2. J. J. Muray, "Electron Beam Processing," in *Beam Processing Technologies*, N. G. Einspruch, S. S. Cohen, and R. N. Singh, eds., Academic Press, San Diego, Calif., 1989.

3. P. Nehmiz, W. Zapka, U. Behringer, M. Kallmeyer, and H. Bohlen, "Electron Beam Proximity Printing," *J. Vac. Sci., Technol.* **B3**:136 (1985).

4. J. Frosien, B. Lischke, and K. Anger, "Aligned Multilayer Structures Generated by Electron Microprojection," in *Proc. 15th Int. Symp. Electron, Photon, Ion Beam Technol.*, p. 1827 (1980).

5. F. J. Hohn, "Development and Use of High Brightness Lanthanum Hexaboride Electron Guns," in *Proc. Scanning Elect. Micr., SEM In. IV*, p. 1327 (1985).

6. Commercially available from FEI Co., Beaverton, Ore.

7. M. Gesley, F. J. Hohn, R. G. Viswanathan, and A. D. Wilson, "A Vector Scan Thermal Field Emission Nanolithography System," *J. Vac. Sci. Technol.* **B6**:2014 (1988).

8. H. Nakazawa, H. Takemura, and M. Isobe, "Thermally Assisted Field Emission Electron Beam Exposure System," *J. Vac. Sci. Technol.* **B6**:2019 (1988).

9. H. C. Pfeiffer, "Variable Spot Shaping for Electron Beam Lithography," *J. Vac. Sci. Technol.* **15**:887 (1978).

10. H. C. Pfeiffer, "Recent Advances in EBL for High Volume Production of VLSI Devices," *IEEE Trans. Electron. Dev.* **ED-26**:663 (1979).

11.   L. Veneklasen, "A High Speed EBL Column," *J. Vac. Sci. Technol.* **B3(1)**; (1985).

12.   F. Hohn, "Electron Beam Lithography, Directions in Direct Write and Mask Making," *Electron-Beam, X-ray, and Ion-Beam Submicrometer Lithographies IX*, in *Proc. SPIE* **1263**:152 (1990).

13.   P. Herriot, R. Collier, D. Alles, and J. Stafford, "EBES, A Practical Electron Lithography Systems," *IEEE Trans. Electron. Dev.* **ED-22**:385 (1975).

14.   T. H. P. Chang, M. Hatzakis, A. D. Wilson, A. Speth, A. Kern, H. Luhn, "Scanning Electron Beam Lithography for Fabrication of Magnetic Bubble Circuits," *IBM J. Res., Dev.* **20**:376 (1976).

15.   L. Veneklasen, "Electron Beam Patterning and Direct Write," in *Handbook of VLSI Microlithography*, Noyes, Park Ridge, NJ, 1991.

16.   T. H. P. Chang and A. D. G. Stewart, "Proximity Effect in Electron Beam Lithography," *J. Vac. Sci. Technol.* **12**:1271 (1975).

17.   C.-Y. Chang, G. Owen, F. R. Pease, and T. Kailath, "A Computational Method for the Correction of Proximity Effect in Electron-Beam Lithography," *Electron-Beam, X-ray, and Ion-Beam Submicrometer Lithographies*, in *Proc. SPIE* **1671**:208 (1992).

18.   P. Vermeulen, R. Jonckheere, and L. Van Den Hove, "Proximity Effect Correction in e-beam Lithography," *J. Vac. Sci. Technol. B* **7**:1556 (1989).

19.   T. H. P. Chang and A. D. G. Stewart, *Electron-Beam, X-ray, and Ion-Beam Submicrometer Lithographies*, SPIE **1671**:97 (1969).

20.   C. R. K. Marrian and E. A. Dobisz, "Scanning Tunneling Microscope Lithography: A Viable Lithographic Technology?," *Electron-Beam, X-ray, and Ion-Beam Submicrometer Lithographies*, SPIE **1671**:166 (1992).

21.   A. N. Broers, W. W. Molzen, J. J. Cuomo, N. D. Wittels, "Electron Beam Fabrication of 80 Å Metal Structures," *Appl. Phys. Lett.* **29** (1976).

22.   A. Gonzales, "Recent Results in the Application of Electron Beam Direct Write Lithography," *Electron-Beam, X-ray, and Ion-Beam Submicrometer Lithographies*, p. 374 (1989).

23.   K. Hara and T. Itoh, "Study of Large-field Beryllium Window for SR Lithography," *Electron-Beam, X-ray, and Ion-Beam Submicrometer Lithographies*, SPIE **1671**:391 (1992).

24.   M. Lepselter, D. S. Alles, H. Y. Levinstein, G. E. Smith, and H. A. Watson, "A Systems Approach to 1 μm NMOS," in *Proc. IEEE* **71**:640 (1983).

25.   S. E. Bernacki and H. I. Smith, "Characteristic and Bremmsstrahlung X-ray Radiation Damage," *IEEE Trans. Electron. Dev.* **22**:421 (1975).

26.   T. Hayasaka, S. Ishihara, H. Kinoshita, and N. Takeuchi, "A Step-and-Repeat X-ray Exposure System for 0.5 μm Pattern Replication," *J. Vac. Sci. Technol.* **B3**:1581 (1985).

27.   W. B. Glendenning and F. Cerrina, "X-ray Lithography," in *Handbook of VLSI Microlithography*, W. B. Glendenning and J. N. Helbert, eds., Noyes, Park Ridge, NJ, 1991.

28.   R. Fedosejevs, R. Bobkowski, J. N. Broughton, and B. Harwood, "keV X-ray Source Based on High Repetition Rate Excimer Laser-produced Plasmas," *Electron-Beam, X-ray, and Ion-Beam Submicrometer Lithographies*, SPIE **1671**:373 (1992).

29.   J. A. Abate, "Solid State Laser Reliability, Productibility, and Maintenance Issues for X-ray Lithography Applications," in *Proc. SPIE* **1223**:37 (1990).

30.   G. K. Celler, J. Frackoviak, R. R. Freeman, C. W. Jurgensen, R. R. Kola, A. S. Novembre, L. F. Thompson, L. E. Trimble, and D. N. Tomes, "Evaluation of a Laser-based Proximity X-ray Stepper," *Electron-Beam, X-ray, and Ion-Beam Submicrometer Lithographies*, SPIE **1671**:312 (1992).

31.   J. B. Murphy, "Electron Storage Rinds as X-ray Lithography Sources: An Overview," *Electron-Beam, X-ray, and Ion-Beam Submicrometer Lithographies*, SPIE **1671**:116 (1990).

32. D. E. Andrews, M. N. Wilson, A. I. Smith, V. C. Kempson, A. L. Purvis, R. J. Anderson, A. S. Bhutta, and A. R. Jorden, "Helios: A Compact Superconducting X-ray Source for Production Lithography," *Electron-Beam, X-ray, and Ion-Beam Submicrometer Lithographies, SPIE* **1671**:124 (1990).

33. W. B. Glendenning and F. Cerrina, "X-ray Lithography," in *Handbook of VLSI Microlithography*, W. B. Glendenning and J. N. Helbert, eds., Noyes, Park Ridge, NJ, 1991.

34. "Soviet X-ray 'Lens' Seen as Promising," *IEEE Inst.*, p. 1, May/June 1992.

35. E. Spiller, "Reflective Multilayer Coatings for the Far UV Region," *Appl. Optics* **15**:2333 (1975).

36. D. W. Kruger, D. E. Savage, and M. G. Lagally, "Diffraction Determination of the Size Distribution of Nanocrystalline Regions in a Crystalline Substrate," *Phys. Rev. Lett.* **63**:402 (1989).

37. D. G. Stearns, N. M. Ceglio, A. M. Hawryluk, and R. S. Rosen, "Multilayer Optics for Soft X-ray Projection Lithography: Problems and Prospects," *Electron-Beam, X-ray, and Ion-Beam Submicrometer Lithographies, SPIE* **1671**:80 (1991).

38. G. Zorpette, "Rethinking X-ray Lithography," *IEEE Spectrum*, p. 33, June 1992.

39. H. Kinoshita, K. Kurihara, Y. Ishii, and Y. Torii, "Soft X-ray Reduction Lithography Using Multilayer Mirrors," *J. Vac. Sci. Technol.* **B6**:1648 (1989).

40. D. G. Stearns, R. S. Rosen, and S. P. Vernon, "Multilayer Mirror Technology," in *Soft-X-Ray Projection Lithography Techn. Dig.*, Opt. Soc. Amer., Washington, DC **8**:44 (1992).

41. W. T. Silfvast and O. R. Wood, II, "Tenth Micron Lithography With a 10 Hz 37.2 nm Sodium Laser," *Microelectron. Eng.* **8**:3 (1988).

42. B. E. Newnam and V. K. Viswanathan, "Development of XUV Projection Lithography at 60–80 nm," *Electron-Beam, X-ray, and Ion-Beam Submicrometer Lithographies, SPIE* **1671**:419 (1992).

43. A. R. Shikunas, "Advances in X-ray Mask Technology," *Solid State Technol. J.* **27**:192 (1984).

44. H. Windischmann, "A 75 mm Diamond X-ray Membrane," *Electron-Beam, X-ray, and Ion-Beam Submicrometer Lithographies, SPIE* **1671**:241 (1990).

45. R. Nachman, G. Chen, M. Reilly, G. Wells, H. H. Lee, A. Krasnoperova, P. Anderson, E. Brodsky, E. Ganin, S. A. Campbell, and F. Cerrina, "X-ray Lithography Processing at CXrL From Beamline to Quarter-Micron NMOS Devices," *Proc. SPIE* (1994).

46. D. L. Spears and H. I. Smith, "X-ray Lithography—A New High Resolution Replication Process," *Solid State Technol. J.* **15**:21 (1972).

47. G. E. Georgiou, C. A. Janoski, and T. A. Palumbo, "DC Electroplating of Submicron Gold Patterns on X-ray Masks," in *Proc. SPIE—Int. Soc. Opt. Eng.* **471**:96 (1984).

48. R. E. Acosta, J. R. Maldonado, L. K. Towart, and J. R. Warlaumont, "B-Si Masks for Storage Ring X-ray Lithography," in *Proc. SPIE—Int. Soc. Opt. Eng.* **448**:114 (1983).

49. J. L. Bartelt, C. W. Slayman, J. E. Wood, J. Y. Chen, C. M. McKenna, C. P. Minning, J. F. Coakley, R. E. Hollman, and C. M. Perrygo, "Mask Ion-beam Lithography: A Feasibility Demonstration for Submicrometer Device Fabrication," *J. Vac. Sci. Technol.* **19**:1166 (1981).

50. D. L. Laird and R. L. Engelstad, "Effects of Mounting Imperfections on an X-ray Lithography Mask," *Electron-Beam, X-ray, and Ion-Beam Submicrometer Lithographies, SPIE* **1671**:366 (1992).

51. K. H. Zaininger and A. G. Holmes-Siedle, "A Survey of Radiation Effects in Metal-Insulator-Semiconductor Devices," *RCA Rev.* **28**:208 (1967).

52. T. P. Ma and P. V. Dressendorfer, eds., *Ionizing Radiation Effects in MOS Devices and Circuits*, Wiley Interscience, NY (1989).

53. F. B. McLean, H. E. Boesch, Jr., and T. R. Oldham, "Electron-Hole Generation and Trapping in SiO_2," in *Ionizing Radiation Effects in MOS Devices and Circuits*, T. P. Ma and P. V. Dressendorfer, eds., Wiley Interscience, New York (1989).

54. C. R. Viswanathan and J. Maserjian, "Model for the Thickness Dependence of Radiation Charging in MOS Structures," *IEEE Trans. Nucl. Sci.* **NS-23**:1540 (1976).

55. N. S. Saks, M. G. Ancona, and J. A. Modolo, "Radiation Effects in MOS Capacitors in Very Thin Oxides at 80 K," *IEEE Trans. Nucl. Sci.* **NS-31**:1249 (1984).

56. J. M. Aitken, "1 μm MOSFET VLSI Technology: Part III—Radiation Effects," *IEEE J. Solid State Circ.* **SC-14**:294 (1979).

57. M. Shimaya, N. Shiono, O. Nakajima, C. Hashimoto, and Y. Sakakibara, "Electron-beam Induced Damage in Poly-Si Gate MOS Structures and Its Effect on Long Term Stability," *J. Electrochem. Soc.* **130**:945 (1983).

58. K. H. Lee, S. A. Campbell, R. Nachman, M. Reilly, and F. Cerrina, "X-ray Damage in Low Temperature Ultrathin Silicon Dioxide," *Appl. Phys. Lett.* **61**:1635 (1992).

59. S. A. Campbell, K. H. Lee, H. H. Li, and F. Cerrina, "Charge Trapping and Device Degradation Induced by X-ray Irradiation in Metal Oxide Semiconductor Field Effect Transistors," *Appl. Phys. Lett.* **63**:1646 (1993).

60. C. C.-H. Hsu, L. K. Wang, D. Zicherman, and A. Acovic, "Effect of Hydrogen Annealing on Hot-Carrier Instability of X-ray Irradiated CMOS Devices," *J. Electron. Mat.* **21**:769 (1992).

Chapter 10

Vacuum Science and Plasmas

Until now most of the unit processes that have been discussed are run at atmospheric pressure. The microelectronic processes that will be covered in the next four chapters however, are run in a vacuum chamber. The first half of this chapter will review vacuum science and technology. It will first discuss the fundamental physics of molecules and atoms in a vacuum chamber. It will then go through some of the basic equipment used to produce, contain, and measure vacuums. Further information can be obtained in several references [1]–[4].

The second half of the chapter is devoted to the physics and technology of plasmas or glow discharges. This type of system is used for the physical and chemical deposition of thin films and for etching. Subsequent chapters will discuss the details of each of these processes, but this chapter will lay the foundations for understanding plasma processes. Finally, a phenomenon will be introduced that occurs in a wide range of processes, including e-beam and x-ray lithography, ion implantation, sputtering, reactive ion etching, and e-beam evaporation: radiation generated by charged particle acceleration. In most cases, the production of radiation is an undesirable side effect of the transport of charged species during the process, but it will be useful to understand their origin.

10.1 The Kinetic Theory of Gasses°

One of the topics that will arise through the next section of the book is the behavior of gasses in a vacuum chamber. In order to appreciate gas for the behavior phase reactions, gaseous beams, heat flow, and surface bombardment, we need to develop a model of gas molecules. One of the most common, the kinetic theory of gasses, treats gaseous molecules as hard spheres. Then the probability distribution of velocities is given by the Maxwell speed distribution (Figure 10.1). For a simple monotonic gas the probability of a molecule having a certain speed is given by

$$P(v) = 4\pi \left[\frac{m}{2\pi kT} \right]^{3/2} v^2 e^{-mv^2/2kT} \tag{10.1}$$

where m is the mass of the molecule, k is Boltzmann's constant, v is the magnitude of the velocity, and T is the temperature in kelvin. The speed, or average magnitude of the velocity, is given by

$$|v| \equiv \bar{c} = \int_0^\infty v P(v) dv = \sqrt{\frac{8kT}{\pi m}} \tag{10.2}$$

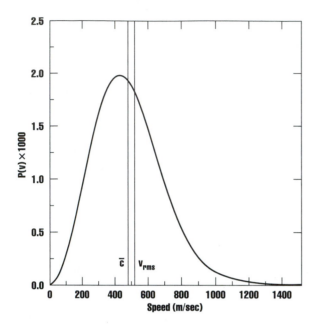

Figure 10-1 A Maxwellian speed distribution of particles. $P(v)$ is the probability that a particular particle will have the indicated magnitude of velocity.

The average component of the velocity in any direction is given by

$$\bar{v}_x = \bar{v}_y = \bar{v}_z = \sqrt{\frac{2kT}{\pi m}} \tag{10.3}$$

and the root mean square velocity is given by

$$v_{rms} = \sqrt{\frac{3kT}{m}} \tag{10.4}$$

The thermal velocity is often large. For example, molecular nitrogen at room temperature has an average x-component of the thermal velocity of about 240 m/sec, or 530 mi/hr. The direction of the thermal velocity is random, and therefore if there are no externally applied forces the average velocity is zero. If a small pressure gradient is imposed across the gas there will be a net macroscopic flow from high pressure to low pressure. On a microscopic level this induced flow velocity vector is added onto the much larger thermal velocity. Individual atoms may be traveling against the pressure gradient at any instant of time, but on average the flow will move from high pressure to low pressure.

One of the primary mechanisms by which gasses at atmospheric pressure change velocity is gas phase collisions. Consider a molecule of diameter d moving randomly in the gas. If another molecule of the same type is within a distance d of the projected path of the first atom, a collision occurs. The molecule then has a collision cross section of πd^2, which in the simple hard sphere approximation, is four times the actual cross section of a single molecule. The probability that a collision will occur over a distance l is given by

$$P = l\pi d^2 n \tag{10.5}$$

where n is the number of gas molecules per unit volume. For simple diatomic molecules like N_2 and O_2, d is commonly taken as 3 Å. If we set $P \approx 1$, the average distance between collisions, commonly called the mean free path λ, can be defined as

$$\lambda \approx \frac{1}{\pi d^2 n} \tag{10.6}$$

A more rigorous statistical treatment will show that

$$\lambda = \frac{1}{\sqrt{2}\pi d^2 n} \tag{10.7}$$

Table 10.1   Various formulas that can be derived from the kinetic theory of gasses

Parameter	Symbol	Equation
Diffusivity	D	$\bar{c}\lambda/3$
Viscosity	η	$Mn\bar{c}\lambda/3$
Thermal conductivity	K	$\overline{C_v}\bar{c}\lambda/3$

Since n is not commonly known, it must be calculated from a macroscopic equation of state. The ideal gas law

$$n = \frac{P}{kT}$$

(10.8)

where P is the pressure of the chamber, is commonly used. Combining equations,

$$\lambda = \frac{kT}{\sqrt{2}\pi d^2 P}$$

(10.9)

A host of useful equations can be derived from this simple model. Table 10.1 lists a few gas characteristics as derived from kinetic theory. In this table, $\overline{C_v}$ is the heat capacity per unit volume of the gas species and $\bar{c}$ is the average speed as defined in Equation 10.2. These formulas are only applicable when $\lambda \ll L$, where L is a characteristic length of the chamber (for example, the chamber diameter). This is called the viscous flow regime. We will for the time being confine our remarks to this pressure regime.

Finally, a topic must be introduced that may seem a bit abstract, but ultimately will prove quite useful. Recall from Chapter 3 that a flux density J_n can be defined as the net flux of molecules through a unit area per unit time. J_n can also be thought of as the number of molecules that strike a surface of unit area per unit time. It is given by

$$J_n = \frac{n\bar{v}_x}{2} = \sqrt{\frac{n^2 kT}{2\pi m}} = \sqrt{\frac{P^2}{2\pi kTm}}$$

(10.10)

Table 10.2   Conversion factors for various pressure units

Units	Conversion Factor
Standard atmospheres (atm)	1.333×10^{-3}
Pounds per square inch (PSI)	1.933×10^{-2}
Torr or mm of Hg (torr or mm)	1.000
Pascal or Nt/m^2 (Pa)	133.3
Microns of Hg (μm)	1.000×10^3
Millitorr (mtorr)	1.000×10^3

To change from torr to the desired unit multiply by the conversion factor. To change to torr from the listed unit divide by the conversion factor.

Once again this can be a very large number. For example, J_n for 1 atm of nitrogen at room temperature is about 6×10^{29} cm^{-2} sec^{-1}.

Before finishing this section, a word is required about units of pressure. Table 10.2 lists some of the most common units and their meanings. The unit of torr (the pressure that would raise a column of mercury 1 mm) has been commonly used in describing vacuum equipment. We have adopted this convention in this book. However, it is not an MKS unit. To use the equations developed in this chapter you must convert to a consistent unit set.

10.2 Gas Flow and Conductance

This section will describe the equations used to calculate pumping speeds and gas flows. In measuring the flow of a simple liquid like water, a common unit might be the volumetric flow rate, say in gallons per hour. While volumetric flow rate is sometimes used in describing gas flows, particularly in describing pumping speed, the problem of using the gas flow in this manner is that gasses are much more compressible than liquids. There are two primary standards that are more commonly used. The first, commonly used in specifying vacuum systems, is the throughput. The mass of a gas in some volume V is

$$G = \rho V \tag{10.11}$$

where in this case ρ is the mass density ($m \times n$). Then the mass flow rate is

$$q_m = \frac{dG}{dt} \tag{10.12}$$

The throughput of a gas Q, which has units of pressure-volume/time, is given by

$$Q = q_m \frac{P}{n} \tag{10.13}$$

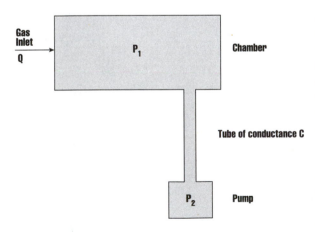

Gas flows are often measured in terms of a standard volume, that is, a volume that an equivalent amount of gas would occupy at 0°C and 1 atm of pressure. For example, one standard liter is the amount of gas that would occupy a liter at 1 atm at 273 K. Since one mole of gas occupies 22.4 l at standard conditions, one standard liter is 1/22.4 moles. A flow of gas then, is measured in standard liters or standard cubic centimeters per minute.

Now consider a simple vacuum system shown in Figure 10.2. Gas flows through the chamber, which is assumed to be at a uniform pressure P_1. A tube connects the chamber to a pump and the inlet of the pump is at a pressure P_2. The conductance C of a vacuum component is given by

Figure 10-2 A simple vacuum system showing a uniform pressure chamber with inlet flow Q, a vacuum pump, and a tube of conductance C.

$$C = \frac{Q}{P_1 - P_2} \tag{10.14}$$

Butterfly valve   **Venetian blind valve**

Figure 10-3   Variable conductance valves used in small and large diameter vacuum lines.

The conductance can be calculated for a variety of geometries or they can be found in tables. Like electrical conductances, conductances in parallel simply add, while conductances in series add as inverses:

$$\frac{1}{C_{total}} = \frac{1}{C_1} + \frac{1}{C_2} + \frac{1}{C_3} + \dots \qquad (10.15)$$

While detailed conductance calculations are beyond the scope of this book, we note in passing that the conductance of a tube of diameter D and length L in the viscous flow regime is

$$C = 0.18 \text{ torr}^{-1} \text{ sec}^{-1} \frac{D^4}{L} P_{av} \qquad (10.16)$$

where P_{av} is the average of P_1 and P_2. Elbows, bends, and narrow sections all reduce the conductance of a system. Flowing a large amount of gas through a vacuum system while keeping the chamber pressure close to the pump pressure requires a vacuum system with a very large conductance. Such a system must therefore be designed with large diameter tubes, and the pump must be placed close to the chamber.

Often it is desired to control the pressure in a vacuum chamber during a process. The volumetric flow rate of a typical pump cannot be independently controlled. Pumps produce a certain vacuum at a given flow. The pressure in the chamber can be set by adjusting the flow rate of the gas in the chamber, but this variable is normally reserved for optimizing other process parameters such as uniformity. The pressure in a vacuum chamber can be easily controlled however, by inserting a variable conductance valve (Figure 10.3) into the pumping line. This can be done by adding a simple vane that can be rotated to partially block the tube, or in large diameter vacuum plumbing, a venetian blind arrangement can be used. A pressure monitor attached to the chamber along with a feedback loop can be used to maintain the pressure in the chamber for a broad range of pumping rates and inlet flows.

Pumps are usually specified in terms of pumping speed, S_p:

$$S_p = \frac{Q}{P_p} = \frac{dV_p}{dt} \qquad (10.17)$$

where P_p is the inlet pump pressure. For example, a pump rated for 1000 liters per minute (slm), will pump 1000 slm at 1 atm inlet pressure. On the other hand, if the inlet pressure is 0.1 atm and the pumping speed at this pressure remains 1000 liters per minute, the maximum gas flow that this same pump could accept is 100 slm. Furthermore, the pumping speed generally is not a constant. It depends on the gas being pumped and the inlet pressure. The next section will discuss the pressure ranges commonly used for microelectronic fabrication and the types of pumps used for these systems.

10.3 Pressure Ranges and Vacuum Pumps

Two pressure ranges corresponding to the mean free path of the gas molecules have been previously identified. The division between viscous flow and molecular flow occurs at about 1 mtorr. The equations for the behavior of the gas in these two ranges are entirely different. This division is based on the physics of the molecule-molecule and molecule-wall interactions. As a practical matter, vacuum regions are also often defined by the technology required to achieve them. The divisions are rather flexible, but let us approximately define them as:

Rough vacuum	0.1 torr–760 torr	10 Pa–10^5 Pa
Medium vacuum	10^{-4} torr–10^{-1} torr	10^{-2} Pa–10 Pa
High vacuum	10^{-8} torr–10^{-4} torr	10^{-6} Pa–10^{-10} Pa
Ultrahigh vacuum	$<10^{-8}$ torr	$<10^{-10}$ Pa

Most of the processing equipment used in semiconductor fabrication operate in the rough or medium vacuum regime. To ensure a pure chamber, however, they are often pumped into the high or ultrahigh vacuum regime before introducing the process gasses. For that reason, we will also discuss the production of high vacuum. The production of ultrahigh vacuum is a difficult task that requires a considerable art. In microelectronics it is used primarily for molecular beam epitaxy (Chapter 14).

Rough vacuum pumps all involve the positive displacement of gas through the mechanical movement of a piston, vane, plunger, or diaphragm. All of these pumps involve three steps: capture of a volume of gas, compression of the captured volume, and gas expulsion. The simplest conceptual picture of such a pump is a piston pump (Figure 10.4). The gas to be pumped is drawn into the cylinder through a valve as the piston is drawn back into the cylinder. During the next part of the cycle both valves are closed and the gas is compressed. Near the end of the stroke the second valve is opened and the gas is expelled to the higher pressure region. Often these valves open automatically in response to a pressure difference. If an ideal gas is used, the pressure differential is just the ratio of the fully expanded to fully compressed volumes. If, for example, the exhaust pressure is 1 atm and

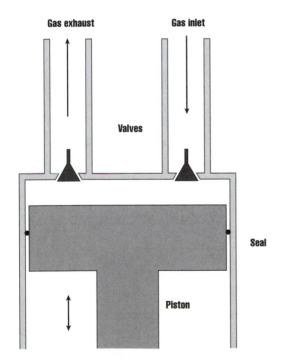

Figure 10-4   A schematic of a single stage two valve piston pump.

Figure 10-5   One of the most common types of pumps for microelectronic processing is the rotary vane vacuum pump.

the compression ratio is 100:1, the lowest pressure that can be achieved in this simple pump is 0.01 atm (7.6 torr). These steps may be run in a series of stages to obtain higher pressure differentials between inlet and outlet.

This simple piston pump is not widely used for microelectronic processing. Instead, a very common pump for rough and medium vacuums is the rotary vane system shown in Figure 10.5. A metal cylinder attached to an electric motor is rotated in a cylindrical chamber about an axis that is displaced from the chamber center. Spring loaded vanes slide along the wall of the chamber, sealing off different areas of the pump. Oil is used to seal the vane and as a lubricant to help with this sliding action. The oil also helps to cool the pump, dissipating the heat generated by friction and by the compression of the gas. As with the piston pump, the rotary vane system works by compressing the gas as the pump rotates. The circular motion of the pump eliminates the need for crankshafts that translate the rotational motion of the motor into the up and down motion required for the piston pump. Furthermore, as the vane sweeps past the outlet, the compressed volume becomes very small. These pumps are commonly used in both single and two stage models, with a variety of throughputs. The ultimate vacuum of the single stage rotary vane pump is about 20 mtorr, while a two stage pump can produce a vacuum of less than 1 mtorr.

One of the problems associated with compression type pumps is the potential for the condensation of vapors. As the gas is compressed, if the partial pressure of a gaseous vapor exceeds the vapor pressure of the corresponding liquid at the gas temperature, it will begin to condense, forming droplets of liquid. These liquids mix with the pump oil and may lead to corrosion. As a simple example, consider water. The vapor pressure of water at room temperature is about 20 torr. If the inlet gas is compressed by a factor of 10^4, water will condense if the partial pressure of water in the chamber is more than 2 mtorr. The problem is most acute when pumping such corrosive condensibles as Cl_2 and the chlorosilanes. To avoid this problem, a small flow (or bleed) of an inert such as N_2 can be injected into the chamber. The use of these gas ballasts limit the ultimate pressure of the pump. Figure 10.6 shows the pumping characteristics of some typical rotary vane pumps with and without the use of gas ballast. The pumping speed is nearly constant over a broad range of inlet pressures, eventually falling as the inlet pressure approaches the ultimate vacuum. Similarly, the throughput is nearly proportional to the inlet pressure over most of the range.

Example 10.1

Assume that the desired chamber for a particular process is 0.1 torr and that there is no pressure drop between the pump and the chamber. Also assume that a gas flow of 1 slm is needed to obtain suitable process results. Would the D65 pump whose characteristics are shown in Figure 10.6 be acceptable if no ballast is used? What is the maximum flow that could be used with this pump? According to Figure 10.6, pumping speed of the D65 pump at 0.1 torr is about 40 cfm (1150 slm). This would allow a maximum flow rate of about 1150 * 0.1/760 or 0.15 slm. Therefore this pump is not adequate for this application.

There are two ways to obtain higher throughputs: Use a larger pump or add a second pump between the rotary vane pump and the chamber. In many cases, the latter is the more economical choice. Pumps that have been designed for this application are called *blowers*. The most common blower in microelectronic fabrication is the Roots blower. As shown in Figure 10.7, the Roots blower is a

Pumping speed curves

Figure 10-6   The pumping characteristics of Leybold D65® and D40® two stage rotary vane vacuum pump with and without gas ballasting *(courtesy Leybold)*.

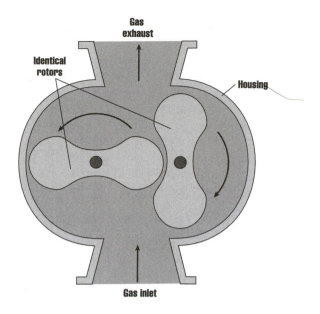

Figure 10-7   Schematic diagram of a Roots blower.

positive displacement pump that consists of two figure eight rotors that revolve at high speed in opposite directions and have very small clearances (<0.1 mm) with respect to each other and the pump walls. There is no mechanical seal between the surfaces. As a result, the compression ratio of these pumps is only about 30:1 (Figure 10.8), but due to their rapid rotation the pumping speed is large. Allowances for thermal expansion to prevent scoring that reduces pump life preclude higher compression ratios. Two stage blowers can also be used for larger compression ratios.

The inlet pressure of the roughing pump can be raised by placing a Roots blower before the rotary vane pump. The increase in the pumping speed of a Roots blower with a zero flow compression ratio of k_o and pumping speed S_R and a rotary vane pump with pumping speed S_{RV} can be calculated using the expression

$$S_{eff} = S_{RB}S_{RV}\frac{k_o}{S_{RB} + k_o S_{RV}} \quad \textbf{(10.18)}$$

Thus, the combination of a 40 cfm rotary vane pump and a 200 cfm blower with a compression ratio of 20 will provide an effective pumping speed of 160 cfm. Taking the previous example where the chamber pressure was 0.1 torr, the inlet pressure of the roughing pump after the blower now rises to 20×0.1 or 2 torr. According to Figure 10.6, if the blower is backed with a D65 rotary vane pump, this increases the maximum permissible chamber flow rate from 0.1 slm to about 3 slm.

Finally, the use of an inert purge or a filter before the roughing pump is also common practice. As the pump operates, the pump oil gets hot. The vapor pressure of pump oil at this temperature is quite high. Unless there is a flow of gas into the pump, the oil vapor flows from the pump into the chamber in a process known as backstreaming. A common example is the use of a roughing pump to evacuate a sealed chamber. As the pressure in the chamber falls, so does the flow of gasses into the pump. Eventually the flow drops low enough to allow backstreaming. The pump oil vapor condenses on and contaminates the cold chamber walls. An inert purge flow in front of the pump will reduce backstreaming, but will also increase the apparent ultimate pressure of the pump. Chemical traps are also available to solve this problem.

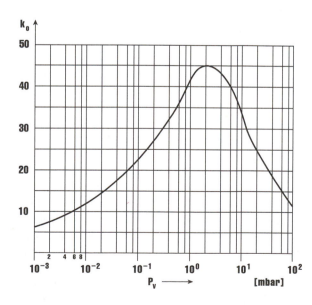

Figure 10-8   Compression ratio as a function of inlet pressure for a typical Roots blower (*courtesy Leybold*).

Figure 10-9 Cutaway view of a diffusion pump *(courtesy Varian).*

Figure 10-10 Cutaway view of a small turbomolecular pump. Notice the change in the blade angle and shape going from the high vacuum (top) to low vacuum (bottom) ends *(courtesy Varian).*

High vacuum pumps for microelectronic fabrication fall into two categories: those that pump the gas by transferring momentum to gaseous molecules and those that trap gaseous molecules. Of these, the former are preferred when pumping corrosive and toxic gasses or when pumping high flows of gasses, and the latter are preferred by pumping small flows of inert gasses or when only using the high vacuum pump for pumping down the chamber before processing. Valves are almost always used to isolate high vacuum trapping pumps, allowing the chamber to be pumped by medium and rough vacuum pumps when first pumping down from an atmosphere and once the process gas begins to flow.

The two most popular types of momentum transfer pumps are diffusion pumps and turbomolecular pumps. Diffusion pumps (Figure 10.9) are extremely simple and robust. They operate by heating an oil at the bottom of the pump. The pump oil vapors rise through the center stack and are ejected through vents at very high speeds. They then strike cooled walls at the top of the pump, condense, and run down the walls. Gasses are primarily pumped by momentum transfer between the vapor stream and the gas molecules. Gas molecules may also be transported by dissolving in the vapor droplets. As the oil is heated again at the bottom of the pump, the gas is given off and removed through a roughing pump. It is possible to obtain compression ratios of 10^8 in these pumps.

Diffusion pumps have high pumping speeds as long as the inlet pressure is in the molecular flow regime. Most diffusion pumps may not be exposed to an atmosphere, or a chemical reaction called cracking may occur between the hot pump oil and the oxygen in the air. Even more serious are concerns that some of the pump oil vapor will not condense in the pump and will backstream into the vacuum system leading to contamination. Diffusion pumped systems may use baffles or cooled traps to remove most of the backstreamed pump oil. Because of this concern, diffusion pumps are generally not desirable when extremely high purity is required.

A turbomolecular (or turbo) pump as shown in Figure 10.10 has a large number of stages in series. Each stage consists of a fan blade that rotates at extremely high speed ($>20,000$ rpm) and a stationary set of blades called a *stator*. The spacing between the stator and rotor is of order 1 mm. Each stage may have a modest compression ratio, but because of the large number of stages the total pump has compression ratios as large as 10^9. The high pressure side of the pump must be attached to a roughing pump since the outlet pressure must be maintained at low pressure. Since the momentum transfer depends on the mass of the gaseous molecule, the compression ratio depends

Figure 10-11   Pumping speed and compression ratio of a typical turbo pump as a function of inlet pressure for several representative gasses *(courtesy Leybold).*

Figure 10-12   Schematic view of a typical cryopump.

strongly on gas being pumped. A typical pump that has a compression ratio of 10^9 for N_2 will have a compression ratio of less than 10^3 for H_2 (Figure 10.11). As a result, turbopumped chambers often have high concentrations of hydrogen. Because pump oil vapors have high mass, however, turbopumped systems are very clean.

When the ultimate in purity is required, gas entrainment pumps are used. One of the most common in use in microelectronic processing is the cryopump. As shown in Figure 10.12, a cryopump consists of a closed cycle refrigerator. The cold head of the refrigerator, which is maintained at about 20 K, is contained in a pump body that attaches to the vacuum system. The cold head is generally constructed of copper or silver and may have coatings of active charcoal to further trap gasses. A radiation shield is often used to minimize the thermal load on the head. All of the chamber gasses except He will condense on the cold head, although gasses with high boiling points are pumped more efficiently than those with low boiling points. Eventually, the cold head becomes saturated with adsorbed gasses whose low thermal conductivity prevent further adsorption. At this point, the cryopump must be isolated from the chamber, heated and pumped to desorb the collected gasses, and then cooled down to return it to service. Cryopumps may not be turned on until the chamber is at a rough vacuum, or their pumping capacity will be greatly diminished.

Another type of gas entrainment pump used for high vacuum is the sorption pump. Sorption pumps operate by chemically or physically absorbing the gas molecules. Early sorption pumps used carbon, but the material in modern pumps is commonly some form of activated Al_2O_3. These alumino-silicates zeolites are permeated by internal cavities interconnected by uniform diameter pores. The pore size varies with the zeolite that is chosen, but is typically between 3 and 15 Å. This type of pump is also called a *molecular sieve* due to the sorbate material. Sorption pumps are often operated by cooling them to liquid nitrogen temperatures and allowing the gas molecules to physically adsorb on the pore walls. The pumping efficiency depends on the size of the gas molecules relative to the size of the pores. N_2, CO_2, H_2O, and heavy hydrocarbons are pumped well by most sorption pumps, but light noble gasses such as He are not. As with cryopumps, sorption pumps can be recycled by heating them while under a vacuum.

The third type of entrainment pump uses replenishable coatings of reactive metal to remove gaseous molecules. In titanium sublimation pumps a titanium containing filament is heated to deposit a thin film of highly reactive metal on the inside surface of the pump. As the inner surfaces become coated with gas molecules, the pumping speed falls. To regenerate the pump, the filament is reheated and a fresh coating of Ti is put down. Sputter ion pumps operate in a very similar manner, but the thin reactive layers are replenished by sputtering (see Chapter 12) instead of sublimation. The inner surface of sputter ion pumps may be shaped to maximize the area being coated. As with sublimation pumps, the most common material deposited to pump the gas is titanium. Both sublimation and sputter ion pumps are extremely clean, robust, and simple to operate.

10.4 Vacuum Seals and Pressure Measurement

In the rough and medium vacuum region elastomers, commonly in the form of O-rings, are used to seal the vacuum chamber. O-rings have the advantages of low cost and ease of reuse. A common application is the wafer introduction and exit doors of the vacuum system. The choice of the elastomer depends primarily on the chemistry of the chamber and on the temperature that the elastomer must withstand. One of the most popular elastomer materials is viton, a vulcanized rubber. Some sort of mechanical support must be built into the joint to prevent the O-ring from being pulled into the chamber by the vacuum. Common practice is to recess the O-ring groove to 70% of the diameter of the O-ring and to make the groove wide enough to prevent the compressed O-ring from filling the groove; 140% of the diameter of the O-ring is often used. This ensures a metal to metal contact between the joining surfaces in the sealed condition, reducing any vibration at the seal and thereby improving the reliability. Two common types of O-ring sealed quick connect flanges are shown in Figure 10.13. Because of reliability concerns regarding O-ring seals, highly toxic gasses may be contained by a double O-ring arrangement where the space between the O-rings is sampled for the presence of the toxic.

Elastomer seals begin to leak at pressures below 10^{-7} torr. For high and ultrahigh vacuum systems, metal to metal seals must be used. In this type of seal, the sealing material must be plastically deformed during the sealing process. Early high vacuum systems used a malleable wire such as gold in much the same manner as an elastomer O-ring. One of the most common high vacuum seals in current use, the Conflat® flange, is shown in Figure 10.13. A flat, soft, 2 mm thick metal ring is pinched between two stainless steel knife edges. The most common gasket material is copper or a copper/silver alloy. A uniform closing force must be applied to ensure a tight seal. This is usually provided by closely spaced bolts around the edge of the flange. To maintain high reliability, the

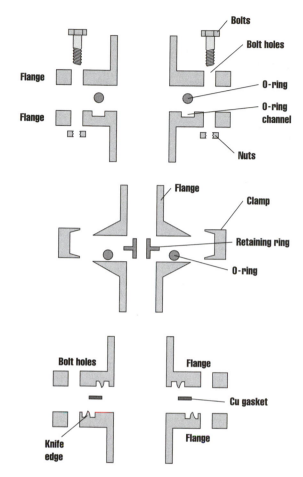

Figure 10-13 Two types of O-ring seals for medium vacuums and the Conflat® flange used for sealing high vacuum systems.

sealing gaskets must be replaced after every use and the knife edges must be protected against damage when the surface is exposed.

Pressure in a vacuum system can be measured with a number of different gauges, the most common of which in microelectronic applications are capacitance manometers, thermal conductivity gauges, and ionization gauges. Capacitance manometers fall into a mechanical class of gauges that depend on a pressure difference, between the chamber to be measured and a reference volume, to cause a mechanical deflection. Capacitance manometers detect the movement of a thin metal diaphragm. While these gauges can be used to detect pressures as low as 1 mtorr, they are most often used at pressures above 1 torr. They are noncontaminating, and since they measure the pressure of the gas directly, they are independent of the gas being measured.

A family of popular vacuum gauges operate by measuring the thermal conductivity of the gas and from that, inferring a pressure. This type of gauge dates back to 1906 [5],[6]. They operate by passing a current through a wire and measuring the temperature of the wire. Wire operating temperatures are kept low (a few hundred degrees centigrade) to prolong wire life, maximize sensitivity, and to ensure that most of the heat transfer is thermal conduction through the gas rather than radiative heat loss. In Pirani gauges, the resistance of the wire is carefully measured by a Wheatstone bridge. The power to the filament is adjusted to balance the bridge and therefore maintain a constant wire temperature. In thermocouple gauges, a thermocouple is spot welded to the heater wire to measure the wire temperature. Thermocouples consist of two dissimilar metals and they produce a small temperature dependent voltage. The measurement of this voltage is easier than balancing a Wheatstone bridge, but the technique has less sensitivity than Pirani gauges. Either type of gauge has a resolution of about 1 mtorr. Care must be taken in using this type of gauge however, in that the thermal conductivity depends on the gas in the chamber.

Neither mechanical deflection nor thermal conductivity gauges are useful in measuring pressures much below 1 mtorr. The types of gauges most commonly used in high and ultrahigh vacuum applications are ionization gauges. They operate by using an electron stream to ionize the gas in the gauge and an electrical field to collect the ions. The ion current produced this way is a function of the pressure in the chamber. The electron stream can be created by heating a filament (hot cathode) or by a plasma (cold cathode). Of these, hot cathode gauges are much simpler to operate, but may present problems in highly corrosive ambients or in ultraclean chambers where filament outgassing is a concern. Furthermore, hot filament gauges usually cannot be operated at pressures above 10^{-4} torr due to short filament life. The ultimate pressure for either type of gauge is ultimately limited by the ability to measure very small ion currents, but some types of ionization gauges can detect pressures as low as 10^{-12} torr.

10.5 The DC Glow Discharge°

Many of the low pressure processes that are described in the next two chapters will involve the use of a glow discharge or plasma. These include various etch processes, chemical vapor deposition, and sputtering. Even the bulbs in many lithography tools are plasma sources. The term "glow discharge" refers to the light given off by the plasma. Plasmas can be used in place of high temperature to crack molecules and so drive some reaction chemistry. It may also be used to create and accelerate ions. In some applications both effects may be important.

A plasma is a partially ionized gas. Assume that the inlet flow of gas contains molecule AB made from atoms A and B. The types of processes that may occur in the glow discharge can be characterized as [7]

Dissociation	$e^* + AB \rightleftarrows A + B + e$
Atomic ionization	$e^* + A \rightleftarrows A^+ + e + e$
Molecular ionization	$e^* + AB \rightleftarrows AB^+ + e + e$
Atomic excitation	$e^* + A \rightleftarrows A^* + e$
Molecular excitation	$e^* + AB \rightleftarrows AB^* + e$

where the superscript "*" refers to a species whose energy is much larger than the ground state. Dissociated atoms or molecular fragments are called *radicals*. Radicals have an incomplete bonding state and are extremely reactive. In Ar or other elemental plasmas, there are no radicals. Ions are charged atoms or molecules such as A^+ and AB^+. They may have more than one positive charge or may even be negatively charged. In typical plasmas, the radicals may constitute 1% of the total plasma; charged species may be less than 0.01%. Although not specifically represented, energetic ionized species are possible as well.

Figure 10.14 shows a simple plasma reactor. Two parallel plates are contained in a vacuum system and attached to a dc supply through vacuum power feedthroughs. A typical pressure for a plasma process is 1 torr. A high voltage source (often a charged capacitor) is connected momentarily to the circuit to start the plasma. The inductor protects the dc supply from the high voltage arc. At 1 torr the required voltage for a 10 cm electrode spacing is about 800 V, while about 500 V is required for a 5 cm spacing.

Until the arc is struck, the gas will not conduct current since it acts as an insulator. Consider what happens as the plasma forms. If the voltage is high enough, the field in the reactor will exceed the breakdown field of the gas, and a high voltage arc will flash between the two electrodes. This arc will create a large number of ions and free electrons. Because of the electric field in the chamber, the electrons will be accelerated toward the positively charged anode, and the ions will be accelerated toward the negatively charged cathode. Due to their small mass, the electrons will be accelerated much more rapidly than the slowly moving ions. The ions travel across the tube and eventually strike the cathode. When they do so, they release a cloud of secondary electrons from the material in the cathode. These electrons are accelerated back toward the anode. If the voltage across the electrodes is large enough, when these high energy electrons collide inelastically with a neutral atom they create more ions. This process of secondary electron release and ion creation sustains the plasma.

Even in this simple 1-D case, the solution for the movement of ions in the plasma is beyond the scope of this text, since the differential equations describing the field and charge distribution are coupled and must be solved simultaneously. Figure 10.15 shows the positive and negative charge densities and the electric field strength as functions of position in the plasma. In the bulk of the plasma, the density of ions and electrons are equal. Since the electrons are rapidly accelerated from the cathode, the electron density near the cathode is much less than the ion density, and the region has a net positive charge. Near the edge of this positively charged region, the electrons have gained

Figure 10-14 A simple parallel plate plasma reactor.

Figure 10-15 Positive and negative charge densities and electric field as a function of position in the plasma.

enough energy to create ions, and so the ion density increases with distance into the plasma. This positive charge shields the remainder of the plasma from the cathode, reducing the field and therefore the ionization rate. As a result, the ion density peaks and then falls to a constant value through the remainder of the plasma.

When a moderate energy electron scatters inelastically off of a neutral atom, it may excite a core level electron to a high energy state. This state is very short lived (of order 10^{-11} sec), and so the excited atom or molecule does not move an appreciable distance between excitation and decay. When the electron decays back to its ground state, it gives off the energy in the form of visible radiation. The light from the glow discharge arises because of this optical emission process. In order for this process to occur, a high concentration of moderate energy electrons is required. Electrons with energies greater than 15 eV primarily ionize the gas molecules rather than excite them.

The requirement for moderate energy electrons precludes optical emission near each of the electrodes. These regions are called *dark spaces*. The region just above the cathode where most of the electrons have very low energies is called the *Crooke's dark space* (Figure 10.16). The anode is a sink for electrons and so the electron density just above the anode is too small for appreciable emission. This dark space is called the *anode dark space*. Finally, there is a region above the cathode where the electrons have been accelerated to very high energies, leading to ionization, and there are few electrons with energies appropriate for emission. This region is called the *Faraday dark space*.

One of the most important aspects of the glow

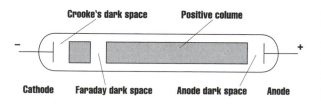

Figure 10-16 Structure of a dc plasma.

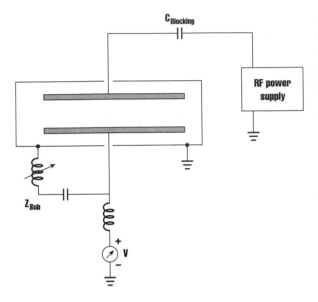

Figure 10-17   Schematic of an RF plasma system.

discharge for fabrication purposes is the large electric field in the Crooke's dark space. Ions that drift and diffuse to the edge of this region are accelerated rapidly toward the cathode. If the cathode is covered with wafers or other materials of interest, we can use this ion bombardment to drive various processes. The width of the dark space depends on the chamber pressure. At low pressures, the mean free path of electrons increases, and so the width of the dark space increases. By controlling the chamber pressure, one can control the energy with which ions strike the surface. This effect limits dc plasmas to pressures greater than about 1 mtorr.

10.6 RF Discharges

In many cases, the material on one or more of the electrodes is insulating. Take as an example an application that will be covered in the next chapter, the etching of silicon dioxide. In this process, wafers with an upper layer of SiO_2 are first patterned lithographically and then placed on a plasma electrode. The exposed materials on the top of the wafers, photoresist and oxide, are both insulators. As ions strike the surface of the wafer and secondary electrons are ejected, these layers become charged. The charge accumulates on the surface and the field is reduced until the plasma is ultimately extinguished. To solve this problem, the plasma can be driven by an ac signal. Sources are in the radio frequency (RF) range, commonly 13.56 MHz, which has been set aside by the FCC for this application. Figure 10.17 shows a typical RF plasma system. A tuning network is used to match the impedance between the plasma and the power source. A blocking capacitor is used to dc isolate the supply from the chamber.

At low frequencies, the plasma follows the excitation, and the width of the dark spaces pulse with the applied signal. When the rate of excitation is greater than 10 kHz, the slow ions in the plasma cannot follow the voltage change. Electrons however, are rapidly accelerated. During alternate half cycles, electrons strike the surface of each electrode, giving both a net negative charge with respect to the plasma. In this case, one dark space exists in the vicinity of each electrode. Figure 10.18 shows the dc voltage as a function of position across the chamber. Superimposed on this dc level is the RF signal. Since the plasma is conductive, the voltage drop across the glow discharge is small. Due to electron depletion however, large dc voltage drops exist between the plasma and the electrodes.

We can label the voltages as

$$V_1 \equiv V_{plasma} - V_{top}$$
$$V_2 \equiv V_{plasma} - V_{bottom} \qquad \textbf{(10.19)}$$

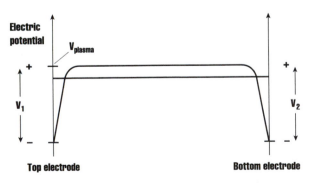

Figure 10-18   Typical plot of dc voltage as a function of position in an RF plasma.

If the electrodes have the same area by symmetry, they must have the same potential difference. By conserving the current through the plasma, it can be shown [8] that for an asymmetric chamber

$$\frac{V_1}{V_2} \approx \left[\frac{A_2}{A_1}\right]^4 \qquad (10.20)$$

where A_1 and A_2 are the areas of the two electrodes. To maximize the voltage difference between the plasma and the lower electrode and therefore the ion bombardment energy on the lower electrode, it is desirable to increase the area of the upper electrode. This can be done by connecting the upper electrode to the walls of the chamber. For safety reasons this point is typically set to dc ground. Now the effective area of the upper electrode is the sum of the areas of the actual electrode and the chamber walls. In practice it is found that the exponent in Equation 10.20 is not a constant, but decreases with increasing area ratio [9]. It is difficult to obtain a voltage ratio much larger than 10:1. It is also possible however, to adjust the dc bias voltage on the grounded chamber and the lower electrode by changing the value of the inductor in the LC circuit shown in Figure 10.17 [10].

10.7 Magnetically Enhanced and ECR Plasmas

As already mentioned, the concentration of ions and radicals in a standard plasma are a small fraction of the total gas. Since it is the ion bombardment and/or chemical reactions that will be of interest for driving the process, it would be highly desirable to find ways to increase their relative concentrations. Two methods that are in widespread use will be discussed here: magnetic confinement and electron cyclotron resonance. If a magnetic field is added to the plasma, the Lorentz force will deflect the motion of the electrons in a direction perpendicular to both the velocity and the magnetic field:

$$\overline{F} = q\overline{v}x\overline{B} \qquad (10.21)$$

If $|v|$ is constant, such a field will induce a circular motion with a radius

$$r = \frac{mv}{qB} \qquad (10.22)$$

Because of their large mass, ions will move through the field with only minor deflections, crossing the discharge long before traveling a full circle, while electrons will follow a helical path of much smaller radius. Effectively, then, the path of the electrons is many times larger in the presence of the field. For a fixed mean free path, the opportunity for impact ionization is much greater with the field "on," and so the ion density and free radical density in such a system will be large. If the magnetic field is configured properly the secondary electrons being accelerated away from the cathode can actually be trapped by the magnetic field and return to the cathode, where the cycle will repeat (Figure 10.19). This geometry is called a magnetron.

A new type of plasma being exploited for micro-

Figure 10-19   In a magnetically confined plasma, electrons ejected from the cathode are confined by the Lorentz force to stay in the cathode dark space.

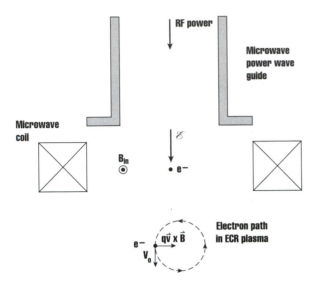

Figure 10-20 In an ECR plasma, an alternating field causes the electrons to move in circular orbits, dramatically increasing the ion density.

electronic processing is electron cyclotron resonance (ECR). This type of plasma uses a perpendicular magnetic field along with an alternating electric field (Figure 10.20). The electric field increases the magnitude of the electron's velocity, and the magnetic field changes the direction of the velocity vector. Consider, for example, an electron on the surface of this page, which is initially at rest. An electric field of $E_o \cos \omega t$, which is directed left to right at time $t = 0$, is switched on. The field will cause the electron to accelerate to the left, then slow down and reverse direction, as it oscillates back and forth. Now switch on a magnetic field that is directed into the page. As the electron moves to the left, the magnetic field will deflect the electron toward the top of the page. If the frequency of oscillation is set to the electron cyclotron resonance frequency

$$\omega = \omega_o = \frac{eB}{m} \qquad (10.23)$$

The amount of deflection caused by the magnetic field is just enough to turn the electron by 180° as the direction of the field changes sign. As a result the electron will move in a circle. This condition is called *electron resonance*.

The electron in resonance gains energy throughout the circle if the mean free path is much larger than the circumference of the circle. This effect increases the power coupling efficiency and reduces the breakdown field [11] to as little as 10 V/cm. As with other magnetically enhanced plasmas, the use of an ECR dramatically increases the density of ions and free radicals in the discharge. If the process to be run depends on the ion flux from the plasma, an ECR will improve the rate at which the process occurs. For reasonable fields, the required cyclotron frequency is large. As an example, for a field of 900 Gauss (0.09 Tesla), ω is of order 1.6×10^{10} rad/sec or about 2.5 GHz. An ECR system, therefore, requires the generation of high power microwave signals that are coupled into the reactor. The microwave energy is typically carried by rectangular waveguides. The power may be transmitted into the chamber through a quartz window [12].

10.8 Radiation from Accelerated Charged Particles°

Many of the processes in microelectronic fabrication involve the acceleration and deceleration of charged particles. Several have already been introduced, including ion implantation and x-ray sources. Plasmas open up new classes of processes, including plasma etching, sputtering, and plasma enhanced deposition. It is known that, whenever a charged particle is accelerated, it radiates an electromagnetic field [13]. Microelectronic fabrication must deal with free nonrelativistic charged particles. In MKS units the power radiated by such a charge is given by

$$P = \frac{\mu_o \varepsilon_o q^2}{[4\pi]^2} \sqrt{\frac{\mu_o}{\varepsilon_o}} \, a^2 \sin^2\theta \qquad (10.24)$$

where c is the speed of light (3×10^8 m/sec), a is the acceleration of the charged particle, and θ is the angle between the line joining the particle and observer and the acceleration vector. Notice that the radiated power is a maximum 90° from the acceleration vector. Integrating over all possible angles gives the total radiated power

$$P = \frac{2}{3} \frac{\mu_o \varepsilon_o q^2}{4\pi} \sqrt{\frac{\mu_o}{\varepsilon_o}} \, a^2 = 5.7 \times 10^{-54} \, \frac{\text{W-sec}^4}{m^2} \, a^2 \qquad (10.25)$$

This power radiated by an accelerated charge is sometimes referred to as Bremsstrahlung radiation. The spectral distribution of this radiation for many microelectronic processes of interest reaches well into the deep UV and sometimes into the soft x-ray region. Some process equipment in which this effect is pronounced must use shielding to protect the operator.

Example 10.2

An electron beam is accelerated to 5 keV and directed into a target where the electrons stop in 2 μm. If the electron beam current is 2 mA and relativistic effects can be ignored, (a) find the power and total energy radiated per electron and (b) find the total power per unit area radiated by the process.

From simple classical mechanics, an electron with an energy of 5000 eV is traveling at a speed of

$$v = \sqrt{\frac{2 \times 5000 \times 1.6 \times 10^{-19}}{9.1 \times 10^{-31}}} = 4.2 \times 10^7 \text{ m/sec}$$

Since the electron must stop in 2 μm, the acceleration is 8.8×10^{16} m/sec^2, and the time required to stop an ion is 2.2×10^{-9} sec. Each electron radiates

$$P = 5.7 \times 10^{-54} \text{ W-sec}^4/m^2 [8.8 \times 10^{16} \text{ m/sec}^2]^2 = 4.4 \times 10^{-20} \text{ W}$$

Since each electron radiates for 2.2 ns, the total energy radiated per electron is 9.7×10^{-29} J. Multiplying by the electron bombardment rate gives

$$P_{tot} = 9.7 \times 10^{-29} \text{ J} \times \frac{0.002 \text{ A}}{1.6 \times 10^{-19} \text{ C}} = 1.2 \times 10^{-12} \text{ W}$$

This would appear to be a rather small effect. In fact, for this situation the x-ray generation caused by this type of effect would be negligible compared to x-rays generated by highly excited core level electrons that radiate when they return to the ground state.

10.9 Summary

This chapter discussed some science and technology that will form the basis of many of the subsequent chapters. Vacuums can be well described by kinetic theory that treats gas molecules as hard spheres. Equations have been developed to describe macroscopic quantities in terms of microscopic variables such as velocity distributions. Rough and medium vacuums typically use rotary vane pumps, o-ring seals, and thermal conductivity gauges. When high flow throughputs are required, a Roots blower may be used before the rotary vane pump. For high vacuum systems, either momentum transfer (turbo or diffusion) or entrainment (cryo, sorption, or sublimation pumps) are used. These systems use metal to metal seals, and pressure is measured with ionization gauges. The second half of the chapter described plasmas or glow discharges. The basic dc plasma reactor can only be used when both electrodes are coated with conducting materials. When insulating layers are exposed, an ac source must be used. The voltage difference between the electrode and the plasma depends on the areas of the electrodes. Finally, it was pointed out that charged particles radiate power when they are accelerated or decelerated.

Problems

1. Air pressure decreases as altitude increases. At 18,000 feet msl, the atmospheric pressure is about half of what it is at sea level (i.e., 760 torr). The temperature also decreases by about 70°C. Does this change the speed distribution of the air molecules? Does it change the mean free path? How much does each change?

2. If we take the transition between the molecular flow and molecular beam regimes to occur when the mean free path is 1 cm, at what pressure will this occur? (Assume $d = 3$ Å.)

3. Thermocouple gauges measure the thermal conductivity of a gas to infer a temperature. Derive a relationship between pressure and thermal conductivity assuming a kinetic model for the gas. H_2 and He have very large thermal conductivities. Explain why.

4. To avoid vibration and improve serviceability, vacuum pumps must sometimes be located on a different surface than the process equipment. In some facilities, this is done by drilling holes through the floor and putting the pumps below the clean room on a separate floor. Discuss the drawback of this approach. If the distance between the pump and the system is 3.0 m, the tubing diameter is 5 cm, and the total gas flow is 1.0 slm, what is the minimum chamber pressure that could possibly be achieved?

5. If a D65 pump is used for the application in the previous problem and no gas ballast is used, estimate the pressures at the pump inlet and in the vacuum chamber.

6. Explain why thermal conductivity type gauges will not work in an ultrahigh vacuum.

7. Why are entrainment pumps generally not used when processing with toxic gasses?

8. It is possible to determine the chemical composition of a plasma by measuring the optical emission spectrum (i.e., optical intensity as a function of wavelength). Explain why.

9. Referring to the material discussed in Section 10.7, if a process relies on ion bombardment of the wafers, would you want to put the wafers on the electrode connected to the chamber walls or on the electrode that is isolated from the chamber walls?

10. Would you expect that adding magnetic confinement to a plasma would increase the x-ray emission? Justify your answer.

11. For a certain process the maximum allowable pressure is 0.20 torr. If a D65 rotary vane pump is used and conductance losses in the vacuum system can be ignored, what is the maximum permissible inlet flow in SCCM?

12. Repeat Problem 10.11 assuming that a blower with $k_o = 20$ and $s = 200$ cfm inserted before the rotary vane pump.

References

1. *Vacuum Technology: Its Foundations, Formulae and Tables*, Leybold AG, Export, PA (1992).
2. A. Roth, *Vacuum Sealing Techniques*, American Institute of Physics, New York (1994).
3. T. A. Delchar, *Vacuum Physics and Techniques*, Chapman and Hall, London (1993).
4. M. H. Hablanian, *High Vacuum Techniques: A Practical Guide*, M. Dekkar, New York (1990).
5. M. Pirani, *Deutsche Phys. Ges. Verk.* **8**:686 (1906).
6. W. Voege, *Phys. Z* **7**:498 (1906).
7. B. Chapman, *Glow Discharge Processes*, Wiley, New York (1980).
8. H. R. Koenig and L. I. Maissel, "Application of R. F. Discharges to Sputtering," *IBM J. Res. Devel.* **14**:168 (1970).
9. C. M. Horwitz, "RF Sputtering-Voltage Division Between the Two Electrodes," *J. Vac. Sci. Technol. A* **1**:60 (1983).
10. J. S. Logan, "Control of RF Sputtered Films Properties Through Substrate Tuning," *IBM J. Res. Dev.* **14**:172 (1970).
11. S. C. Brown, "Breakdown in Gases: Alternating and High Frequency Fields," in *Handbuch der Physik* **22**, S. Flugge, ed., Springer-Verlag (1956).
12. S. Matsuo and M. Kiuchi, "Low Temperature Deposition Apparatus Using an Electron Cyclotron Resonance Plasma," in *Proc. Symp. VLSI Sci., Technol.*, Electrochemical Soc., Pennington, NJ (1982), p. 83.
13. Consult any E + M text such as J. D. Jackson, *Classical Electrodynamics*, Wiley, New York (1975) or R. H. Good, Jr. and T. J. Nelson, *Classical Theory of Electric and Magnetic Fields*, Academic Press, New York, 1971.

Chapter 11

Etching

After a photoresist image has been formed on the surface of a wafer, the next process often involves transferring that image into a layer under the resist by etching. The chapter will begin with simple wet chemical etching processes where the wafer is immersed in a solution that reacts with the exposed film to form soluble byproducts. Often the photoresist mask is highly resistant to attack by the etching solution. While still widely used for noncritical processes, wet chemical etching is difficult to control, is prone to high defect levels due to solution particulate contamination, cannot be used for small features, and produces large volumes of chemical waste. The chapter will therefore go on to discuss dry or plasma etch processes.

It is useful to begin the discussion of etching by identifying the appropriate figures of merit. The primary one is the etch rate, which has dimensions of thickness per unit time. A high etch rate is generally desirable in a manufacturing environment. Too high an etch rate however, may render a process difficult to control. Common desired etch rates are hundreds or thousands of angstroms per minute. When a batch of wafers are etched simultaneously, the etch rate can be less than the rate required for a single wafer etch process. Several related figures are equally important. Etch rate uniformity is measured in terms of percent variation of the etch rate. It is quoted across a wafer and from wafer to wafer. Selectivity is the ratio of the etch rates of various materials, such as the photoresist or the underlying layer, referenced to the etch rate of the film being patterned. A particular process may be quoted as having a selectivity of 20 to 1 for polysilicon over oxide, meaning that polysilicon etches 20 times faster than oxide.

Undercut is the lateral extent of the etch under the photoresist mask. It can be quoted in two forms. The first is an undercut distance per side. For example, a particular etching process may be found to produce 0.8 μm lines when the patterned photoresist line is 1.0 μm wide. The process bias is 0.1 μm per side. As shown in Figure 11.1, the etched sidewalls are not always vertical. The amount of undercut therefore depends on how it is measured. Most electrical measurements of the line are sensitive to its cross-sectional area, providing an average undercut. If the etch process attacks the resist pattern, this attack will also produce an etch bias. Some processes are designed to intentionally do this. For now however, assume that the resist mask does not change during the etching.

The second method of describing undercut is to quote the etch anisotropy. Anisotropy is given by

$$A = 1 - \frac{R_L}{R_V}$$

(11.1)

Figure 11-1 Typical isotropic etch process showing the etch bias.

where R_L and R_V are the lateral and vertical etch rates. A process is said to be perfectly anisotropic ($A = 1$) if the lateral etch rate is zero. On the other hand, $A = 0$ would imply that the lateral and vertical etch rates are identical.

All of the preceding metrics can be expressed quantitatively. There are several other metrics that are somewhat more difficult to quantify. The first is the substrate damage. As an example, p-n junctions have been shown to degrade under certain types of plasma etching. The amount of degradation depends not only on the etch process, but on the depth and type of the junction. Finally, the process must be safe, both to the operator and to the environment. Chlorine plasmas in particular, are known to produce hazardous byproducts that must be neutralized before they are exhausted into the air. Many current generation plasma etch processes are based on Freon chemistries that are known to degrade the environment. Replacing these etches is an area of current research.

Etches can be done by physical damage, chemical attack, or some combination of the two. One can define a scale of the etch mechanism that describes the range of etch processes (Figure 11.2). Ion milling, for example, is done with an energetic beam of inert atoms in a very low pressure chamber, such that the mean free path of ions is much longer than the chamber diameter. It is at the extreme of purely physical etching. These processes are characterized by a high degree of anisotropy, but the etch rate is nearly independent of the substrate material. The selectivity of ion beam milling therefore, is close to 1. Wet etching is at the other extreme of the spectrum with no physical attack. This type of process is usually characterized by low anisotropy, but may have high selectivity.

Figure 11-2 Types of etch processes on a chamber pressure scale.

11.1 Wet Etching

Wet etching is a purely chemical process that can have serious drawbacks: a lack of anisotropy, poor process control, and excessive particle contamination. However, wet etching can be highly selective and often does not damage the substrate. As a result, while the process is much less popular than it once was, it continues to be used for a wide range of ''noncritical'' tasks. Comprehensive reviews of wet etching can be found elsewhere [1].

Since the reactive species is normally present in the etchant solution, wet chemical etching consists of three processes: movement of the etchant species to the surface of the wafer, a chemical reaction with the exposed film that produces soluble byproducts, and movement of the reaction products away from the surface of the wafer. Since all three steps must occur, the slowest one, called the *rate limiting step*, determines the etch rate. Since it is generally desirable to have a large, uniform, well controlled etch rate, the wet etch solution is often agitated in some manner to assist in the

Figure 11-3   Resist scumming occurs when the photoresist is incompletely developed. The residual resist may serve as an etch mask to prevent a complete etch process.

movement of etchant to the surface and the removal of the etch product. Some wet etch processes use a continuous acid spray to ensure a fresh supply of etchant.

For most wet etch processes, the film to be etched is not directly soluble in the etchant solution. It is usually necessary to change the material to be etched from a solid to a liquid or a gas. If the etching process produces a gas, this gas can form bubbles that can prevent the movement of fresh etchant to the surface. This is an extremely serious problem since the occurrence of the bubbles cannot be predicted. The problem is most pronounced near pattern edges. In addition to assisting the movement of fresh etchant chemicals to the surface of the wafer, agitation in the wet chemical bath will also reduce the ability of the bubbles to adhere to the wafer. Even in the absences of bubbles, however, small geometry features may etch more slowly, due to the difficulty in removing all of the etch products. This phenomenon has been shown to be related to microscopic bubbles of trapped gas [2]. Another common problem for wet etch processes is undetected resist scumming. This occurs when some of the exposed photoresist is not removed in the develop process (Figure 11.3). Common causes are incorrect or incomplete exposures and insufficient developing of the pattern. Due to their high selectivity, even a very thin layer of resist residue is sufficient to completely block the wet etch process.

In the last few years, wet etching has enjoyed something of a resurgence. Automated wet etch benches have been developed that allow the operator to precisely control the etch time, bath temperature, degree of agitation, bath composition, and the degree of misting in spray etches [3]. Increased use of filtration, even in hot, very aggressive, compounds has helped control particle deposition concerns. Even with these improvements, however, wet etching is still not regarded as practical for most features smaller than 2 μm.

One recent area of strong interest in wet etching is chemical mechanical polishing. As will be discussed in Chapter 15, there is a need to planarize the surface of the wafer to improve the lithography and metallization step coverage. Increasingly, this is being done by mechanically polishing the wafer (Figure 11.4) in a chemical slurry [4] that may contain both etchants and abrasives. This can be thought of as adding a physical component to the etching, but unlike most plasma processes, this physical damage is directed parallel to the surface of a wafer. Considerable work has been done to make the process manufacturable, and its use is expected to increase dramatically [5].

One of the most common etching processes is the wet etching of SiO_2 in dilute solutions of hydrofluoric acid (HF) [6]. Common etchants are 6:1, 10:1, and 20:1, meaning 6, 10, or 20 parts (by volume) of water to one part HF. A 6:1 HF solution will etch thermal silicon dioxide at about 1200 Å/min. Deposited oxides tend to etch must faster. The ratio of the deposited film etch rate in HF to that of thermal oxides is often taken as a measure of its density. Doped oxides such as phosphosilicate glass and borophosphosilicate glass etch faster yet, as the etch rate increases with impurity con-

Figure 11-4   Electromechanical polishing of the surface of a partially processed wafer done to achieve a high degree of global planarization.

centration. HF solutions are extremely selective of oxide over silicon [7]. Some etching of silicon does occur, since the water will slowly oxidize the surface of the silicon and HF will etch this oxide. Selectivities are commonly better than 100:1. Wet etching of oxide in HF solutions is however, completely isotropic.

The exact reaction pathway is complex and depends on the ionic strength, the solution pH, and the etchant solution [8]. The overall reaction for etching SiO_2 is

$$SiO_2 + 6HF \rightarrow H_2 + SiF_6 + 2H_2O \tag{11.2}$$

Since the reaction consumes HF, the reaction rate will decrease with time. To avoid this, it is common to use HF with a buffering agent (BHF) such as ammonium fluoride (NH_4F) which maintains a constant concentration of HF through the dissolution reaction

$$NH_4F \rightleftharpoons NH_3 + HF \tag{11.3}$$

where NH_3 (ammonia) is a gas. Buffering also controls the pH of the etchant, which minimizes photoresist attack.

Silicon nitride is etched very slowly by HF solutions at room temperature. A 20:1 BHF solution at room temperature, for example, etches thermal oxide at about 300 Å/min, but the etch rate for Si_3N_4 is less than 10 Å/min [9]. More practical etch rates of silicon nitride can be obtained in H_3PO_4 at 140 to 200°C [10]. A 3:10 mixture of 49% HF (in H_2O) and 70% HNO_3 at 70°C can also be used, but is much less common. Typical selectivities in the phosphoric etch are 10:1 for nitride over oxide and 30:1 for nitride over silicon. If the nitride layer is exposed to a high temperature oxidizing ambient, therefore, a dip in BHF is often done before the nitride wet etch to strip any surface oxide that may have grown on top of the nitride.

Wet etching has also been widely used to pattern metal lines. Since the metal layers used in ICs are often polycrystalline, the lines produced this way sometimes have ragged edges. A common aluminum etchant is (by volume) 20% acetic acid, 77% phosphoric acid, and 3% nitric acid. Most of the metal interconnect in silicon technologies is not elemental but rather dilute alloys. In many cases, these impurities are much less volatile in the bath than the base material. In particular, silicon and copper additions to aluminum are often difficult to completely remove in standard aluminum wet etch solutions.

While several techniques have been used for wet etching silicon, most of them use a strong oxidant to chemically oxidize the silicon and HF to etch the oxide. A common etchant solution is a combination of HF and nitric acid (HNO_3) in water. The overall reaction is given by

$$Si + HNO_3 + 6HF \rightarrow H_2SiF_6 + HNO_2 + H_2 + H_2O \tag{11.4}$$

Often acetic acid (CH_3COOH) is used as a diluent rather than water. Figure 11.5 shows a diagram of the etch rate of silicon in HF and HNO_3 [11]. Notice that the three axes are not independent. To find the etch rate, draw lines from the percentages of HNO_3 and HF. They should intersect at a point on the line corresponding to the remaining percentage of diluent. At low HNO_3 concentrations, the etch rate is controlled by the oxidant concentration. At low HF concentrations, the etch rate is controlled by the HF concentration. The maximum etch rate for this solution is 470 μm/min. A hole can be etched completely through a typical wafer at this rate in about 90 sec.

The most common wet etchants for GaAs [12],[13] are H_2SO_4-H_2O_2-H_2O [14], Br_2-CH_3OH [15], $NaOH$-H_2O_2 [16], and NH_4OH-H_2O_2-H_2O [17]. Figure 11.6 shows the etch contours for the

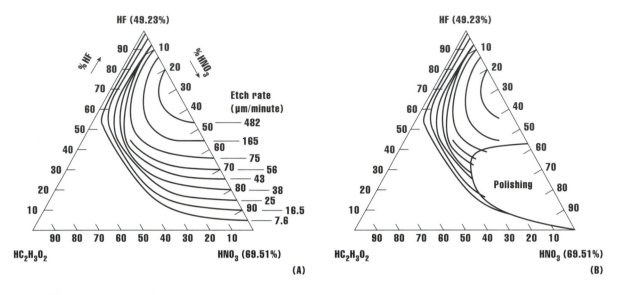

Figure 11-5   The etch rate of silicon in HF and HNO$_3$ *(after Schwarz and Robbins, reprinted by permission of the publisher, The Electrochemical Society Inc.).*

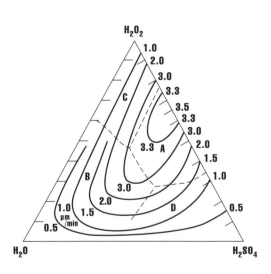

Figure 11-6   The etch rate of GaAs in H$_2$SO$_4$, H$_2$O$_2$, and H$_2$O. The bottom leg is the concentration of H$_2$SO$_4$, the left leg is H$_2$O, and the right leg is H$_2$O$_2$. All scales increase in the clockwise direction *(after Iida and Ito, reprinted by permission of the publisher, The Electrochemical Society Inc.).*

H$_2$SO$_4$-H$_2$O$_2$-H$_2$O system [18]. The curves are qualitatively similar to those of Figure 11.5. At high sulfuric acid concentrations, the solution is quite viscous, and so the etch may be limited by diffusion of fresh etchant to the wafer. Etches are uneven in these solutions. As a result, the acid content is usually kept to less than 30%. As with all etching solutions using H$_2$O$_2$, the etch rate decreases with time.

GaAs etches that are selective to Al$_x$Ga$_{1-x}$As are often required since many advanced devices consist of very thin layers of one material on top of the other. Selective etches are needed to make electrical contact to the various layers. The first reports of a selective etch used a 30% solution of H$_2$O$_2$ diluted with NH$_4$OH or H$_3$PO$_4$ to control the pH [19]. Selectivities as large as 30:1 have been reported in this system [20]. Redox solutions such as I$_2$-Kl, K$_3$Fe(CN)$_6$-K$_4$Fe(CN)$_6$ and C$_6$H$_4$O$_2$-C$_4$H$_6$O$_2$ are now more commonly used due to their high etch rates and good selectivities. Tijburg and van Dongen [21] reported selective etching of GaAs over AlGaAs for K$_3$Fe(CN)$_6$-K$_4$Fe(CN)$_6$ solutions with pH values greater than 9 and selective etching of AlGaAs over GaAs for the same solutions with pH values between 5 and 9. NH$_4$OH-H$_2$O$_2$-H$_2$O and K$_3$Fe(CN)$_6$-K$_4$Fe(CN)$_6$ have also been found to be highly selective etches for GaAs and AlGaAs over InGaAs [22].

It is sometimes useful to be able to etch deep patterns in semiconductor substrates with some degree of directional con-

SEM TOP VIEW (100) DI ETCH

SEM CROSSECTIONAL VIEW (100) DI ETCH

1000X

160 μm DEEP

Figure 11-7 (100) silicon wafers after directional etching in KOH, isopropyl alcohol, and water. The upper photo shows a 50 μm deep etch. The lower photographs are of 80 μm deep trenches etched at 10 μm pitch on (110) and 10° off (110) *(after Bean, ©1978 IEEE).*

trol. An example of this type of structure is a deep trench approximately 0.5 μm wide etched several microns into the wafer, sometimes used to isolate devices from each other in high density silicon ICs. While directional etching will be discussed in much more detail later in the chapter, this section will point out that some solutions will etch much faster along certain crystal directions. As a result, these etches produce sharp facets with well controlled angles in the single crystal substrates [23]. In the zinc-blende structure, the density of bonds in the (111) directions is much larger than in the (100) or (110) directions. Some etchants will therefore etch more slowly in the (111) directions.

Common directional wet etchants for silicon are mixtures of KOH, isopropyl alcohol, and water. A 23.4:13.5:63 mixture etches 100 times faster in the (100) directions than in the (111) [24]. Since this etchant contains no HF, a simple thermal oxide can be used as the masking layer. Figure 11.7 shows a cross-sectional view of (100) wafers after directional etching. Closely spaced features will result in a characteristic V shape with sidewalls at 54.7° to the surface normal. More widely spaced features will etch deeper into the substrate, exposing more of the (111) surface. If the wafers are etched long enough, these facets can extend completely through the wafer. Br_2-CH_3OH is a common directional etchant for GaAs [25].

A variety of special purpose wet etches have also been developed for selectively etching certain regions of the wafers. This section will review two: doping selective and defect selective. Doping selective etches were developed primarily for junction staining, providing steps that can be used to show the depth of a p-n junction (see Section 3.6). More recently, these techniques have been applied to the fabrication of etch stop layers for sensors and other special structures. One of the most common doping selective etches is a 1:3:8 mixture of HF/HNO_3/CH_3COOH. The etch rate of either type of heavily doped ($>10^{19}$ cm^{-3}) layers of silicon is about 15 times larger than the etch rate of lightly doped layers [26]. An ethylene-diamine-pyrocatechol-water mixture, however, etches lightly doped silicon, but does not attack heavily doped p-type layers [27].

Defect selective etching is done to decorate (highlight) defects in the wafer. Defects that may be invisible before etching are often readily observable by simple optical microscopy after preferential defect staining. By counting the defects after staining, a defect density can be derived. While some defect stain etches have been developed for GaAs, the most common (summarized in Table 11.1) have been used in silicon.

Table 11.1   Wet chemical etch solutions used to delineate defects in (100) and (111) silicon

(100)	Schimmel [28]	2:1; HF/CrO_3 (1 molar)	Works well on lightly doped layers (>0.6 Ω-cm)
	Modified Schimmel [28]	4:3:2; $HF/H_2O/CrO_3$ (1 molar)	Works well on heavily doped layers
	Secco [29]	2:1; $HF/K_2Cr_2O_7$ (0.15 molar)	Delineates oxidation induced stacking faults
(111)	Sirtl [30]	1:1; HF/CrO_3 (5 molar)	Requires agitation
	Dash [31]	1:3:10; $HF/HNO_3/CH_3COOH$	Etch rates are slow and concentration dependent

11.2 Basic Regimes of Plasma Etching

Etching in a plasma environment has several significant advantages when compared to wet etching. Plasmas are much easier to start and stop than simple immersion wet etching. Furthermore, plasma etch processes are much less sensitive to small changes in the temperature of the wafer. These two factors make plasma etching more repeatable than wet etching. Most important for small features, plasma etches may have high anisotropies. Plasma environments may also have far fewer particles than liquid media. Finally, a plasma etch process produces less chemical waste than wet etching.

As shown in Figure 11.2, there is a wide variety of dry etch processes with differing amounts of physical and chemical attack. Overlaid on this is the variety of etch chemistries that is used in each type of etch system. Table 11.2 lists some of the most common. Clearly, a complete review of this topic would be complicated and lengthy. The remainder of the chapter will review a representative sample of these processes. A few of the typical etch chemistries for the most common etch process will also be presented. For further information of plasma etching, the reader is referred to any of several references [32]–[34].

The preceding chapter introduced the concept of a glow discharge. For a plasma etch process to proceed, six steps must occur. A feed gas introduced into the chamber must be broken down into chemically reactive species by the plasma. These species must diffuse to the surface of the wafer and be adsorbed. Once on the surface, they may move about (surface diffusion) until they react with the exposed film. The reaction product must be desorbed, diffused away from the wafer, and be transported by the gas stream out of the etch chamber. As with wet etching the etch rate is determined by the slowest of these steps.

Table 11.2   Typical etch chemistries

Si	CF_4/O_2, CF_2Cl_2, CF_3Cl, $SF_6/O_2/Cl_2$, $Cl_2/H_2/C_2F_6/CCl_4$, C_2ClF_5/O_2, Br_2, SiF_4/O_2, NF_3, ClF_3, CCl_4, CCl_3F_5, C_2ClF_5/SF_6, C_2F_6/CF_3Cl, CF_3Cl/Br_2
SiO_2	CF_4/H_2, C_2F_6, C_3F_8, CHF_3/O_2
Si_3N_4	$CF_4/O_2/H_2$, C_2F_6, C_3F_8, CHF_3
Organics	O_2, CF_4/O_2, SF_6/O_2
Al	BCl_3, BCl_3/Cl_2, $CCl_4/Cl_2/BCl_3$, $SiCl_4/Cl_2$
Silicides	CF_4/O_2, NF_3, SF_6/Cl_2, CF_4/Cl_2
Refractories	CF_4/O_2, NF_3/H_2, SF_6/O_2
GaAs	BCl_3/Ar, $Cl_2/O_2/H_2$, $CCl_2F_2/O_2/Ar/He$, H_2, CH_4/H_2, $CClH_3/H_2$
InP	CH_4/H_2, C_2H_6/H_2, Cl_2/Ar
Au	$C_2Cl_2F_4$, Cl_2, $CClF_3$

Cotler and Elta [35].

11.3 High Pressure Plasma Etching

The earliest plasma etch equipment, introduced into IC fabrication facilities in the early 1970s, was based on high pressure, low power plasmas, where the mean free path of the species in the plasma is much less than the chamber size. The plasma in such a process is used to start and stop the chemical reaction or etching. It does this by producing a reactive species from an inert precursor. Since the energy of ions in the plasma is quite low, the etching process depends primarily on the chemistry of the plasma.

Plasma chemistry is extremely complex. To begin to understand the chemistry of the glow discharge, it is useful to start from a prototype system. The most widely studied plasma etch chemistry is that produced by carbon tetrafluoride (CF_4). Assume that a flow of CF_4 gas is established in a chamber, where a pressure of 500 mtorr (a high pressure plasma) is maintained. Silicon wafers with a photoresist mask are in contact with the plasma. The process intent is to etch the substrate silicon. This choice does not mean that only CF_4 or even fluorinated plasmas are run at high pressure. Chlorine and other species may also be used in high pressure plasmas, but fluorine is more common. By the same token, fluorinated species are sometimes used for low pressure reactive ion etching.

One can begin with a simple energy balance argument as discussed in Morgan [33]. This model suggests that etching can occur if a chemical reaction that produces a gaseous or high vapour pressure liquid or solid, is energetically favored. Applied to the etching of silicon, the basic idea is that volatizing silicon with a halogen requires the replacement of Si-Si bonds with Si-halogen bonds. The amount of energy necessary to break a C-F bond in CF_4 is 105 kcal/mole. The amount of energy required to break a Si-Si bond is 42.2 kcal/mole. In order for CF_4 to etch silicon, the sum of these two energies (147 kcal/mole) must be less than the energy of the Si-F bond (130 kcal/mole).

$$C \veebar F + Si \veebar Si = Si - F + 17 \text{ kcal/mole} \tag{11.5}$$

where $\veebar$ represents a bond breaking event. Since there is a net positive energy required for the reaction, CF_4 will not etch silicon directly. In a plasma however, collisions with high energy electrons will dissociate (crack) some of the CF_4 molecules, producing free fluorine atoms and molecular radicals. The first term of Equation 11.5 is now ignored from an energy balance point of view. The formation of SiF compounds is then energetically favored. Furthermore by choosing the feed gas, chamber pressure, and plasma power, one can increase the density of species that are energetically favored to etch the film.

In a typical plasma, most of the species are unreacted gas molecules. The density of these molecules at the prototype chamber pressure is about 3×10^{16} cm^{-3}. After the unreacted feed gas, the next most common (5 to 10%) species in the plasma are neutral radicals from the precursor. CF_3, CF_2, C, and F will be found in this CF_4 plasma. These radical species are extremely reactive. Assuming a simple kinetic theory gas one can use Equation 10.10 to roughly estimate the flux of radicals striking the surface:

$$J_n = \sqrt{\frac{n^2 kT}{2\pi M}} \tag{11.6}$$

Assuming a radical species temperature of 500 K, the radical bombardment rate due to simple diffusion is of order 10^{23} m^{-2}-sec^{-1}. Also assuming that every radical that strikes the surface etches one silicon atom, and that the radical flux is the rate limiting step, the resultant etch rate would be about 1000

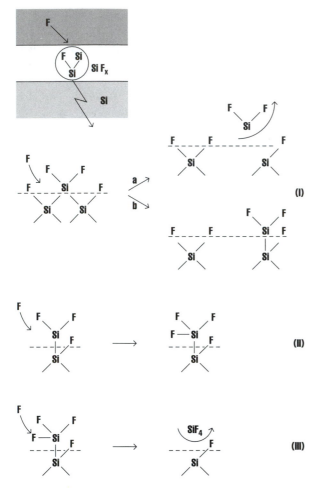

Figure 11-8 Proposed mechanism of plasma etching of silicon in CF_4. A 1–5 atom thick SiF_x layer forms on the surface. A silicon atom on the upper level is bonded to two fluorine atoms. An additional fluorine atom may remove the silicon as SiF_2. Much more likely, however, is that additional fluorine atoms bond to the silicon atom until SiF_4 forms and desorbs *(after Manos and Flamm, reprinted by permission, Academic Press)*.

Å/min. Thus, one could conceive of such a purely chemical process that would provide ample etch rates. In practice, not all of the radical species will etch all of the substrate materials. Some of the reaction byproducts such as carbon do not react with silicon or do not form volatile species. As these materials coat the surface of the wafer they slow the etching reaction. Furthermore, not all reactant species that strike the surface of the wafer will stick.

Typical plasma power densities in this prototype reactor create an ion concentration of order 10^{10} cm^{-3}. In a CF_4 plasma, the most abundant ion species is CF_3^+. Because of the small ion concentration, most of the etching will not be directly due to ions in the plasma. Rather, the constant bombardment of ions incident on the surface creates damage in the form of unsatisfied bonds that are exposed to the reactive radicals. The neutral radicals diffuse to the surface due to the concentration gradient in the plasma. On the surface they quickly react, form volatile products, and are pumped away. Referring back to Equation 11.5, ion bombardment of the surface eliminates most of the energy penalty associated with the Si-Si bond breaking event. This makes the forward reaction even more energetically favored.

CF_4 plasmas can be used to selectively etch silicon on silicon dioxide or silicon dioxide on silicon. To understand how, start with a picture of what happens at the surface of the wafer. It is believed that in a fluorine plasma, the surface silicon atoms are bonded to two F atoms, forming a fluorinated skin that may be several atoms thick. Both SiF_2 and SiF_4 are volatile species, but SiF_2 will not readily desorb since it is chemically bonded to the wafer. As shown in Figure 11.8, the arrival of additional F atoms will reduce the number of bonds between the surface silicon atom and the substrate until eventually the silicon fluoride molecule can be released at a minimum of energy. The primary sources of these atoms are atomic fluorine, F_2, and radicals of the form CF_x, where $x \le 3$.

Without ion bombardment, F_2 radicals will etch SiO_2 very slowly since there is a net energy loss in the reaction

$$F \veebar F + Si \veebar O = Si - F + -5 \text{ kcal/mole} \qquad \textbf{(11.7)}$$

while CF_3 radicals will etch SiO_2 much more aggressively. In a pure fluorine beam at low energy, the selectivity of Si of SiO_2 is almost 50:1 when the wafer is held at room temperature and about 100:1 at $-30°C$. As a result, plasma processes for etching silicon selectively over silicon dioxide are those that produce high atomic fluorine concentrations. While it is possible to use F_2 as the feed gas,

Figure 11-9 Species concentrations in a CF₄ plasma as a function of the amount of oxygen in the feed gas *(after Smolinsky and Flamm, reprinted by permission, AIP).*

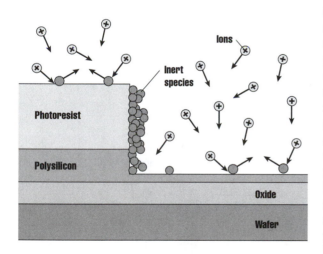

Figure 11-10 Schematic diagram of a high pressure anisotropic etch showing the formation of sidewall passivating films.

it is not desirable due to its particularly high toxicity. The preferred species are CF_4, C_2F_6, and SF_6 which produces very large concentrations of free F. It is found experimentally that the addition of small concentrations of oxygen to a CF_4 feed gas will increase the etch rate of both silicon and silicon dioxide [36]. It is believed that the oxygen reacts with the carbon atoms to produce CO_2. This removes (scavenges) the carbon from the plasma, thereby increasing the F concentration. These plasmas are called fluorine rich. Figure 11.9 shows plots of some of the species concentrations in a CF_4 plasma as a function of the amount of oxygen in the feed gas [37]. The F concentration can be increased by an order of magnitude by the addition of 12% O_2 to a CF_4 plasma producing about an order of magnitude increase in the Si etch rate, while the increase in the etch rate of SiO_2 is considerably less. At higher oxygen concentrations, the selectivity of silicon to oxide drops sharply as the molecular and atomic oxygen chemisorb on the silicon surface making it appear more like SiO_2 [38].

A typical application of a selective etch of Si to SiO_2 is patterning a polysilicon gate for an MOS transistor over a thin gate oxide. While the addition of O_2 to CF_4 can increase the selectivity of the etch, the processes are not necessarily anisotropic. It is possible to etch silicon anisotropically in a high pressure fluorine plasma by encouraging the formation of nonvolatile fluorocarbons that deposit on the surface of the wafer. The deposit can only be removed by physical collisions with incident ions. Figure 11.10 shows a schematic of an anisotropic high pressure plasma etch process. Fluorocarbon films deposit on all surfaces [39], but the ion velocity, which follows the electric field, is nearly vertical. As a result, as the etching proceeds there is little ion bombardment of the sidewalls and the fluorocarbon film accumulates. (Due to collisions with the ions, neutral species may also strike the surface of the wafer, but they are not shown in this figure.) The nature of the film is a sensitive function of plasma conditions [40]. If it is not reactive, ion bombardment of the horizontal surfaces drives reactions with the underlying substrate [41],[42]. Under steady state conditions an approximately 15 Å layer of fluorinated silicon is formed under the fluorocarbon [43]. The process of producing nonvolatile species that reduce the etch rate is known as *polymerization*. The film is said to passivate the sidewall, preventing lateral etching.

One way to encourage the formation of these films is to add hydrogen to the plasma. The hydrogen scavenges the fluorine, creating a carbon rich plasma. The excess carbon can then form these nonvolatile products. The same thing happens if C_2F_6 is used as the feed gas instead of CF_4. In many cases, the etch products from the resist which are also hydrocarbons, may also participate in the polymer formation process. Scavenging the fluorine however, does the opposite of what was done when oxygen was added to the plasma. As a result, the silicon to oxide selectivity drops in such an etch process. One can therefore obtain anisotropy or selectivity of Si to SiO_2 in such a plasma, but not both simultaneously.

The gasses in a plasma reactor however, can be changed during the etch process. In the case of MOS technologies, to obtain good dimensional control of polysilicon gate electrodes, one might start the etch process in a CF_4 plasma with H_2. When the etch is most of the way through the polysilicon one could shut off the H_2 and instead add O_2. There may be some undercutting during this part of the process, but before undercutting can begin, the plasma must first erode the accumulated polymer. As a result, these hybrid etch processes, while using very simple equipment and relatively innocuous feed gasses, can produce adequate etch profiles for features down to about 1 μm.

As just described, adding a small amount H_2 to a CF_4 plasma causes the etch rate of both silicon and silicon dioxide to decrease in good agreement with the nonvolatile species model described above. At moderate H_2 concentrations, the etch rate of SiO_2 exceeds that of Si. The chemistry can therefore be used to selectively etch SiO_2 over Si. In this case, the hydrogen reacts with the fluorine radicals to form HF, which etches silicon dioxide but not silicon [44]. The plasma in this situation is said to be fluorine deficient. At moderate H_2 concentrations, the nonvolatile fluorocarbon film deposition process that increases anisotropy further increases the selectivity. On the surface of the SiO_2, the oxygen produced during the ion bombardment will react with the carbon to form CO and CO_2, both of which are volatile and so are pumped away. No such reaction occurs over the silicon and so the etch selectivity of SiO_2 over Si increases sharply with the addition of H_2 to the plasma (Figure 11.11). Eventually, the carbon deposition rate swamps the ability of the plasma etch process to remove it,

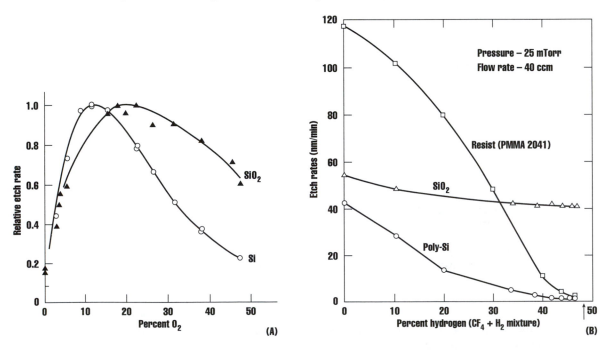

Figure 11-11   Etch rate of Si and SiO_2 in (a) CF_4/O_2 plasma *(after Mogab et al., reprinted by permission, AIP)*, and (b) CF_4/H_2 plasma *(after Ephrath and Petrillo, reprinted by permission of the publisher, The Electrochemical Society Inc.)*.

5% NF$_3$

20% NF$_3$

100% NF$_3$

├─────┤
1 μm

Figure 11-12 Etch profiles of SiO$_2$ with increasing concentrations of NF$_3$ *(after Donnelly et al., reprinted by permission, AIP).*

and instead of etching, there may be a net deposition.

Coburn and Kay recognized that for fluorocarbon systems the onset of polymerization depends on the fluorine to carbon ratio [45],[46]. After taking into account the formation of HF, CO, CO$_2$, and other scavenging compounds, polymerization proceeds when the remaining carbon concentration in the feed gas mixture is greater than half of the fluorine concentration. Examples of gasses at the point of polymerization include C$_2$F$_4$, CHF$_3$, and a 1:1 mixture of CF$_4$ and H$_2$. Highly selective plasma etch processes operate close to this point and can achieve etch selectivities of over 20 to 1 for etching SiO$_2$ over Si. The hydrogen can be supplied as H$_2$ gas. It can also be incorporated into the fluorinated precursor as CHF$_3$ or even CH$_2$F$_2$. It can also be added in the form of methane (CH$_4$), ethane (C$_2$H$_6$), or other simple hydrocarbons. In the latter case, one must take into account that this not only adds hydrogen to the plasma, but carbon as well.

As was just pointed out, the concept of depositing a nonvolatile polymeric film during the plasma etch process has an important application in producing etch selectivity. As was the case for silicon etching, sidewall passivation can produce highly anisotropic etch profiles when etching SiO$_2$; however, there is no tradeoff of selectivity required. As a result, it is relatively easy to use fluorinated plasmas to produce nearly vertical etching of SiO$_2$ over Si. Figure 11.12 shows etch profiles of SiO$_2$ with increasing concentrations of NF$_3$ [47]. At the highest concentrations, there is a large etch rate, but the polymeric formation is suppressed and the etch is isotropic with substantial undercut of the mask.

Selective etching of silicon nitride to both silicon dioxide and silicon cannot be achieved simultaneously due to its intermediate position in terms of bond strength and electronegativity. For selectivity of nitride to oxide, silicon nitride can be etched in plasma processes similar to those used to etch silicon, that is, plasmas rich in atomic fluorine. A typical example is a CF$_4$-O$_2$ plasma [48]. Etching of silicon nitride selective to silicon can be done in a typical silicon dioxide etching process such as a CF$_4$-H$_2$ plasma.

In some applications, the plasma etch rate for a process is measured for a given set of process conditions, and the time to etch is inferred. Such etches provide poor reproducibility since subtle changes in the etch process or in the film properties can result in unacceptable under or over etch. One such subtlety is loading effect, the tendency for the etch rate in many plasma etches to decrease with an increase in the area of the exposed film being etched. Usually, this is caused by a depletion of the reactive species in the plasma. Loading effect can be described by the equation

$$R = \frac{R_o}{1 + kA} \tag{11.8}$$

where R_o is the empty chamber etch rate, A is the area of the exposed film to be etched, and k is a constant [49]. k can often be reduced by fixing the reactor pressure and increasing the flow of the etch gasses, but this approach is subject to pumping limitations.

Example 11.1

A plasma etch process is found to produce an etch rate of 300 Å/min when etching a single wafer. When a second wafer is added to the reactor the etch rate falls to 240 Å/min. What etch rates would you expect for three and four wafers?

Denote the area to be etched on each wafer as A_o and the etch rates as R_n where n is the number of wafers. Then from Equation 11.8,

$$\frac{R_1}{R_2} = 1.25 = \frac{1 + 2kA_o}{1 + kA_o} \tag{11.9}$$

Solving this equation for the kA_o product gives a value of 1/3. Then, inserting back into Equation 11.8 using either of the two known etch rates gives $R_o = 400$ Å/min. Finally, for three and four wafers, $R_3 = 200$ Å/min, and $R_4 = 171$ Å/min.

If a more repeatable process is required, some method of detecting either the remaining thickness of the film to be etched or the presence of etch byproduct in the plasma can be used to determine the etch endpoint. If the film being etched is transparent, it is possible to use the same constructive and destructive interference effect described in Chapter 4 for measuring oxide thickness. The sum of the reflected and transmitted optical beams will oscillate with film thickness until the transparent layer is removed. This endpoint method however, requires a substantial area of unpatterned film to focus and align the laser beam.

Alternatively, one can monitor the chemistry of the plasma either by using mass spectrometry of the etch effluent or by using the peaks of the optical emission spectra. In either case one can look for the extinction of the etch product of the desired film or the emergence of an etch product of the substrate. It is also possible to detect changes in the reactive species. If a substantial part of the gas is consumed by the reaction, the endpoint will be seen as a sudden increase in the concentration of the etchant. Table 11.3 lists some of the most common etch processes and the associated detection wavelengths. Figure 11.13 shows a commercial parallel plate plasma etcher. The system contains two etch chambers fed by a central load lock and robot. Windows are provided at the front of each chamber for possible spectral analysis of the glow discharge.

Table 11.3 Characteristic optical emission wavelengths used for endpoint detection of plasma etch processes

Film	Etchant	Wavelength (Å)	Emitter
Al	CCl_4	2614	AlCl
		3962	Al
Resist	O_2	2977	CO
		3089	OH
		6563	H
		6156	O
Si	CF_4/O_2	7037	Fl
		7770	SiF
	Cl_2	2882	Si
Si_3N_4	CF_4/O_2	3370	N_2
		7037	F
		6740	N
SiO_2	CHF_3	1840	CO
W	CF_4/O_2	7037	F

After Manos and Flamm.

Figure 11-13 Photograph of a computer controlled, dual chamber, parallel plate plasma etch system *(courtesy Plasma Therm)*.

11.4 Ion Milling

Ion milling is at the opposite extreme of the spectrum of etch processes compared to high pressure plasma etches. Pure ion milling or ion beam etching involves no chemical reactions with the etch species, since it uses noble gasses such as argon. It is a strictly mechanical process, sometimes called the micromechanical analog of sandblasting. The physics of the etch process is entirely analogous to sputtering. The reader is referred to Sections 12.7 and 12.8 for a review of the physical interactions between incident ions and solid targets. Ion milling has two significant advantages compared to high pressure plasmas: directionality and applicability. The directionality of the erosion is due to the fact that the ions in the beam are accelerated by a strong vertical electric field, and the chamber pressure is so low that atomic collisions are extremely unlikely. As a result, the ions' velocity as they impinge on the surface of the wafer is almost completely vertical. The second advantage of ion milling is that it can be used to pattern a wide variety of materials including compounds, even if there are no suitable volatile etch products. The erosion rate of the target does not vary by much more than a factor of two from material to material. Thus, ion milling is popular for patterning YBaCuO, InAlGaAs, and other ternary and quaternary systems.

The ability of ion milling to pattern a wide range of materials is also one of its most serious liabilities. The process selectivity to photoresist and underlying layers is generally close to 1:1 unless a chemical component is added to the process. The other drawback with ion milling is throughput. Most ion sources are no more than 200 mm in diameter. Thus, for large silicon wafers, ion milling is a one wafer at a time process. Combined with the low erosion rates and the need for high vacuums, this makes ion milling impractical for high volume manufacturing in silicon based technologies. For III–V technologies however, the smaller wafer sizes and reduced number of wafers per lot make ion milling quite viable.

The most popular source for ion milling is the Kaufman source (Figure 11.14). This basic source was originally developed as a space based rocket engine [50]. Unlike most plasma processes, it has direct and independent control over both the ion bombardment energy and ion flux [51],[52]. A Kaufman source consists of an electron filament heated using the supply V_f. The filament is held at a voltage V_a below the anode potential. Electrons boil off of the filament and are accelerated toward the anode grid. V_a must be large enough so that the accelerated electrons impact the neutral gas atoms with enough energy to ionize them. Typically, this is about 40 V. Much higher voltages are not desirable since high energy ions that strike the inside of the chamber walls will etch the chamber and so contaminate the system. In order to maintain the plasma, the source is held at about 10^{-3} torr.

The source body is perforated on one side. A grid held at a voltage V_g below the target just below this side of the source body accelerates the ejected ions toward the target. Typical accelerating potentials are 500 to 1000 V. Ions arrive at the target surface with an energy given by

Figure 11-14   Cross section schematic of a Kaufman ion source.

$$E_{ion} = q|V_p - V_g| = q|V_a + V_{pa} - V_g| \quad \textbf{(11.10)}$$

where V_p is the plasma voltage with respect to ground, and V_{pa} is the plasma voltage with respect to the anode. This latter number is determined by the ion flux and the geometry of the discharge. As discussed in Chapter 10, V_{pa} is positive. Because of the large anode area however, it is only a few volts. The extracted ion energy then is simply controlled by changing V_a or V_g.

In order to increase the ion density, a magnetic field is applied to the source as well. Field strengths are about 100 Gauss [53]. (The use of magnetic fields to enhance the ion density in a plasma was discussed in Section 10.6.) Typical Kaufman sources can produce ion currents of almost an ampere over areas as large as 300 mm in diameter. This maximum current is limited by the electric field produced by a high density ion stream. The maximum current is approximately

$$ j_{max} \approx K \sqrt{\frac{q}{m} \frac{V_t^{3/2}}{I_g^2}} \qquad (11.11) $$

where K is a constant for a given chamber, q/m is the charge to mass ratio of the ions, V_t is the potential difference between the screen and accelerator grids, and I_g is the distance between the grids [54]. Typical values of I_g are 1 to 2 mm. These systems produce beams with divergences of 5° to 7°.

Example 11.2

Calculate the maximum flux from a Kaufman argon source, the potential difference between the grids is 500 V, the grid spacing is 1 mm, and $K = 2 \times 10^{-15}$ sec/Ω-m. Assume singly ionized argon.

Substituting $q = 1.6 \times 10^{-19}$ C and $m = 40 \times 1.67 \times 10^{-27}$ kg into Equation 11.10, j_{max} is directly calculated as 35 mA/m^2. This corresponds to 2.2×10^{17} ions/m^2-sec.

Recently electron cyclotron resonance (ECR) and other advanced plasma sources have been developed for etching. The basic operation of an ECR plasma is described in Section 10.6. An advantage of such a source for ion milling is the elimination of the filament [55]. This reduces source heating and therefore contamination, and improves system reliability. This advantage is most notable if a reactive species is used rather than argon (see CAIBE, below). Unlike plasma or reactive ion etching, an ECR plasma has nearly independent control of ion flux and ion energy. Such a system may have less substrate damage than reactive ion etchers [56]. The disadvantages of these systems include increased cost and complexity for the source and the need for large and expensive pumping systems. Since the flux density in an ion mill is limited by Equation 11.11, substantially higher erosion rates are not obtained with these sources [57] unless modified extraction methods are used.

The ions that are ejected from the source enter the target chamber. In order to maintain a highly directional etch, the target chamber is pumped to as low a pressure as the incident flux and the pumping speed will allow. This minimizes collisions between the ions and residual gas molecules. Cryopumps or turbomolecular pumps are typically used to evacuate the target chamber. It is normally maintained at a pressure 10 to 100 times lower than the source. Ions travel in a straight line to the target electrode,

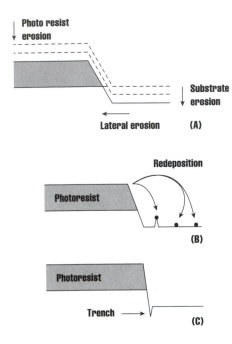

Figure 11-15 Problems that may occur during ion milling: (a) mask taper transfer, (b) redeposition from the mask, and (c) trenching.

which contains the wafers to be milled. Since the ions are positively charged, a voltage will build up on the surface unless the wafer and the film being etched are conductive. To avoid charging effects, an electron flood gun is used. Exactly the same principle is used for maintaining accurate dosimetry in ion implantation (Chapter 5).

Figure 11.15 shows some of the problems that may occur during ion milling. Since the process erodes the mask, any taper in the masking layer will be transferred to the pattern. After the etch is done and the photoresist mask has been removed, the resultant pattern is broadened and appears very similar to an undercut. Since the eroded material from the target is not volatile, some of it will redeposit on the surface of the wafer. This can lead to an uneven etch and a considerable amount of organic residue from the photoresist mask. The latter problem is particularly pronounced on the sidewalls of the features. Finally, some milling cross sections show an enhanced erosion rate at the edges of the pattern called *trenching*. This most often occurs when mask erosion causes the sidewalls of the pattern to be tapered at a steep angle. Some of the low angle ions will reflect off of the tapered surface toward the pattern edge, where they cause trenches. Commercial ion beam systems often allow the target to be tilted and rotated with respect to the incident beam during milling to minimize these effects [58].

It is necessary to use reactive species to increase selectivity in ion milling. The mixture of ions producing physical damage along with chemical attack will also be exploited in reactive ion etching, but the name chemically assisted ion beam etching (CAIBE, pronounced kay-bee) or reactive ion milling will be reserved for those processes in which the wafer is in a broad area, well collimated ion beam, but not in the plasma chamber itself. An inert species such as argon can be mixed with small amounts of oxygen to reduce the erosion rate of many metals in an ion mill. This has particular applications in the etching of GaAs over $Al_xGa_{1-x}As$. It has been shown that a 2% O_2 dilution of Ar at 100 eV etches GaAs at hundreds of Å/min, while the etch rate of $Al_{0.5}Ga_{0.5}As$ is not measurable [59].

The difficulty with introducing some reactive species, of course, is that they will tend to attack the source components. In particular, the hot filament is readily attacked. Oxygen is known to simply shorten filament life, but more aggressive precursors such as the halogens may be completely unusable for this reason. Some species may also have problems with gas phase polymerization that leads to coating of the inside of the source. It is therefore necessary to design ion source to accommodate a very reactive plasma. As previously discussed, ECR sources are desirable for this application.

Another way to avoid this problem is to add a reactive bleed near the wafer surface in a conventional ion mill (Figure 11.16). This type of etch, sometimes called an ion assisted chemical etch, proceeds primarily by adsorption of the precursor gas on the wafer surface, simultaneously with ion bombardment. The ion stream leads to both adsorbed species decomposition and substrate damage to drive a chemical reaction. Since the chemical process depends on ion bombardment however, this type of process cannot achieve the level of selectivity that can be obtained in a purely chemical process.

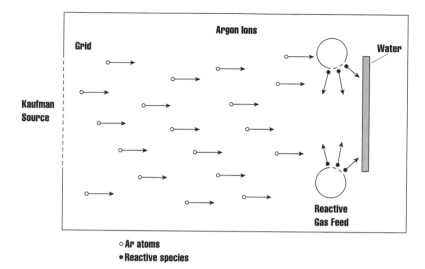

Figure 11-16   The use of a reactive bleed near the wafer surface in a conventional ion mill to introduce a chemical component to the etch process.

11.5 Reactive Ion Etching

Reactive ion etching (RIE) was developed because there was a very strong need for an anisotropic etch with much higher selectivity than can be achieved in ion milling. The name is something of a misnomer since the ions are not the primary etching species in this process. A more appropriate name that is sometimes used is ion assisted etching, although RIE (pronounced as either R I E or rye) is far more common.

Two common systems for RIE are shown in Figure 11.17. Unlike high pressure plasma etchers, the wafers rest on the powered electrode. In the parallel plate reactor, the neutral electrode is attached to the chamber wall to enlarge its effective area. In the hexode, the neutral electrode is the chamber wall. As discussed in Chapter 10, the effect of this arrangement is to increase the potential difference from the plasma to the powered electrode and so the energy of the ion bombardment. In order for the parallel plate arrangement to be effective, the plasma must contact the chamber walls. As pressures increase (>1 torr) the plasma contracts and loses contact with the walls. RIE, however, is done in low pressure plasmas, where the mean free path in the plasma is at least of order millimeters. In this region, the plasma remains in good contact with the walls, and a large potential appears between the plasma and the powered electrode.

Chlorine based plasmas are commonly used to anisotropically etch silicon, GaAs, and aluminum based metallizations. Although corrosive, chlorinated precursors such as CCl_4, BCl_3, and Cl_2 have high vapor pressures, and both the precursors and etch products are easier to deal with than the comparable bromides or iodides. Fluorinated plasmas can also be used under RIE conditions, but the chlorine system will serve as a model for this section.

It is relatively easy to develop a basic understanding of the anisotropic etching of silicon in a chlorine RIE. Undoped silicon etches very slowly in a Cl or Cl_2 ambient without the addition of ion bombardment. Heavily n-type doped silicon or polysilicon, however, spontaneously etches at high rates without bombardment in the presence of Cl, but not in the presence of Cl_2. The doping enhance-

Figure 11-17 Top and side views of parallel plate and hexode batch RIE systems. Typical conditions for either is 50 mtorr and 5 kW/m^2.

ment, which can be as much as a factor of 25, depends on the carrier concentration in the film rather than the chemical identity of the dopant [60],[61].

This very pronounced doping effect implies that the chlorine etching process involves electron transfer from the substrate. The model that has developed postulates that in a chlorine plasma, atomic chlorine chemisorbs on the silicon but does not break the underlying Si-Si bonds. The absorption of more chlorine atoms are impeded by stearic hindrance (Figure 11.18). Once the surface chlorine becomes negatively charged, however, it can bond ionically with the substrate. This frees additional chemisorption sites and greatly increases the probability that the chlorine atoms will penetrate the surface and produce volatile silicon chlorides.

Chlorine penetration of the surface is also dramatically increased by ion bombardment. As a result, those surfaces subject to ion bombardment etch much more rapidly than those that are not. In particular, the vertical sidewalls receive very little ion bombardment. Because of these effects, the etch profile of undoped polysilicon or single crystal silicon in a Cl$_2$ RIE is almost completely aniso-tropic. Unfortunately, the charge transfer mechanism described above produces isotropic etch profiles in heavily doped layers such as polysilicon gates [62] and aluminum metallizations. In these structures, it is necessary to obtain anisotropic etch profiles through sidewall polymerization schemes. This is commonly done by adjusting the relative concentrations of Cl$_2$ and an inhibitor forming gas such as BCl$_3$, CCl$_4$, or SiCl$_4$. It can also be done by combining fluorinated precursors and Cl$_2$. A common

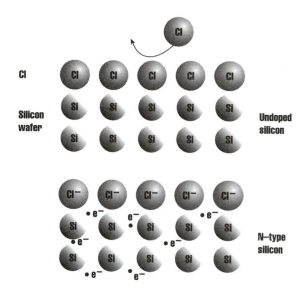

Cl

Silicon
wafer

Undoped
silicon

N-type
silicon

Figure 11-18 Once a monolayer of chlorine atoms
builds on the surface, it impedes any chlorine addition.

mixture is 90% C_2F_6 with 10% Cl_2. By controlling the ratio of Cl to F, one can control the degree of undercut and obtain a range of etch profiles [63]. This is particularly useful in etching multiple layer films such as silicides on polysilicon with acceptable profiles [64]. As with other passivation schemes, the etch rate must be traded off against the etch selectivity.

The lower half of Figure 11.19 shows a silicon line etched with BCl_3, Cl_2, and O_2 in a sidewall passivation technique [65]. In this micrograph the sidewall film has begun to peel off of the line. Notice that the top of the sidewall film is thicker than the bottom because of the difference in the deposition time. This produces a slight taper in the etch profile. The sidewall film, which is more clearly shown in the upper film, is almost pure SiO_2. The other features in the lower micrograph are sometimes called grass or black silicon [66]. They are due to the unintentional deposition of nonvolatile species and incomplete removal by ion bombardment. Grass appears in a microscope as a darkened area and often occurs when operating a sidewall passivation RIE process under heavy deposition conditions. Grass formation is suppressed by reducing the polymer formation by decreasing the flow of BCl_3, increasing the O_2 concentration, or by increasing the ion bombardment by reducing the pressure or increasing the power.

Reliable and repeatable etching of aluminum and aluminum based compounds requires considerable attention to detail. Aluminum readily oxidizes to form Al_2O_3, which etches very slowly in most Cl plasmas. To avoid problems with etch uniformity and to maintain proper etch timing, a short sputter etch in Ar or an appropriate chemical RIE can be used before beginning the Cl based RIE. At the end of the etch, it is also necessary to remove any adsorbed chlorinated species from the surface of the wafer. When the wafers are removed from the vacuum chamber, these compounds can react with moisture in the air to form hydrochloric acid, leading to metal corrosion. Several methods have been developed to avoid this problem. The simplest is a dip in deionized water immediately after the etch. Other post etch treatment can be done inside the vacuum chamber. These include an inert sputter etch, an oxygen plasma, or a fluorinated plasma to displace the adsorbed chlorine atoms [67].

A problem related to etching aluminum compounds is the volatilization of silicon and/or copper additives to the metal. Copper residues are especially difficult to remove due to the low vapor pressures of these copper chlorides at room temperature. Furthermore, the copper residues are even more prone to corrosion than their aluminum counterparts. Etching compounds with greater than 2% copper generally require increased ion bombardment and/or heated substrates. Another way to etch these compounds is the use of sacrificial aluminum in the etch chamber to provide a sufficient amount of $AlCl_3$, an etch product that helps volatize copper chlorides.

Finally, a serious problem in chlorinated RIE is resist erosion due to Cl attack. This is particularly aggravated in aluminum RIE because of the thickness of typical interconnect layers, the severe topology that may exist, and because of the presence of the etch product $AlCl_3$, which also accelerates resist attack. Some resist vendors supply chlorine resistant resists to address this problem, but even with these resists, erosion remains a concern.

Most anisotropic etching of GaAs is also done in chlorinated reactive ion etchers. Purely chemical, isotropic etching occurs in high pressure plasmas, while anisotropic etching occurs for chamber

Figure 11-19   SEM micrographs of sidewall passivation films created in a HCl/O₂/BCl₃ plasma. In the upper micrograph, the photoresist and silicon posts have been removed by subsequent wet etching. The lower photo shows the sidewall film is partially peeled from the etched pattern *(after Oehrlein et al., reprinted by permission, AIP)*.

pressures of less than 10 mtorr under strong ion bombardment [68]. It is believed that the primary group III byproduct, GaCl₃, accumulates on the surface of the wafer [69] even though the compound is reported to have a high vapor pressure. Due to differences in the etch rates of group III and group V halides, GaAs etches more rapidly along some crystal planes than others. In low power, high pressure Cl₂ plasmas, significant faceting can occur. To avoid this problem, compounds may be added to form polymers in the plasma and passivate the sidewalls.

Another chemistry that can be used for anisotropically etching GaAs is hydrogen based. Arsenic, of course, forms several hydrides, the most stable of which is AsH₃. To volatize gallium, 5 to 25% methane must be added [70]. The maximum etch rate occurs at about 500 K [71]. As will be discussed in Section 14.9, both the long lived radical monomethyl gallium and the stable compound trimethyl gallium (Ga(CH₃)₃) are volatile products. At high methane concentrations, however, excessive polymerization occurs and etching stops. Anisotropic etching with very high etch rates has also been demonstrated with chloromethanes [72].

In fluorinated RIE as in chlorinated RIE, etch anisotropy may be due to either sidewall passivation, physical damage of the horizontal surfaces, or some combination of the two. The physical damage mechanism is believed to be described by the chemical sputtering model [73]. In this model, the ion bombardment supplies energy, increasing the mobility and reactivity of surface species. The development of anisotropic etch profiles due to sidewall passivation was discussed in a previous section. In such a process, some ion bombardment of the surfaces is required to allow any etching at all [74]. In reactive ion etching, there is a substantial ion bombardment suggesting that these tools may also be used with fluorinated chemistries. The high energy bombardment in RIE, however, damages the substrate, particularly when etching SiO₂ over Si. This will be discussed in the next section.

11.6 Damage in Reactive Ion Etching⁺

One of the limitations of reactive ion etching is residual damage left in the substrate after the etch. Typical ion fluences of 10^{15} ions/cm² are delivered at energies of 300 to 700 eV in a RIE. Both substrate damage [75]

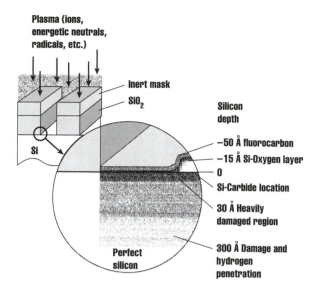

Plasma (ions,
energetic neutrals,
radicals, etc.)

Inert mask

SiO₂

Si

Silicon
depth

~50 Å fluorocarbon

~15 Å Si-Oxygen layer

0

Si-Carbide location

30 Å Heavily
damaged region

300 Å Damage and
hydrogen
penetration

Perfect
silicon

Figure 11-20 A cross section schematic of the results of a typical etch of SiO₂ down to Si using CF₄/H₂ *(after Oehrlein, Rembetski, and Payne, reprinted by permission, AIP).*

and chemical contamination are serious issues. The latter is particularly a concern in polymerization etches, which are known to leave behind residual films. Gas phase particle deposition is also a significant problem [76]. Furthermore, one can often find metallic impurities including Fe, Ni, Na, Cr, K, and Zn on the surface of the wafer after the etch [77] due to sputtering of the electrodes, chamber, and fixtures in contact with the plasma. Techniques developed for the removal of these impurities include O₂ plasma treatments followed by wet acid cleaning [78],[79] and H₂ plasma treatments [80]. The drawback of these post etch treatments is the increased complexity of the process.

Physical damage and drive in of impurities is a second class of problem with RIE. After a typical etch in a carbon containing RIE, the top 30 Å is heavily damaged with an extensive concentration of Si-C bonds, while significant damage is as deep as 300 Å [81]. RIE processes done in a hydrogen containing ambient also have Si-H defects that can be observed electrically as deep as 400 Å [82] and are very difficult to remove [83]. Hydrogen may actually penetrate several microns into the surface where it can deactivate dopants in the substrate [84]. Figure 11.20 shows a cross section schematic of the results of a typical etch of SiO₂ down to Si [65]. Removal of this damage requires an initial clean followed by annealing at temperatures of over 800°C [85]. It is also possible to design RIE processes without hydrogenated species [86].

11.7 Magnetically Enhanced Reactive Ion Etch (MERIE) Systems⁺

As discussed above, in a typical RIE system, ions reach the surface of the wafer with 300 to 800 eV. This is far in excess of the energy required to drive the surface chemistry and sputter nonvolatile species, but operating at much lower energies also reduces the ion flux. A type of system currently being developed to anisotropically etch layers with high selectivity and low substrate damage is Magnetically Enhanced Reactive Ion Etch (MERIE). As discussed in Chapter 10, MERIE uses a magnetic field applied perpendicular to the direction of ion travel from anode to cathode [87] (Figure 11.21). The field causes the electrons to move in helical paths, greatly increasing the ionization probability and increasing the fraction of ions in the plasma by as much as a factor of 50. Because the electron bombardment of the cathode is reduced, the electrode bias is significantly reduced. Typical MERIE biases are less than 100 V. As a result, ion bombardment damage is greatly reduced [88]. Compared to traditional RIE, MERIE has a much higher ion flux at much lower energies. MERIE is especially appropriate, therefore, in etching vertical features down to substrates that have low damage tolerance

+

Anode

B field
on

B field
off

e⁻

e⁻

Bᵢₙ

Bᵢₙ

Cathode

−

Figure 11-21 Schematic of a magnetically enhanced RIE system.

and cannot be annealed at sufficiently high temperature after the etch to remove all of the damage. This includes etching SiO_2 down to shallow junctions in silicon [89] and etching AlGaAs down to thin layers of GaAs [90].

11.8 Liftoff

Most GaAs technologies have been developed around liftoff rather than etching. The sequence for liftoff is shown in Figure 11.22. A thick layer of resist is spun and patterned. Next, a thin layer of metal is deposited using evaporation (Chapter 12). One characteristic of evaporation is its difficulty in covering high aspect ratio features. If a re-entrant profile is obtained in the resist, a break in the metal is virtually assured. Next, the wafer is immersed in a solution capable of dissolving the photoresist. The metal lines that were deposited directly on the semiconductor remain, while the metal deposited on the resist lifts off of the wafer as the resist dissolves. Etch damage to the substrate is avoided and the lines patterned with infinite selectivity with no undercut. Since the process in its simplest form requires only a wet bench and perhaps ultrasonic agitation, it is widely used in research laboratories.

Methods that have been used to form a re-entrant resist profile generally harden the surface of the resist. This can be done to some extent by promoting cross linking by deep UV exposure, in a suitable plasma environment, or by ion implantation. Another solution is the use of multiple layers of resist such as a DQN resist on PMMA. After PMMA coating, the upper layer of resist is spun, baked, exposed, and developed as normal with a UV source. The patterned upper layer can then be used as a mask for a deep UV exposure of the PMMA, which is then overdeveloped using a solution that does not attack the upper resist. The result is a pronounced ledge that is very difficult to cover. Due to the complexity of multilayer resist processing, the most popular method of producing re-entrant profiles is soaking a single layer DQN resist after softbake in chlorobenzene [91],[92] or similar compounds. Typical soak times are 5 to 15 min. The soak process reduces the dissolution rate of the upper surface of the resist [93]. After developing the pattern, therefore, a ledge appears (Figure 11.23).

Figure 11-22   Process sequence for a liftoff operation *(after Hatzakis et al., ©1980 International Business Machines Corporation).*

Figure 11-23   Liftoff profiles after various resist treatments *(after Hatzakis et al., ©1980 International Business Machines Corporation.)*

The thickness of the ledge depends on the soak time, the temperature of the chlorobenzene bath, and the resist prebake cycle [94].

Liftoff processes have several shortcomings. The first is that the surface topology must be very smooth, since the metal deposition step is designed to have poor step coverage. The technology, therefore, must either be limited to one layer of metallization, or each layer must be planarized before liftoff patterning. This effectively prevents the use of sputtering. The other serious problem is that the metal lifted off remains solid and floats in the bath. Pieces of it are very likely to redeposit on the surface of the wafer. Unless the patterns are very simple, liftoff has serious yield impacts.

11.9 Summary

This chapter reviewed a variety of etch techniques that ranged from the purely physical (ion milling) to the purely chemical (wet etching). In between the two extremes are chemically assisted ion beam etching, reactive ion etching, and high pressure plasma etching. Generally, the greater the degree of ion bombardment, the higher the etch anisotropy, but the lower the selectivity. Sidewall passivation was introduced as a technique for retaining high etch anisotropy with good selectivity. A wide variety of etches is used. Two of the most common, fluorine and chlorine plasma, have been described in some detail. Liftoff processes were developed to avoid the need for etching altogether. In these processes, the resist is patterned before metal deposition.

Problems

1. A solution consisting of four parts 70% HNO_3, four parts 49% HF, and two parts $HC_2H_3O_2$ is used to etch silicon. If the solution is held at room temperature, what etch rate would you expect? If the compound is to remain at two parts $HC_2H_3O_2$, what mixture of these same chemicals would be suitable to etch silicon at a rate of approximately 10 μm/min?

2. An etch engineer only has a high pressure plasma etch and an ion mill at her disposal. Select which you would use in the following applications. Be sure to justify your answer.
    (a) Etching a 5000 Å polysilicon layer that serves as the upper electrode of a large square capacitor. The capacitor dielectric is 50 Å of SiO_2.
    (b) Recessing the channel of a GaAs FET. For this application the residual etch damage must be minimized.
    (c) Anisotropic patterning of a thin layer of $YBa_2Cu_3O_7$ on a thick insulating layer.

3. A cylindrical parallel plate RF etch chamber is constructed with 12 in. diameter electrodes. The chamber diameter is 18 in. and the height of the chamber is 6 in. One of the electrodes is attached to ground. The dc voltage between the electrodes is measured to be 20 V when a plasma is established. Assuming that the plasma is in contact with the chamber walls, calculate the potential difference between the plasma and each electrode. Explain the significance of the result.

4. Describe the differences between high pressure plasmas and reactive ion etch systems. Explain when each is the preferred process.

5. Describe the difference between chemically assisted ion beam etching (CAIBE) and ion assisted chemical etching.

References

1. W. A. Kern and C. A. Deckert, "Chemical Etching," in *Thin Film Processing*, J. L. Vossen, ed., Academic, New York, 1978.
2. K. McAndrews and P. C. Subanek, "Nonuniform Wet Etching of Silicon Dioxide," *J. Electrochem. Soc.* **138**:863 (1991).
3. P. Burggraaf, "Wet Etching: Alive, Well, and Futuristic," *Semiconductor Int.*, p. 58, July 1990.
4. C. Kaanta, W. Cote, J. Cronin, K. Holland, P.-I. Lee, and T. Wright, "Submicron Wiring Technology with Tungsten and Planarization," in *Proc. 1987 IEDM*, IEEE, p. 209 (1987).
5. H. Landis, P. Burke, W. Cote, W. Hill, C. Hoffman, C. Kaanta, C. Koburger, W. Lange, M. Leach, and S. Luce, "Integration of Chemical-Mechanical Polishing into CMOS Integrated Circuit Manufacturing," *Thin Solid Films* **220**:1 (1992).
6. W. Kern, "Chemical Etching of Dielectrics," in *Etching for Pattern Definition*, Electrochem. Soc., Pennington, NJ, 1976.
7. S. M. Hu and D. R. Kerf, "Observation of Etching of n-type Silicon in Aqueous HF Solutions," *J. Electrochem. Soc.* **114**:414 (1967).
8. J. S. Judge, in *Etching for Pattern Definition*, H. G. Hughes and M. J. Rand, eds., Electrochem. Soc., Princeton, NJ, 1976.
9. L. M. Loewenstein and C. M. Tipton, "Chemical Etching of Thermally Oxidized Silicon Nitride: Comparison of Wet and Dry Etching Methods," *J. Electrochem. Soc.* **138**:1389 (1991).
10. J. T. Milek, *Silicon Nitride for Microelectronic Applications, Part 1—Preparation and Properties*, IFI/Plenum, New York, p. 1, 1971.

11. B. Schwartz and H. Robbins, "Chemical Etching of Silicon: Etching Technology," *J. Electrochem. Soc.* **123**:1903 (1976).

12. For a comprehensive listing of III–V etching solutions see Kern and Deckert [1] and references therein.

13. R. E. Williams, *Gallium Arsenide Processing Techniques*, Artech, Dedham, Mass., 1984.

14. S. Adache and K. Oe, "Chemical Etching Characteristics of (001) GaAs," *J. Electrochem. Soc.* **130**:2427 (1983).

15. Y. Tarui, Y. Komiya, and Y. Harada, "Preferential Etching and Etched Profiles of GaAs," *J. Electrochem. Soc.* **118**:118 (1971).

16. D. W. Shaw, "Enhanced GaAs Etch Rates Near the Edges of a Patterned Mask," *J. Electrochem. Soc.* **113**:958 (1966).

17. J. J. Gannon and C. J. Nuese, "A Chemical Etchant for the Selective Removal of GaAs Through SiO_2 Masks," *J. Electrochem. Soc.* **121**:1215 (1974).

18. S. Iida and K. Ito, "Selective Etching of Gallium Arsenide Crystals in the H_2SO_4-H_2O_2-H_2O System," *J. Electrochem. Soc.* **118**:768 (1971).

19. R. A. Logan and F. K. Reinhart, "Optical Waveguides in GaAs-AlGaAs Epitaxial Layers," *J. Appl. Phys.* **44**:4172 (1973).

20. J. J. LePore, "Improved Technique for Selective Etching of GaAs and $Ga_{1-x}Al_xAs$," *J. Appl. Phys.* **51**:6441 (1980).

21. R. P. Tijburg and T. van Dongen, "Selective Etching of III–V Compounds with Redox Systems," *J. Electrochem. Soc.* **123**:687 (1976).

22. D. G. Hill, K. L. Lear, and J. S. Harris, Jr., "Two Selective Etching Solutions for GaAs on InGaAs and GaAs/AlGaAs on InGaAs," *J. Electrochem. Soc.* **137**:2912 (1990).

23. K. E. Bean, "Anisotropic Etching of Si," *IEEE Trans. Electron. Dev.* **ED-25**:1185 (1978).

24. S. Wolf and R. N. Tauber, *Silicon Processing for the VLSI Era*, Vol. 1, Lattice Press, Sunset Beach, Calif., 1986.

25. P. D. Greene, "Selective Etching of Semi-Insulating Gallium Arsenide," *Solid-State Electron.* **19**:815 (1976).

26. H. Muraoka et al., "Controlled Preferential Etching Technology," in H. R. Huff and R. R. Burgess, eds., *Semiconductor Silicon 73*, Electrochemical Society, Princeton, NJ, p. 327, 1973.

27. J. C. Greenwood, "Ethylene Diamine-Catechol-Water Mixture Shows Preferential Etching of p-n Junction," *J. Electrochem. Soc.* **116**:1325 (1969).

28. D. G. Schimmel, "Dry Etch for <100> Silicon Evaluation," *J. Electrochem. Soc.* **126**:479 (1979).

29. F. Secco d'Aragona, "Dislocation Etch for (100) Planes in Silicon," *J. Electrochem. Soc.* **119**:948 (1972).

30. E. Sirtl and A. Adler, *Z. Metallk* **52**:529 (1961).

31. W. C. Dash, "Copper Precipitation on Dislocations in Silicon," *J. Appl. Phys.* **27**:1193 (1956).

32. D. M. Manos and D. L. Flamm, *Plasma Etching, An Introduction*, Academic Press, Boston, 1989.

33. R. A. Morgan, *Plasma Etching in Semiconductor Fabrication*, Elsevier, Amsterdam, 1985.

34. A. J. van Roosmalen, J. A. G. Baggerman, and S. J. H. Brader, *Dry Etching for VLSI*, Plenum, New York, 1991.

35. T. J. Cotler and M. Elta, "Plasma-Etch Technology," *IEEE Circuits, Dev. Mag.*, p. 38, July 1990.

36. V. M. Donelly, D. I. Flamm, W. C. Dautremont-Smith, and D. J. Werder, "Anisotropic Etching of SiO_2 in Low-Frequency CF_4/O_2 and NF_3/Ar Plasmas," *J. Appl. Phys.* **55**:242 (1984).

37. G. Smolinsky and D. L. Flamm, "The Plasma Oxidation of CF_4 in a Tubular, Alumina, Fast-Flow Reactor," *J. Appl. Phys.* **50**:4982 (1979).

38. C. J. Mogab, A. C. Adams, and D. L. Flamm, "Plasma Etching of Si and SiO_2—The Effect of Oxygen Additions to CF_4 Plasmas," *J. Appl. Phys.* **49**:3796 (1978).

39. J. W. Coburn, "*In-situ* Auger Spectroscopy of Si and SiO_2 Surfaces Plasma Etched in CF_4-H_2 Glow Discharges," *J. Appl. Phys.* **50**:5210 (1979).

40. R. d'Agostino, F. Cramarossa, F. Fracassi, E. Desimoni, L. Sabbatini, P. G. Zambonin, and G. Caporiccio, "Polymer Film Formation in C_2F_6-H_2 Discharges," *Thin Solid Films* **143**:163 (1986).

41. M. Shima, "A Study of Dry-Etching Related Contaminations of Si and SiO_2," *Surf. Sci.* **86**:858 (1979).

42. S. Joyce, J. G. Langan, and J. I. Steinfeld, "Chemisorption of Fluorocarbon Free Radicals on Si and SiO_2," *J. Chem. Phys.* **88**:2027 (1988).

43. Ch. Cardinaud and G. Turban, "Mechanistic Studies of the Initial Stages of Si and SiO_2 in a CHF_3 Plasma," *Appl. Surf. Sci.* **45**:109 (1990).

44. M. M. Millard and E. Kay, "Difluocarbene Emission Spectra from Fluorocarbon Plasmas and Its Relationship to Fluorocarbon Polymer Formation," *J. Electrochem. Soc.* **129**:160 (1982).

45. J. W. Coburn and E. Kay, "Some Chemical Aspects of Fluorocarbon Plasma Etching of Silicon and Its Compounds," *IBM J. Res. Dev.* **23**:33 (1979).

46. J. W. Coburn and E. Kay, "Some Chemical Aspects of Fluorocarbon Plasma Etching of Silicon and Its Compounds," *Solid State Technol.* **22**:117 (1979).

47. V. M. Donnelly, D. E. Ibbotson, and D. L. Flamm, in *Ion Bombardment Modification of Surfaces: Fundamentals and Applications*, O. Auciello and R. Kelly, eds., Elsevier, New York, 1984.

48. F. H. M. Sanders, J. Dieleman, H. J. B. Peters, and J. A. M. Sanders, "Selective Isotropic Dry Etching of Si_3N_4 over SiO_2," *J. Electrochem. Soc.* **129**:2559 (1982).

49. C. J. Mogab, "The Loading Effect in Plasma Etching," *J. Electrochem. Soc.* **124**:1262 (1977).

50. B. A. Heath and T. M. Mayer, in *VLSI Electronics Microstructure Science 8, Plasma Processing for VLSI*, N. G. Einspruch and D. M. Brown, eds., Academic, New York, 1984.

51. J. M. E. Harper, "Ion Beam Techniques in Thin Film Deposition," *Solid State Technol.* **30**:129 (1987).

52. R. E. Lee, "Ion-Beam Etching (Milling)," in *VLSI Electronics Microstructure Science 8, Plasma Processing for VLSI*, N. G. Einspruch and D. M. Brown, eds., Academic, New York, 1984.

53. H. R. Kaufman, J. J. Cuomo, and J. M. E. Harper, "Techniques and Applications of Broad-Beam Ion Sources Used in Sputtering—Part 1. Ion Source Technology," *J. Vac. Sci. Technol.* **21**:725 (1982).

54. J. M. E. Harper, "Ion Beam Etching," in *Plasma Etching: An Introduction*, D. M. Manos and D. L. Flamm, eds., Academic, New York, 1989.

55. S. Matsup and Y. Adachi, "Reactive Ion Beam Etching Using a Broad Beam ECR Ion Source," *Jap. J. Appl. Phys.* **21**:L4, (1982).

56. A. S. Yapsir, G. S. Oehrlein, F. Wiltshire, and J. C. Tsang, "X-ray Photoemission and Raman Scattering Spectroscopic Study of Surface Modifications of Silicon Induced by Electron Cyclotron Resonance Etching," *Appl. Phys. Lett.* **57**:590 (1990).

57. C. Keqiang, A. Erli, W. Jinfa, Z. Hansheng, G. Zuoyao, and Z. Bangwei, "Microwave Electron Cyclotron Resonance Plasma for Chemical Vapor Deposition and Etching," *J. Vac. Sci. Technol.* **A4**:828 (1986).

58. R. E. Lee, "Microfabrication by Ion-Beam Etching," *J. Vac. Sci. Technol.* **16**:164 (1979).

59. H. Kinoshita, T. Ishida, and K. Kaminishi, "Surface Oxidation of GaAs and AlGaAs in Low-Energy Ar/O_2 Reactive Ion Beam Etching," *Appl. Phys. Lett.* **49**:204 (1986).

60. S. Berg, N. Nender, R. Buchta, and H. Norstrom, "Dry Etching of n-type and p-type Polysilicon: Parameters Affecting the Etch Rate," *Vac. Sci. Technol.* **A5**:1600 (1987).

61. H. Okano, Y. Horiike, and M. Sekine, "Photo-Excited Etching of Poly-Crystalline and Single-Crystalline Silicon in Cl_2 Atmospheres," *Jpn. J. Appl. Phys.* **24**:68 (1985).

62. G. C. Schwartz and P. M. Schaible, "Reactive Ion Etching of Silicon," *J. Vac. Sci. Technol.* **16**:410 (1979).

63. D. L. Flamm, D. N. K. Wang, and D. Maydan, "Multiple-Etchant Loading Effects and Silicon Etching in $CClF_3$ and Related Mixtures," *J. Electrochem. Soc.* **129**:2755 (1982).

64. L. Peters, "Plasma Etch Chemistry: The Untold Story," *Semicond. Int.*, p. 67, May 1992.

65. G. S. Oehrlein, J. F. Rembetski, and E. H. Payne, "Study of Sidewall Passivated and Microscopic Silicon Roughness Phenomena in Chlorine-Based Reactive Ion Etching of Silicon Trenches," *J. Vac. Sci. Technol.* **B8**:1199 (1990).

66. G. K. Herb, D. J. Rieger, and K. Shields, "Silicon Trench Etching in a Hex Reactor," *Solid State Technol.* **30**:109 (1987).

67. Y. T. Fok, *Electrochem. Soc. Proc.* **80**:301, Electrochemical Society, Pennington, NJ, 1980.

68. D. E. Ibbotson and D. L. Flamm, "Plasma Etching for III–V Compound Devices: Part 1," *Solid State Technol.* **31**:77, Oct. 1988.

69. M. Balooch, D. R. Orlander, and W. J. Siekhaus, "The Thermal and Ion-Assisted Reactions of GaAs (100) with Molecular Cl," *J. Vac. Sci. Technol.* **B4**:794 (1986).

70. R. Cheug, S. Thomas, S. P. Beamont, G. Doughty, V. Law, and C. D. W. Wilkinson, "Reactive Ion Etching of GaAs Using a Mixture of Methane and Hydrogen," *Electronics Lett.* **23**:16 (1987).

71. J. M. Villalvilla, C. Santos, and J. A. Vallés-Abarca, "Temperature Dependence of Reactive Ion Beam Etching of GaAs with CH_4/H_2," *Vacuum* **43**:591 (1992).

72. V. J. Law and G. A. C. Jones, "Chloromethane-Based Reactive Ion Etching of GaAs and InP," *Semicond. Sci. Technol.* **7**:281 (1992).

73. T. J. Tu, T. J. Chang, and H. F. Winters, "Chemical Sputtering of Fluorinated Silicon," *Phys. Rev.* **B23**:823 (1981).

74. L. M. Ephrath and E. J. Petrillo, "Parameter and Reactor Dependence of Selective Oxide RIE in CF_4 and H_2," *J. Electrochem. Soc.* **129**:2282 (1982).

75. G. S. Oehrlein, "Dry Etching Damage of Silicon: A Review," *Mat. Sci. Eng.* **B4**:441 (1989).

76. G. S. Selwyn, J. Singh, and R. S. Bennett, "*In-situ* Laser Diagnostic Studies of Plasma-generated Particulate Contamination," *J. Vac. Sci. Technol.* **A7**:2758 (1989).

77. S. J. Fonash, "Overview of Dry Etching Damage and Contamination Effects," *J. Electrochem. Soc.* **137**:3885 (1990).

78. X.-C. Mu, S. J. Fonash, G. S. Oehrlein, S. N. Chakravarti, C. Parks, and J. Keller, "A Study of $CClF_3/H_2$ Reactive Ion Etch Damage and Contamination Effects in Silicon," *J. Appl. Phys.* **59**:2958 (1986).

79. J. P. Gambino, M. D. Monkowski, J. F. Shepard, and C. C. Parks, "Junction Leakage Due to RIE—Induced Metallic Contamination," *J. Electrochem. Soc.* **137**:976 (1990).

80. J. P. Simko, G. S. Oehrlein, and T. M. Mayer, "Removal of Fluorocarbon Residues on CF_4/H_2

Reactive Ion Etched Silicon Surfaces With a Hydrogen Plasma,'' *J. Electrochem. Soc.* **138**:277 (1991).

81. G. J. Coyle and G. S. Oehrlein, ''Formation of a Silicon-Carbide Layer During CF$_4$/H$_2$ Dry Etching of Silicon,'' *Appl. Phys. Lett.* **47**:604 (1985).

82. G. S. Oehrlein, R. M. Tromp, Y. H. Lee, and E. J. Petrillo, ''Study of Silicon Contamination and Near-surface Damage Caused by CF$_4$/H$_2$ Reactive Ion Etching,'' *Appl. Phys. Lett.* **45**:420 (1984).

83. G. S. Oehrlein, J. G. Clabes, and P. Spirito, ''Investigation of Reactive Ion Etching Related Fluorocarbon Film Deposition onto Silicon and a New Method for Surface Residue Removal,'' *J. Electrochem. Soc.* **133**:1002 (1986).

84. J. M. Heddleson, M. W. Horn, and S. J. Fonash, ''Evolution of Damage, Dopant Deactivation, and Hydrogen Related Effects in Dry Etched Silicon as a Function of Annealing History,'' *J. Electrochem. Soc.* **137**:1960 (1990).

85. S. J. Fonash, X.-C. Mu, S. Chakravarti, and L. C. Rathbun, ''Recovery of Silicon Surfaces Subjected to Reactive Ion Etching Using Rapid Thermal Annealing,'' *J. Electrochem. Soc.* **135**:1037 (1988).

86. J. P. Simko and G. S. Oehrlein, ''Reactive Ion Etching of Silicon and Silicon Dioxide in CF$_4$ Plasmas Containing H$_2$ and C$_2$F$_4$ Additives,'' *J. Electrochem. Soc.* **138**:2748 (1991).

87. J. A. Thornton, ''Magnetron Sputtering: Basic Physics and Applications to Cylindrical Magnetrons,'' *J. Vac. Sci. Technol.* **15**:171 (1978).

88. M. Engelhart, ''Evaluation of Dry Etch Processes with Thermal Waves,'' *Solid State Technol.* **33**(4):151 (1990).

89. D. C. Hinson, I. Lin, W. H. Class, and S. Hurwitt, ''Magnetron-Enhanced Plasma Etching of Silicon and Silicon Dioxide,'' *Semicond. Int.* **6**:103, Oct. 1983.

90. M. Namaroff, J. Sasserath, M. Meyyappan, G. McLane, H. S. Lee, and W. M. Cole, ''Magnetron Enhanced Ion Etching for GaAs IC Manufacturing Technology,'' in *Proc. GaAs IC Symp.*, p. 297 (1991).

91. B. J. Canavello, M. Hatzakis, and J. M. Shaw, ''Single Step Optical Lift-off Processes,'' *IBM Technol. Disc. Bull.* **19**:4048 (1977).

92. M. Hatzakis, B. J. Canavello, and J. M. Shaw, ''Single-Step Optical Lift-off Process,'' *IBM J. Res. Dev.* **24**:452 (1980).

93. R. M. Halverson, M. W. MacIntyre, and W. T. Motsiff, ''The Mechanism of Single-Step Lift-off with Chlorobenzene in a Diazo-Type Resist,'' *IBM J. Res. Dev.* **26**:590 (1982).

94. G. G. Collins and G. W. Halsted, ''Process Control of the Chlorobenzene Single-Step Liftoff Process with a Diazo-Type Resist,'' *IBM J. Res. Dev.* **26**:596 (1982).

Part IV

Unit Processes 3: Thin Films

A well-designed apparatus will do its work without effort. It will cost considerably less than one improperly planned and in whose design established principles have been overlooked. It will also perform with greater economy of steam, which is always an important consideration.[1]

T he previous chapters have discussed the processes involved in growing oxides, diffusing impurities, and transferring a pattern through photolithography and etching. These are primarily processes used to fabricate the transistor. This section of the book will discuss processes that can be used for depositing thin films. These are a very important set of processes since all of the layers above the surface of the wafer must be deposited. Generally the techniques used to deposit metals are physical, that is, they do not involve a chemical reaction. Processes used to deposit semiconducting and insulating layers often involve chemical reactions. This distinction however, is changing. One of the areas currently being developed is the chemical deposition of metals.

This section begins with the physical process of evaporation, which has been applied primarily to III–V fabrication. Evaporated films are unable to cover sudden steps and other severe topology on the surface of the wafer. Since many III–V technologies use liftoff for patterning, evaporation is ideally suited for this process. Chapter 12 will also cover a second physical deposition process: sputtering. Sputtering has been used extensively in silicon technologies. It is able to deposit a wide range of alloys and compounds and has a moderate ability to cover surface topology. The reasons for the choice of materials to be used for metallization will not be emphasized here but will be covered in more detail in Chapter 15. Chapter 13 will cover chemical vapor deposition, which has the best ability to cover surface topology of the three methods and produces the least substrate damage. Because the chemistry of each process is unique, Chapter 13 will cover the deposition of several representative materials. The use of these materials is described in Chapter 15.

---

[1]A. L. Weber, M.E., *Evaporation*, The Chemical Catalog Co., New York, 1926.

Chapter 14 will extend the deposition discussion of the first two chapters into the area of epitaxial growth. The chapter first covers chemical vapor deposition of silicon from silanes and chlorosilanes. These processes are widely used in silicon production. The chapter will then cover processes that have been developed to grow epitaxial layers with atomic level control over the thickness. These processes have been applied primarily to III–V technologies. One of them, molecular beam epitaxy, is essentially an extension of simple evaporation. Another, MOCVD, is an extension of chemical vapor deposition to lower growth temperatures using organometallic sources.

Chapter 12

Physical Deposition: Evaporation and Sputtering

Figure 12-1 A simple diffusion pumped evaporator showing vacuum plumbing and the location of the charge containing crucible and the wafers.

The metal layers for all of the early semiconductor technologies were deposited by evaporation. While still widely used in research applications and for III–V technologies, evaporation has been displaced by sputtering in most silicon technologies for two reasons. The first is the ability to cover surface topology, also called the *step coverage*. As the lateral dimensions of transistors have decreased, the thickness of many of the layers has remained nearly constant. As a result, the topology that the metal must cover has become more severe. Evaporated films have very poor ability to cover for these structures, often becoming discontinuous on the vertical walls. It is also difficult to produce well controlled alloys by evaporation. Since many modern silicon technologies require alloys to form reliable contacts and/or metal lines, it would be very difficult to use evaporation for a silicon technology in a manufacturing environment. Many III–V technologies use the poor step coverage of evaporation to advantage. Rather than depositing and etching metal layers, the film is deposited on top of a patterned photoresist layer. The films naturally tend to break at the edges of the resist so that when the resist is subsequently dissolved the layer on top of the resist is easily lifted. Rather than trying to form alloys, these technologies use thin layers of different metals. Such a stack would be difficult to etch, but this problem is circumvented with liftoff.

A simple evaporator is shown in Figure 12.1. The wafers are loaded into a high vacuum chamber that is commonly pumped with either a diffusion pump or a cryopump. Diffusion pumped systems commonly have a cold trap to prevent the backstreaming of pump oil vapors into the chamber. The charge or material to be deposited is loaded into a heated container called the *crucible*. It can be heated very simply by means of an embedded resistance heater and an external power supply. As the material in the crucible becomes hot, the charge gives off a vapor. Since the pressure in the chamber is much less than 1 mtorr, the atoms of the vapor travel across the chamber in a straight line until they strike a surface where they accumulate as a film. Evaporation systems may contain up to four crucibles to allow the deposition of multiple layers without breaking vacuum and may contain up to 24 wafers suspended in a frame above the crucibles. Furthermore, if an alloy is desired, multiple crucibles can be operated simultaneously. To help start and stop the deposition abruptly, mechanical shutters can be used in front of the crucibles.

12.1 Phase Diagrams: Sublimation and Evaporation°

Section 2.1 discussed phase diagrams for multicomponent systems such as Al-Si. Phase diagrams are essentially maps in 2-D for binary mixtures and in 3-D for ternary mixtures showing regions of stability. For evaporation, one is usually interested in the behavior of single component systems. Often this type of phase diagram is presented as a function of temperature and pressure. Evaporation only involves the pressure regime well below 1 torr. This is such a small range that we can ignore any pressure effects as well as the existence of multiple solid phases as they have little effect on evaporation.

As the temperature of the sample is raised, the material typically goes through the solid, liquid, and gas phases. At every temperature there exists an equilibrium pressure P_e of vapor above the material. When the sample is below the melting temperature this process is called *sublimation*. When the sample is molten it is called *evaporation*. The semiconductor process called evaporation typically involves molten samples, but this is simply because this region of operation has high vapor pressures and so produces acceptable deposition rates. The distinction between molten and solid samples is of secondary importance for evaporation. Figure 12.2 shows the equilibrium vapor pressure as a function of temperature for a variety of elements [1]. A wide range of pressures is accessible simply by changing the temperature of the sample.

When the charge is liquid, the vapor pressure is given by [2]

$$P_e = 3 \times 10^{12}\sigma^{3/2}T^{-1/2}e^{\Delta H_v/NkT} \qquad \textbf{(12.1)}$$

where σ is the surface tension of the metal, N is Avogadro's number, and ΔH_v is enthalpy of evaporation. In practice, even small errors in the determination of the enthalpy will produce dramatic errors in the vapor pressure, so the data in Figure 12.2 is generally determined experimentally.

To obtain reasonable deposition rates the sample vapor pressure must be at least 10 mtorr. From Figure 12.2 it is obvious that some materials must be heated to much higher temperatures than others to obtain the same vapor pressure. The class of materials known as refractory metals, which includes Ta, W, Mo, and Ti, have very high melting temperatures and therefore have low vapor pressures at moderate temperatures. Tungsten re-

Figure 12-2   Vapor pressure curves for some commonly evaporated materials (*data adapted from Alcock et al.*).

quires a temperature in excess of 3000°C to obtain a vapor pressure of 10 mtorr. The evaporation of these refractory materials entails special equipment problems. Al, on the other hand, has this same vapor pressure at 1250°C. Some ionic contaminants such as K have significant vapor pressure even at room temperature.

12.2 Deposition Rates

According to Equation 10.10, the number of gas molecules crossing a plane per unit time is given by

$$J_n = \sqrt{\frac{P^2}{2\pi kTM}} \tag{12.2}$$

where P is the pressure in the chamber and M is the atomic mass. While this expression has been used so far to describe the flux of atoms striking the surface of a vacuum system, it can also be used to describe the loss rate of atoms from an evaporative source. Multiplying this flux rate by the mass of the molecule produces the Langmuir expression for the mass evaporation rate [3]:

$$R_{ME} = \sqrt{\frac{M}{2\pi kT}}\, P_e \tag{12.3}$$

where P_e is the equilibrium vapor pressure of the crucible material.

Using this expression one can readily calculate the mass loss rate of the crucible as

$$R_{ML} = \int \sqrt{\frac{M}{2\pi kT}}\, P_e dA = \sqrt{\frac{M}{2\pi k}} \int \frac{P_e}{\sqrt{T}}\, dA \tag{12.4}$$

where the integral is done over the surface of the charge in the crucible. Equation 12.4 can be quite formidable unless two approximations are made. If the charge is completely molten it is common to assume that natural convection and thermal conduction will keep the temperature of the charge nearly constant across the crucible. This is analogous to the rapid stirring approximation made when discussing Czochralski growth in Chapter 2. If it is also assumed that the opening of the crucible has a constant area, A

$$R_{ML} = \sqrt{\frac{M}{2\pi k}} \frac{P_e}{\sqrt{T}}\, A \tag{12.5}$$

To find the deposition rate on the surface of the wafer, one needs to determine the fraction of that material leaving the crucible that accumulates on the surface of the wafer. The ultrahigh vacuum chamber makes the determination of this constant of proportionality relatively easy. The pressure is low enough so that material ejected from the crucible travels in a straight line to the surface of the wafer. If it is assumed that all of the material that arrives at the wafer sticks and remains there, the arrival rate then is governed by simple geometry. That is, the constant of proportionality is just the fraction of the total solid angle subtended by the wafer as seen from the crucible. This proportionality constant is the same view factor discussed in Chapter 6. It is given by [4]

$$k = \frac{\cos\theta\,\cos\phi}{\pi R^2} \tag{12.6}$$

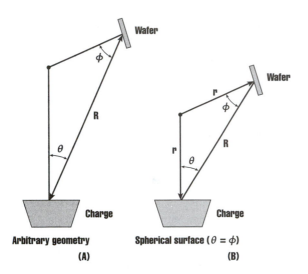

Figure 12-3   The geometry of deposition for a wafer (a) in an arbitrary position and (b) on the surface of a sphere.

where R is the distance between the surface of the crucible and the surface of the wafer and θ and ϕ are the angles between R and the surface normals of the crucible and wafer, respectively (See Figure 12.3a).

Equation 12.6 suggests that the deposition rate in the evaporator will depend on the location and orientation of the wafer in the chamber. Wafers directly above the crucible will be coated more heavily than wafers off to the side. Furthermore, θ, ϕ, and R actually vary across the surfaces of the crucible and the wafer so that film uniformity is a concern. One method commonly used to obtain good uniformity is to place the crucible and wafers on the surface of a sphere (Figure 12.3b). Then

$$\cos\theta = \cos\phi = \frac{R}{2r} \qquad (12.7)$$

where r is the radius of the sphere. Combining Equations 12.5, 12.6, and 12.7, and recognizing that the deposition rate is just the mass arrival rate per unit area divided by the mass density of the film (ρ),

$$R_d = \sqrt{\frac{M}{2\pi k\rho^2}} \frac{P_e}{\sqrt{T}} \frac{A}{4\pi r^2}. \qquad (12.8)$$

In this equation, the first term depends only on the material to be evaporated, the second term depends on the temperature (and therefore the equilibrium vapor pressure), and the third term is determined by the geometry of the chamber.

To accommodate a spherical geometry, wafers are mounted in a hemispherical cage called a *planetary*. The wafers are flat, and so there will be small variations of the deposition rate across the surface of the wafer, but to a good approximation it can be ignored. Because all of the wafers are on the surface of the sphere, not only will the deposition occur at a uniform rate, the rate can be monitored at a single point in the chamber.

To obtain large deposition rates, evaporators are often operated with very high crucible temperatures. Then the vapor pressure in the region immediately above the crucible is high enough so that this region is in the viscous flow regime. Operation in this regime can lead to condensation of the evaporated materials into droplets. If these droplets travel to the surface of the wafer and adhere, the film can have very poor surface morphology. Operation in this regime also affects the deposition uniformity by creating a virtual source (Figure 12.4) at some distance above the crucible. To obtain the best uniformity, the evaporator must be run at low rates [5], although operation at these low rates requires extremely high vacuums to avoid contamination of the film.

The deposition rate is commonly measured using a quartz crystal rate monitor [6],[7]. This device is a resonator plate that is allowed to oscillate at the resonance frequency, which is then measured. The resonance frequency shifts due to the additional mass as material is deposited on top of the crystal. When enough material has been added, the resonance frequency will shift by several

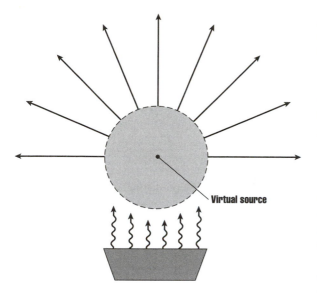

Virtual source

Figure 12-4   At high deposition rates, the equilibrium vapor pressure of the charge puts the region just above the crucible into viscous flow, creating a virtual source up to 10 cm above the top of the crucible.

percent and the oscillator will no longer show a sharp resonance. The sensing elements are quite inexpensive, however, and are easily replaced. By linking the output of the frequency measurement system to the mechanical shutters, the thickness of the deposited layers can be well controlled over a wide range of deposition rates. Furthermore, the time rate of change of the deposited thickness can be fed back to the crucible temperature to maintain a constant deposition rate, if so desired. Crucibles must be constructed with tapered walls to avoid a progressive shadowing of part of the planetary as the charge is depleted. However, this decreases the surface area of the crucible as the charge is consumed. Again, this is not a serious concern if a crystal film thickness monitor is used to control the process.

Since the chamber does not represent a perfect vacuum, gasses other than the vapor may also be incorporated in the deposited film. This deposition may be a simple physical mixing, or it may involve a chemical reaction. Because of the long mean free path in the chamber, if a chemical reaction occurs it will probably happen at the surface of the wafer. The incorporation of gaseous species may be intentional (reactive evaporation) or may be an unintentional incorporation of gasses due to vacuum leaks or incomplete chamber evacuation.

Example 12.1

An evaporator is used to deposit aluminum. The aluminum charge is maintained at a uniform temperature of 1100°C. If the evaporator planetary has a radius of 40 cm and the diameter of the crucible is 5 cm, what is the deposition rate of aluminum? If the chamber also has a background pressure of 10^{-6} torr of water vapor and the water vapor is assumed to be at room temperature, determine the ratio of the arrival rates of the aluminum atoms and the water molecules. The mass density of solid aluminum is 2.7 gm/cm^3 (2700 kg/m^3).

According to Figure 12.2, at 1100°C the vapor pressure of aluminum is about 1×10^{-3} torr. The area of the crucible is 19.6 cm^2. The atomic mass of aluminum is 27. Using Equation 12.8 with MKS units,

$$R = 8.45 \times 10^{-6} \, \frac{m^2\text{-sec}}{kg} \, \sqrt{K} \times 3.59 \times 10^{-3} \, \frac{kg}{m\text{-sec}^2\text{-}\sqrt{K}} \times 9.8 \times 10^{-4}$$

$$R = 2.9 \times 10^{-11} \, \frac{m}{sec} = 17.8 \, \frac{Å}{min}$$

The arrival rate of the aluminum atoms is just the growth rate times the number density of aluminum or

$$J_{Al} = R \times \frac{\rho}{M} = 1.78 \times 10^{18} \frac{atoms}{m^2\text{-sec}}$$

According to Equation 10.10 and 12.2, the arrival rate of the water (atomic mass 18) is

$$J_{H_2O} = \sqrt{\frac{P_{H_2O}^2}{2\pi k T_{H_2O} M_{H_2O}}} = 4.8 \times 10^{18} \frac{molecules}{m^2\text{-sec}}$$

Since the arrival rate of water is larger than that of aluminum, it would appear that the aluminum film will be heavily oxygen contaminated. In reality, the freshly evaporated aluminum film rapidly getters the residual oxygen and water vapor from the chamber. Most of this occurs before the shutter is opened. Such low deposition rates would still produce films with some oxygen contamination, however. This oxygen comes from the small amount of air that leaks past the chamber vacuum seals during the deposition. It is common practice therefore, to deposit aluminum with crucible temperatures of about 1200°C where the vapor pressure is above 10^{-2} torr and the growth rate is several hundred Å/min.

12.3 Step Coverage

As already mentioned one of the primary limitations of evaporation is step coverage. Figure 12.5 shows a schematic of the geometry of a film evaporated over a step. In this case, the step is the cross section of a contact etched through an insulating layer down to the substrate. Over this scale of distance ($\sim$1 μm), the incoming material beam can be considered nondivergent. Assuming that the incident atoms are immobile on the surface of the wafer, the topology will cast well defined shadows, and the film will frequently be discontinuous on one side of the contact. As the deposition proceeds, the growing film on top of the insulator moves the edge of the shadow upward. The problem of covering steps is particularly acute for metallization layers because these are the last steps in the process and the accumulated topology may be quite severe. Some limited improvement can be made by optimizing the wafer orientation [8].

One frequently used method of improving the step coverage is to rotate the wafer in the evaporated beam. Thus, the hemispherical cages used to hold the wafers in the evaporator are designed so as to roll the wafer around the top of the evaporator. The deposition rate on the sidewalls is still less than the rate on the flat surface but is axially uniform. The aspect ratio of the contact is defined as

Low adatom mobility and nonrotated

(A)

Heated and rotated

(B)

Figure 12-5 (a) Time evolution of the evaporative coating of a feature with aspect ratio of 1.0, with little surface atom mobility (i.e., low substrate temperature) and no rotation. (b) Final profile of deposition on rotated and heated substrates.

$$AR = \frac{\text{step height}}{\text{step diameter}} \qquad \textbf{(12.9)}$$

Standard evaporation cannot be used to form a continuous film over features with aspect ratios greater than 1.0 and is very marginal for aspect ratios between 0.5 and 1.0.

A second method to improve step coverage is to heat the wafer. Many evaporators use banks of infrared lamps or low intensity refractory metal coils behind the planetaries to do this. Atoms that reach the wafer can diffuse across the surface before they chemically bond and become part of the growing film. When shadowing leads to a concentration gradient, this random motion will result in a net movement of material into the low rate deposition regions. In a manner similar to bulk diffusion discussed in Chapter 3, one can define a surface diffusivity which, like bulk diffusivity, follows a simple Arrhenius function

$$D_s = D_o e^{-E_s/kT} \qquad \textbf{(12.10)}$$

The surface activation energy is much smaller than that of the bulk diffusivity so that significant diffusion can occur at temperatures of a few hundred degrees centigrade. If the mean time before incorporation is T, the characteristic surface diffusion length is

$$L_s = \sqrt{D_s T} \qquad \textbf{(12.11)}$$

Because of the exponential dependence of D_s on temperature, heating the wafer above room temperature can dramatically increase L_s. It is not unusual for L_s to be much larger than the feature sizes of interest on the wafer. Many groups have demonstrated filling high aspect ratio structures using this technique [9]. One concern with heating the substrate when applied to alloy deposition is that the surface diffusivities of the constituent atoms may be very different. Thus, the composition of the film at the bottom of a contact may be different than the composition at the top of the structure. A second concern is that the increased substrate temperature can affect the film morphology, often leading to large grains. This can be avoided by using an ion beam after evaporation in order to redistribute the deposit [10]. The ions strike the surface and transfer energy to the evaporated film. The atoms in the film may redistribute by diffusion or by sputtering (to be discussed later in the chapter). Few evaporators are equipped with this capability, however. As a result, evaporation does not generally have any capability for independent control of morphology and step coverage.

12.4 Evaporator Systems: Crucible Heating Techniques

There are three types of crucible heating systems: resistive, inductive, and electron beam systems. A resistively heated system is the simplest type of source. Given a high vacuum chamber with power feedthroughs, it is possible to construct a simple evaporator with only a small coil of wire and a simple variable transformer. The charge in such a system is a small solid bar laid over the heated element (Figure 12.6a). The input power is adjusted to prevent the charge from becoming molten and dripping through the coil. More practical arrange-

Figure 12-6 Resistive evaporator sources. (a) Simple sources including heating the charge itself and using a coil of refractory metal heater coil and a charge rod. (b) More standard thermal sources including a dimpled boat in a resistive media.

Figure 12-7   Example of an inductively heated crucible used to create moderately charged temperatures.

ments are shown in Figure 12.6b. The charge is contained in a crucible that is heated with a coil of wire.

Since the filament wire must be at least as hot as the material to be evaporated, one of the problems with resistively heated crucibles is evaporation and outgassing from the wire. If a material such as aluminum is to be deposited, adequate vapor pressures can be obtained with only moderate power input. On the other hand, if a refractory metal must be deposited, there is often no suitable resistive heating element. One way to achieve at least moderate charge temperatures is the use of inductively heated crucibles. As shown in Figure 12.7, a solid charge is placed in a crucible, typically made of boron nitride (BN). A metal element is wound around the crucible and RF power is run through the coil. The RF induces eddy currents in the charge causing it to heat. The coil itself can be water cooled to keep its temperature below 100°C, effectively eliminating any loss of material from the coil.

While inductive heating can be used to raise the crucible temperature high enough to evaporate refractory materials, contamination of the charge from the crucible itself remains a serious problem. This effect can be avoided by heating only the charge and cooling the crucible. A common method for accomplishing this is electron beam (e-beam) evaporation. A simple low flux electron beam system (Figure 12.8a) consists of a loop of heated tungsten wire surrounding a thin rod of material held at a high bias with respect to the wire. Electrons boiling off of the wire impact the rod, raising the temperature at the end of the rod and creating an atomic beam [11].

In most e-beam evaporators, an electron gun under the crucible ejects an intense, high energy beam. The location of the filament minimizes deposition of the filament material, typically W, on the surface of the wafer. A strong magnetic field bends the beam through 270° causing it to be incident

Figure 12-8   Electron beam evaporative sources. (a) A simple low flux source using a hot wire electron source and a thin movable rod. (b) A popular source using a 270° source arc in which the beam can be rastered across the surface of the charge. The magnet must be larger than shown to achieve the full 270° of arc.

Figure 12-9   Photograph of a commercial evaporator. Insert shows a planetary *(photographs courtesy of CHA Industries).*

on the surface of the charge (Figure 12.8b). The beam can be rastered across the charge to melt a significant fraction of the surface. The hot portion of the charge is then effectively self contained by the cooler portion of the charge.

Because of their ability to easily deposit a wide range of materials, e-beam evaporators (Figure 12.9) are commonly used in GaAs technologies. When thermionic emission electron guns are used (see Section 9.2), the hot electron filaments remain a source of contamination in the chamber. Special care must be taken in e-gun design when operating these systems at very high vacuums [12]. A more serious concern, particularly for silicon based technologies, is radiation damage. As discussed in Chapter 10, accelerated charged particles radiate. The radiation due to highly excited electrons in the material being evaporated decaying back to core levels is far more intense. Since x-rays will damage the substrate and the dielectrics, e-beam evaporators cannot be used in MOS or other technologies that are sensitive to this type of damage unless later thermal annealing steps are adequate to remove the damage.

It is often convenient to have multiple sources in the evaporator, even if only one of them is to be used at a time. Such an arrangement allows the deposition of different materials without opening the high vacuum chamber. For a resistive evaporator, a high current switching box can be used, with each sample crucible having its own heating coil. E-beam evaporators are particularly well suited to this application since the electron beam is easily steered between small charges using an electrostatic potential and perhaps a change in the magnetic field. Alternatively, a different charge can be moved into the beam mechanically by mounting them on a wheel that can be rotated into the irradiated position.

12.5 Multicomponent Films

Often it is desirable to deposit alloys and compounds of materials. Figure 12.10 shows three possible methods for depositing alloy films using evaporation. It is possible to simply evaporate materials with very similar vapor pressures such as Al and Cu [13] from a properly prepared compound target. In some applications, such as the formation of ohmic contacts to GaAs, the vapor pressure of the constituent species is reasonably close and the variation in composition of the alloy is acceptable [14]. If, however, we place a solid sample of TiW with the desired composition in a crucible and evaporate it, the primary evaporated material may not be TiW, but rather some other combination of Ti and W. For example, at a crucible temperature of 2500°C, the vapor pressure of Ti is about 1 torr, while the vapor pressure of W is only 3×10^{-8} torr. The

Figure 12-10   Methods for evaporating multicomponent films include: (a) single source evaporation, (b) Multisource simultaneous evaporation, and (c) Multisource sequential evaporation.

vapor that comes off initially is almost pure Ti. As the remaining melt composition changes due to this evaporation, the composition of the deposited film would slowly drift.

The basic problem with this process is the difference in vapor pressure of the various components. Controlling the composition with any reasonable accuracy is difficult; controlling it in many compounds is impossible. In coevaporation, multiple sources are run simultaneously in order to deposit an alloy structure. To deposit TiW, for example, two crucibles, one containing W and one containing Ti, are run at different temperatures. While this is a significant improvement over a single crucible process, the problem remains that the vapor pressure and ultimately the deposition rate are extremely sensitive functions of the temperature of the charge. In a single component film, the absolute deposition rate is not that important, since a deposited film thickness monitor can be used to open and close the shutter. Unless multiple rate monitors are used in different regions of the evaporator [15], this type of system has no ability to control the flux of each component separately. During coevaporation therefore, even reasonable control of the film composition requires extraordinary temperature control.

An alternate technique for depositing multicomponent films is to preform a sequential deposition [16]. This is readily accomplished in a multiple source system by opening and closing shutters. After the deposition is complete, an alloy can be formed by elevating the sample temperature and allowing the components to interdiffuse. This can be assisted by depositing a large number of very thin layers with alternating compositions. The process requires the wafer to tolerate a subsequent high temperature step. The integrating time of the crystal rate monitor however, prevents the use of this method for extremely thin layers without using very low deposition rates. Alternatively, the deposition can be controlled with the use of a mass spectrometer [17] or by measuring the mass of a deposited film on a sensor in the beam of each constituent with a microbalance [18].

12.6 An Introduction to Sputtering

Sputtering is the primary alternative to evaporation for metal film deposition in microelectronics fabrication. First discovered in 1852 [19], sputtering was developed as a thin film deposition technique by Langmuir in the 1920s [20]. It has better step coverage than evaporation, induces far less radiation damage than electron beam evaporation, and is much better at producing layers of compound materials

Figure 12-11   Chamber for a simple parallel plate sputtering system.

and alloys. These advantages have made sputtering the metal deposition technique of choice for most silicon based technologies.

A simple sputtering system, as shown in Figure 12.11 is very similar to a simple reactive ion etch system, a parallel plate plasma reactor in a vacuum chamber. In a sputtering application however, the plasma chamber must be arranged so that a high density of ions strikes a target containing the material to be deposited. (The requirements for obtaining a large ion flux were discussed in Chapter 11.) In sputtering, the target material, not the wafers, must be placed on the electrode with the maximum ion flux. To collect as many of these ejected atoms as possible, the cathode and anode in a typical sputtering system are closely spaced, often less than 10 cm. An inert gas is normally used to supply the chamber. The gas pressure in the chamber is held at about 0.1 torr. This results in a mean free path of order hundreds of microns.

Due to the physical nature of the process, sputtering can be used for depositing a wide variety of materials. In the case of elemental metals, simple dc sputtering is usually favored. When depositing insulating materials such as SiO_2, an RF plasma must be used [21]. If the target material is an alloy or compound, the stoichiometry of the deposited material may be slightly different than the target material (Table 12.1). It has been shown however, that the material with a lower sputter yield will accumulate on the surface of the target until the composition of the deposited film is approximately that of the bulk of the target [22]. (This is only true if the target temperature is kept sufficiently low to prevent solid state diffusion.) This makes sputtering very attractive for depositing not only elements, but a very wide range of materials.

12.7 Physics of Sputtering°

The topic of glow discharges was covered in the latter half of Chapter 10. To briefly review, a plasma is initiated by applying a large voltage across a gap containing a low pressure gas. The required breakdown voltage is given by Paschen's law:

$$V_{bd} \propto \frac{P \times L}{\log P \times L + b} \tag{12.12}$$

Table 12.1   Composition of aluminum alloy films sputtered from composite targets

Aluminum Alloy Material	Target	Film
Si	0.5%–1.0%	0.86%
Si	2%	2.8%
Cu	3.9%–5.0%	3.81%
(Al + Si)Si	2%	2% ± 0.1%
(Al + Cu + Si)Si	4%	3.4%

See Wilson and Terry [23].

Table 12.2   First and second ionization potentials of some common gasses

Atom	First Ionization (eV)	Second Ionization (eV)
Helium	24.586	54.416
Nitrogen	14.534	29.601
Oxygen	13.618	35.116
Argon	15.759	27.629

See Konuma [24].

where P is the chamber pressure, L is the electrode spacing, and b is a constant. Once a plasma is formed, ions in the plasma are accelerated toward the negatively charged cathode. When they strike the surface, they release secondary electrons, which are accelerated away from the cathode. They may collide with neutral plasma species while crossing from cathode to anode. If the energy transfer is less than the ionization potential of the gaseous species, the atom can be excited to an energetic state (Table 12.2). The atom decays from this excited state through an optical transition, providing the characteristic glow. If the energy transfer is high enough, however, the atom will ionize and be accelerated toward the cathode. The bombardment of the cathode in this ion stream gives rise to the process of sputtering.

When an energetic ion strikes the surface of a material, four things can happen. Ions with very low energies may simply bounce off the surface. At energies of less than about 10 eV, the ion may also adsorb to the surface, giving up its energy to phonons (heat). At energies above about 10 keV, the ion penetrates into the material many atomic layer spacings, depositing most of its energy deep into the substrate, where it changes the physical structure. These high energies are typical for ion implantation. Between these two extremes, both energy transfer mechanisms occur. Part of the ion energy is deposited in the form of heat. The remainder goes into a physical rearrangement of the substrate. At this low energy, nuclear stopping at the surface is quite effective. Most of the energy transfer occurs within several atomic layers. When this happens, substrate atoms and clusters of atoms will be ejected from the surface of the substrate. The atoms and atomic clusters ejected from the cathode escape with energies of 10 to 50 eV. This is about 100 times the energy of evaporated atoms. This additional energy provides sputtered atoms with additional surface mobility for improved step coverage relative to evaporation. At typical sputtering energies about 95% of the ejected material is atomic. Most of the remainder is diatomic molecules [25].

At high energies like those used in implantation, chemical bonding processes can be largely ignored and the target can be considered as simply a collection of atoms. At very low energies, no disruption of the target occurs, and a chemical model can be readily developed. At sputtering energies however, the physics of the material removal is quite complicated, involving the coupled effects of bond breaking and physical displacement. Figure 12.12 shows some of the processes that may occur when an ion strikes a surface. A simple

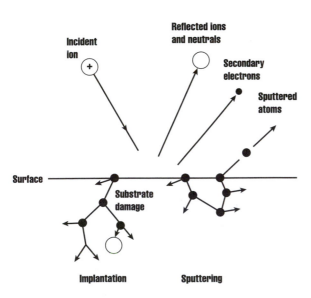

Figure 12-12   Possible outcomes for an ion incident on the surface of a wafer.

model developed by Wehner [26], ignoring chemical effects and treating the substrate atoms as hard spheres, provides at least a qualitative picture of sputtering. An ion incident on the target surface may travel several atomic layers into the target until it strikes an atom with a small impact parameter and is deflected through a large angle. This near head on collision may also liberate a target atom that has a large momentum directed at a significant angle with respect to the surface normal. During this process, many of the bonds in the top layers of the target will be broken. If several of these large angle collisions occur, the incident atom or the recoiled target atom may develop a significant velocity component parallel to the surface of the wafer. A subsequent collision can then eject an atom or small cluster of atoms.

12.8 Deposition Rate: Sputter Yield

The sputter deposition rate depends on the ion flux to the target, the probability that the impact of an incident ion will eject a target atom, and the transport of the sputtered material across the plasma to the substrate. The ion flux in a dc plasma can be approximated by the Langmuir-Child relationship:

$$J_{ion} \propto \sqrt{\frac{1}{m_{ion}}} \frac{V^{3/2}}{d^2} \tag{12.13}$$

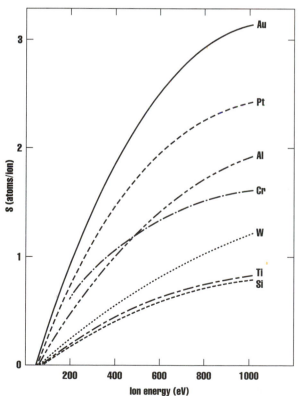

where V is the voltage difference from the target electrode to the wafer, d is the dark space thickness, and m_{ion} is the mass of the ion. The value of the voltage exponent is not always found to be 1.5.

The sputter yield S is the ratio of the number of target atoms ejected from the target to the number of ions incident on the target. It depends on the ion mass, the ion energy, the target mass, and the target crystallinity. Figure 12.13 shows the sputter yield as a function of the ion energy for a variety of materials in an argon plasma [27]. For each target material, there exists a threshold energy, below which no sputtering occurs. This energy is typically in the range of 10 to 30 eV [28]. Anderson and Bay [27] have shown that for target atomic masses greater than five times the incident ion atomic mass, the threshold energy is given by

$$E_{th} = \frac{U_o}{\gamma(1 - \gamma)} \tag{12.14}$$

where U_o is the heat of vaporization of the target material and γ is the energy transfer parameter given by

$$\gamma = \frac{4M_1 M_2}{(M_1 + M_2)^2} \tag{12.15}$$

Figure 12-13 Sputter yield as a function of ion energy for normal incidence argon ions for a variety of materials *(after Anderson and Bay, reprinted by permission).*

The threshold energy reaches a maximum when $M_1 \approx M_2$.

Figure 12-14 Sputter yield as a function of the bombarding ion atomic number for 45 keV ions incident on silver, copper, and tantalum targets *(after Wehner, reprinted by permission, AIP).*

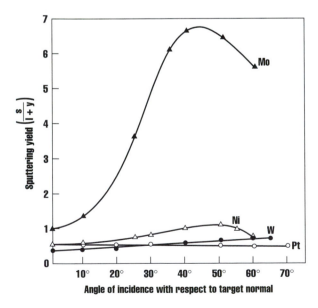

Figure 12-15 Typical angular dependence of the sputter yield for several different energies. The sputter profiles follow a cosine distribution *(after Wehner, reprinted by permission, AIP).*

For ions whose energy is slightly greater than threshold the sputter yield increases as the square of the energy up to about 100 eV, then linearly with energy up to about 750 eV. Above 750 eV, the yield increases only slightly until the onset of implantation [29]. The maximum sputter yield typically occurs at about 10 keV. The sputter yield is also a weak function of the plasma composition, generally increasing with ion mass (Figure 12.14) [30].

The angular dependence of the sputter yield is a strong function of the target material and the incident ion energy. Materials such as Au, Pt, and Cu, which have high sputter yields, generally have little angular dependence. Materials such as Ta and Mo, which have low sputter yields, have a pronounced angular dependence at low ion energies. The maximum in the sputter yield occurs at about an 80° incidence angle [31]. Figure 12.15 shows typical angular distributions for several materials [32]. At low energies a minimum exists at near normal incidence in a profile known as an under cosine. At high energies the sputter yield approaches [33]

$$S \propto \frac{M_{gas}}{M_{target}} \frac{\ln E}{E} \frac{1}{\cos \theta}$$

(12.16)

where θ is the angle between the target normal and the velocity vector of the incident ion. At these energies the angular dependence of the net deposition rate approaches a simple cosine similar to that of evaporation. Crystalline targets however, may have less regular distributions with maxima in the directions along the low Miller indices of the target.

The transport across the discharge is a complicated function involving computation fluid dynamics, including ion drift and diffusion. The normalized net thickness distribution, including transport across the electrode spacing, for a simple circular planar target of radius S is given by [6]

$$\frac{d}{d_o}(r) = \left[\frac{h^2}{2S^2} + \frac{1}{2}\right] - \frac{1 + \left[\frac{r}{h}\right]2 - \left[\frac{S}{h}\right]^2}{\sqrt{\left[1 - \left[\frac{r}{h}\right] + \left[\frac{S}{h}\right]^2\right]^2 + 4\left[\frac{r}{h}\right]^2}}$$

(12.17)

where the electrode separation is h and the distance from the center of the plate is r. To improve the uniformity, many sputtering systems mechanically scan the wafers past a target. In some systems, the wafers may also be rotated.

12.9 Magnetron Sputtering

As discussed in Section 10.7, the application of a magnetic field in a plasma causes the electrons to spiral around the direction of the magnetic field lines. The radius of this orbital motion is given by

$$r = \frac{mv}{q} B$$

(12.18)

Figure 12-16 Planar and cylindrical magnetron sputtering systems T: target; P: plasma; SM: solenoid; M: magnet; E: electric field; B: magnetic field *(after Wasa and Hayakawa, reprinted by permission, Noyes Publications).*

The orbital motion of the electrons increases the probability that they will collide with neutral species and create ions. This increased ion density decreases the Crooke's dark space and increases the rate of ion bombardment of the target. Typical ion densities in normal sputtering systems are 0.0001%, while in magnetron systems it often approaches 0.03%. The use of a magnetron also allows the formation of a plasma at lower chamber pressures, typically 10^{-5} to 10^{-3} torr.

Figure 12.16 shows two geometries that have been used for magnetron sputtering. A planar magnetron starts with the basic parallel plate reactor shown in Figure 12.11. Either a solenoid coil [34] or a set of magnets is added behind the target in such a way as to create magnetic field lines parallel to the surface of the target. Circular planar magnetron targets can be made by arranging the bar magnets radially from the center of the target. A cylindrical magnetron starts with a cylindrical plasma chamber with a target electrode at the center. The wafer holding electrode is the vertical wall of the chamber. The magnetron is added by inserting the chamber in an electromagnetic winding that generates a vertical field [35]. This type of system is of less interest to microelectronic production due to the difficulty with placing large diameter wafers on the walls of a cylindrical chamber. A typical field for either type of system is a few hundred Gauss.

Figure 12.17 shows a detailed cross section of a planar magnetron target. The target material is typically hot pressed sintered disks. Most of the power in the plasma is dissipated in the form of heat in the target. To avoid excessive heating, the target is typically water cooled. Deionized water is often used to prevent an electrical short. The target material is bonded to the backing plate using conductive epoxy or is held mechanically. To avoid sputtering the edge of the target or backing material, the edge of the target is surrounded by a ground shield. One of the concerns with the simple magnetron target, as shown in the lower half of Figure 12.17, is that the material in some parts of the target will erode much more rapidly than the material in other areas, resulting in a low utilization efficiency, commonly 25 to 35%. Other geometries have been shown to yield much better efficiencies [36]. Sputter utilization efficiency is especially a problem in modern systems where wafer load locks are used to avoid opening the chamber for long periods of time. These systems may be unusable for several days after a target change while the main chamber is pumped to high vacuum.

One interesting target geometry [37], shown in Figure 12.18, is the S-Gun[2]. As in other magnetron systems, high deposition rates are achieved through crossed electric and magnetic fields. The anode and cathode are built into a single assembly. The I–V characteristics of a magnetron system are shown in Figure 12.19 [38],

Figure 12-17   Detailed cross section of a rectangular planar magnetron target using permanent magnets to supply the field *(after Wasa and Hayakawa, reprinted by permission, Noyes Publications)*.

Figure 12-18   Cross section diagram of an S-Gun® sputtering source *(courtesy Varian Associates)*.

---

[2]S-Gun is a registered trademark of Varian Associates.

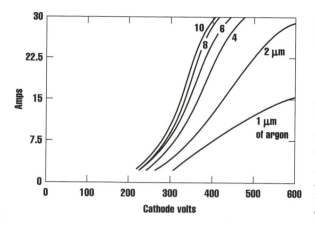

Figure 12-19   Current voltage characteristics of a magnetron system with chamber pressure as a parameter. The lines are for 1, 2, 4, 6, 8, and 10 μm of Ar *(after Van Vorous, reprinted by permission, Solid State Technology)*.

[39]. Such curves are characteristic of all magnetron sputtering and are of the form [40]

$$I = k(V - V_o)^n \qquad (12.19)$$

where V_o is the minimum voltage necessary to maintain the plasma, and n is a positive number between 1.5 and 8, whose value depends on the geometry of the target and the pressure of the chamber. With an S-Gun, the wafers may then be mounted in a planetary cage, similar to the geometry of an evaporator. (One of the major applications of S-Guns is to upgrade an evaporator to have sputter deposition capability.) Typical deposition rates are hundreds of angstroms per minute.

12.10 Morphology and Step Coverage

Because the chamber pressure during most sputtering processes is high, ejected atoms will suffer many collisions before reaching the surface of the wafer. As a result, the deposition rate on the surface of the wafer will depend on the solid angle of the plasma at that point on the wafer. Once they reach the surface of the wafer, the adatoms will diffuse along the surface until they form nuclei of critical size. Once stable nuclei are formed, they capture more adatoms, forming islands. If the surface mobility is high, the islands merge while still very thin, forming a smooth, continuous film.

The morphology of the deposited film has been described by a zone model first introduced by Movchan and Demchishin [41] and later modified by Thornton [42]. Figure 12.20 shows a schematic of the regions of film morphology as a function of substrate temperature and incident ion energy. The substrate temperature is normalized to the melting temperature of the film. While the model is broadly applicable, the region boundaries vary somewhat from material to material. At the lowest temperature and ion energy, the film will be an amorphous, highly porous solid with a low mass density. This is the first zone of the diagram. It is caused by the low adatom mobility of the growing film. Metal films deposited in this region can readily oxidize when exposed to air and so may also have high resistivities. If the chamber pressure is lowered or the substrate temperature is raised, the deposition process enters the ''T'' zone. Films deposited in this region are highly specular and have very small grains. For many microelectronic applications, this is the most desirable region of operation. Increasing the temperature and/or impinging energy further causes the grain size to increase. The second zone has tall narrow columnar grains that grow vertically from the surface. The grains end in facets. Finally, in Zone 3, the film has large 3-D grains. The surfaces of the films in the second and third zones are moderately rough, and the films appear milky or hazy.

Figure 12-20   The three zone model of film deposition as proposed by Movchan and Demchishin *(after Thornton, reprinted by permission, AIP)*.

Figure 12-21   Cross section of the time evolution of the typical step coverage for unheated sputter deposition in a high aspect ratio contact.

The most common microelectronic application of sputtering is the deposition of metal interconnect layers for ICs. Typically these layers are applied over thick insulating films such as SiO_2. To contact the device, holes are first etched in the SiO_2. As discussed earlier in the chapter, a crucial feature of the deposition process is the ability of the films to maintain an adequate thickness, even over high aspect ratio structures. The step coverage of films deposited by magnetron sputtering has been calculated by many authors [43],[44]. Figure 12.21 shows the development of a cross section over a typical high aspect ratio contact. On the top surface and near the upper corner the deposition rate is high. A more moderate deposition rate occurs on the sidewalls. The sidewall thickness tapers toward the bottom. At the bottom corner of the step a pronounced notch or crack may occur. This tendency increases with the aspect ratio of the feature unless substrate heating is done [45]. Compared to evaporated films, however, even low temperature deposited sputtered films have better step coverage, both because of the higher pressure and because of the incident energy of the deposited species. Sputtered films also tend to form a cusp, or protrusion, at the edge of an isolated step, due to the increased view factor at this position.

As with evaporation, the application of substrate heat will dramatically improve the step coverage due to surface diffusion. The wafer containing electrode is often water cooled as well as resistively heated to provide a wide range of available temperatures. This is a very active area of thin film deposition research. Very hot metal can completely fill contacts whose aspect ratio is greater than one, but if aluminum is used, a very robust barrier must be used between the silicon and the aluminum to avoid undesirable intermixing (see Chapter 15 on contact spiking). Furthermore, the temperature of the surface of the wafer during plasma deposition is difficult to control. Radiative heating from the target and bombardment by high energy secondary electrons present a large heat load to the wafer. RF sputtering systems are the most prone to uncontrolled substrate heating. Figure 12.22 shows a typical plot of substrate temperature versus plasma power for an SiO_2 target [46].

Step coverage remains a serious problem for high density interconnect in purely sputtered interconnect technologies. Sufficient heating of the substrate during deposition in order to obtain the desired step coverage may produce unacceptably large grains or interdiffusion effects (see Al/Si/Cu deposition below). A second technique for improving the step coverage in sputtering is to apply an RF bias to the wafers. If the bias is sufficiently large, the wafer will begin to be bombarded by energetic ions. This will tend to re-deposit the sputtered material, somewhat improving step coverage.

Figure 12-22   The temperature rise of substrates as a function of the plasma power for an argon RF plasma with an SiO_2 target *(after Wasa and Hayakawa, reprinted by permission, Noyes Publications).*

12.11 Sputtering Methods

All exposed surfaces in an RF plasma develop a negative potential with respect to the plasma due to the higher mobility of electrons than ions. In a typical sputtering system, most of the voltage drop is on the target electrode, but the bias on the substrate electrode leads to a bombardment on ions of the wafers as well. The bombardment leads to a removal of material from the surface of the wafer. This effect can be controlled by adjusting the dc bias on the electrode with respect to the plasma. This has two major applications in microelectronics: sputter cleaning and bias sputtering. The removal of all surface contaminants from the wafer before film deposition has been studied extensively for low temperature epitaxial growth. That topic will be covered in a later chapter. This section will restrict its attention to a purely physical method of cleaning: sputter etching of the substrates. A typical example of the use of a sputter deposition process is the deposition of an aluminum alloy to form an ohmic contact to heavily doped silicon. The contacts are patterned and etched through a thick insulating layer of SiO_2. Immediately before loading into the sputtering system, the wafers are dipped in a dilute mixture of hydrofluoric acid and water (1:100) to remove any oxide that has regrown on the silicon after the contact etch. The wafers are rinsed in deionized water, spun dry, and loaded directly into the vacuum system for deposition. The brief exposure to a water rinse and the air however, can allow a regrowth of a thin, patchy native oxide. To obtain a reliable low resistance contact, it is desirable to remove this very thin oxide layer before metal deposition.

By reversing the electrical connections, it is possible to sputter from the substrates rather than from the target. This is frequently done for a short time before deposition to remove the native oxide and any residual contaminants from the surface of the wafer [47]. Sputter etching has serious problems, however. Sputtered material from the substrate electrode or from the oxide coated regions of the wafer may deposit on the surface, leading to more contamination rather than less. The contamination may have serious consequences if it contains heavy metal impurities that cause junction leakage [48]. Organic contaminants, such as those obtained from condensed pump oil vapors, may be polymerized, making them very difficult to remove [49],[50].

Sputter etched silicon layers show damage extending 40 to 110 Å into the wafer [51]. The surface of the silicon may contain up to 20 atomic percent Ar, depending on the bias condition [52]. Material removal from the surface of the wafer may also be nonuniform, leading to the formation of steep cones and etch valleys [53]; however, for a typical sputter etch the desired etch depth is less than 100 Å, which is typically less than 10% of the junction depth. The development of a sputter preclean step however, must be optimized experimentally for each sputtering system by measuring the effects of the sputter cleaning step on the contact resistance, contact reliability, and the junction leakage.

It is possible to adjust the bias on the substrate with respect to the plasma if the substrate and the deposited films are conductive. By placing a negative bias on the substrate, the ion bombardment of the substrate is increased. By controlling the bias, one can change the rate of deposition independent of the rate of sputter etching of the growing film. Since the sputter etched film may redeposit on the wafer if sputtered at low bias, a net improvement in the step coverage may be achieved (Figure 12.23). Vossen has shown that this technique can be used to actually produce a higher deposition rate on the sidewalls of a contact than on the surface [54]. High sputter etch rates, however, may lead to damage of the underlying substrate and faceting of the deposited layers. At lower bias voltages, the incident ion energy may also improve step coverage by increasing the adatom mobility [55].

Figure 12-23   In bias sputtering, the ions incident on the surface of the wafer redistribute the deposited film to improve step coverage.

It is also possible and often desirable to clean the target before beginning the deposition. To presputter, the plasma is ignited before opening the shutter so that the material on the

Figure 12-24 Photograph and schematic of a commercial sputtering system used for silicon IC fabrication *(courtesy Balzers)*.

top of the target is deposited on the back side of the shutter instead of on the substrates. A common application of presputtering is the removal of the native oxides formed on metal targets. For this application, the change in the glow discharge current can be used to determine the removal of the oxide [56]. The surface oxides generally have high secondary electron emission rates, and so the discharge current falls to a steady state value once the target is clean. Presputtering of these reactive species also getters the residual reactive gasses such as O_2, H_2O, and N_2 from the chamber, both by gas phase reactions and by coating the back surface of the shutter with a highly reactive film.

Figure 12.24 shows a photograph and schematic of a Balzers LLS 802 production sputtering system. The top of the process chamber has been cut away in the photograph to reveal two of the planar magnetron sources. The wafers are introduced into the process area (2) through the loadlock (1) by opening the gate valve (3). Also contained in the loadlock area is an RF sputter station (8) that can be used to preclean the wafers before deposition. In the process chamber, the wafers are mounted on a drum that rotates in the process chamber, bringing the wafers before each target (5) sequentially.

A shutter (9) can be opened in front of the targets to be used. The process chamber also contains a heater (6) to adjust the substrate temperature during deposition.

12.12 Sputtering of Specific Materials

Many reports have been given regarding the utility of aluminum based sputtered metallizations for silicon based ICs [57],[58]. Pure aluminum has been replaced by aluminum silicon alloys to increase the reliability of ohmic contacts formed to shallow junctions. Typical silicon concentrations are 0.5 to 2.0 atomic percent. The addition of 0.5 to 1.0 atomic percent copper also reduces the tendency of the metal film to form hillocks [59] and dramatically improves the ability of lines formed from these films to pass high currents without electromigration degradation or stress induced voiding. Each of these effects will be discussed in Chapter 15. The first part of this section will review the sputter deposition of aluminum alloy films.

To obtain large sputter deposition rates, most aluminum is deposited in planar dc magnetron systems. Since all of these films are alloys, control of the stoichiometry of the film is a primary concern. As discussed previously, the deposited film composition is usually close to the bulk target composition. At moderate substrate temperature where re-evaporation of the deposited material can be neglected, the exact composition of the film is controlled by the transport properties of the constituents in the plasma [60]. At low chamber pressures, for example, sputtering from an AlCu target will result in slightly higher Cu concentrations in the film than in the target. The high copper composition is related to the ability of the Ar gas to thermalize the very light aluminum atoms, while collisions with the much heavier copper atoms have little effect [61]. This result, called *selective thermalization* [62], is most pronounced with sputtering materials like AlCu or TiW, in which one material has a much larger atomic mass than the other.

One way to obtain better control of the stoichiometry is to have multiple targets. By adjusting the power to each target, the composition of the deposited layer can be altered. A second method of controlling composition that doesn't require a second power supply is the use of composite targets that have different regions of concentration. It is possible, for example, to construct a target with cylindrically symmetric regions of two different materials. By changing the electrical properties of the plasma, one can control the composition of the deposited layer [63]. In most production sputtering however, the target composition is simply chosen so as to give the desired film composition for a particular process.

Film resistivity is a sensitive function of the base pressure and gas composition in the chamber. Aluminum reacts readily with oxygen, nitrogen, and water. The resistivity rises sharply for nitrogen partial pressures above 10^{-6} torr [64]. If the primary source of nitrogen is an incomplete pump out, much of the nitrogen will be gettered during the deposition. In this case, a presputter may substantially clean the chamber. More serious are small vacuum leaks that contribute nitrogen through the bulk of the deposit. To reduce the film resistivity, many production systems designed for aluminum deposition use a load lock to avoid the necessity of venting the main chamber to air. Wafers may be baked at temperatures greater than 400°C or may be exposed to deep UV sources to desorb most of the adsorbed species in the load lock prior to high vacuum chamber entry. A low temperature stage called a Meissner trap [65] is often used to condense residual gasses and so to reduce the pumpdown time.

Another variable of primary importance for IC fabrication is the deposited film reflectance. Films that have a low reflectance often appear hazy or milky. The large grain size of these films make photolithography difficult, both because of an inability to see the alignment marks from previous layers and because of stray light scattered off of the aluminum grains. The minimum reflectivity

Figure 12-25   Reflectance of Al-Si layers at 400 nm as a function of the partial pressures of N_2, H_2, and H_2O *(after McLeod and Hartsough, reprinted by permission, AIP).*

required for most UV photolithography is about 0.6. The factors that influence the specularity of Al based metallizations include the substrate temperature [66], film thickness [67], and residual gasses in the chamber [68]. It is found that the reflectivity obeys the relationship

$$R \propto e^{-[4\pi\sigma/\lambda]^2} \qquad \textbf{(12.20)}$$

where λ is the wavelength of light and σ is the root mean square (RMS) surface roughness, as long as $\lambda \gg \sigma$ [69]. Since the material on which the aluminum is deposited is also known to affect the morphology of the deposited film [70], this may also affect the film specularity.

Figure 12.25 shows the percent reflectance of Al-Si layers deposited at a constant rate as a function of the partial pressures of N_2, H_2, and H_2O. Generally, the higher the deposition rate, the less the effects of the residual gasses. Once again, nitrogen has the most pronounced effect, seriously degrading the reflectivity if its partial pressure is greater that 10^{-6} torr. While H_2 requires a partial pressure almost ten times higher than N_2 for the same effect, hydrogen pumping is much less efficient. In a turbomolecular pump, hydrogen has the lowest compression ratio because of its light mass. In a cryopump, the pumping rate of hydrogen is low due to its low boiling point. Production systems may therefore use a getter pump, such as a titanium sublimation pump [71], to remove most of the hydrogen during the pumpdown. State of the art sputter systems commonly operate with base pressures of about 1×10^{-8} torr.

Finally, as discussed above, step coverage is a crucial issue in sputter deposited conductors. The primary variable in determining step coverage is the wafer surface temperature. This temperature may be partially controlled by adjusting the temperature of the substrate. In most cases however, the aluminum is being deposited over a thick layer of SiO_2. Power is deposited in the film through radiation transfer, secondary electron bombardment, thermal excitation of plasma species, and the heat of condensation of the aluminum vapor. Due to the low thermal conductivity of SiO_2 and a poor contact between the wafer and the substrate, the temperature of the aluminum may be 50 to 200°C higher than that of the substrate. As a result, optimization of the deposition process must be done for each technology. The film temperature generally desired is 300–500°C. Much less than this and step coverage will be poor. Much larger than this and the contacts will not be reliable, due to interdiffusion of the substrate silicon and the aluminum.

When compounds whose components have very different sputter yields are to be deposited and where one component can be obtained in a gaseous precursor, it may be preferable to deposit the film reactively. Reactive sputtering is a process in which the normally inert sputter gas is replaced by a inert/reactive mixture. The composition of the deposited film can be controlled by varying the partial pressure of the reactive species in the plasma. The deposition rate is also affected by changing the partial pressure of the reactive species. Using reactive sputtering, the film composition can be smoothly controlled over a wide range of compositions. Because of the range of energies available to drive chemical reactions in the plasma, many compounds may be produced. The task is to select deposition and post deposition annealing conditions so as to preferentially form the desired compound or phase.

Figure 12-26   Resistivity and composition of reactively sputtered TiN as a function of the N$_2$ flow in the sputtering chamber *(after Tsai, Fair, and Hodul, reprinted by permission, The Electrochemical Society, and Molarius and M. Orpana, reprinted by permission, Kluwer Academic Publishing).*

Figure 12-27   Cross section electron micrograph of a high aspect ratio contact that has been sputter deposited with TiN *(after Kohlhase, Mändl, and Pamler, reprinted by permission, AIP).*

As will be discussed in Section 15.7, there is a requirement in modern devices to form ohmic contacts to very thin junctions. To prevent an interdiffusion of aluminum and silicon, a thin barrier metal is sometimes used. One of the most popular barrier metals is TiW; however, sputtering of TiW from a compound target produces high particle counts and so can be a serious problem for yielding dense ICs. As an alternate diffusion barrier, there has been a great deal of work in TiN films [72], both produced by chemical vapor deposition and by sputtering [73]. TiN is unusual in its low resistivity, high thermal and chemical stability, and extreme hardness. (It has also received considerable attention as a material for coating machining workpieces for reduced wear.) The deposition of TiN will therefore be used as a vehicle to describe the process of reactive sputtering.

The primary variable for controlling the composition of the deposit is the nitrogen partial pressure in the sputtering system. Figure 12.26 shows the resistivity and composition of reactively sputtered TiN$_x$ films as a function of nitrogen flow. A sharp jump in resistivity is seen at the onset of TiN formation [74]. Since it is difficult to maintain exact stoichiometry [75], the deposition is usually done with a nitrogen rich ambient. Sharp corners or edges of the steps to be covered should be avoided to minimize stress cracking of the barrier metal (Figure 12.27) [76]. Films deposited in a nitrogen rich ambient however, have excess gas molecules that can passivate any microcracks that might develop in the film after deposition [77]. These films are known as stuffed barriers in the literature [78]. The films may also be annealed in a nitrogen or ammonia [79] ambient after deposition to increase the nitrogen packing and decrease the stress in the film. The residual gas partial pressures of H_2O and O_2 in the chamber should be less than 10^{-6} torr to avoid significant contamination of the film [80]. Application of bias increases the film density for unheated depositions (zone 1 to zone T in Figure 12.20) [81]. Finally, it has been shown that a sputter cleaning of the substrate prior to reactive TiN sputtering may be necessary to obtain a low leakage diode when depositing on n-silicon [82].

12.13 Stress in Deposited Layers

A thin film deposited on a substrate can be either in tensile stress, in which the film would relax by contracting, or it may be in compressive stress. If the stress in the film is too large, the film may peel from the surface of the wafer. The film stress and structure play important roles in the performance, and particularly in the reliability, of the metal lines. Large stress may give rise to void formation upon subsequent thermal cycling where the strain state provides a driving force for grain boundary diffusion.

One component of the stress is caused by thermal expansion mismatch of the film with the substrate. This stress is seen when the deposition is not done at room temperature. If E is Young's modulus for the material and v is Poisson's ratio, and it is assumed that E and v are temperature independent [83], this stress is given by

$$\sigma_{th} = \frac{E_{film}}{1 - v_{film}} \int_{T_o}^{T_{dep}} (\alpha_{film} - \alpha_{sub}) dT \qquad \textbf{(12.21)}$$

Intrinsic stress is less well understood. As already noted, films may form single crystal grains during the deposition process. For polycrystalline films, stress can arise when the deposition is carried out at high temperature. Under these conditions, the larger grains may grow at the expense of the smaller grains. The increase in the crystalline order will lead to an overall stress. Thus, the intrinsic stress can depend on such variables as the substrate temperature, deposition rate, film thickness, and background chamber ambient. Generally, if the deposition is carried out at low temperature, the intrinsic stress is small. Taking into account both sources of stress,

$$\sigma = \sigma_{th} + \sigma_{bi} \qquad \textbf{(12.22)}$$

Stress in a film is commonly measured using the change in the bow of the wafer before and after film deposition (Figure 12.28). Then the film stress is given by

Tensile stress   **Compressive stress**

$$\sigma = \frac{\delta}{t} \frac{E}{1 - v} \frac{T^2}{3R^2} \qquad \textbf{(12.23)}$$

Figure 12-28   The change in wafer deflection may be used to measure the stress in a deposited layer. This is typically measured using a reflected laser beam.

where δ is the change in the deflection of the center of the wafer, t is the film thickness, R is the radius of the wafer, and T is the thickness of the wafer.

12.14 Summary

Evaporation is an ultrahigh vacuum technique for depositing thin films. The material to be evaporated is heated in a crucible. The vapor of the material travels in a straight line to the substrates. Crucible heating can be done resistively, inductively, or with an electron beam. The latter technique is particularly useful for materials that require very high temperatures for deposition. The deposition rate depends on the vapor pressure of the charge and the geometry of the reactor. To improve the uniformity and reduce the effects of shadowing, the wafers are often mounted on a planetary that causes them to rotate during deposition. Substrate heating may also be used. The primary problems with evaporation include step coverage, alloy formation, and in electron beam systems, radiation damage.

Sputtering is another purely physical process that uses a glow discharge to remove material from a target. The ejected material diffuses to and collects on the surface of the wafer. The primary advantages of sputtering compared to evaporation are improved step coverage and ease of deposition of alloys and compounds. Some compounds may also be deposited by reactively sputtering the target with a dilute mixture of argon and a reactive species. As a result of these advantages, sputtering is used almost universally for metallization in silicon technologies. High aspect ratio features require high substrate temperatures and/or substrate biases to obtain acceptable step coverage. These temperatures however, promote the growth of large grains in the film. For the student who would like to investigate sputtering further, there are a variety of good texts. Stuart has written a short, qualitative introduction [84]. In addition, chapters on sputtering are included in Chapman and Mangano [85] and Vossen and Kern [86]. Excellent books that focus on sputtering include Konuma [24] and Wasa and Hayakawa [46].

Problems

1. In Example 12.1, the atomic arrival rate of aluminum is compared to the molecular flux of water vapor. For the same evaporator, what crucible temperature would be required to set the aluminum atomic flux equal to the water vapor flux calculated in the example? What would be the growth rate of the aluminum film?
2. The last problem suggests that very high rate depositions may be desirable in reducing the film contamination. What drawbacks might exist for such a process?
3. It is desirable to deposit a mixture of Ga and Al using a single source evaporator. If the deposition temperature is 1100°C, the mixture in the crucible is initially 50% molar, and both components have a sticking coefficient of one, what would the initial film composition be? How will this change with time?
4. It has been suggested that directing a broad beam laser could improve the step coverage of evaporated films. Discuss possible mechanisms for this improvement.
5. Suggest why the sputter yield might increase with ion mass for a fixed ion energy. (Hint: This may require a discussion of inelastic collisions.)
6. Discuss why the sputter yield has a maximum at some energy.
7. Why would the reflectivity of a metal film be a concern?
8. Explain why the stress in a film depends on the temperature of the film when it is measured.

References

1. C. B. Alcock, V. P. Iktin, and M. K. Horrigan, *Canadian Metallurgical Quarterly* **23**:309 (1984).

2. T. Iida, Y. Kita, H. Okano, I. Katayama, and T. Tanaka, "Equation for the Vapor Pressure of Liquid Metals and Calculations of Their Enthalopies of Evaporation," *J. High Temp. Mat., Proc.* **10**:199 (1992).

3. L. I. Maissel and R. Glang, eds., *Handbook of Thin Film Technology*, McGraw-Hill, New York, 1970.

4. L. Holland, *Vacuum Deposition of Thin Films*, Wiley, New York, 1956.

5. M. Aceves, J. A. Hernández, and R. Murphy, "Applying Statistics to Find the Causes of Variability in Aluminum Evaporation: A Case Study," *IEEE Trans. Sem. Man.* **5**:165 (1992).

6. K. Chopra, *Thin Film Phenomena*, McGraw-Hill, New York, 1969, p. 91.

7. C. Lu and A. W. Czanderna, *Applications of Piezoelectric Quartz Crystal Microbalances*, Elsevier, Amsterdam, 1984.

8. I. A. Blech, B. D. Fraser, and S. E. Haszko, "Optimization of Al Step Coverage Through Computer Simulation and Scanning Electron Microscopy," *J. Vac. Sci. Technol.* **15**:13 (1978).

9. Y. Homma and S. Tsunekawa, *J. Electrochem. Soc.* **132**:1466 (1985).

10. M. A. Lardon, H. P. Bader, and K. J. Hoefler, "Metallization of High Aspect Ratio Structures with a Multiple Cycle Evaporation/Sputter Etching Process," in *Proc. IEEE VLSI Multilevel Intercon. Conf.*, p. 212 (1986).

11. B. T. Jonker, "A Compact Flange-Mounted Electron Beam Source," *J. Vac. Sci. Technol.* **A8**:3883 (1990).

12. J. Bloch, M. Heiblum, and J. J. O'Sullivan, "Ultra-High Vacuum Compatible Electron Source," *IBM Tech. Disc. Bull.* **27**:6789 (1985).

13. L. C. Hecht, "Use of Successive Dilution for Reproducible Control of Al-Cu Alloy Evaporation," *J. Vac. Sci. Technol.* **14**:648, Jan/Feb 1977.

14. M. L. Kniffin and C. R. Helms, "The Synthesis and Properties of Low Barrier Ag-Ga Intermetallic Contacts to n-type GaAs," *J. Appl. Phys.* **68**:1367 (1990).

15. J. Villalobos, R. Glosser, and H. Edelson, "A Quartz Crystal-Controlled Evaporator for the Study of Metal Film Alloy Hydrides," *Meas. Sci. Technol.* **1**:365 (1990).

16. L. Esaki, in *Novel Materials and Techniques in Condensed Matter*, G. W. Crabtree and P. Vashishta, eds., North-Holland, Amsterdam, 1982.

17. W. Sevenhans, J.-P. Locquet, and Y. Bruynseraede, "Mass Spectrometer Controlled Electron Beam Evaporation of Multilayer Materials," *Rev. Sci. Instr.* **57**:937 (1986).

18. N. Uetake, T. Asano, and K. Suzuki, "Measurement of Vaporized Atom Flux and Velocity in a Vacuum Using a Microbalance," *Rev. Sci. Instr.* **62**:1942 (1991).

19. W. R. Grove, *Philos. Trans. Faraday Soc.*, p. 87, 1852.

20. I. Langmuir, *General Electric Rev.* **26**:731 (1923).

21. G. K. Wehner, *Adv. in Electron., Electron Phys.* **VII**:253 (1955).

22. M. L. Tarng and G. K. Wehner, "Alloy Sputtering Studies with *in-situ* Auger Electron Spectroscopy," *J. Appl. Phys.* **42**:2449 (1971).

23. R. L. Wilson and L. E. Terry, "Application of High-Rate ExB or Magnetron Sputtering in the Metallization of Semiconducting Devices," *J. Vac. Sci. Technol.* **13**:157 (1976).

24. M. Konuma, *Film Deposition by Plasma Techniques*, Springer-Verlag, Berlin, 1992.

25. L. I. Maissel and R. Glang, eds., *Handbook of Thin Film Technology*, McGraw-Hill, New York, 1970, p. 3–23.

26. G. K. Wehner and G. S. Anderson, "The Nature of Physical Sputtering," in *Handbook of Thin Film Technology*, L. I. Maissel and R. Glang, eds., McGraw-Hill, New York, 1970.

27. H. H. Anderson and H. L. Bay, *Sputtering by Particle Bombardment I*, R. Behrisch, ed., Springer-Verlag, Berlin, 1981.

28. R. V. Stuart and G. K. Wehner, "Sputtering Yields at Very Low Bombarding Ion Energies," *J. Appl. Phys.* **33**:2345 (1962).

29. G. K. Wehner, *Phys. Rev.* **102**, "Controlled Sputtering of Metals By Low-Energy Hg Ions," p. 690 (1956); ibid **108**, "Sputtering Yields for Normally Incident Hg+ Ion Bombardment at Low Ion Energy," p. 35 (1957); ibid **112**, "Low-Energy Sputtering Yields in Hg," p. 1120 (1958).

30. G. K. Wehner, D. Rosenberg, "Hg Ion Beam Sputtering of Metals at Energies 4–15 keV," *J. Appl. Phys.* **32**:1842 (1962).

31. H. H. Anderson and H. L. Bay, in *Sputtering by Particle Ion Bombardment I*, R. Behrisch, ed., Springer-Verlag, Berlin, 1981, p. 202.

32. G. K. Wehner and D. L. Rosenberg, "Angular Distribution of Sputtered Material," *J. Appl. Phys.* **31**:177 (1960).

33. O. Almen and G. Bruce, "Collection and Sputtering Experiments with Noble Gas Ions," *Nucl. Instr. Meth.* **11**:257, 279 (1961).

34. K. Wasa and S. Hayakawa, Jpn. Patent 642,012, assigned to Matsushita Electric Ind. Corp. (1967).

35. F. M. Penning, U.S. Patent 2,146,025 (Feb. 1935).

36. R. S. Rastogi, V. D. Vankar, and K. L. Chopra, "Simple Planar Magnetron Sputtering Source," *Rev. Sci. Instr.* **58**:1505 (1987).

37. P. J. Clarke, U.S. Patent 3,616,450 (1971).

38. T. Van Vorous, "Planar Magnetron Sputtering: A New Industrial Coating Technique," *Solid State Technol.*, p. 62, Dec. 1976.

39. D. B. Fraser, "The Sputter and S-Gun Magnetrons," in *Thin Film Processes*, J. L. Vossen and W. Kern, eds., Academic Press, New York, 1978, p. 115.

40. J. A. Thorton and A. S. Penfold, "Cylindrical Magnetron Sputtering," in *Thin Film Processes*, J. L. Vossen and W. Kern, eds., Academic Press, New York, 1978.

41. B. A. Movchan and A. V. Demchishin, "Study of the Structure and Properties of Thick Vacuum Condensates of Nickel, Titanium, Aluminum Oxide, and Zirconium Dioxide," *Phys. Met. Metallogr.* **28**:83 (1969).

42. J. A. Thornton, "Influence of Apparatus Geometry and Deposition Conditions on the Structure and Topology of Thick Sputtered Coatings," *J. Vac. Sci. Technol.* **11**:666 (1974).

43. R. J. Gnaedinger, "Some Calculations of the Thickness Distribution of Films Deposited from Large Area Sputtering Sources," *J. Vac. Sci. Technol.* **6**:355 (1969).

44. I. A. Blech and H. A. Vander Plas, "Step Coverage Simulation and Measurements in a DC Planar Magnetron Sputtering Systems," *J. Appl. Phys.* **54**:3489 (1983).

45. W. H. Class, "Deposition and Characterization of Magnetron Sputtered Aluminum and Aluminum Alloy Films," *Solid State Technol.* **22**:61 (June 1979).

46. K. Wasa and S. Hayakawa, *Handbook of Sputter Deposition Technology*, Noyes, Park Ridge, NJ, 1992.

47. P. Burggraaf, "Sputtering's Task: Metallizing Holes," *Semicond. Int.*, p. 28, Dec. 1990.

48. J. L. Vossen, J. J. O'Neill, K. M. Finlayson, and L. J. Royer, "Backscattering of Materials Emitted from RF-Sputtering Targets," *RCA Rev.* **31**:293 (1970).

49. L. Holland, "The Cleaning of Glass in a Glow Discharge," *Br. J. Appl. Phys.* **9**:410 (1958).

50. J. L. Vossen and E. B. Davidson, "The Interaction of Photoresist with Metals and Oxides During RF Sputter Etching," *J. Electrochem. Soc.* **119**:1708 (1972).

51. G. W. Sachse, W. E. Miller, and C. Gross, "An Investigation of RF-Sputter Etched Silicon Surfaces Using He Ion Backscattering," *Solid-State Electron.* **18**:431 (1975).

52. J. C. Bean, G. E. Becker, P. M. Petroff, and T. E. Seidel, "Dependence of Residual Damage on Temperature During Ar$^+$ Sputter Cleaning of Silicon," *J. Appl. Phys.* **48**:907 (1977).

53. M. J. Whitcomb, "Sputter Etch Profiles of Spheres, Cylinders, and Slab-like Silica Targets," *J. Mat. Sci.* **11**:859 (1976).

54. J. L. Vossen, "Control of Film Properties By RF-Sputtering Techniques," *J. Vac. Sci. Technol.* **8**:512 (1971).

55. Y. H. Park, F. T. Zold, and J. F. Smith, "Influence of DC Bias on Aluminum Films Prepared with a High Rate Magnetron Sputtering Cathode," *Thin Solid Films* **129**:309 (1985).

56. J. E. Houston and R. D. Bland, "Relationships Between Sputter Cleaning Parameters and Surface Contamination," *J. Appl. Phys.* **44**:2504 (1973).

57. L. D. Hartsough and P. S. McLeod, "High-Rate Sputtering of Enhanced Aluminum Mirrors," *J. Vac. Sci. Technol.* **14**:123 (1977).

58. T. N. Fogarty, D. B. Fraser, and W. J. Valentine, "MOS Metallization Via Automatic 'S-Gun' Planetary Deposition System," *J. Vac. Sci. Technol.* **15**:178 (1978).

59. D. S. Herman, M. A. Schuster, and R. M. Gerber, "Hillock Growth on Vacuum Deposited Aluminum Films," *J. Vac. Sci. Technol.* **9**:515 (1972).

60. S. M. Rossnagel, "Deposition and Redeposition in Magnetrons," *J. Vac. Sci. Technol.* A **6**:3049 (1988).

61. S. M. Rossnagel, I. Yang, and J. J. Cuomo, "Compositional Changes During Magnetron Sputtering of Alloys," *Thin Solid Films* **199**:59 (1991).

62. F. J. Cadieu and N. Cheneinski, "Selective Thermalization in Sputtering to Produce High T_c Films," *IEEE Trans. Magnet.* **11**:227 (1975).

63. S. Kobayashi, M. Sakata, K. Abe, T. Kamei, O. Kasahara, H. Ohgishi, and K. Nakata, "High Rate Deposition of $MoSi_2$ Films by Selective Co-Sputtering," *Thin Solid Films* **118**:129 (1984).

64. V. Hoffman, "High Rate Magnetron Sputtering for Metallizing Semiconductor Devices," *Solid State Technol.* **19**:57 (Dec. 1976).

65. D. R. Denison, "Sputtering System Design for Optimum Deposited Film Quality," *Microelectron. Man., Testing*, p. 12, July 1985.

66. P. S. McLeod and L. D. Hartsough, "High Rate Sputtering of Aluminum for Metallization of Integrated Circuits," *J. Vac. Sci. Technol.* **14**:263 (1977).

67. K. Kamoshida, T. Makino, and H. Nakamura, "Preparation of Low Reflectivity Al-Si Films Using DC Magnetron Sputtering and its Application to Multilevel Metallization," *J. Vac. Sci. Technol.* B **3**:1340 (1985).

68. R. S. Nowicki, "Influence of Residual Gasses on the Properties of DC Magnetron Sputtered Al-Si," *J. Vac. Sci., Technol.* **17**:384 (1980).

69. R. J. Wilson and B. L. Wiess, "The Sputtered Reflectivity of DC Magnetron Sputtered Al-1%-Si Films," *Vacuum* **42**:987 (1991).

70. R. J. Wilson and B. L. Wiess, "The Structure of DC Magnetron Sputtered Al-1%-Si Films," *Thin Solid Films* **203**:147 (1991).

71. J. Visser, *Supplement a la Revue, Le Vide*, number 157.

72. J. E. Sundgren, "Structure and Properties of TiN Coatings," *Thin Solid Films* **128**:21 (1985).

73. K. Wasa and S. Hayakawa, *Microelectron. Rel.* **6**:213 (1967).

74. W. Tsai, J. Fair, and D. Hodul, "Ti/TiN Reactive Sputtering: Plasma Emission, X-Ray Diffraction and Modeling," *J. Electrochem. Soc.* **139**:2004 (1992).

75. J. M. Molarius and M. Orpana, "Titanium Nitride Process Development," in *Issues in Semi-conductor Materials and Processing Technologies*, S. Coffa, F. Priolo, E. Rimini, and J. M. Poate, eds., Kluwer, Dordrecht, 1991.

76. A. Kohlhase, M. Mändl, and W. Pamler, "Performance and Failure Mechanisms of TiN Diffusion Barrier Layers in Submicron Devices," *J. Appl. Phys.* **65**:2464 (1989).

77. I. Suni, M. Mäenpää, M. A. Nicolet, and M. Luomajärvi, "Thermal Stability of Hafnium and Titanium Nitride Diffusion Barriers in Multilayer Contacts to Silicon," *J. Electrochem. Soc.* **130**:1215 (1983).

78. M. A. Nicolet, "Diffusion Barriers in Thin Films," *Thin Solid Films* **52**:415 (1978).

79. T. Hara, A. Yamanoue, H. Ito, K. Inoue, G. Washidzu, and S. Nakamura, "Properties of Titanium Nitride Films for Barrier Metal in Aluminum Ohmic Contact Systems," *Jpn. J. Appl. Phys.* **30**:1447 (1991).

80. S. Berg, N. Eguchi, V. Grajewski, S. W. Kim, and E. Fromm, "Effect of Contamination Reactions on the Composition and Mechanical Properties of Magnetron Sputtered TiN Coatings," *Surface, Coatings Technol.* **49**:127 (1991).

81. P. Jin and S. Maruno, "Bias Effect on the Microstructure and Diffusion Barrier Capability of Sputtered TiN and TiO$_x$N$_y$ Films," *Jpn. J. Appl. Phys.* **31**:1446 (1992).

82. S. S. Ang, H. M. Le, and W. D. Brown, "Sputtering-etching and Plasma Effects on the Electrical Properties of Titanium Nitride Contacts on n-Type Silicon," *Solid-State Electron.* **33**:1387 (1990).

83. G. J. Kominiak, "Silicon Nitride by Direct RF Sputter Deposition," *J. Electrochem. Soc.* **122**:1271 (1975).

84. R. V. Stuart, *Vacuum Technology, Thin Films, and Sputtering, An Introduction*, Academic Press, New York (1983).

85. B. Chapman and S. Mangano, in *Handbook of Thin-Film Deposition Processes and Techniques*, K. K. Schuegraf, ed., Noyes, Park Ridge, NJ, 1988.

86. J. L. Vossen and W. Kern, *Thin Film Processes II*, Academic Press, Boston, 1991.

Chapter 13

Chemical Vapor Deposition

The last two chapters discussed physically based methods for depositing thin films: evaporation and sputtering. They are called physical processes because the techniques did not involve chemical reactions. Rather, they produce a vapor of the material to be deposited by heating (evaporation) or by energetic ion bombardment (sputtering). While most metal films for silicon ICs are deposited using these methods, they have major problems associated with step coverage. This is a particular concern with submicron technologies where very small contacts require the coverage of high aspect ratio features. Furthermore, these techniques are not well suited to the deposition of insulating or semiconducting films. This chapter will discuss methods of thin film deposition based on chemical reactions. The chemicals are normally supplied through the use of gaseous compounds. By definition, the chemical bonding state of the final product is different than the precursor gasses.

Chemical vapor deposition (CVD) has become extremely popular and is the preferred deposition method for a wide range of materials. Thermal CVD also forms the basis for most epitaxial growth in IC manufacturing. Modifications of simple thermal CVD processes provide alternate energy sources such as plasmas or optical excitation to drive the chemical reactions, allowing the deposition to occur at low temperature. For a comprehensive review of CVD for IC fabrication, the reader is referred to Sherman [1]. Unfortunately, CVD does not lend itself to simple analytic explanation. Both the gas flow in the reactor and the chemical reactions require a detailed numerical analysis that is both reactor and process dependent. This chapter will begin by introducing a simple CVD system that will be used in the following two sections to discuss the equations that would have to be solved in order to understand the chemical reactions and flows in the chamber. The remainder of the chapter will discuss various types of deposition systems and specific gas chemistries used to deposit materials of interest.

13.1 A Simple CVD System for the Deposition of Silicon

To begin to understand CVD processes, consider the simple reactor shown in Figure 13.1. The reactor consists of a tube with a rectangular cross section. The walls of the tube are maintained at a temperature T_w. A single wafer rests on a heated susceptor in the center of the tube. This susceptor is maintained at T_s, where in many cases $T_s \gg T_w$. In order to discuss a simple but representative process we will use the decomposition of silane gas (SiH_4) to form polycrystalline silicon. Assume that the gas flows through the tube from left to right. Since the silane will begin to decompose when it approaches the

Figure 13-1   A simple prototype thermal CVD reactor.

hot susceptor, the concentration of silane, and therefore the deposition rate, may decrease along the length of the tube. To improve the uniformity of the deposition, the silane can be mixed in an inert carrier gas. A common diluent for silane is molecular hydrogen (H_2). Assume the chamber is fed a mixture of 1% SiH_4 in H_2. Not only are the use of diluents common practice in real systems, it also avoids further complications in the chemistry of the reaction, since at typical deposition conditions very little of the hydrogen can decompose. Finally assume that the temperature of the gas as it enters the tube is the same as the wall temperature. The reaction products and any unreacted silane flows out of the tube at the right. The flows in the chamber will be slow enough (see Section 9.3) that the pressure in the chamber can be considered uniform.

The overall reaction that must occur is

$$SiH_4(g) \rightarrow Si(s) + 2H_2(g) \tag{13.1}$$

where the quantity in parenthesis is (g) for gas phase and (s) for solid. The detailed process by which this overall reaction occurs is much more complex. One of the distinctions drawn regarding CVD is the location of the reaction that liberates a solid atom or cluster of atoms from a gaseous source. If that reaction occurs spontaneously in the gas above the wafer, it is called a *homogeneous process*. Such processes are generally undesirable. Using the deposition of silane as an example, excessive homogeneous reactions will result in large silicon particles in the gas phase that gradually accumulate on the wafer. The result is a deposit with poor surface morphology and inconsistent properties. In real systems, such deposits have a poorly controlled composition and may have significant contamination from residual gasses in the chamber. This chapter will therefore emphasize processes that are heterogeneous. That is, processes that operate in such a manner as to greatly favor the formation of solids only at surfaces. Even for processes that are run this way, homogeneous reactions are still important. For example, in the deposition of silane, the homogeneous production of silylene (SiH_2) is a crucial process because it is generally believed it is the silylene rather than silane itself that adsorbs on the surface of the wafer and produces the solid silicon. The distinction here is that the homogeneous reaction produces a gaseous, not a solid, product. The chapter will focus initially on the simplest type of heterogeneous reaction, which is run in a cold wall chamber (like the one shown in Figure 13.1), where all deposition reactions occur at the surface of the wafer.

Generally, the steps that occur during a chemical vapor deposition process include: (1) the transport of the precursors from the chamber inlet to the proximity of the wafer, (2) reaction of these gasses to form a range of daughter molecules, (3) transport of these reactants to the surface of the wafer, (4) surface reactions to release the silicon, (5) desorption of the gaseous byproducts, (6) transport of the byproducts away from the surface of the wafer, and (7) transport of the byproducts from the reactor. Even if the discussion is limited to thermal CVD in this very simple deposition system, understanding each of these steps is a formidable task. To simplify matters, the problem is often divided in half. The next section will focus on the chemical reactions that occur in the reactor, both in the gas phase and at the surface of the wafer. The following section will discuss the flow of gasses in the reactor. Choosing to study a system that contains only a small concentration of the reactant gas (1% SiH_4 in H_2) allows this separation to be fairly realistic. The thermal and mechanical properties of the gas are relatively unaffected by any chemical reaction in the small, active component.

13.2 Chemical Equilibrium and the Law of Mass Action°

Focusing for the moment on CVD processes that involve long times and many collisions between molecules, the chemical composition at each point in the reactor approaches equilibrium. To understand chemical equilibrium consider a unit volume of the gas somewhere in the chamber (Figure 13.2). Assume that the volume is small enough that the temperature and chemical composition in this volume is uniform. One of the reactions that one might think of is

$$SiH_4(g) \leftrightarrows SiH_2(g) + 2H(g) \tag{13.2}$$

The double arrow indicates that the reaction proceeds in both directions. Chemical equilibrium is reached when the concentration of each species is constant, even if the gasses take an arbitrarily long time traversing this unit volume. (Actually this reaction is unlikely, but it is instructive.)

Assume for the moment that this is the only reaction that occurs. The law of mass action says that

$$K_p(T) = \frac{p_{SiH_2}p_H^2}{p_{SiH_4}} \tag{13.3}$$

where p refers to the partial pressure of the subscripted species, and $K_p(T)$ is a reaction equilibrium constant that depends only on the temperature. The atomic hydrogen term is squared because of the 2 in front of the atomic hydrogen term of the species balance equation (Equation 13.2). The equilibrium constant generally follows an Arrhenius function

$$K_p(T) = k_o e^{\Delta G/kT} \tag{13.4}$$

where ΔG is the change in the Gibbs free energy in the reaction.

Assume that $K_p(T)$ is known for this process. There are three unknowns (the three partial pressures) and only one equation. Solving for them requires two more equations. The total pressure of the reactor P is a constant whose value is normally known. It is the sum of the partial pressures

$$P = p_{SiH_4} + p_{SiH_2} + p_H + p_{H_2}. \tag{13.5}$$

Assuming, for example that the chamber is run at atmospheric pressure, the partial pressure of H_2 is the same as its inlet partial pressure (0.99P) since it is assumed to be inert. The final equation comes from the inlet flow. One can use the Si/H ratio as

$$\frac{Si}{H} = \frac{f_{SiH_4}}{4f_{SiH_4} + 2f_{H_2}} = \frac{p_{SiH_4}}{4p_{SiH_4} + 2p_{H_2} + p_H} \tag{13.6}$$

where the f terms are the inlet flows, also assumed to be known.

This expression has not taken any other reactions into account. For example, in actual CVD, silicon is consumed

Figure 13-2   A volume element dV at some point in the gas above the surface of the wafer.

from the gas phase. In that case, the partial pressure of silicon cannot be completely determined by the inlet flows. Instead, one must consider the inlet flow to be a source of silicon containing molecules and the deposition surface to be a sink. Then, the silicon containing molecule flux, which depends on the flow field and diffusion, must be calculated.

To begin to develop a more realistic picture, some of the reactions that would have to be included are [2]:

$$SiH_4(g) \rightleftharpoons SiH_2(g) + H_2(g) \tag{13.7}$$

$$SiH_4(g) + SiH_2(g) \rightleftharpoons Si_2H_6(g) \tag{13.8}$$

$$Si_2H_6(g) \rightleftharpoons HSiSiH_3(g) + H_2(g) \tag{13.9}$$

Other reactions are, of course, possible, and *a priori* one cannot decide which reactions to include. Instead, one must find the equilibrium constant for each possible reaction and ignore only those reactions for which the $K_p(T)$ values are negligibly small. Finding the equilibrium partial pressures, therefore, requires an equilibrium constant for each of the three reactions listed above and the solution of a set of coupled algebraic equations.

Example 13.1

Assume that the gas AB is introduced into a reactor and that the only chemical reaction that occurs in the chamber is:

$$AB \rightleftharpoons A + B \tag{13.10}$$

If the process is run at 1 atm (760 torr) and a temperature of 1000 K and the process reaches chemical equilibrium, calculate the partial pressure of each species. The equilibrium constant for this reaction is given by

$$K(T) = 1.8 \times 10^9 \text{ torr } e^{-2.0eV/kT} \tag{13.11}$$

At 1000 K the equilibrium constant is calculated to be 0.15 torr. Then

$$0.15 = \frac{p_A p_B}{p_{AB}} \tag{13.12}$$

and the total pressure P is the sum of the partial pressures:

$$P = p_A + p_B + p_{AB} \tag{13.13}$$

We have three unknowns but only two equations. Since A and B are both created by the dissociation of the inlet gas, there must be an equal number of each. It would be reasonable to assume that the partial pressures are equal. Then

$$p_A^2 + 0.3p_A - 0.15 \cdot P = 0 \tag{13.14}$$

Setting $P = 760$ torr and solving gives $p_A = p_B = 10.5$ torr and $p_{AB} = 739$ torr.

The discussion thus far makes a crucial approximation: that all species are in chemical equilibrium. To understand the limitation of this approximation, consider what happens as the pressure of the chamber is reduced. At a low enough pressure, the mean free path of the molecules approaches the width of the chamber (see Section 10.1). If these gas phase collisions do not occur, the species in the gas often do not reach thermal equilibrium and therefore cannot achieve chemical equilibrium. Furthermore, since the gas molecules have a distribution of energies, a large number of collisions must occur in each unit volume for the gasses to potentially reach chemical equilibrium. Therefore, a characteristic chamber length such as the distance between the gas injector and the susceptor must be at least a few orders of magnitude larger than the mean free path. Depending on the particular reaction equations involved, some processes may reach equilibrium, others will not. Processes that do not reach chemical equilibrium are called *kinetically controlled processes*. Typically low pressure CVD is kinetically controlled, while atmospheric CVD may be in equilibrium.

To begin to understand this class of deposition, reconsider Reaction 13.7. This is a prototypical reaction in silicon CVD. When the reaction is kinetically controlled, it is written as

$$\text{SiH}_4(g) \underset{k_r}{\overset{k_f}{\rightleftharpoons}} \text{SiH}_2(g) + \text{H}_2(g) \tag{13.15}$$

where k_f and k_r are the forward and reverse reaction rate coefficients.

By writing a similar expression for each of the chemical reactions, it is possible to construct a differential equation for the time rate of change of the concentration (or partial pressure) of all of the chemical species. For example, if only Reactions 13.7 through 13.9 are considered, the time rate of change of silane is given by

$$\frac{d}{dt} C_{\text{SiH}_4} = -k_{f1} C_{\text{SiH}_4} + k_{r1} C_{\text{SiH}_2} C_{\text{H}_2} - k_{f2} C_{\text{SiH}_4} C_{\text{SiH}_2} + k_{r2} C_{\text{Si}_2\text{H}_6} \tag{13.16}$$

where the subscript 1 on the first two terms refers to the reaction given in Equation 13.7, and the subscript 2 refers to the reaction listed in Equation 13.8. Equation 13.9 does not involve silane and so does not enter into Equation 13.16. One can construct similar equations for the other species. If the $k(T)$'s are known, we are left with a set of coupled first order differential equations that can be solved for the rate of change of each chemical species in the unit volume. If one also takes into account diffusion of chemical species due to concentration gradients, the residence time of the gasses in the unit volume are known (through the flow velocities), and the value of the temperature for each of the unit volumes is known, one could begin to solve for a map of the chemical species in the chamber. While it is possible to solve for the species balance in very simple systems, real CVD systems stretch the capabilities of even supercomputers [3]. Furthermore, many of the important rate coefficients, even in the most commonly used chemical systems, also are not presently known accurately enough to produce meaningful information. For that reason, we will set aside attempts at meaningful quantitative discussion of the CVD reactions and use this introduction as a basis for a more qualitative discussion.

While the chemical reactions in the gas phase are at least qualitatively understood, the situation at the surface is much less clear. Part of the problem are the tools that can be used to investigate the gas sample over some finite volume. When these same techniques are applied to the surface, there is insufficient signal. Methods that have been used to develop a detailed understanding of the surface during high vacuum growth cannot be used in chambers under typical CVD conditions. Although new techniques have emerged in the last few years, the picture that we have is still tentative and qualitative.

The last two sections suggest that for the deposition of silicon from silane, several types of silicon containing species may strike the surface of the wafer. In the boundary layer model, one can readily calculate the flux of the molecules if a boundary condition is established for the concentration at the surface. It is common practice to define a phenomenological parameter called the *sticking coefficient*, which varies from zero for molecules that are completely reflected from the surface to unity for molecules that irreversibly adsorb. In the case of silylene, it is commonly assumed that the sticking coefficient is one. On the other hand, it has been shown that the probability of an incident silane molecule sticking and reacting is given by [4]

$$\gamma_{SiH_4} = 0.054 e^{-0.81 eV/kT}. \tag{13.17}$$

At typical deposition conditions, the sticking coefficient is only about 10^{-6}; however, the flux of silane to the surface is much larger than the flux of silylene, so that both molecules may contribute to the deposition rate.

Once the molecule is on the surface, a chemical reaction must occur to remove the silicon atom and free the hydrogens. Taking silylene as an example, the molecule first adsorbs:

$$SiH_2(g) \rightleftarrows SiH_2(a) \tag{13.18}$$

The overall surface reaction must take the form

$$SiH_2(a) \rightleftarrows Si(s) + H_2(g) \tag{13.19}$$

where (a) refers to adsorbed species and (s) refers to atoms that have been incorporated into the solid. Because of the high concentration of H_2 in the gas, it is believed that the surface is covered with either physically adsorbed H_2 (at low temperature) or chemically adsorbed H (at high temperature). These surface species must be desorbed to allow Reaction 13.19 to proceed. The desorption process also follows an Arrhenius behavior, so that the surface will have a concentration of vacancies, whose density increases with increasing temperature. The adsorbed silylene can diffuse across the passivated surface (Figure 13.3) until it finds such a vacancy at which point it will bond and eventually lose its hydrogen atoms. This diffusion across the surface plays an important role in CVD processes. When the surface diffusion is large (of order millimeters), the deposition is very uniform. When the surface diffusion length is short, a less uniform deposit will result. As with physical deposition processes, surface diffusion increases exponentially with temperature, and so film uniformity can generally be improved by heating the wafer.

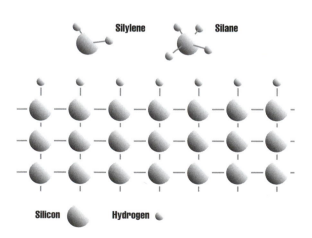

Figure 13-3   A simple model of the surface of the wafer during silane CVD includes adsorbed SiH_4 and SiH_2.

Figure 13-4   Flow development in a tubular reactor. The gas enters with a simple plug flow on the left and exits with a fully developed parabolic flow.

13.3 Gas Flow and Boundary Layers°

The second area that needs to be understood for CVD is gas flow dynamics. The gas flow in the reactor is important because it determines the transport of the various chemical species in the chamber, and it plays a significant role in the temperature distribution in the gas in many reactors. Furthermore, the temperature distribution will also affect the flow. If the mean free path of the gas is much smaller than the chamber geometries, the gas can be treated as a viscous fluid. Furthermore, if the flow velocities are much less than the speed of sound (low Mach numbers) the gas can be considered incompressible. Nearly all CVD systems operate in pressure and flow regimes that make these approximations valid. Finally, as a starting point assume that the gas velocity is low enough that the gas flows along the contours of the chamber. The flow is said to be laminar and it can be well described by the mechanical properties of the gas.

If the reactor is a circular tube and all surfaces are at the same temperature, the problem can be simplified considerably. Assume that the gas is introduced with a uniform velocity U_∞ at the left end of the tube (Figure 13.4). One important feature of gas flows is that the gas velocity must be zero at all surfaces. Because of the finite gas viscosity, the flow velocity must vary smoothly from zero at the walls to some maximum value at the center. This change from a uniform or plug flow, to a fully developed tube flow occurs over a distance z_v:

$$z_v \approx \frac{a}{25} N_{Re}$$ (13.20)

where a is the radius of the tube, and N_{Re} is a dimensionless quantity known as the Reynolds number. The Reynolds number (N_{Re}) is given by

$$N_{Re} = U_\infty \frac{L}{\mu} = U_\infty \frac{L\rho}{\eta}$$ (13.21)

where L is a characteristic length of the chamber (such as the radius a), μ is the kinematic viscosity, ρ is the mass density of the gas, and η is the dynamic viscosity of the gas [5]. When Re is low, the flow in the tube is dominated by the finite viscosity effects and so is parabolic across the chamber. The velocity then is given by

$$v(r) = \frac{1}{4\eta} \frac{dp}{dz} (a^2 - r^2)$$ (13.22)

where dp/dz is the pressure gradient across the tube, which is assumed to be small. At very large Re, the gas cannot support the large velocity gradients required for fully developed laminar flow, and so the flow becomes turbulent. The transition between laminar and turbulent flows depends on the gas. For example, when $Re > 2300$ in H_2, the flow is turbulent [6].

Example 13.2

An LPCVD tube operating at 10 torr has an inlet gas flow of 1000 sccm (standard cubic centimeters per minute) of nitrogen. At the reactor temperature of 1000 K, the dynamic viscosity of nitrogen is 0.04 gm/cm-sec. The reactor is 200 mm in diameter. Estimate the length required for fully developed flow and calculate $v(r)$ after the flow is fully developed.

According to 13.20 and 13.21,

$$z_v \approx \frac{a^2}{25} U_\infty \frac{\rho}{\eta}$$

The mass density can be calculated using the ideal gas law

$$\rho = m_{N_2} \frac{P}{kT} = 5.8 \times 10^{-6} \text{ gm/cm}^3$$

If U_∞ is taken as the plug flow velocity,

$$U_\infty \approx 1000 \text{ cm}^3/\text{min} \frac{1 \text{ min}}{60 \text{ sec}} \frac{1000 \text{ } K}{273 \text{ } K} \frac{760 \text{ torr}}{10 \text{ torr}} \frac{1}{\pi(10 \text{ cm})^2} = 15 \text{ cm/sec}$$

and

$$z_v \approx \frac{(10 \text{ cm})^2}{25} 15 \text{ cm/sec} \frac{5.8 \times 10^{-6} \text{ gm/cm}^3}{0.04 \text{ gm/cm-sec}} = 0.0085 \text{ cm}$$

Since the total flow must be constant,

$$\int 2\pi r v(r) dr = 15 \text{ cm/sec } \pi 100 \text{ cm}^2$$

one can readily show that

$$v(r) = 30 \text{ cm/sec} \left[1 - \frac{r^2}{a^2} \right]$$

To begin to understand the flow fields of a more complicated chamber, return to Figure 13.1, but for the moment keep the system at a uniform temperature. The wafers will rest on the bottom surface of the chamber. The previous discussion suggests that the gas flow velocity must go to zero at the surface of the wafer. The standard textbook picture has the height of the chamber large enough to have a large N_{Re} and so a broad region of uniform gas velocity approximately equal to U_∞. To further simplify the behavior of flow, it is customary to approximate the parabolic fall off of the gas velocity in the vicinity of the wafer as a boundary layer of width $\delta(z)$ where, for a flat surface whose normal is perpendicular to the flow direction

$$\delta \approx 5 \sqrt{\frac{\mu z}{U_\infty}} \tag{13.23}$$

In this model, the gas flow in the boundary layer is zero, while outside the boundary layer the flow velocity is U_∞ (Figure 13.5).

    If deposition is occurring at the surface of the wafer, the deposition gasses must diffuse through the stagnant boundary layer. The boundary layer thickness, therefore, may play a critical role in determining the deposition rate. Notice that, according to Equation 13.23, for a flat surface the boundary layer thickness increases as $z^{1/2}$. To maintain a uniform boundary layer thickness, CVD systems in which gas transport plays a significant role in the deposition rate often tilt the deposition surface with respect to the flow direction. The wafers then rest on a wedge shaped susceptor. The tilt angle of the wedge must be optimized to obtain the best uniformity for a particular CVD process.

    Gas phase diffusivities are much less temperature sensitive than bulk diffusivities. One common form is given by Hammond [7]

$$D_e \propto T^{3/2} \frac{P_g}{P} \tag{13.24}$$

where P_g and P are the partial pressure of the diffusing species and the total pressure, respectively, at the edge of the stagnant layer. One of the ways to distinguish CVD processes that are limited by diffusion through the gas is to measure the temperature dependence of the deposition rate. If P_g can be considered independent of temperature, the deposition rate of such a mass transport limited reaction would increase only weakly with increasing temperature.

    The gas flow in CVD reactors is often nonlaminar and may involve recirculations and roll cells. One of the most common sources of these flows is natural convection. As the gas flows past hot surfaces it expands. The expansion is described by an equation of state such as the ideal gas law:

$$\rho = \frac{nm}{V} = \frac{Pm}{kT} \tag{13.25}$$

where m is the molecular mass. In so doing, the mass density decreases. (Recall that the pressure is fixed.) The hot gas tends to float or rise in the reactor with respect to cooler gasses. This effect, called *natural con-*

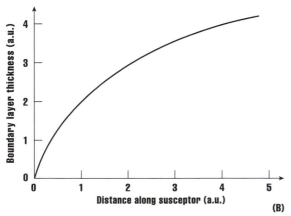

Figure 13-5 (a) Gas flow as predicted by a parabolic flow model and in the stagnant layer approximation. (b) Stagnant layer thickness versus position assuming a plug flow inlet.

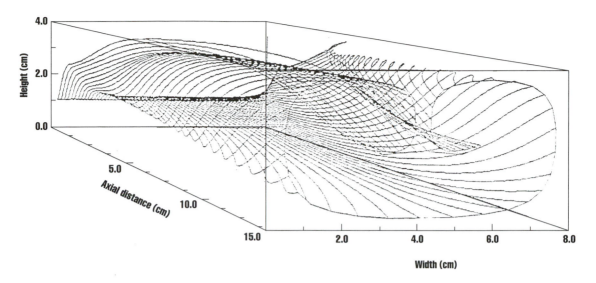

Figure 13-6a   Flow field for a square cross section CVD reactor showing roll cells *(after Moffat).*

vection, must also be considered when calculating the flow of gasses in a real chamber under deposition conditions [8]. The effects are most pronounced when heavy molecules are used at near atmospheric pressures. Conversely, natural convection has almost no effect for low pressure H_2 ambients.

Figure 13.6a shows the calculated flow fields for a horizontal reactor with a square cross section. The calculations were done by discretizing the reactor in 3-D and solving for momentum conservation in the form of the Navier-Stokes equation over this grid. Gasses are injected at the back (upper left) end of the box and flow toward the front (lower right) end of the box. The lower surface is a hot susceptor. The hot susceptor causes the gas at the center of the reactor to rise, leading to transverse

Figure 13-6b   Roll cells in hemispherical tubes at 760 torr (left) and 160 torr (right) *(after Takahashi et al. reprinted by permission of the publisher, The Electrochemical Society).*

High flow/low pressure   **Low flow/high pressure**

Figure 13-7   Buoyancy driven recirculation cells.

roll cells that develop along the length of the reactor. Again, a quantitative discussion of the exact flow in any reactor is well beyond the scope of this text, but these principles can be used to develop a qualitative understanding of a variety of chamber geometries and CVD processes. Figure 13.6b shows TiO_2 particles in a carrier gas as they flow through semicircular tubes. The left-hand figure shows the flow at 760 torr and the right-hand picture shows the flow at 160 torr [8]. Obviously the roll cells are greatly diminished at low pressure.

In other reactors, natural convection shows up as recirculation cells in which part of the gas flow has a significant velocity component in the opposite direction of the intended flow (Figure 13.7). Recirculation cells may also arise when the cross-sectional area of the chamber changes abruptly. These cells can trap gas components that would otherwise be swept from the growth system. The effect can lead to an incomplete flushing of the chamber. If the composition of the deposit changes, this change will be much less abrupt than if the reactor and process are designed to avoid recirculation cells.

13.4 Evaluation of the Simple CVD System

Now examine what happens if the deposition rate in our example CVD reactor is measured as a function of temperature. The prediction of the growth rate and growth rate uniformity requires extensive fluid dynamics and chemical concentration calculations. Instead we will work backward by developing a qualitative understanding of experimental results. The deposition rate of the silicon containing precursor (silane) will be used as a growth parameter. Figure 13.8 shows the results. At low wafer temperature the deposition rate increases exponentially with decreasing inverse temperature. In this regime the limiting step of the process is some reaction rate. This may be in the gas phase or at the surface. The information provided is insufficient to determine whether the gas is in chemical equilibrium or is kinetically controlled. Processes operated in this region are called *reaction rate limited*. CVD systems that operate in this regime must have excellent temperature control and temperature uniformity. These systems are primarily large batch systems where many wafers are processed at relatively low rates. Flow dynamics in this chamber are less of a concern, except to the extent that they contribute to temperature nonuniformity across the wafer. For that reason, this type of reactor often heats not only the wafer but the walls as well. It is therefore called a *hotwall batch CVD reactor*.

At high wafer temperature, the growth is limited by the arrival rate of the growth species. It is often assumed that the limitation is diffusion across the boundary layer. As already pointed out, the temperature dependence of the gas phase diffusivity is much lower than the activation energy of the chemical reaction. Processes operating in this regime are referred to as mass transport limited. Then the concentration of the deposition gas or gasses controls the deposition rate. Remember that the rate of production of these gasses from the precursors may also have a temperature dependence. CVD systems that operate in mass transport limited regime must have excellent control of the gas flows, and the chamber geometry must be designed so as to ensure a uniform transport to all parts of all wafers. As a result, these systems often are single wafer or small batch systems.

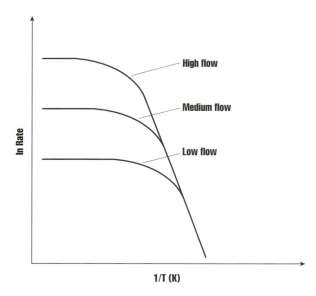

Figure 13-8 Typical deposition rates for CVD as a function of the temperature with flow rate as a parameter.

In addition to deposition rate and uniformity, CVD films must be examined with regard to stress, step coverage, and composition. Just as is the case for sputtered films, CVD layers with large compressive or tensile stress may crack, particularly when they cover steps. The step coverage of CVD films is usually very good even for high aspect ratio features. Plasma enhanced CVD films however, may be somewhat re-entrant. This will be discussed later in the chapter. One of the major concerns with CVD layers is chemical composition. In the silane decomposition process that has been discussed so far, the deposited silicon may have a high concentration of hydrogen. This leads to low density films that may etch more quickly than pure silicon films. Residual gasses in the chamber, such as oxygen or water vapor, may also react with the silicon to form SiO_x layers with high resistivities. In many processes, compounds such as SiO_2 or Si_3N_4 are the desired result. In that case, not only are contaminants a possibility, but also the stoichiometry of the deposited films may differ from the ideal. For example, it is common to anneal a CVD oxide in oxygen at high temperature to densify the film. This step adds oxygen and moves the film closer to stoichiometric SiO_2. Plasma enhanced CVD (PECVD) films are particularly prone to stoichiometry concerns. One common and easy method to check the density of deposited oxides is to measure the wet chemical etch in dilute HF solutions. As deposited, PECVD films frequently have etch rates ten times those of thermal oxides.

13.5 Atmospheric CVD of Dielectrics

Some of the earliest CVD processes were done at atmosphere pressure (APCVD) because of the large reaction rates and the simplicity of the CVD system, particularly for the deposition of dielectrics.

Figure 13-9 Simple continuous feed atmospheric pressure reactor (APCVD).

While the deposition of silicon from silane, as discussed earlier in the chapter, has been run at atmospheric pressure, the uniformity is poor. It is easy to obtain good uniformity at low pressure, and so APCVD is generally reserved for thick dielectrics, where deposition rates in excess of 1000 Å/min make the process very attractive.

Figure 13.9 shows a simple continuous feed atmospheric CVD reactor. Wafers travel from cassette to cassette on a heated chain track. Wafer temperatures may be anywhere from 240 to 450°C [9]. The gasses are injected from a showerhead above the wafers. When the oxygen to silane gas flow ratio is at least 3:1, stoichiometric SiO_2 will result. SiH_4 and O_2 is a very volatile mixture, which reacts readily. Without a sufficient flow of a diluent such as N_2, this reaction proceeds in the gas phase, resulting in poor morphology.

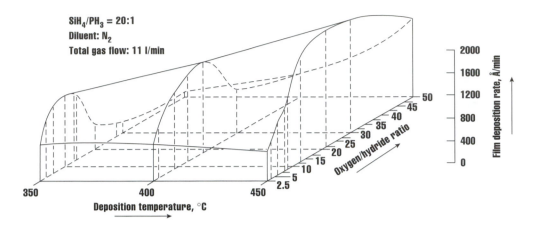

SiH$_4$/PH$_3$ = 20:1
Diluent: N$_2$
Total gas flow: 11 l/min

Figure 13-10   Deposition rate of PSG in an APCVD system *(after Kern and Rosler, ®1977, AIP).*

As will be discussed in Chapter 15, it is often desirable to deposit silicon dioxide films with 4 to 12% phosphorus. These phosphosilicate glasses (PSG) soften and reflow at moderate temperatures, smoothing wafer topology and gettering many impurities. PSG can be formed in an atmospheric process by adding phosphine (PH$_3$). Figure 13.10 shows a typical plot of the deposition rate of PSG versus temperature and oxygen to hydride flow rates. For high oxygen concentration ambients (30:1), the deposition rate increases sharply with temperature and is probably reaction rate limited. For low oxygen containing ambients (2.5:1), the growth rate actually decreases slightly with increasing temperature. The phosphorus content of the film can be controlled by changing the phosphine to silane ratio. Due to the toxicity of phosphine and silane, APCVD systems designed for PSG deposition are generally housed in a vented cabinet. To improve uniformity and step coverage, many PSG and borophosphosilicate glass (BPSG) processes now use organometallic sources such as TEOS (tetraethyloxysilane or Si(OC$_2$H$_5$)$_4$). TEOS and ozone can also be used to deposit SiO$_2$ at about 400°C [10]. TEOS is supplied as a stable, inert, high vapor pressure liquid that is used in a bubbler (see Section 3.8). One of the advantages of using TEOS is the elimination of the need for some of the hazardous chemical handling. The lines from the bubbler must be heated to prevent deposition on the walls of the tubing. Various alternate organometallics have also been investigated. Hexamethyldisiloxane ((CH$_3$)$_3$-Si-O-Si-(CH$_3$)$_3$), a linear disiloxane, has also shown excellent characteristics comparable to TEOS [11], but the deposition rate was found to depend on the substrate material. Similar films have also been deposited from hydridospherosiloxanes [12] with wet oxygen at about 500°C.

The major drawback of APCVD is particle formation. While particle formation in the gas phase can be controlled by adding a sufficient amount of N$_2$ or another inert gas, heterogeneous deposition can also oc-

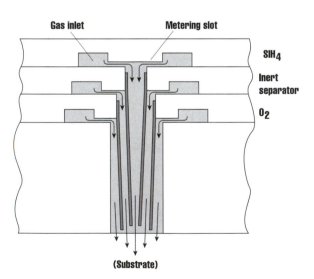

Figure 13-11   Showerhead design used to minimize deposition at the nozzle by maintaining an inert curtain between the reactants.

Figure 13-12   Common LPCVD reactor geometries.

Figure 13-13   Photograph of a standard bank of LPCVD tubes.

cur at the gas injectors. Even if the growth rate of these particles is low, after a number of wafers the particles will become large enough to flake off and fall on the wafer surface. To avoid this problem, the showerhead may be segmented so as to keep the reactant gasses separated until they are injected into the chamber. Figure 13.11 shows a showerhead arrangement designed to reduce this problem.

13.6 Low Pressure CVD of Dielectrics and Semiconductors in Hot Wall Systems

A variety of system geometries have been used for low pressure CVD (LPCVD). Figure 13.12 shows a sample of some of the most common. We can divide the reactors into hot and cold wall systems. Hot wall systems have the advantages of uniform temperature distributions and reduced convection effects. Cold wall systems are able to reduce deposition on the walls. These deposits can lead to depletion of the deposition species and the formation of particles, which may flake off the walls and fall on the wafers. Deposits on the walls also lead to memory effects: the deposition on the wafer of material previously deposited on the walls. For that reason hot wall reactors must be dedicated to the growth of a particular film.

Virtually all polycrystalline silicon and a considerable amount of dielectric deposition is done in horizontal flow hot wall systems. Instead of using an inclined susceptor, the wafers are close packed vertically, like wafers in thermal oxidation systems. In order to achieve reasonable deposition uniformity in such a system, the process must be designed so as to keep the reaction strictly controlled by the deposition kinetics [13]. Instead of a diluent gas, the use of low pressures (0.1 to 1.0 torr) reduces gas phase nucleation. This process is commonly called LPCVD.

Figure 13.13 shows a photograph of an LPCVD system. Like furnaces, they generally are built in banks of four tubes. In this bank, the upper tube has been removed to show the heater coils. The gasses are controlled at the back of the tube using mass flow controllers and routed to the front of the furnace. Depending on the process to be run, the gasses are either injected through a ring at the front of the tube, or they are plumbed to a tube that runs the length of the furnace and injects gasses uniformly across the load. Due to the amount of deposition on the walls, most production systems have soft landing cantilever loaders to minimize particulate formation and flaking during the load/unload process. After loading the furnace is closed with an o-ring sealed door. The tube is flushed with an inert gas such as N_2 and pumped to a medium vacuum. The furnace is ramped to the deposition temperature if it is not already idling at that temperature, and the deposition gasses are switched on. The deposition is allowed to proceed for a predetermined time, then the furnace is again flushed with N_2, the pressure raised to an atmosphere, and the wafers unloaded.

A recent innovation in the LPCVD area is the introduction of vertical chambers. Similar to vertical oxidation/diffusion tubes, these systems have several advantages over standard tubes. Since the wafers are all held by gravity, the wafer to wafer spacing in the reactor is more uniform. Convective effects are more uniformly distributed across the wafer. As a result of these advantages, vertical LPCVD systems can routinely achieve uniformities of better than 2% in the deposition of undoped poly and silicon nitride [14]. Vertical CVD systems are more easily integrated into automated factories since the wafers do not have to be tipped to the vertical allowing easier robotic handling. Perhaps the most important advantage of vertical LPCVD systems is the reduced particle counts. The cost of these systems is considerably higher than conventional LPCVD systems however.

Another newer type of hot wall LPCVD system that has widespread application to silicon IC manufacturing is the hot wall cross-flow reactor. In this reactor, the wafers are placed vertically in closely spaced cassettes, arranged so that fresh gas flows past each wafer. This reduces particle formation and improves uniformity. The system requires extensive quartzware maintenance, however.

Most LPCVD poly is done with silane in furnaces at temperatures ranging from 575 to 650°C. The activation energy for poly deposition is about 1.7 eV. It is believed that the deposition rate is limited by H_2 desorption from the silicon surface. Typical polysilicon deposition rates are 100 to 1000 Å/min, so common deposition times are tens of minutes. When the gas is injected at the front of the tube, the furnace may be programmed with a small temperature gradient (25°C) from front to back. This allows the higher reactivity at the back of the tube to compensate for silane depletion. Temperature ramping has the undesirable effect of producing wafers with larger poly grains at the back of the tube than at the front; however, after high temperature annealing the crystal structure of the films is indistinguishable [15]. It is common to obtain thickness uniformities of 5% across a batch of 100 large diameter wafers [16].

Figure 13.14 shows the deposition rate of poly as function of pressure for a range of temperatures [17]. The morphology of the deposit is a sensitive function of the deposition conditions. When the deposit is performed at temperatures below 600°C, it is usually amorphous [15], but may be polycrystalline if the deposition rate is low enough. These layers can be crystallized into poly layers when annealed at low temperatures [18].

LPCVD polysilicon can be N^+ doped by implantation, solid or $POCl_3$ diffusion, or in situ doping

Figure 13-14 Deposition rate of polysilicon as functions of temperature and silane flow rate *(after Voutsas and Hatalis, reprinted by permission of the publisher, The Electrochemical Society).*

through the addition of arsine (AsH_3) or PH_3. Impurity concentrations approaching 10^{21} cm^{-3} can be achieved in poly layers doped in this manner [19]. It is common to achieve resistivities of less than 1 mΩ-cm in both in situ doped and diffused polysilicon. The major difficulty with in situ doping is that the addition of PH_3 degrades the deposition uniformity, particularly near the edge of the wafer. Recent modeling efforts have suggested that this may be the result of an increased silylene concentration [20]. It is also critical to include the effects of temperature variations across the load, particularly when considering the first and last few wafers [21].

Aside from particle control, one of the major limitations of traditional LPCVD furnaces is the difficulty of integrating them into automated, cluster tool environments. The most common example is polysilicon. It would be desirable to grow a gate oxide, deposit the poly, and dope the poly in a single cluster tool. Such systems can be made by using a number of simple single wafer reactors that can be fed by a central robot arm. In Applied Material's Precision 5000®, for example, wafers can be shuttled among up to five chambers (Figure 13.15). The wafer enters the load locked area, where it is transferred to a rapid thermal chamber for oxidation. After oxidation, it is transferred to one of two poly deposition chambers. After deposition, it travels to another rapid thermal chamber for phosphorus diffusion using a $POCl_3$ source.

Various chemistries have been used for hot wall deposition of oxide in LPCVD reactor including silane and oxygen, dichlorosilane ($SiCl_2H_2$ or DCS) and nitrous oxide (N_2O), and the decomposition of TEOS. The silane and oxygen process can be run at substrate temperatures of less than 500°C. This means that it can often be used to deposit a layer of oxide over an aluminum layer. Films deposited at these temperatures are found to contain significant quantities of silanol (SiOH), hydride

Figure 13-15   A CVD cluster tool showing the central robot and one of the single wafer processing stations *(photo courtesy Applied Materials).*

Figure 13-16 BPSG deposition rate and the boron and phosphorus concentrations as a function of phosphine flow in the TEOS/TMB system *(after Becker et al. ©1986, AIP)*.

(SiH), and water [22]. As with the corresponding APCVD process, the major limitations are particulate generation and low deposition rates [12]. Furthermore, due to the low substrate temperature, the step coverage of these films is often unacceptable.

The DCS and nitrous oxide process [23] must be run at temperatures of about 900°C. Uniformity and step coverage are excellent, and etch rates are close to those of thermal SiO_2. Although these films are virtually free of hydrogen, they do contain measurable amounts of chlorine, which can lead to etching of underlying poly-silicon layers. The DCS and nitrous oxide deposition process has a strong nonlinear pressure dependence. Deposition nonuniformities across the tube are common. Special injectors and temperature gradients can be used to attempt to provide a more uniform deposition.

A popular LPCVD oxide process is the thermal decomposition of TEOS. Figure 13.16 shows the deposition rate of oxide as a function of TEOS partial pressure [24]. Typical deposition temperatures of 650 to 750°C are low enough that redistribution of dopants in the substrate is not a concern, but the temperature is still too high to use over aluminum layers. Lower temperature depositions have been reported with alternate organometallics such as diacteoxyditertiarybutoxysilane, which decomposes at temperatures as low as 450°C [25] and provides excellent step coverage, but is not yet used extensively.

The deposited stress in undoped LPCVD oxides are about $1–3 \times 10^9$ dyne/cm^{-2}. Low temperature depositions are tensile, while high temperature depositions are compressive. The refractive index of thermal SiO_2 is 1.46 [26], while deposited oxides are generally higher. It is often found that the higher refractive index of deposited oxides correlates with a low mass density and high etch rates in buffered HF. As with APCVD, dilute mixtures of phosphine or phosphine and diborane [27] in hydrogen can be added to the TEOS LPCVD process [28] or the SiH_4 and O_2 process [29] to produce doped oxide layers. The addition of phosphine generally increases the deposition rate, but degrades the deposition uniformity, while adding diborane to form BPSG sometimes increases the deposition rate and has little effect on the uniformity.

While silicon nitride can be deposited from silane and a nitrogen containing precursor, LPCVD silicon nitride is most commonly deposited from mixtures of DCS and ammonia (NH_3). Typical deposition temperatures are 700 to 900°C [12]. The activation energy for this process is 1.8 eV. The deposition rate increases with the DCS flow rate. Due to DCS depletion effects, a ramped profile is usually required for this process.

Two common techniques to check the composition of the films are the measurement of the refractive index with ellipsometry and the measurement of the etch rate in buffered HF. The refractive index is usually between 1.8 and 2.2, with 2.0 being the ideal value. High refractive indices indicate a silicon rich film. Low refractive indices indicate the presence of oxygen, often due to a vacuum leak, a contaminated gas, or an incomplete pumpdown. The presence of oxygen is also indicated if the etch rate of the film exceeds 1 nm/min in 49% HF. Other common impurities in Si_3N_4 include hydrogen and approximately 0.4% Cl [30]. The authors of this work also found that these films have a 10 to 20 Å layer of SiO_2. LPCVD silicon nitride films often have high tensile stress (10^{10} dyne/cm^{-2}). Increasing the concentration of DCS in the growth ambient can decrease the stress, but results in films with a considerable amount of excess silicon. Due to the high concentration of silicon containing species (primarily $SiCl_2$), gas phase nucleation is likely during the LPCVD of Si_3N_4, leading to a high concentration of particles formed in the gas.

Figure 13-17 Basic PECVD geometries: cold wall parallel plate, hot wall parallel plate, and after glow.

13.7 Plasma Enhanced CVD of Dielectrics

In many applications, it is necessary to deposit films at very low substrate temperatures. The deposition of SiO_2 over aluminum and the deposition of Si_3N_4 capping layers over GaAs are two common examples. To accommodate these lower substrate temperatures, an alternate energy source must be applied to the gaseous and/or adsorbed molecules. While photo-enhanced deposition has been experimentally demonstrated, and even seen some limited use in production, the primary nonthermal energy source used to drive CVD reactions is the RF plasma. Plasma enhanced chemical vapor deposition (PECVD) systems have the added advantage of using ion bombardment of the surface to provide energy to the adspecies to allow them to diffuse further along the surface, without a high substrate temperature. As a result, the process is very good at filling small features. Chapter 10 introduced the concept of plasmas. The application of glow discharges to sputtering and etching have already been discussed. This section will discuss the use of plasmas to enhance the deposition rate in CVD processes. Since the deposition of insulating layers is of primary interest, only RF discharges need to be considered.

There are three basic types of PECVD systems as shown in Figure 13.17. In each system, the RF frequency chosen is normally less than 1 MHz, although PECVD oxide can be deposited at 13.56 MHz. The first PECVD systems were cold wall parallel plate reactors. Gasses are either injected at the edge or through an upper electrode showerhead, and exhausted through a port at the center, or the gas is injected at the center and exhausted around the edges. As the wafer diameter has increased, the low throughput and marginal uniformity of these systems has obviated their use for silicon IC production. Due to the small wafer diameters and the limited number of wafers per batch however, this style of reactor is often preferred for GaAs technologies.

For silicon IC manufacturing with large diameter wafers, the currently preferred technique is a parallel plate hot wall system. Similar in appearance to an LPCVD tube, the wafers are mounted vertically on conductive graphite electrodes of alternating polarity. The substrate temperature is controlled as in any furnace, although it is much cooler than it would be for a comparable LPCVD process. Although the throughput of this type of reactor is better than the parallel plate plasma reactor, it is much less than that of a standard LPCVD system.

Hot wall batch PECVD systems suffer from the same types of gas depletion/uniformity and particle problems as their thermal counterparts. For that reason, there has been a resurgence of interest in cold wall PECVD systems. To increase the throughput, a number of deposition stages may be put into a single vacuum system, or several single wafer chambers may be run in parallel with a robot arm to feed the chambers. One manufacturer does the deposition sequentially in five stages. Not only does this improve throughput, uniformity of close to 1% is seen as well [31].

To deposit high quality layers at low substrate temperatures, remote plasmas have recently been introduced. These reactors use a conventional or electron cyclotron resonance (ECR) plasma [32] to dissociate or crack one or more of the precursors. These sources were covered in Chapter 10. One application of ECR is to dissociate N_2 to form atomic nitrogen that readily reacts with silane to form Si_xN_y with virtually no ion bombardment of the substrate. The silane can be introduced outside the plasma [33]. Because of the high reactivity of the atomic species, a large substrate temperature is not necessary to drive the reaction and obtain dense films [34]. Good silicon dioxide films have also been demonstrated at temperatures as low as 120°C [35].

The low pressure of an ECR plasma (about 0.1 torr) results in long mean free paths and, therefore, poor step coverage. If the system is designed to allow a significant amount of ion bombardment of the surface however, the deposited species will be continuously sputtered, allowing the fill of high aspect ratio features. One of the primary limitations of ECR deposition systems is the high concentration of particles produced in the plasma. This has been addressed recently by creating particle traps and/or particle absorbent chamber surfaces. To improve the low deposition rates of ECR, production systems can be fabricated with a large number of the remote plasma injectors in parallel in the same vacuum chamber.

The PECVD of silicon nitride has been used in GaAs implantation for many years. In silicon technologies, this process has also been used as a final passivation or scratch protect layer as one of the final steps of the process [36]. The process is run at 300 to 400°C using mixtures of a diluent gas such as Ar or He, SiH_4, and either NH_3 or N_2. Both hot wall and cold wall reactors have been used (Figure 13.17).

Figure 13.18 shows plots of the deposition rate, mass density, and atomic composition as a function of the fraction of ammonia in the gas flow [37]. The deposition rate is relatively insensitive to gas composition. The maximum in the density occurs at a Si/N ratio of 0.75, the correct value for stoichiometric silicon nitride. Increasing ammonia flow however, increases the hydrogen concentration, which at about 20%, is typical for PECVD Si_3N_4. Increasing the substrate temperature decreases the hydrogen content in the film. The use of N_2 rather than NH_3 also decreases the hydrogen content [38].

One of the major applications for deposited dielectrics is to form the insulator between metal interconnect levels in an IC. Particularly when aluminum is used for the metal, PECVD provides the

Figure 13-18 Deposition rate, density, and stoichiometry in the PECVD of Si_3N_4 *(after Claasen et al., reprinted by permission of the publisher, The Electrochemical Society).*

necessary low temperature deposition. A problem with PECVD nitrides in this application is the relatively large dielectric constant of Si_3N_4. When used as an insulator between two metal layers, this will result in a large node capacitance and therefore, a slower circuit speed. To improve circuit performance, the nitride can be replaced with a PECVD oxide. PECVD silicon dioxide processes can be run using silane and an oxidizer. O_2 can be used, but the reaction between silane and O_2 does not need a plasma to drive it. As a result, considerable homogeneous nucleation occurs in the inlet nozzles and in the gas above the wafer resulting in high particle counts and poor morphology. CO_2 can be used but N_2O is the preferred oxidant to avoid carbon incorporation. It has also been reported that the addition of He as a diluent improves deposition uniformity and reproducibility [39].

Oxides deposited with this technique have high concentrations (1 to 10%) of hydrogen [40]. It is also common to find substantial amounts of water and nitrogen [41]. The exact composition depends critically on the chamber power and the gas flows. As shown in Figure 13.19, increasing the plasma power increases the deposition rate but also reduces the density [42]. As a result of the ease of silicon oxidation reaction, low plasma power densities provide large deposition rates. It has also been shown that if the power density of the plasma is large enough to ensure a sufficient concentration of atomic oxygen in the plasma to completely oxidize the silane, good films with a high dielectric strength are observed [43]. Post deposition high temperature bakes may be used to reduce the hydrogen content and densify the film. These bakes can also be used to control film stress [44] but often PECVD processes are chosen precisely because such high temperature steps cannot be tolerated.

One of the interesting features of PECVD films is that it is possible to change the composition of the films continuously from oxide to nitride by changing the gas flow. By adding increasing amounts of N_2O to a mixture of SiH_4, NH_3, and He in a 13.56 MHz cold wall PECVD system, films were deposited whose refractive index varies smoothly from that of nitride to that of oxide [45]. This allows the technologically interesting possibility of stacks and graded composition films.

PECVD of silicon dioxide may also be done using a TEOS source [46],[47]. One of the motivations for this approach is the danger associated with silane use. At sufficiently large O_2 to TEOS inlet flow ratios, the residual carbon contamination in the deposited films can be made extremely small [48], and reasonable values of the index of refraction and dielectric constant [49] can be obtained. Substrate temperatures are well below 400°C. As with thermally deposited TEOS films, the stress of the layer can be controlled over a broad range [50] by a post deposition bake. Doped layers such as PSG and BPSG are also increasingly looking toward the use of organometallics such as tetrabutylphosphine and trimethylborane to reduce the use of hydrides.

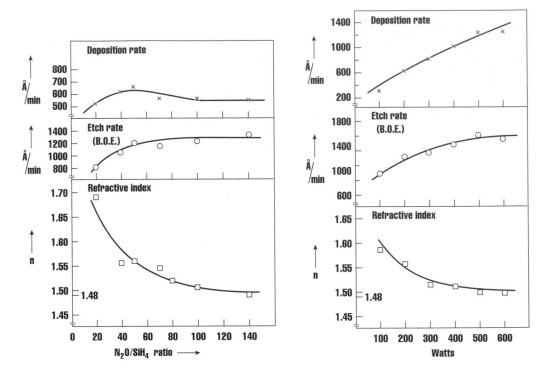

Figure 13-19   Results for the PECVD of SiO₂ *(after Van de Ven, ©1981, Solid State Technology).*

13.8 Metal CVD⁺

The last sections described the attributes of CVD: excellent step coverage and the potential for low substrate temperature depositions. One of the most serious step coverage concerns is that of metal as it goes into contacts. Particularly for deep submicron devices, the step coverage of sputter deposited films into increasingly high aspect ratio features has become unacceptable unless very high temperatures are used. Furthermore, in order to ensure coverage of the metal over the contact, sidewalls must be carefully tapered during the etch process. To accommodate this widening of the contact, typical metal lines must have caps (Figure 13.20). These caps can significantly reduce wiring density.

Finally, the topology created by tapering the contact accumulates so that coincident contacts or "plugs" are not allowed. If, on the other hand, metal CVD is available, vertical contact structures can be used, filling the contact and decreasing the surface topology. No cap is required, and step coverage is much less of a concern. It is therefore highly desirable to develop processes for metal CVD. Of the various metals that have been attempted, the greatest success has been achieved with tungsten [51].

Much of the early work in tungsten CVD was done in standard horizontal LPCVD tubes [52],[53].

Figure 13-20   The use of caps versus plug filled contacts.

Tungsten was found not to adhere to the tube walls and so particulation was a serious problem. Even more of a concern however, was the fact that thin layers of tungsten could effectively block the IR radiation from the coils that heat the wafers. This effect resulted in poor uniformity and reproducibility. As a result of these problems, most tungsten CVD is now carried out in cold wall reactors [54]. Keeping the chamber walls below about 150°C is critical to the success of the process due to the high reactivity of the precursors. Tungsten sources include WCl_6 [55],[56]; $W(CO)_6$ [57],[58]; and WF_6 [59]. Of these, only WF_6 is a liquid at room temperature. The others are high vapor pressure solids. Most tungsten CVD is therefore run using WF_6 with a H_2 carrier gas. Deposition temperatures are usually less than 300°C.

Figure 13-21   A filled contact showing the CVD tungsten morphology *(after Schmitz, reprinted by permission, Noyes).*

The simplest type of tungsten deposition process is the blanket CVD of tungsten. Since blanket CVD W films do not adhere well to the oxide, a thin adhesion layer must first be deposited. One of the most commonly used layers is sputtered TiW, which is often used as a barrier metal under aluminum interconnect. TiN has also received considerable attention, particularly since recent demonstrations of the CVD of these layers [60],[61]. It has been shown however, that the deposition of W over TiN has a substantial initiation time, during which no film is formed unless SiH_4 is used during the growth initiation phase [62].

Although typical deposition rates are rather slow, the best step coverage for W CVD is obtained using WF_6 and H_2 [63]. This is a major concern since one of the primary applications for CVD tungsten films is the filling of high aspect ratio contacts. If the films have poor step coverage, a void will be left in the center of the contact as the top closes. Even a small taper of the contact will ensure complete filling (Figure 13.21). The overall reaction for this process is

$$WF_6 + 3H_2 \rightleftarrows W + 6HF$$

The tungsten layer deposited in this manner may be left in place and used for the interconnect level, or it may be etched back leaving only the filled contacts or plugs. In the latter case, a sacrificial nitride layer may be used on top of the oxide to avoid roughening the surface of the oxide during the etchback (Figure 13.22).

To avoid the etchback altogether, it is also possible to deposit tungsten selectively in the exposed contact windows. Selective tungsten is expected to fill any aspect ratio structure, since the deposition occurs from the bottom of the structure up, rather than from the sides. This has the additional advantage of reducing the cost of the deposition, which is considerable. The problem with these processes is that maintaining selectivity is extremely difficult [64]. Deposits tend to form on the dielectric and may also creep up the walls of the contact.

Aside from tungsten CVD, there are two primary areas of metal CVD research: silicides for gate electrodes, local interconnect, and to serve as W adhesion layers, and alternate interconnect metals such as aluminum and copper to replace the use of sputtered AlSi and AlCuSi altogether. In the former

(A)

(B)

(C)

Figure 13-22   The use of a sacrificial nitride layer with tungsten CVD to avoid surface roughening *(after Schmitz, ©1992, Noyes).*

category, the primary material is $TiSi_2$. $TiSi_2$ is deposited from a low pressure mixture of SiH_4 and $TiCl_4$ in a H_2 carrier. Typical deposition temperatures are 600 to 750°C and deposition rates are several thousand Å/min, making the deposition process suitable for use with single wafer systems [65]. Selective deposition is also possible. This is attractive for a salicide replacement (see Chapter 16), since the very shallow junctions employed by deep submicron CMOS do not tolerate much silicon consumption. This process can even be run in standard LPCVD tubes [66], but as with tungsten the use of such systems for metal deposition leads to eventual problems with thermal uniformity. Furthermore, the process window for selective deposition is very narrow [67]. If the process is carried out in an PECVD reactor, the films deposited over silicon tend to be rough unless subsequently annealed at high temperature [68].

The two films that have received the most attention as potential replacements for sputtered aluminum are CVD copper and CVD aluminum. Because of its current use in IC interconnect, aluminum was one of the first films studied [69]. Due to the grain structure, electromigration [70] and junction spiking [71] are much less severe in CVD aluminum than in sputtered films. The most commonly used precursor is tri-isobutyl-aluminum (TIBA), because of the low carbon content in LPCVD deposited layers. Typical deposition temperatures are 200 to 300°C, and deposition rates are hundreds of Å/min. It is also possible to deposit these films selectively as well [72]. Of course, the deposition of the desired alloys AlSi and AlCuSi are much more difficult.

The most popular precursor for copper CVD are the Cu β-diketones. Deposition temperatures are typically 100 to 200°C [73]. Although the films have very low resistivity, there is currently no acceptable way to anisotropically etch copper. As a result, much of the copper CVD work has been directed toward selective deposition. Here the concept would be to deposit and pattern a thin seed layer, then selectively deposit the copper on top of that. To achieve good selectivity, the deposition temperature must be kept low and the surface must be very clean [74].

13.9 Summary

This chapter reviewed the basic chemistry and fluid mechanics of chemical vapor deposition. Although an accurate quantitative description of any reaction is highly process and reactor dependent, this introductory material can be used to develop a qualitative understanding of the deposition process. Atmospheric pressure CVD (APCVD) is only widely used for SiO_2 deposition. The major problem with these processes is particle formation either at the injector nozzles or through homogeneous nucleation. Low pressure CVD (LPCVD) of polysilicon and silicon nitride is widely used in silicon technologies. Standard furnace tube arrangements provide good uniformity and high throughput. To improve the uniformity further, cross-flow reactors have been developed. The deposition of oxide can be done with silane and oxygen or nitrous oxide, but the currently preferred method is the thermal decomposition of TEOS. Plasma enhanced CVD is also popular when low deposition temperatures are required. These films tend to have poor stoichiometry and suffer from high etch rates unless subsequent anneals are done. To obtain high density PECVD films, after glow reactors have been developed. The CVD of metals is a new and important area that allows the fabrication of high density interconnect. Currently, tungsten is the most popular CVD metal system.

Problems

1. Repeat the calculations done in Example 13.1 if the reaction of interest is

$$AB_2 \rightleftharpoons A + 2B$$

Assume that the equilibrium constant is unchanged.
2. Repeat the calculations done in Example 13.2 assuming that the gas is hydrogen ($\eta = 30$ gm/cm-sec).
3. Briefly describe the advantages and disadvantages of APCVD.
4. Assume that you wanted to deposit NaCl (table salt) on the wafer. What precursors might you use? What problems could you foresee?
5. A particular process is reaction rate limited at 700°C and the activation energy is 2 eV. At this temperature the deposition rate is 1000 Å/min. What would you guess that it would be at 800°C? If the measured deposition rate at 800°C is much less than this prediction, what might your conclusion be? How would you prove it?
6. A process is run in a standard horizontal LPCVD tube, where the wafers are loaded on edge in a standard slotted boat. What factors might explain a reduction in the deposition rate from the front of the tube to the back? From the edge of each wafer to the center? What would you do to try to improve the uniformity in each case?
7. A post-deposition anneal called a *densification step* is often used to reduce the etch rate of CVD SiO_2. The densification is typically run at 900–1000°C. The step is not normally done for PECVD films although they would benefit from the anneal. Explain why the process is not done for these films.

References

1. A. Sherman, *Chemical Vapor Deposition for Microelectronics: Principles, Technology, and Applications*, Noyes, Park Ridge, NJ, 1987.
2. J. Dzarnoski, S. F. Rickborn, H. E. O'Neal, and M. A. Ring, ''Shock Induced Kinetics of the Disilane Decomposition and Silylene Reactions with Trimethylsilane and Butadiene,'' *Organometallics* **1**:1217 (1982).
3. H. K. Moffat, *Numerical Modeling of Chemical Vapor Deposition Processes in Horizontal Reactors*, Thesis, Univ. Minn., 1992.
4. M. E. Coltrin, R. J. Lee, and J. A. Miller, ''A Mathematical Model of Silicon Vapor Deposition Further Refinements and the Effects of Thermal Diffusion,'' *J. Electrochem. Soc.* **133**:1206 (1986).
5. For the viscosities of common carrier gasses, see L. J. Giling, in *Crystal Growth of Electronic Materials*, E. Kaldis, ed., Elsevier Science Publ., 1985, p. 71.

6. G. B. Stringfellow, *Organometallic Vapor-Phase Epitaxy Theory and Practice*, Academic, Boston, 1989.

7. M. L. Hammond, ''Introduction to Chemical Vapor Deposition,'' *Solid State Technol.* **22**:61 (Dec. 1979).

8. R. Takahashi, Y. Koga, and K. Sugawara, ''Gas Flow Patterns and Mass Transfer Analysis in a Horizontal Flow Reactor for Chemical Vapor Deposition,'' *J. Electrochem. Soc.* **119**:1406 (1972).

9. W. Kern and R. S. Rosler, ''Advances in Deposition Processes for Passivation Films,'' *J. Vac. Sci. Technol.* **14**:1082 (1977).

10. K. Shankar, ''Thermal CVD Equipment: Competition Heats Up,'' *Solid State Technol.*, p. 47 (April 1991).

11. K. Fujino, Y. Nishimoto, N. Tokumasu, and K. Maeda, ''Low Temperature, Atmospheric Pressure CVD Using Hexamethylsilixane and Ozone,'' *J. Electrochem. Soc.* **139**:2282 (1992).

12. S. B. Desu, C. H. Peng, T. Shi, and P. A. Agaskar, ''Low Temperature CVD of SiO_2 Films Using Novel Precursors,'' *J. Electrochem. Soc.* **139**:2682 (1992).

13. R. S. Rosler, ''Low Pressure CVD Production Processes for Poly, Nitride, and Oxide,'' *Solid State Technol.* **20**:63 (1977).

14. R. Iscoff, ''Hotwall LPCVD Reactors: Considering the Choices,'' *Semicond. Int.*, p. 60 (June 1991).

15. T. I. Kamins, ''Structure and Properties of LPCVD Silicon Films,'' *J. Electrochem. Soc.* **127**:686 (1980).

16. A. C. Adams, ''Dielectric and Polysilicon Film Deposition,'' in *VLSI Technology*, S. M. Sze, ed., McGraw-Hill, New York, 1988.

17. A. T. Voutsas and M. K. Hatalis, ''Structure of As-Deposited LPCVD Silicon Films at Low Deposition Temperatures and Pressures,'' *J. Electrochem. Soc.* **139**:2659 (1992).

18. M. K. Hatalis and D. W. Greve, ''Large Grain Polycrystalline Silicon by Low-Temperature Annealing of Low Pressure Chemical Vapor Deposited Amorphous Silicon Films,'' *J. Appl. Phys.* **63**:2260 (1988).

19. A. Baudrant and M. Sacilotti, ''The LPCVD Polysilicon Phosphorus Doped in situ as an Industrial Process,'' *J. Electrochem. Soc.* **129**:2620 (1982).

20. P. Duverneuil and J.-P. Couderc, ''Two Dimensional Modeling of Low Pressure Chemical Vapor Deposition Hot Wall Tubular Reactors,'' *J. Electrochem. Soc.* **139**:296 (1992).

21. T. A. Badgwell, T. F. Edgar, I. Tractenberg, and J. K. Elliott, ''Experimental Verification of a Fundamental Model for Multiwafer Low-pressure Chemical Vapor Deposition of Polysilicon,'' *J. Electrochem. Soc.* **139**:524 (1992).

22. W. A. Pliskin, ''Comparison of Dielectric Films Deposited by Various Methods,'' *J. Vac. Sci. Technol.* **14**:1064 (1977).

23. K. Watanabe, T. Tanigaki, and S. Wakayama, ''The Properties of LPCVD SiO_2 Film Deposited by SiH_2Cl_2 and N_2O Mixtures,'' *J. Electrochem. Soc.* **128**:2630 (1981).

24. F. S. Becker, D. Pawlik, H. Schafer, and G. Standigl, ''Process and Film Characterization of Low Pressure Tetraethylorthosilicateborophosilicate Glass,'' *J. Vac. Sci. Technol.* **B4**:732 (1986).

25. G. Smolinsky, ''The Low Pressure Chemical Vapor Deposition of Silicon Oxide Films in the Temperature Range 450°C to 600°C From a New Source: Diacteoxyditertiarybutoxysilane,'' *Proc. 1986 Symp. VLSI Technol.*, San Diego, May, 1986.

26. W. A. Pliskin, ''Comparison of Properties of Dielectric Films Deposited by Various Methods,'' *J. Vac. Sci. Technol.* **14**:1064 (1977).

27. C. L. Ramiller and L. Yau, "Borophosphosilicate Glass for Low Temperature Reflow," *Semicon West Technol. Proc.* **5**:29 (1982).

28. A. C. Adams and C. D. Capio, "The Deposition of Silicon Dioxide Films at Reduced Pressure," *J. Electrochem. Soc.* **126**:1042 (1979).

29. A. J. Learn, "Phosphorus Incorporation Effects in Silicon Dioxide Grown at Low Pressure and Temperature," *J. Electrochem. Soc.* **132**:405 (1985).

30. F. H. P. M. Habraken, A. E. T. Kuiper, A. V. Oostrom, and Y. Tamminga, "Characterization of Low Pressure Chemical Vapor Deposited and Thermally Grown Silicon Nitride Films," *J. Appl. Phys.* **53**:404 (1982).

31. R. D. Compton, "PECVD: A Versatile Technology," *Semicond. Int.*, p. 60 (July 1992).

32. S. Matsuo and M. Kiuchi, "Low Temperature Deposition Apparatus Using an Electron Cyclotron Resonance Plasma," *Proc. Symp. VLSI Sci., Technol.*, Electrochem. Soc., Pennington, NJ, 1982, p. 83.

33. P. D. Lucovsky, D. V. Richard, S. Tsu, Y. Lin, and R. J. Markunas, "Deposition of Silicon Dioxide and Silicon Nitride by Remote PECVD," *J. Vac. Sci. Technol.* **A4**:681 (1986).

34. V. Herak and D. J. Thomson, "Effects of Substrate Temperature on the Electrical and Physical Properties of Silicon Dioxide Films Deposited from Electron Cyclotron Resonance Microwave Plasmas," *J. Appl. Phys.* **67**:6347 (1990).

35. F. Plais, B. Agius, F. Abel, J. Siejka, M. Puech, G. Ravel, P. Alnot, and N. Proust, "Low Temperature Deposition of SiO_2 by Distributed Electron Cyclotron Resonance Plasma Enhanced Chemical Vapor Deposition," *J. Electrochem. Soc.* **139**:1489 (1992).

36. R. S. Rosler, W. C. Benzing, and J. A. Baldo, "A Production Reactor for Low Temperature Plasma Enhanced Silicon Nitride Deposition," *Solid State Technol.* **19**:45 (1976).

37. W. A. P. Claasen, W. G. J. N. Valkenburg, M. F. C. Willemsen, and W. M. v. d. Wijgert, "Influence of the Deposition Temperature, Gas Pressure, Gas Phase Composition, and RF Frequency on Composition and Mechanical Stress of Plasma Silicon Nitride Layer," *J. Electrochem. Soc.* **132**:893 (1985).

38. R. Chow, W. A. Lanford, W. Ke-Ming, and R. S. Rosler, "Hydrogen Content of a Variety of Plasma-Deposited Silicon Nitrides," *J. Appl. Phys.* **53**:5630 (1982).

39. J. Batey and E. Tierney, "Low-Temperature Deposition of High-Quality Silicon Dioxide by Plasma Enhanced Chemical Vapor Deposited," *J. Appl. Phys.* **60**:3136 (1986).

40. A. C. Adams, F. B. Alexander, C. D. Capio, and T. E. Smith, "Characterization of Plasma Deposited Silicon Dioxide," *J. Electrochem. Soc.* **128**:1545 (1981).

41. A. Sherman, *Chemical Vapor Deposition for Microelectronics: Principles, Technology, and Applications*, Noyes, Park Ridge, NJ, 1987.

42. E. P. G. T. van de Ven, "Plasma Deposition of Silicon Dioxide and Silicon Nitride Films," *Solid State Technol.* **24**:167 (April 1981).

43. D. L. Smith and A. S. Alimonda, "Chemistry of SiO_2 Plasma Deposition," *J. Electrochem. Soc.* **140**:1496 (1993).

44. H.-J. Schliwinski, U. Schnakenberg, W. Windbracke, H. Neff, and P. Lange, "Thermal Annealing Effects on the Mechanical Properties of Plasma Enhanced Chemical Vapor Deposited Silicon Oxide Films," *J. Electrochem. Soc.* **139**:1730 (1992).

45. S. Nguyen, S. Burton, and P. Pan, "The Variation of Physical Properties of Plasma-Deposited Silicon Nitride and Oxynitride with Their Compositions," *J. Electrochem. Soc.* **131**:2348 (1984).

46. D. R. Secrist and J. D. Mackenzie, "Deposition of Silica Films by the Glow Discharge Technique," *J. Electrochem. Soc.* **113**:914 (1966).

47. U. Mackens and U. Merkt, ''Plasma Enhanced Chemical Vapor Deposition of Metal-Oxide-Semiconductor Structures on InSb,'' *Thin Solid Films* **97**:53 (1982).

48. F. Fracassi, R. d'Agostino, and P. Favia, ''Plasma-Enhanced Chemical Vapor Deposition of Organosilicon Thin Films From Tetraethoxysilane-Oxygen Feeds,'' *J. Electrochem. Soc.* **139**:2636 (1992).

49. W. J. Patrick, G. C. Schwartz, J. D. Chapple-Sokol, R. Carruthers, and K. Olsen, ''Plasma Enhanced Chemical Vapor Deposition of Silicon Dioxide Films Using Tetraethoxysilane and Oxygen: Characterization and Properties of Films,'' *J. Electrochem. Soc.* **139**:2604 (1992).

50. K. Ramkumar and A. N. Saxena, ''Stress in Thermal SiO_2 Films Deposited by Plasma and Ozone Tetraethylorthosilicate Chemical Vapor Deposition Processes,'' *J. Electrochem. Soc.* **139**:1437 (1992).

51. John E. J. Schmitz, *Chemical Vapor Deposition of Tungsten and Tungsten Silicides for VLSI/ULSI Applications* Noyes, Park Ridge, New Jersey (1992).

52. E. K. Broadbent and C. L. Ramiller, ''Selective Low Pressure Chemical Vapor Deposition of Tungsten,'' *J. Electrochem. Soc.* **131**:1427 (1984).

53. Y. Pauleau and P. Lami, ''Kinetics and Mechanism of Selective Tungsten Deposition by LPCVD,'' *J. Electrochem. Soc.* **132**:2779 (1985).

54. J. E. J. Schmitz, *Chemical Vapor Deposition of Tungsten and Tungsten Silicides*, Noyes, Park Ridge, NJ, 1992.

55. C. M. Melliar-Smith, A. C. Adams, R.-K. Kaiser, and R. A. Kushner, ''Chemical Vapor Deposition of Tungsten for Semiconductor Metallizations,'' *J. Electrochem. Soc.* **121**:298 (1974).

56. N. Hashimoto and Y. Koga, ''The Si-WSi_2-Si Epitaxial Structure,'' *J. Electrochem. Soc.* **114**:1189 (1967).

57. L. Kaplan and F. d'Heurle, ''The Deposition of Molybdenum and Tungsten Films from Vapour Decomposition of Carbonyls,'' *J. Electrochem. Soc.* **117**:693 (1970).

58. M. Diem, M. Fisk, and J. Goldman, ''Properties of Chemically Vapor Deposited Tungsten Thin Films on Silicon Wafers,'' *Thin Solid Films* **107**:39 (1983).

59. R. Hogle and K. Aitcheson, *Tungsten Workshop I*, 1985, p. 225.

60. S. R. Kurtz and R. G. Gordon, ''Chemical Vapor Deposition of Titanium Nitride at Low Temperatures,'' *Thin Solid Films* **140**:277 (1986).

61. A. Sherman, ''Growth and Properties of LPCVD Titanium Nitride as a Diffusion Barrier for Silicon Device Technology,'' *J. Electrochem. Soc.* **137**:1892 (1990).

62. M. Iwasaki, H. Itoh, T. Katayama, K. Tsukamoto, and Y. Akasaka, *Tungsten Workshop V*, 1990, p. 187.

63. J. E. J. Schmitz, R. C. Ellwanger, and A. J. M. van Dijk, *Tungsten Workshop III*, 1988, p. 55.

64. P. Singer, ''The Multiple Challenges of Tungsten CVD,'' *Semicond. Int.* p. 39, (Aug. 1992).

65. A. Bouteville, J. C. Remy, and C. Attuyt, ''$TiSi_2$ Selective Growth in a Rapid Thermal Low Pressure Chemical Vapor Deposition System,'' *J. Electrochem. Soc.* **139**:2260 (1992).

66. A. Bouteville, A. Royer, and J. C. Remy, ''LPCVD of Titanium Disilicide,'' *J. Electrochem. Soc.* **134**:2080 (1987).

67. V. Ilderem and R. Reif, ''Very Low Pressure Chemical Vapor Deposition Process for Selective Titanium Silicide Films,'' *Appl. Phys. Lett.* **53**:687 (1988).

68. J. Lee and R. Reif, ''Plasma Deposited Chemical Vapor Deposition of Blanket $TiSi_2$ on Oxide Patterned Wafers,'' *J. Electrochem. Soc.* **139**:1159 (1992).

69. T. Amazawa and Y. Arita, ''A 0.25 μm via Plug Process Using Selective CVD Aluminum for Multilevel Interconnect,'' *IEDM Tech. Dig.*, p. 265 (1991).

70. R. A. Levy, M. L. Green, and P. K. Gallagher, "Characterization of LPCVD Aluminum for VLSI Processing," *J. Electrochem. Soc.* **131**:2175 (1984).

71. K. Tsubouchi, K. Masu, K. Sasaki, and N. Mikoshiba, "Selective Al CVD on Hydrogen-Terminated Si Surfaces," *IEDM Tech. Dig.*, p. 269 (1991).

72. K.-I. Lee, Y.-S. Kim, and S.-K. Joo, "Effect of Substrate Temperature on the Selectivity in Low Pressure Chemical Vapor Deposition of Aluminum," *J. Electrochem. Soc.* **139**:3578 (1992).

73. S. K. Reynolds, C. J. Smart, E. F. Baran, T. H. Baum, C. E. Larsen, and P. J. Brock, "Chemical Vapor Deposition of Copper from 1,5-cyclooctadiene Copper (I) Heafluoroacetylacetonate," *Appl. Phys. Lett.* **59**:2332 (1991).

74. L. H. Dubois and B. R. Zegarski, "Selectivity and Copper Chemical Vapor Deposition," *J. Electrochem. Soc.* **139**:3295 (1992).

Chapter 14

Epitaxial Growth

Many techniques have been used to deposit single crystal silicon and GaAs [1]. A partial list would include a wide variety of CVD variations, including plasma enhanced, rapid thermal, metallorganic, ultrahigh vacuum, laser, optical, and x-ray assisted. Non-CVD methods include molecular beam, ion beam, and clustered ion beam epitaxy, to name just a few. All of these methods have the capability to grow single crystal layers. Very few however, have demonstrated the capability to grow material of sufficient quality to economically fabricate high density ICs. This chapter will first concentrate on thermal chemical vapor deposition or vapor phase epitaxy (VPE), the most commonly used approach for silicon epitaxy in IC production. In this process, heat provides the energy necessary to drive the chemical processes. Later, the chapter will discuss a VPE method for growing epitaxial layers of GaAs, as well as advanced forms of VPE, including metallorganic VPE (MOCVD) and rapid thermal VPE (RTCVD). The chapter will conclude with a discussion of molecular beam epitaxy (MBE).

A prototype VPE system is shown in Figure 14.1. Much like a low pressure CVD reactor, the wafer is in a vacuum chamber resting on a heated susceptor. This heating is usually done either by radiation from filament lamps or by inductively heating the graphite susceptor. The system is usually pumped, although some silicon epitaxial growth systems have no vacuum capability at all. The gas flows are controlled with mass flow controllers and pneumatic valves.

At commercial VPE temperatures the dopants diffuse significantly, making very thin layers and abrupt transitions impossible. As the temperature is reduced the diffusivity decreases more rapidly than the growth rate, making lower temperature epitaxy very desirable. A great deal of progress has been made in the last few years in reducing the temperature of silicon epitaxy. These improvements have generally involved reducing the impurity concentrations in the growth chamber. Small loading chambers called *load locks* that are attached to, and can be pumped independent of, the main chamber, allow the introduction of new wafers without venting the growth chamber after each run. Also, better pumps and seals are now being used that improve the quality of the vacuum in the chamber. Finally filters are being used to remove chemical contaminants from the gas feed. Some production growth is now done at temperatures below 1000°C. Research has demonstrated high quality growth in single wafer systems at temperatures of about 800°C [2]. Ultrahigh vacuum growth from the vapor has been done at much lower temperatures. Given this progress and the high defect density of the alternative growth techniques the future of these VPE approaches look very attractive for the growth of thick layers. Their largest liability for growing very thin structures is their lack of in situ, real time growth diagnostics [3].

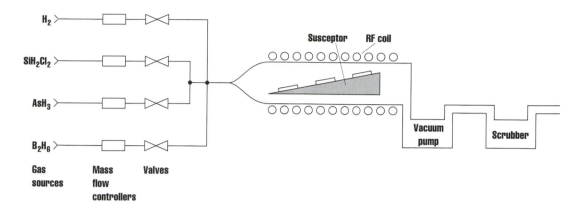

Figure 14-1   A simple VPE system. The susceptor in this chamber is inductively heated using RF power in the external coil.

Current mainstream silicon manufacturing however, uses epitaxial growth only for thick layers (1 to 10 μm). Devices may be fabricated in these layers after their growth, but the basic device regions are not formed by the epitaxy. In contrast, virtually all high speed III-V technologies require the use of advanced epitaxial growth processes as an integral part of the device. Future silicon technologies may also rely more heavily on these growth methods. These processes have the capability to control the vertical dimension in the device to the atomic scale, and they allow the fabrication of nearly perfect interfaces of dissimilar semiconductors. The use of these heterojunction layers provide a new degree of freedom in transistor design. Critical to their application however, is the ability of epitaxial techniques to simultaneously provide excellent control of composition, doping, and thickness, as well as near zero defect density.

14.1 Wafer Cleaning and Native Oxide Removal

Before growth, the substrate must be cleaned; that is the native oxide and any residual impurities must be removed. This is a common requirement for many processes [4]. The ideal goal of this step would be to produce a perfect silicon surface with no impurities or change in atomic position. This is impossible to achieve even if all impurities are removed, since the surface itself has unsatisfied bonds. The atoms on the surface move or reconstruct in order to minimize the density of these bonds. Furthermore, in actual semiconductor fabrication the wafer is often exposed to air as it is moved from the preclean station to the growth chamber. During this exposure the surface may oxidize and often collects carbon contamination from the air. Thus, even a pure reconstructed surface is normally only achieved in ultrahigh vacuum (UHV) equipment.

The types of impurities likely to exist on the surface of the wafer include organics from photoresist residues or polymeric materials deposited during etches, metallic particles deposited in plasma chambers such as implanters and etchers, and thin chemical oxides. The removal of all types of impurities often requires a sequential operation. To isolate the operations, each step normally ends with a rinse in deionized water. The purity of the water is measured by the resistivity and the total organic content (TOC). Typical resistivities are 14 to 18 MΩ-cm and TOC counts are a few parts per million. One of the most popular wet chemical preclean steps for silicon is the RCA clean [5], commonly used when only bare silicon or silicon and thick oxides are exposed on the surface of the wafer. This process will be used as a standard example. When mixed in the proper proportions and held at the correct temperatures, the RCA clean will not significantly attack the silicon surface. The simplest arrangement for an RCA clean is a series of baths for immersion of the wafers. To ensure a fresh supply of chemicals, the solutions may also be plumbed into a spray cleaning system. These cleaning systems also use less chemicals and ensure a more reproducible cleaning environment.

Organic residues are most often removed in an oxidation/reduction solution. A typical solution is a 5:1:1 mixture (by volume) of ammonium hydroxide, hydrogen peroxide, and water ($NH_4OH/H_2O_2/H_2O$) [6], although more recently, lower ammonium hydroxide concentrations have been found to be as effective in reducing contamination and produce less surface roughening. To increase the rate of chemical attack the bath temperature is held at 70 to 80°C. This solution also dissolves elements from the group IB and IIB columns of the periodic table and other metals including gold, silver, nickel, cadmium, zinc, cobalt, and chromium [7]. Because of the depletion of the reactants, the mixture should be prepared just before using it. Other acids including dilute mixtures of HCl, HNO_3, and CH_3COOH are also effective in reducing the concentration of some types of metallic impurities [8].

Since hydrogen peroxide is a strong oxidant, the wafer surface will oxidize during this clean. To remove this chemical oxide, the wafers are dipped in a dilute HF or buffered HF solution. The time required for HF removal of the chemical oxide depends on the solution strength, but is usually less than 30 sec. Complete removal of a surface oxide can be detected by dewet. This is the tendency of water to rapidly sheet off of the surface of the wafer when bare silicon is exposed. When an oxide remains, the surface stays wet for several seconds as the water slowly flows off of the wafer. On patterned wafers, the bare silicon areas will dewet if they are large enough and if there is a clear downward path to the edge of the wafer. Typically these are the scribe lines or streets between the die.

Finally, the heavy alkali ions and cations are stripped in a halogen containing solution. A common mixture is 6:1:1 of $H_2O:HCl:H_2O_2$ heated to 75 to 80°C. As with the organic step, typical immersion times are 10 to 15 min and end with a DI rinse. A drying step using compressed N_2 gas and/or a spin dryer completes the process. It is important in designing an RCA clean to keep the ammonium hydroxide solution and the hydrochloric solution well separated. The vapors of these solutions can react to form ammonium chloride, a salt that acts as a particle contamination to the bath.

The wet chemical clean described often leaves the wafer with a thin chemical oxide that is free of metallic and organic impurities. This oxide must be removed if good epitaxial growth is to be achieved. Virtually all VPE includes as its first step a thermal preclean of this native oxide. This can be conducted in H_2, an H_2/HCl mixture, or in vacuum. A 1% HCl/H_2 mixture etches silicon nonpreferentially and is preferred when the subsequent growth is to be done using silane or disilane without chlorine [9]. There is no direct measurement in most VPE that precleaning has been done well. The grower must simply standardize on a process that achieves good films.

The prebake must be carried out in a temperature oxidizing ambient partial pressure regime that makes the native oxide thermodynamically unstable. Figure 14.2 shows a plot of data taken from Ghidini and Smith [10],[11] and extrapolated to lower temperatures. For oxygen, a clear boundary exists between regions that are thermodynamically stable. At high temperature and low oxygen partial pressure, the native oxide on the wafer forms volatile suboxides such as SiO that desorb [12]. The process is believed to proceed nonuniformly with pin holes opening up first, allowing the semiconductor to evaporate and react with the oxide [13]. This leads to some surface roughening, suggesting that it is desirable to minimize the thickness of the oxide before prebake. In the presence of water

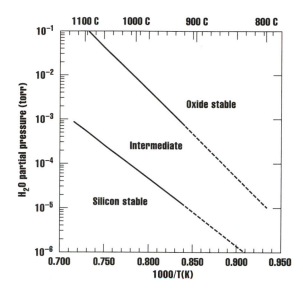

Figure 14-2 Stability diagram for the surface of silicon and silicon dioxide as functions of the temperature and the partial pressure of oxygen and water vapor. The curve for water vapor has an intermediate region where both surfaces can coexist. (*Redrawn using data from Ghidini and Smith, and Smith and Ghidini*).

vapor, there is an intermediate region where both Si and SiO_2 may exist on the surface of the wafer. If the wafer is held in this region, significant surface roughening can occur.

Example 14.1

A preclean must be done using H_2 contaminated with 10 ppm of O_2. If the anneal is done at 1.0 atm, what is the minimum anneal temperature? What if it is done at 0.1 atm?

If the hydrogen prebake is done at 1 atm (760 torr), the H_2 gas contains 10 ppm of O_2 and no other impurities are present, according to Figure 14.2 the wafer must be heated to at least 1070°C to desorb the native oxide. Given a sufficient amount of time, all of the oxide will desorb from the surface if the wafer is held at or above this temperature. If the chamber pressure is reduced to 76 torr using the same gas, the preclean could be done at temperatures as low as 980°C.

The limit to this gas purity example is that the chamber contributes impurities as well, particularly when the wafer is at temperature. As a result, both gas purity and chamber base vacuum are important concerns in epitaxial growth. This careful attention to gaseous and chamber impurities and the ability to heat the wafers to prebake temperatures for relatively short times (of order minutes) are two of the primary differences between epitaxial growth and other CVD equipment.

To allow lower temperature precleans, the chamber may be pumped to a high base vacuum before the preclean. The hydrogen source can also be scrubbed to remove any residual gasses. Hy-

Figure 14-3   Schematic diagram of a palladium diffuser used for purifying hydrogen. Hydrogen diffuses through the high temperature palladium barrier, while the impurities are swept out the vent line.

drogen is usually purified through palladium diffusion (Figure 14.3). In this method, a stream of hydrogen gas is delivered to a canister containing a heated (400°C) palladium tube that has been sealed on one end. Due to the high diffusivity of hydrogen in palladium, the hydrogen diffuses into the tube and from there is delivered to the growth system. Gaseous impurities that cannot diffuse through the tubing wall are swept out the vent line by the hydrogen flow. Palladium diffusers can lower the residual water and oxygen partial pressures to 1 ppb. For other gasses, chemically active resins must be used. Often combined with particle filtration, these units contain a readily oxidizable species that getters oxidants from the gas stream.

There is considerable interest in improving silicon precleaning processes [14]. This research involves both variations in the wet chemical cleans and the in situ part of the preclean. Motivation for this work includes a reduction in the production of waste chemicals, a reduction in particle contamination, and a reduction in the prebake time and temperature to minimize dopant diffusion in the semiconductor. For applications in which even moderate temperatures cannot be tolerated, the surface can be cleaned in situ by sputtering in an argon [15] or hydrogen [16] ambient. A preclean chamber designed for this cleaning can be incorporated into the load lock itself or it may be in line between the chamber and the load lock. Alternately, an Ar$^+$ beam can be set up to impinge off of the surface of the wafer while it is in position for growth. The process is carried out at room temperature at low power densities (1 mA/cm^2). A major problem with the technique is residual sputter damage that is very difficult to remove by thermal annealing. This damage can be reduced by using very low energy ions such as those formed in ECR plasmas [17]. Sputter damage and metallic contamination cannot be completely eliminated, however.

Another method that has received considerable attention in recent years is the surface passivation of silicon through the use of an HF dip [18]. This may be done after a standard RCA clean by removing the final chemical oxide in HF. This process is believed to leave a monolayer of silicon monohydride (SiH) on the (100) silicon surface [19] that leaves the surface resistant to oxidation in air [20]. The Si-F bond is believed to be highly polar due to the large electronegativity difference between the Si and F atoms. This polarization in turn induces a polarization in the Si-Si back bond when a fluorine atom attaches to the surface. The Si-Si polarization allows the bond to be attacked by HF [21], which results in the release of SiF$_x$ and the replacement of the surface Si-Si bond with a stable Si-H bond. Exposure of the wafer to UV light before immersion in HF helps to produce very low levels of residual carbon on the surface. It has also been shown that a brief deionized water rinse can be tolerated if the dissolved oxygen concentration in the water is kept low [22] and the wafers are lightly doped. This is an important consideration since rinses of only a few minutes are sufficient to remove most of the residual F left on the surface after an HF etch [23].

Once the wafer is hydrogen terminated, it can be transferred into the growth chamber and growth may be commenced immediately without a high temperature preclean step. The hydride layer desorbs at about 500°C. This type of process then is preferred primarily for low temperature processes such as MBE. The problem of H desorption during the ramp up to deposition temperatures make this method unattractive for standard thermal VPE. This process is also not preferred for cleaning the surface before thin oxide growth because of a slight surface roughening that reduces the breakdown characteristics of the MOS capacitor [24].

14.2 The Thermodynamics of Vapor Phase Growth

The simplest model that can be used to describe the VPE process is that of Deal [25]. It is essentially the same model as described in Chapter 3 for oxidation. The flux of the deposition species across a gaseous boundary layer is set equal to the flux of reactant consumed by the growing surface:

$$F = h_g(C_g - C_s) = k_s C_s \qquad (14.1)$$

where h_g is a mass transport coefficient that depends on the flow in the chamber, k_s is a surface reaction rate, and C_g and C_s are the concentrations of the growth species in the gas and on the surface of the wafer, respectively. Just as was done for oxidation, one can solve for the growth rate R:

$$R = \frac{k_s h_g}{k_s + h_g} \frac{C_g}{N} \qquad (14.2)$$

where N is the number density of silicon (5×10^{22} cm^{-3}) divided by the number of silicon atoms in the growth molecule. The mass transport limited regime occurs when $k_s \gg h_g$, and the reaction rate limited regime when $h_g \gg k_s$.

While the Deal model provides a semiquantitative model for the growth, it vastly oversimplifies the growth process. Unlike oxidation, the growth species is not the same as the feed gas. Instead, many chemical reactions may occur in the gas phase and on the surface of the wafer. To further complicate matters, a number of processes may be going on simultaneously that assist or compete with the growth process. In the Si-H-Cl system, for example, the silicon containing molecules on the surface of the wafer may be $SiCl_2$, $SiCl_4$, SiH_2, or Si. At low chamber pressure or low inlet gas flows, the production of these reactive species limit the growth rate. Silicon atoms that are adsorbed on the surface and the substrate atoms themselves may be etched by reaction with Cl. The Deal model has no mechanism for taking this complicated chemistry into consideration. It should therefore be considered a simple parameterization of a very complicated process.

Just as for CVD in the previous chapter, a more detailed model for the growth of a layer in a VPE process can begin by dividing the process into a sequence of steps (Figure 14.4), each of which can potentially determine the growth rate. The precursor gas enters the chamber where it partially decomposes to several more reactive daughter species. These growth species must move through the chamber until they reach the vicinity of the wafer. Near the wafer the growth species diffuse through the stagnant or boundary layer until they reach the surface. Here they adsorb, diffuse across the surface, decompose further to atomic silicon and volatile reaction products, which must then desorb and be pumped away. The first part of this section will describe the gas phase reactions of some of the common homogeneous processes and popular silicon epitaxy chemistries. The following section will discuss the reactions that occur on the wafer surface.

The simplest chemical system for the growth of silicon is the pyrolytic decomposition of silane. The overall process is:

$$SiH_4(g) \rightarrow Si(s) + 2H_2(g) \qquad (14.3)$$

Figure 14-4   VPE steps include: (1) gas phase decomposition and (2) transport to the surface of the wafer. At the surface the growth species must: (3) adsorb, (4) diffuse, and (5) decompose, and (6) the reaction byproducts desorb.

This reaction is irreversible and can be driven at temperatures as low as 600°C. As discussed in the last chapter, this process is often used for polysilicon deposition. While silane can be used to epitaxially grow silicon at low temperature (600 to 800°C), its tendency to decompose to form particles in the gas (homogeneous nucleation) means that it is difficult to obtain high quality growth except at very low pressures. Homogeneous nucleation occurs continuously in the gas phase for all pressures; however, the nucleation rate increases sharply with the partial pressure of silane in the vapor. If the size of the particle created in the vapor is small, its surface energy is large enough to make the particle energetically unfavorable. The particle must therefore achieve a minimum size, or it will shrink and ultimately disappear. This critical size is given by [26]:

$$r^* = \frac{2uv}{kT \ \ln(\sigma_o)} \qquad\qquad (14.4)$$

where u is the surface interfacial free energy, v is the atomic volume, and σ_o is the ratio of the pressure of the growth species to the equilibrium pressure of the growth species. This ratio, called the *degree of saturation*, will be discussed later in the chapter. The production of these particles sets a limit on the maximum pressure of SiH_4 that is allowable for a given temperature. Most epitaxial processes dilute the growth species with H_2 to form 1 to 5% growth gas mixtures. Even with this constraint, epitaxial silicon is not commonly grown from silane. Silicon will deposit on all warm surfaces in the reactor. These films may peel off and particulate during movement of the fixtures or during pumpdown and heating. Some of these particles deposit on the wafer leading to high defect densities and potentially poor growth. Of course, epi processes are far more sensitive to these defects than standard CVD because the films will be made into active device instead of simple insulators or conductors.

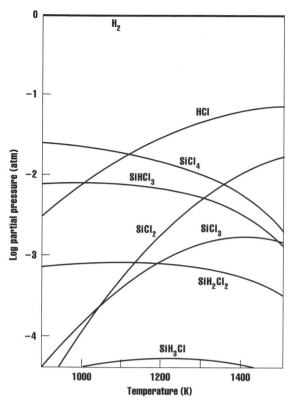

Figure 14-5   Equilibrium partial pressures in the Si-Cl-H system at 1 atm and a Cl to H ratio of 0.06 (*after Bloem and Claassen, reprinted by permission, Philips*).

Nearly all silicon epitaxy for IC fabrication is done by the reduction of chlorosilanes (SiH_xCl_{4-x}, where x = 0, 1, 2, or 3), heavily diluted in hydrogen. The smaller the number of chlorine atoms in the precursor molecule, the lower the temperature required for the same growth rate. The first commonly used precursor was silicon tetrachloride ($SiCl_4$). To obtain useful growth rates with $SiCl_4$, however, the substrate must be held at temperatures in excess of 1150°C. Because of extreme dopant redistribution at this temperature, growth from $SiHCl_3$, SiH_2Cl_2, and SiH_3Cl have all been developed. Dichlorosilane (SiH_2Cl_2 or DCS) is currently the most commonly employed source [27].

The chlorosilanes all follow similar reaction paths. Figure 14.5 shows the equilibrium partial pressures for various species for $SiCl_4$ at one atmosphere and 1500 K [28], [29]. As discussed in the last chapter, these calculations come from a minimization of the total free energy of the system. In real epitaxy systems, the gasses may not actually reach equilibrium [30], but the tendencies remain the same. For low atomic percentages of feed gas, the primary components at the growth temperature are H_2, HCl, and dichlorosilylene ($SiCl_2$). $SiCl_2$ is believed to be the primary growth species in most chlorosilane VPE processes. As shown in Figure 14.6, in the reaction rate limited regime, the growth rate is reasonably well fit to [31],[32]

$$R = c_1 p_{SiCl_2} - c_2 p_{HCl}^2 \qquad (14.5)$$

where c_1 and c_2 show Arrhenius behavior. The negative term on the right side is due to chlorine etching of the substrate. The etch mechanism depends on the square of the HCl partial pressure because the etching reaction requires two HCl molecules:

$$Si(s) + 2HCl(g) \rightleftarrows SiCl_2(g) + H_2(g) \qquad (14.6)$$

where $Si(s)$ refers to silicon atoms already incorporated into the crystal. Adding gaseous HCl slows the net deposition rate by increasing the etch rate.

One can define σ as the degree of supersaturation in the growth ambient given by

$$\sigma = \left[\frac{p_{Si}}{p_{Cl}}\right]_{Feed} - \left[\frac{p_{Si}}{p_{Cl}}\right]_{Eq} \qquad (14.7)$$

where p_{Si} and p_{Cl} are the partial pressures of the silicon and chlorine, respectively. The supersaturation is just the saturation (σ_o) minus one. The first term in Equation 14.7 refers to the ratio of the feed gasses, and the second is the equilibrium ratio given as a function of temperature in Figure 14.7 [33]. To find the equilibrium ratio, first determine the ratio of chlorine to hydrogen in the chamber. Next, read across Figure 14.7 to the chamber temperature to obtain the equilibrium ratio of silicon to

Figure 14-6 Growth rate as a function of the $SiCl_4$ flow. At high concentrations, the chlorine in the chamber leads to etching *(after Theuerer, reprinted by permission of the publisher, The Electrochemical Society).*

Figure 14-7 The equilibrium ratio of silicon to chlorine at 1 atm *(after Arizumi, reprinted by permission, Elsevier Science).*

chlorine. Finally, subtract this ratio from the feed gas ratio to obtain the degree of supersaturation. If σ is positive, the vapor is said to be supersaturated, and under equilibrium conditions growth will result. If σ is negative, on the other hand, the system is undersaturated and etching will result.

A problem arises if σ is very large. Unlike standard CVD, the films grown in VPE are single crystal. The ability of the incident atoms to be correctly incorporated into lattice sites is essential to obtaining high quality growth. The picture of the surface often used is the terrace-kink model. A quantitative description of the model will be covered later in the chapter. For now, assume that the model predicts the existence of two planes (terraces). Molecules landing on the lower plane diffuse to the edge of the terrace where they dissociate and incorporate into the crystal. This type of growth is called *two dimensional* (2-D) since it proceeds primarily across the surface of the wafer. If the arrival rate is too large, the admolecules will agglomerate in the center of the plane, forming a growth island that never reaches the terrace edge. This type of growth is called *islanding* or *three dimensional* (3-D) *growth*. It produces a lower quality film than 2-D growth. The degree of saturation then is an important first order description of the growth process.

Example 14.2

For the data shown in Figure 14.6, calculate the supersaturation as a function of $SiCl_4$ concentration. Compare this value to the growth rate. Does the model predict the peak in the growth rate? Does it predict the transition to etching?

Since the growths use $SiCl_4$, the ratio of Si/Cl in the feed gas is 0.25 for all growths. The equilibrium ratio of Si/Cl can be found using Figure 14.7 by first calculating the Cl/H ratio. For

Table 14.1

SiCl$_4$/H$_2$	[Si/Cl]$_{Feed}$	[Cl/H]$_{Feed}$	[Si/Cl]$_{Eq}$	σ	Growth Rate (μm/min)
0.05	0.25	0.11	0.14	+0.11	+3.7
0.10	0.25	0.22	0.21	+0.04	+5.0
0.20	0.25	0.50	0.24	+0.01	+2.1
0.30	0.25	0.86	0.28	−0.03	−0.6

a 5% mixture, the chlorine to hydrogen ratio is 0.05*4/0.95*2 = 0.11. According to Figure 14.7, at the growth temperature (1270°C = 1543 K), this produces an equilibrium Si/Cl ratio of 0.14. Continuing for increasing SiCl$_4$ concentrations produces Table 14.1. The model does indeed predict a change from growth to etching between 20 and 30% SiCl$_4$ due to the large amount of Cl released. It does not predict the maximum shown at 10% because the small growth rate at low concentrations is due to mass transport limitations. The model does not take this into account.

14.3 Surface Reactions

An atomistic picture of what is happening at the surface of the wafer during VPE is not very clear. The fact that the epitaxial growth of silicon from all of the precursors in the Si-H-Cl system have the same activation energy in the reaction rate limited regime (see Figure 14.8) suggests that the same rate limiting process may be involved for each reaction [34]. This is generally believed to be the desorption of hydrogen from the surface of the wafer. In this model, a large part of the surface is hydrogen terminated. These hydrogen atoms must be desorbed before silicon atoms can be incorporated. The model has been supported by the observations of Claassen and Bloem that the growth rate of silicon in dichlorosilane saturates with increasing DCS flow in a hydrogen ambient, but does not saturate in a nitrogen ambient. Coon et al. [35] have used temperature programmed desorption to measure the surface coverage of the various species. Their results suggest that HCl, and not H, desorption is the rate limiting step at low temperature.

A second important clue about the surface chemistry was obtained by Aoyama and coworkers [36]. Their calculations indicated that the primary growth species on the surface of the wafer is SiCl$_2$. Adsorbed molecules such as SiCl$_2$ are not chemically bonded, but instead are only physically adsorbed (physisorbed) to this passivated surface in VPE. As such there is a constant flux of SiCl$_2$ to and from the surface. Chernov modeled the (111) surface of silicon in a typical

Figure 14-8 Arrhenius behavior of a variety of silicon containing growth species *(after Eversteyn, reprinted by permission, Philips).*

Table 14.2 The major surface components for a (111) silicon surface for Cl/H = 0.01 at 1500 K

Species	% Coverage
H	63
Cl	20
$SiCl_2$	16
Vacancies	1.5
H_2	0.01

After Chernov.

epitaxial growth environment and found that under equilibrium conditions, the major surface components are H, Cl, and $SiCl_2$ (Table 14.2) [37].

In order to develop a model of the surface during VPE, one generally assumes that the gas phase reactions only produce growth species that are then adsorbed on the surface of the wafer where they react and become incorporated into the growing film. It is also assumed that a limited number of sites exist on the surface of the wafer where these molecules can adsorb. θ_x is defined as the fraction of these sites occupied by the admolecule x. Finally, assume that these fractions are proportional to their partial pressure in the gas phase, which in turn is proportional to the mass flow rate of the parent precursor:

$$\theta_x \propto p_x \propto \text{Flow}_x \qquad (14.8)$$

For small fractions of dichlorosilane in a hydrogen carrier the growth rate can then be approximated by [31]:

$$R = k_1 \frac{p_{H_2} p_{SiCl_2}}{1 + k_2 p_{H_2}} \theta - k_3 \frac{p_{HCl}^2}{p_{H_2}} \qquad (14.9)$$

where θ is the fraction of free sites given by:

$$\theta = \frac{1}{1 + k_4 p_{SiCl_2} + k_5 \dfrac{p_{HCl}}{p_{H_2}^2} + k_6 p_{H_2}^2} \qquad (14.10)$$

where the k terms are constants. As with Equation 14.5, the negative term in Equation 14.9 corresponds to etching of the substrate. For most growth conditions, the gas is primarily H_2. Then, the partial pressure of hydrogen is approximately the same as the chamber pressure. The $SiCl_2$ partial pressure is proportional to the dichlorosilane inlet flow rate. The HCl partial pressure, on the other hand, is proportional to the square of the dichlorosilane inlet flow rate, since each dichlorosilane molecule contributes two chlorine atoms.

14.4 Dopant Incorporation

Both intentional and unintentional dopants may be incorporated during epitaxial growth. The primary unintentional sources are solid state diffusion from the substrate and gas phase autodoping. Solid state diffusion dominates the profile near the epi/substrate interface. It follows a complimentary error function dependence if the growth rate obeys the relation

$$R > 2\sqrt{\frac{D}{t}}$$

where D is the diffusivity of the impurity and t is the growth time. The diffusion coefficients used in this calculation may be somewhat different than those calculated in Chapter 3 due to the effects of the growth process on the local concentration of self interstitials and vacancies.

Gas phase autodoping occurs when impurities desorb from the wafer travel through the gas, and readsorb elsewhere on the wafer. It generally follows a doping profile of

$$C(x) = fN_{os}e^{-x/x_m} \tag{14.11}$$

where f is the trapping density, N_{os} is the number of trapping sites on the surface, and x_m is a transition width, which depends on growth rate and the desorption coefficient.

For silicon epitaxy, the most common gaseous dopant sources are diborane (B_2H_6), arsine (AsH_3), and phosphine (PH_3). The doping concentration in the epitaxial layer is controlled by metering the flow of these gasses into the reaction chamber. To achieve low partial pressures for lightly doped layers, the dopant gas is often diluted with hydrogen. One can define a segregation coefficient K as

$$K_{eff} = \frac{P_{Si}}{P_{Dopant}}\frac{C_{Dopant}}{5 \times 10^{22}\text{cm}^{-3}} \tag{14.12}$$

where P_{Si} and P_{Dopant} are the partial pressures of the silicon growth species and dopant respectively, and C_{Dopant} is the concentration of the dopant in the growing film due to intentional doping. When K_{eff} is less than 1, the growing film rejects the dopant. For dilute flows, the pressure ratios can be approximated by the ratios of the inlet flows multiplied by the dilution of the dopant gas. Then, the doping concentration in the film is just linearly proportional to the dopant gas flow rate.

At high dopant gas concentrations, the incorporation rate falls below a simple linear dependence on the dopant gas flow. In the case of PH_3, this occurs due to the parasitic gas phase reaction:

$$2PH_3 \rightarrow P_2 + 3H_2 \tag{14.13}$$

Thus, at low concentrations the phosphorus incorporation is proportional to the partial pressure, but at high concentrations it is proportional to the square root of the partial pressure. Boron doping shows similar effects [38]. The concentration at which these competing gas phase reactions becomes important is a sensitive function of temperature. Lower growth temperatures generally allow the growth of more heavily doped films. High concentration doping has also been shown to affect the growth rate. Arsenic doping at high concentration, in particular, is known to dramatically reduce the growth rate by poisoning the surface. That is, arsenic occupies a high concentration of surface sites, preventing silicon incorporation in the normal 2-D growth process, but not becoming incorporated itself.

14.5 Defects in Epitaxial Growth

Epitaxial growth can introduce defects. If these defects are in the active region of the wafer where the transistors are fabricated, they will often lead to device failures. These failures can be caused directly by electronic states associated with the defects, which lead to excessive leakage. Failures may also be less direct. During processing, the defects may trap other impurities in the wafer that contribute to these electronic states. The defects may also lead to excessive impurity diffusion during processing, which changes the physical transistor structure.

The number and density of defects is a sensitive function of the growth conditions, including substrate temperature, chamber pressure, growth precursor, and wafer precleaning procedure. The most common type of defect in epitaxial silicon layers is the stacking fault. As discussed in Section 2.3, stacking faults are an extra plane of atoms inserted into the crystal or a missing plane of atoms in the crystal. Stacking faults in silicon normally occur in the <111> directions. Figure 14.9 shows an electron micrograph of a stacking fault in a (111) wafer, along with four scans of the surface morphology of the defect [39]. At one end of the defect is a pit, while at the other side, the stacking fault extends above the surface of the wafer. With (100) wafers the stacking fault appears as a line along the <110> directions. In either case, an obvious crystal orientation is seen.

Spikes are protrusions from the epitaxial layer that show little or no alignment with the crystal directions. Spikes may be related to the onset of 3-D growth [40]. Stacking faults and spikes often originate from a defect on the original wafer surface [41]. These original defects include oxygen, metallic impurities, oxidation induced stacking faults in the wa-

Figure 14-9   An electron micrograph of a stacking fault in an epitaxial layer grown on a (111) wafer. The lower plot shows mechanical scans of the surface topology at four sites on the fault *(reprinted by permission, Academic Press)*.

Figure 14-10   A schematic diagram of pattern shift. Alignment marks are etched into the wafer, but nonuniform growth rates cause the mark position to shift after growth.

fer, and particles deposited on the surface of the wafer. Improvements in cleaning procedures have dramatically reduced stacking fault densities in production silicon VPE.

Dislocations are the 2-D analog of the stacking fault. Dislocations are much less obvious on the surface of the wafer but are still a serious yield concern. Dislocations may simply propagate from substrate dislocations. Dislocations may also form in the epitaxial layer due to plastic deformation caused by thermal nonuniformity and excessive growth rates. Subtle defects such as dislocations are often measured by selective etching. As discussed in Chapter 11, these methods are destructive and employ stains that etch more rapidly at defect sites.

Often the reason for growing the epitaxial layer is to reduce the parasitic resistance. This is commonly done by growing a lightly doped epi layer on top of a heavily doped buried region. The transistor is formed in the epi and the heavily doped region is essentially a buried contact to the bottom of the transistor. For bipolar transistors it is necessary to be able to align the upper device layers with this buried layer. To do this alignment marks may be etched into the substrate before epitaxial growth. Pattern shift is the tendency for position of the alignment marks to appear to move after the growth (Figure 14.10). The cause of pattern shift is the dependence of the growth rate on the exposed crystal orientation. Pattern shift is much less pronounced in (100) wafers than in (111). Generally, the lower chlorine content precursors (SiH_2Cl_2 and $SiHCl_3$) also show less pattern shift as does growth at higher temperature.

For relatively thick epitaxial layers, the preferred method of measuring the film thickness is Fourier transform infrared (FTIR) spectroscopy. As shown in Figure 14.11, an infrared source is sent through a beam splitter to the surface of the wafer and to a moveable mirror. The reflected radiation from both surfaces are added and sent to a detector. The distance of the mirrored path is swept, and the intensity of the reflected beam as a function of the position of the mirror is monitored. The separation between these peaks is proportional to the thickness of the epitaxial layer.

The other popular methods for determining the epitaxial thickness are electrical. All of these techniques depend on measuring the differences in the doping levels of the substrate and the epitaxial layer. The quickest measurement is a simple four point probe of the resistivity of the epitaxial layer. More detailed information can be obtained by measuring the doping profile. This can be done using

Figure 14-11   A Fourier transform infrared spectroscopy system. Infrared light is used since silicon is nearly transparent in this region.

C-V analysis, spreading resistance, or staining techniques as discussed in Chapter 3. The accuracy of these techniques however, is limited by autodoping.

14.6 Selective Growth[+]

It is possible to grow selectively from the vapor. That is, to grow the epitaxial layers in some regions of the wafer and not to deposit any material in other regions. The process is known as selective epitaxial growth (SEG). SEG is not currently in widespread production use, although it is extremely attractive in some applications, including device isolation, planarization of contacts, trench isolation filling, and the formation of elevated source/drain structures for MOS transistors [42]. An overview of the history of the process has been presented by Borland [43].

SEG capitalizes on the fact that the nucleation rate for silicon growth increases with substrate type in the order $SiO_2 < Si_3N_4 < Si$. For example, if an attempt is made to deposit polysilicon over thermal oxide using dichlorosilane, the deposition rate will be very slow until a layer of silicon is able to accumulate over the surface of the oxide. If a thermal oxide is grown over the surface of the wafer and holes are etched to the substrate (Figure 14.12), silicon can be made to grow selectively in these holes. To perform selective epitaxial growth, nucleation in the gas phase and on the surface of the masking layer must also be actively suppressed by adding enough chlorine to the growth chemistry. By adjusting the Si/Cl ratio, the process goes from nonselective growth to selective growth to etching. It is believed that higher concentrations of HCl in the gas also improves the selectivity by etching any small nodules that begin to grow on the oxide surface [44]. The best SEG films of silicon are grown at low pressure in SiH_2Cl_2 and HCl mixtures [45].

Problems with selective growth include faceting (a nonplanarity of the surface due to lower growth rates along the various crystal planes) [46], the formation of defects along the semiconductor/oxide interface, and the dependence of the growth rate on the size and distribution of windows. The interface defects are particularly important if a doping junction is bounded by the oxide/semiconductor interface. They can be dramatically reduced if the windows are oriented in the <100> directions [47].

One interesting application of SEG is extended lateral overgrowth (ELO) [48]. If the epitaxy is allowed to continue past the height of the hole, it will begin to spread across the surface of the oxide. If several holes are placed close together, the approaching growth fronts coalesce, and a region of single crystal silicon is produced over the oxide. If the seed windows are along the <100> directions (i.e., the same as the wafer orientation), the growth rates in the vertical and horizontal directions are identical. Devices fabricated in these overgrowth regions have lower parasitic capacitance and can be easily isolated from one another. The problem with the method is defects. In addition to defects along the interface with the oxide, twins will occur where the growth front changes direction [49]. Defects will also be present at the line where the two growth fronts meet. New applications for ELO include structures that do not require the growth fronts to meet. Instead, the lateral overgrowth regions are used to form low capacitance contacts to the active device formed in the growth window.

Figure 14-12 A cross-sectional view of selective epitaxial growth.

Figure 14-13   A halide transport system for growing GaAs epitaxially. Gaseous gallium chlorides are created in the chamber and are transported to the wafer surface.

14.7 Halide Transport GaAs Vapor Phase Epitaxy

GaAs VPE is done somewhat differently than the growth of silicon. AsH_3 is a stable, though very toxic, gaseous arsenic source. Gallium however, does not form stable chlorohydrides. All of the major gallium chlorides are solids at room temperature. It is known that GaCl and $GaCl_3$ can be made gaseous at high temperature. This fact was the basis of one of the early and very successful GaAs growth techniques, halide transport [50],[51] (Figure 14.13). In this method, a hot (700 to 750°C) solid serves as the gallium source. A dilute mixture of HCl in H_2 flows over the hot Ga or GaAs to form GaCl. The carrier gas picks up the gaseous GaCl and transports it to the wafer surface where it forms GaAs heterogeneously on the surface. Some practitioners also react solid As with H_2 to form AsH_3 in situ so that large tanks of AsH_3 are not required. In that case, the solid arsenic is maintained at about 800 to 850°C. Substrate temperatures for halide transport growth tend to be in the 650 to 800°C range.

There are two primary liabilities of the halide transport growth of GaAs. The first is that the substrate temperatures required for device quality growth is high enough that abrupt junctions are difficult to achieve. The second problem is that it is difficult to control the growth thickness for thin structures due to the nature of the process. Halide transport is still widely used for the large scale fabrication of light emitting diodes. For other applications however, these two factors, combined with the emergence of organometallic gallium precursors, have decreased the popularity of the process.

14.8 Incommensurate and Strained Layer Heteroepitaxy

The first half of this chapter discussed processes that are sometimes collectively called *homoepitaxy*, the growth of the same material as the substrate. The rest of the chapter will discuss heteroepitaxy, the growth of dissimilar materials. The various approaches to epitaxial growth can be classified on a pressure temperature chart (Figure 14.14). The older methods (VPE and LPE) are done at high temperature at or near 1 atm. These methods will be briefly discussed here in connection with the growth of thick layers of silicon on insulating substrates. The remainder of the chapter will deal with techniques used to grow thin layers for the active regions of the device. In that application, it is very desirable to be able to grow layers at low temperature to limit diffusion. Molecular beam epitaxy (MBE) has demonstrated the ability to grow device quality layers at very low temperature, but requires ultrahigh vacuum (UHV). The extreme vacuum requirements of MBE make its use in large scale IC manufacturing difficult. Because of the diagnostics available in a UHV environment however, MBE is probably the best understood epitaxy process. Other device growth techniques, such as metallorganic CVD (MOCVD), rapid thermal CVD (RTCVD), and ultrahigh vacuum CVD (UHVCVD), while less well understood, can be viewed as extensions of the VPE processes that have already been described. The remainder of the chapter will therefore begin with these advanced VPE approaches and then proceed to MBE.

Advanced growth processes would not be of great technological interest if not for their capability to fabricate high quality heterostructures; that is, the growth of epitaxial layers of one material on top

Figure 14-14 Temperature and pressure regimes of various epitaxial growth techniques.

of another. There are three types of heteroepitaxial growth (Figure 14.15). It is common to say that the growth is commensurate if the substrate has the same crystal structure and lattice constant as the epi layer. All homoepitaxial growth is by definition commensurate. Incommensurate growth is the production of thick layers that are not lattice matched to the substrate. In this type of growth, the misfit between the two crystals must be accommodated by defects at or near the interface. Many of these defects may propagate into the epi layer, making the successful manufacture of even simple circuits extremely difficult. A third possibility is pseudomorphic growth. This occurs when growing thin layers that are not lattice matched. Instead of forming defects at the interface, the epi layer strains to match the substrate. The critical thickness or maximum thickness that can be grown in this manner before the strain is released in the form of dislocations depends on the lattice mismatch and the mechanical properties of the layers. For a lattice mismatch of a few percent, it is typically of order hundreds of angstroms. Strictly speaking, under this definition all heteroepitaxial growth must be pseudomorphic since one cannot have perfect lattice matching with dissimilar materials. A more practical description suggests that commensurate growth occurs when the adatom/adatom interaction energy is small compared to the adatom/substrate energy. Incommensurate growth then occurs when the adatom/substrate energy is small compared to the adatom/adatom energy. Pseudomorphic growth occurs when the energies are comparable.

Most of the earliest heteroepitaxial growth was incommensurate. A popular example of incommensurate growth is silicon on sapphire (SOS). Experiments by Cadoff and Bicknell [52] and later work by Manasevit and Simpson [53] demonstrated that silicon, which has cubic symmetry with a lattice parameter of 5.43 Å, can be grown successfully on sapphire (Al_2O_3), which is rhombohedral with lattice parameters of 4.75 Å (a-axis) and 12.97 Å (c-axis). Sapphire is insulating, and so circuits

Figure 14-15 Epitaxial growth processes can be divided into (a) commensurate, (b) strain relaxed incommensurate, and (c) incommensurate but pseudomorphic.

Figure 14-16   Defect density as a function of the epitaxial layer thickness for the growth of Si on sapphire (Al$_2$O$_3$) *(after Vasudev).*

built in SOS substrates have high speed and excellent device isolation. Device isolation can be done simply by removing the conducting silicon between the transistors, etching down to the insulating substrate. A common application of SOS is radiation hardened circuits.

Most SOS is done using VPE from SiH$_4$ diluted in H$_2$. Although chlorosilanes may be used, Cl attacks Al and produces heavily doped p-type films. Traditionally, the deposition temperatures are around 1000°C, although more recently some success has been achieved at lower temperatures. Of course, supersaturation and homogeneous nucleation are significant problems for SOS growth. The most basic problem for SOS however, is defects in the epitaxial layers caused by the sapphire substrate. These defects arise both because of the lattice mismatch and because the thermal expansion coefficient of Al$_2$O$_3$ is more than twice that of silicon. The mismatch at the interface creates defects that will propagate up into the growing film. One way to reduce this problem is by growing thick films. Defects may propagate upward, but do not necessarily penetrate the entire layer. Figure 14.16 shows a plot of defect density per unit length as measured by cross-sectional transmission electron microscopy (TEM) [54]. The defects fall off sharply with increasing thickness. Thick films are therefore usually required for incommensurate growth. It has also been found that high dose Si implantation can be used to amorphize part of the epi layer. This can be regrown by solid phase annealing. By repeating this process for different depths, the material quality can be further improved. SOS quality is currently suitable for LSI and some high-cost VLSI circuits, but is strongly challenged by the recent advances in SIMOX technologies (Section 5.7), and bonded SOI (Section 15.4).

The other primary example of incommensurate epitaxy is GaAs on silicon [55],[56]. GaAs and silicon share the same type of crystal structure, but the GaAs lattice constant is about 4% larger. As with SOS, thick epitaxial layers are required in order to reduce the defect densities [57] enough to allow IC fabrication. It has also been reported that a large number of very thin layers with alternating composition called a *strained layer superlattice* can be used to distribute the strain more gradually [58]. GaAs on silicon technology is attractive for three reasons: The silicon substrates are inexpensive and large. It is therefore possible to fabricate GaAs ICs on 6 or 8 in. rather than 3 or 4 in. substrates. The silicon substrate has a higher thermal conductivity than GaAs. This is an important consideration in high power ICs and devices where the performance is often limited by the maximum junction temperature. Finally, it is possible that GaAs devices could be integrated with silicon on the same chip. In this application a silicon technology such as CMOS would be fabricated first. GaAs would be grown selectively in oxide windows etched down to the silicon substrate. A small number of high speed GaAs devices might be used in critical paths to improve the overall system performance. This type of integration has been demonstrated on simple systems but has not seen appreciable commercial application.

In virtually all incommensurate growth, the interface must be far from the active device region since it contains so many defects. From a device design standpoint, it would be much more desirable to grow layers that can be used as part of the device structure itself. Figure 14.17 shows a plot of the

Figure 14-17 The bandgap and lattice parameter of a variety of semiconductor compounds and alloys.

Figure 14-18 For pseudomorphic GeSi on Si there are three regions: mechanically stable, metastable, and unstable *(after People and Bean).*

bandgap of various ternaries as a function of the lattice parameter. By drawing a line vertically, one can predict which useable heterojunctions might be feasible. GaAs for example is lattice matched to the proper mole fraction of $Ga_xIn_{1-x}P$. The most heavily studied commensurate growth system is GaAs on $Al_xGa_{1-x}As$. Although a very small lattice mismatch exists, these two compounds can generally be regarded as perfectly lattice matched. While gross defects such as dislocations are relatively rare in these lattice matched systems, a number of interface states exist. These states may serve as recombination sites that degrade the performance of the heterostructure.

An example of pseudomorphic growth is Ge_xSi_{1-x} layers grown epitaxially on silicon. This system is interesting because it suggests the possibility of combining the high speed of heterostructures with the density of silicon ULSI. Figure 14.18 shows a plot of the critical thickness of GeSi as a function of the germanium mole fraction [59]. As long as the film thickness lies to the left and below this line, the GeSi layer will accept the strain and be thermodynamically stable. A variety of devices both electronic and optoelectronic have been fabricated in the GeSi system.

14.9 Metal Organic Chemical Vapor Deposition (MOCVD)

Metal organic chemical vapor deposition (MOCVD or OMVPE) began in the late 1960s with the research of Manasevit [60]. MOCVD is sometimes regarded as a competitor to MBE. Both are capable of growing high quality III-V devices with atomically abrupt or nearly atomically abrupt interfaces. The process is not generally used for the epitaxial growth of silicon since suitable inorganic sources are readily available.

Classic MOCVD processes were developed using group III organic compounds and group V hydrides. Historically this occurred because the organic molecules had been developed for other applications and were therefore available, although not in the purity that is now required. Figure 14.19 shows examples of some of the relevant precursors. While a wide variety of organic compounds have been used, this chapter will restrict its attention to methyl and ethyl compounds, which are the most commonly used sources. They are designated with the acronyms M and E. For example, TMG is one of the most common gallium sources, trimethylgallium ($Ga(CH_3)_3$).

To understand the bonding arrangements in the organometallics, start from the valence arrangement of the metal atom [61]. Gallium has three valence electrons. In the ground state, gallium has two s electrons and one p electron. In TMG, the carbon atom from each of the methyl groups bonds covalently with the gallium forming a hybridizing sp^2 bond. The three methyl groups form in a plane with the Ga-C bonds 120° from each other. This leaves an empty p orbital perpendicular to the plane of the methyl groups. This unfilled orbital easily accepts electrons. The group V hydrides have 2 s electrons and 3 p electrons in the valence. The three As-H bonds in arsine form covalently with bond angles of about 109.5°. A single pair of unbonded electrons sit on top of the molecule away from the hydrogen atoms. One of these electrons is easily lost or donated. The ability of the molecules to easily accept and lose electrons plays an important role in the chemical reactions that occur during growth.

AsH_3 is a typical group V hydride source. Unfortunately these hydrides are extremely toxic. Exposure to arsine leads to red blood cell destruction, renal failure, and ultimately death. Since III-V growth uses these chemicals not as dopants but as the growth species themselves, they must be stored in large quantities. The Immediately Dangerous to Life and Health (IDLH) concentration for arsine is 6 ppm. A cylinder of AsH_3 at 1500 psi will contaminate about 10^9 liters of air to the IDLH concentration. This corresponds to roughly one square block. Of course, in the event of an accident, personnel close to the release are in much greater danger than those further removed. The cost and discipline of the proper gas handling for such toxic precursors has led to the investigation of less toxic alternatives, typically group V organometallics [62]. Potential group V sources must have a sufficiently high vapor pressure. They must not have parasitic reaction paths with each other or the group III precursor. They must be stable. They must be extremely pure. On reaction they must not introduce impurities into the epitaxial films. It is generally found that replacing H bonds with C bonds decreases the toxicity, but increases the amount of unintentional doping in the epitaxial layers. One of the most promising OM arsenic sources is tertiarybutylarsine [63] (TBA), which is nearly as toxic as AsH_3. Since it is stored as a liquid rather than a high pressure gas however, it is generally regarded as safer than AsH_3.

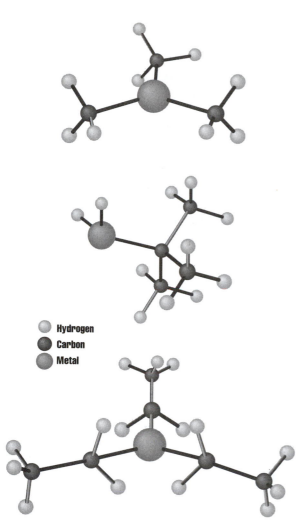

- ○ **Hydrogen**
- ● **Carbon**
- ● **Metal**

Figure 14-19   Examples of common organometallics used in MOCVD include (from top to bottom): trimethylgallium, tetrabutylarsine, and triethylgallium.

Figure 14-20   A typical OM source for MOCVD includes valves that isolate and bypass the bubbler during change-out allowing cycle purging of the process line with nitrogen.

An MOCVD system must be capable of the controlled growth of thin, atomically abrupt, extremely perfect epitaxial layers. The growth system can be divided into three subcomponents: gas delivery, reaction chamber, and gas exhaust. The gas delivery system is the easiest to specify. It is desired that the growth chamber be provided a well mixed, well controlled flow of a number of gasses. For growing GaAs, this is typically H_2, TMG or TEG, AsH_3, and a dopant, typically DMZ for p-type doping or SiH_4 for n-type doping [64].

The OM sources consist of a bubbler immersed in a controlled temperature bath of glycerin and water (Figure 14.20). The OM vapor is obtained by heating the liquid in a constant temperature bath and bubbling H_2 through the bath. The temperature of the bubbler must be carefully controlled since the vapor pressure depends exponentially on the temperature of the bath. The back pressure regulator is used along with the temperature of the bath to establish the mole fraction of the OM in the gas that is sent to the reactor. Figure 14.21 shows the vapor pressure of several commonly used OM sources. The OM source must also include valves so that the source can be isolated from the line and flushed with N_2 when it is to be changed.

The details of the gas delivery system depend on the particular organometallic being used. Some OM compounds such as triethylaluminum (TEAl) require substantial heating in order to obtain a sufficient vapor pressure for deposition. The vapor of these compounds can recondense in the tubing before it reaches the chamber. This results in poorly controlled deposition conditions unless the tubing and valves also are heated. Other compounds such as dimethylzinc (DMZ) which is commonly used as a p-type dopant in GaAs MOCVD, has a very high vapor pressure at room temperature and may actually have to be chilled to obtain the desired vapor pressure. DMZ pyrolyses at only slightly above room temperature, however. To prevent a reaction in the supply tubes, particularly the part of the lines close to the reactor, they may have to be cooled. Other compounds such as TEIn, which can be used to grow InGaAs and InP, are unstable and so will begin to decompose in the liquid cylinder.

The hydride sources are delivered as high pressure gasses just as in standard CVD systems. The gas pressure is reduced to a few atmospheres and the gas is plumbed to a valve and mass flow controller. Due to the toxicity of the hydrides, these lines are typically double contained, leak checked, and continually sniffed for leaks. Furthermore, only the most reliable gas seals are typically allowed on such a system. The design of the actual gas introduction system is one of the most critical parts of the chamber. The simplest method is to simply merge the various lines at a manifold near the entrance to the growth chamber. If one desires the ability to grow thin, abrupt layers, this arrangement

Figure 14-21 Vapor pressure curves for some common organometallics *(after Stringfellow).*

is not satisfactory since the growth during a switching transient is very poorly controlled. The solution is to include valves as an integral part of the manifold. These valves are designed so as to have the minimum dead volume, the volume between the chamber and the valves. The second concern with gas switching transients is that a flow burst or bolus occurs when a valve is switched on. To avoid this problem, gas manifolds for MOCVD may be designed in a run/vent configuration (Figure 14.22). The manifold of such a system uses three way valves that either supply gas to the growth chamber or to a bypass. This redirection provides a much more stable flow than can be achieved by turning the gas on and off. Of course, the amount of gas consumed by this process may be considerably larger, but this is considered an acceptable cost for many applications.

The gas exhaust must allow the system to operate at a controlled pressure. Most of the early MOCVD systems were run at 1 atm. It has since been found that growth at reduced pressure (10 to 100 torr) can improve uniformity and layer abruptness. Most MOCVD systems now include vacuum components to maintain the chamber pressure. The exhaust gasses must also be scrubbed prior to release. This is done in either a high temperature furnace in which the hydrides are oxidized or in wet chemical scrubbers where the effluent reacts with caustic solutions to form water soluble salts that can be diluted and flushed.

A great deal of research has gone into the design of the reactor itself. Figure 14.23 shows cross sections of several typical chambers. The two most common are the wedge shaped horizontal reactor, similar to those described earlier, and the vertical impinging jet (also called the *stagnation point flow*) reactor. For low flow velocities, it is usually assumed that the gas enters the chamber in a fully developed flow. The gas velocity is zero at the walls, and it increases parabollically toward the center of the inlet tube. It is of critical importance to the success of the MOCVD process that the flow through the chamber be laminar, with no turbulence or recirculation. Nonlaminar and recirculative

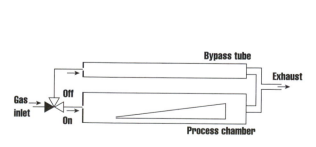

Figure 14-22 Run/vent gas switching used to avoid large gas transients.

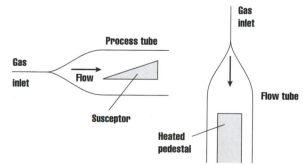

Figure 14-23 Common MOCVD chambers include the horizontal flow and the stagnation point flow reactors.

Figure 14-24   Growth rate of GaAs during MOCVD as a function of temperature *(after Tischler, ©1990 IBM).*

flows trap gasses in the chamber. This concern was also considered during the discussion of CVD, but is much more important in MOCVD, since typical growth runs involve multiple thin layers with different compositions and dopings. Turbulence usually arises at high gas velocities due to shear forces. Recirculation cells arise from two sources. The first is the sudden change in the cross-sectional area of the chamber. The second source of recirculation cells, mentioned previously, is buoyancy forces caused by nonuniform temperatures. The effect of recirculation cells can be reduced by reducing the mass density of the gas. This is a common reason for using H_2 as the carrier gas. By also reducing the pressure one to two orders of magnitude below 1 atm, the effect can be virtually eliminated. Another method for overcoming the effects of convective flow is to rotate the substrate at several thousand rpm. This transfers momentum to the molecules near the surface of the susceptor that opposes the recirculation flow.

A plot of the growth rate of GaAs in MOCVD as a function of the substrate temperature [65] is shown in Figure 14.24. At low temperature the growth rate increases exponentially with temperature. The activation energy for this process has been measured to be between 13 and 22 kcal/mole (0.55 to 0.95 eV). At high temperatures the growth rate begins to drop. One reason for this reduction is the emergence of a parasitic reaction pathway that scavenges the TMG. It is believed that homogeneous pyrolysis of TMG produces CH_3 radicals. These radicals in turn attack both the AsH_3 and H_2. In the case of H_2, the remaining atomic hydrogen attacks the TMG, freeing another methyl group:

$$Ga(CH_3)_3 \rightleftarrows Ga(CH_3)_2 + CH_3$$

$$CH_3 + H_2 \rightleftarrows CH_4 + H$$

$$Ga(CH_3)_3 + H \rightleftarrows GaH(CH_3)_2 + CH_3 \tag{14.14}$$

This reaction path causes gallium to deposit on the walls of the chamber and deplete the TMG source. It is also possible however, that the reduction in the reaction rate is related to desorption of AsH_x admolecules from the surface of the wafer.

Most GaAs growth is done in the intermediate temperature regime where the growth rate is determined by the arrival rate of the TMG. The reaction process for TMG and AsH_3 in this regime was first postulated by Schlyer and Ring [66] and later supported by others [67]. While it has not yet been proven, it is the most widely accepted. In this model, at temperatures less than 650°C very little homogeneous reaction occurs at any pressure. As a result, both TMG and AsH_3 independently adsorb on the surface of the wafer. Once on the surface, TMG appears to react spontaneously to form monomethylgallium ($GaCH_3$) and two methyl groups. AsH_3 reacts with a CH_3 group to form AsH_2 and methane (CH_4). The latter is quickly desorbed and pumped away:

$$AsH_3 + CH_3 \rightleftarrows AsH_2 + CH_4 \tag{14.15}$$

AsH$_2$ and monomethylgallium diffuse independently on the surface until they come into contact. The methyl group reacts with the hydride yielding GaAs, atomic hydrogen, and methane, which also desorb from the surface. It now appears that the atomic hydrogen produced by AsH$_2$ dissociation plays an important role in removing the methyl groups from the surface of the wafer.

One of the first major problems of MOCVD was carbon contamination. This results in unintentional p-type doping. Early MOCVD films [68] had hole concentrations of 10^{19} cm^{-3}. Although they can be counter doped, the films have very low mobilities. It was found however, that by increasing the group V to group III gas flow ratio, the unintentional doping could be dramatically decreased. High group V partial pressures are needed because AsH$_3$ dissociation is much less efficient than dissociation of the organometallics at normal growth temperatures. By the late 1970s, material quality was comparable to MBE with heterojunction undoped liquid nitrogen temperature mobilities routinely reported to be in excess of 100,000 cm^2/V-sec. It has also been shown that the use of TEG rather than TMG also produces layers with little carbon incorporation [69] because the ethyl organometallic pyrolyses to form ethylene (C$_2$H$_4$) which, unlike CH$_3$, is nonreactive.

Within the last few years it has been found that this early liability of carbon contamination can be used to considerable advantage. When heavily doped p-type layers are required for a particular device, the films can be made by choosing the V/III ratio and the temperature so as to incorporate carbon [70] or by adding an appropriate carbon containing gas [71]. Carbon doping is preferred to Be and Zn for many applications, since carbon has a lower diffusion coefficient [72], and the source is not toxic.

Al$_x$Ga$_{1-x}$As can be grown by adding an aluminum containing organic such as trimethylaluminum (TMAl) to the gallium containing organic. The concentration of aluminum in the film is roughly proportional to the ratio of the injected gasses [73] and is nearly independent of temperature. When growing AlGaAs, additional care must be taken because of the extreme reactivity of aluminum, particularly for oxygen. To obtain high quality films in MOCVD systems, the substrate temperature must be heated to about 750°C [74]. It is believed that in this temperature range, the aluminum oxide more readily desorbs from the growing film, and so the oxygen contamination drops sharply. It is also critically important in growing these layers to obtain a high base vacuum, palladium diffuse the H$_2$ source, and use any other method possible to getter out residual O$_2$ and H$_2$O. Liquid metal bubblers have been found to be effective for gettering oxygen from hydride source gasses [75].

14.10 Advanced Silicon Vapor Phase Epitaxial Growth Techniques

The first part of the chapter discussed the epitaxial growth of silicon from the vapor. The application for these processes are the growth of thick layers above a substrate with a different dopant type and/or concentration. The process must have a low defect density, there must be good control over thickness and uniformity, and it must be done so as to minimize the dopant redistribution during the growth. Recall that one can define a characteristic diffusion length as $\sqrt{Dt}$, where D is the diffusivity and t is the time at temperature. There are two approaches to solving this problem. The first uses rapid thermal processing to extend standard VPE using chlorosilanes to very short and well controlled growth times. The second uses ultrahigh vacuum systems with a silane chemistry to grow at very low temperature, limiting the diffusivity.

Figure 14.25 shows a typical time/temperature cycle for growing 1 μm of epitaxial silicon in a VPE reactor. The wafer and susceptor must be heated to the preclean temperature, then held at that temperature for some period to allow the native oxide to desorb. Next, the wafer is ramped to the epitaxial growth temperature, the dichlorosilane turned on and the growth proceeds. After the desired growth time the growth gas is turned off, the chamber is purged with an inert gas, and the wafer is

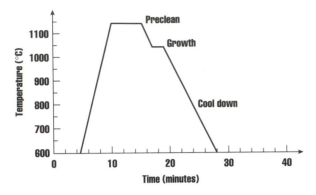

Figure 14-25   A typical time/temperature cycle for growing 1 μm of epitaxial silicon in a VPE reactor.

ramped to room temperature. If one wants to grow a 1000 Å layer instead of 1 μm in order to be able to directly fabricate devices, the thickness of the layer is reduced by a factor of ten, but because of the overhead associated with precleaning the wafer and ramping the temperature, the overall time at temperature may only be reduced by a factor of two. As a result, the layer doping will be dominated by diffusion from the lower layers and will be very poorly controlled.

Gibbons et al. [76] were the first to combine rapid thermal processing and epitaxial growth in a technique called limited reaction processing (LRP). Other groups have called similar techniques rapid thermal vapor phase epitaxy (RTVPE) [77] or rapid thermal chemical vapor deposition (RTCVD) [78]. The chamber for doing this work like the quartz walled RTP systems described in Chapter 6, although a load lock may be used to reduce residual gas impurities. A cross section of a system is shown in Figure 14.26a. The wafer rests on three quartz pins in a quartz flow tube. Gasses are injected at one side, flow across the wafer, and exit at the other end. The wafer is heated by lamps situated above and below the flow tube. Because of the rapid thermal response of the wafer in such a system, the actual growth time for thin layers can now be the majority of the overall time at temperature.

A wide variety of structures can be grown in an RTCVD system [79], and it is an excellent, low cost method of developing advanced device structures. From a manufacturing standpoint, however, the technique has several important liabilities. As with all VPE, there is no direct control over the thickness. Instead the growth rate must be assumed from some previous calibration run and the growth time calculated. Furthermore, in standard VPE the growth temperature is measured using a

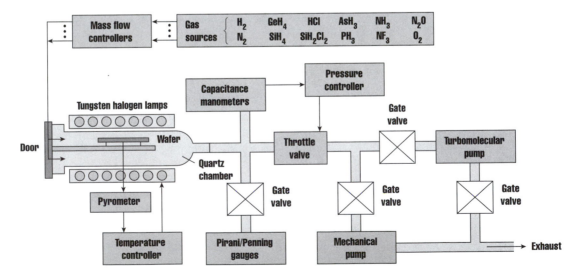

Figure 14-26a   Schematic of a rapid thermal epitaxial growth system *(after Hsieh et al., reprinted by permission of the publisher, The Electrochemical Society).*

Figure 14-26b   Commercial rapid thermal epitaxial growth system showing load and unload stations, robot arm, and rapid thermal flow sleeve *(courtesy of ASM Corp.)*.

thermocouple embedded in the susceptor. In an RTCVD system the growth temperature is most often inferred using pyrometry, a method with significant reproducibility problems. This is particularly a concern during growth when deposited layers may coat the quartz surfaces, changing the optical path to the sensor. It is also well known that the edges of a wafer are 20 to 50°C cooler than the center of the wafer in many RTP systems at growth temperatures [80]. Additional power can be focused on the edge of the wafer to help compensate, but unless dynamic control is used, the temperature uniformity can only be optimized for a single process condition.

To help overcome the temperature uniformity and control problems of RTP, one can run the epitaxial growth process in a mass transport limited regime. Once again this presents difficulties. Recall that in a standard horizontal CVD reactor the susceptor is tilted to provide a uniform stagnant layer thickness. This is not possible in RTCVD since there is no susceptor. An impinging jet system such as those used for MOCVD can be employed, but it is difficult to design such a system that is radiatively heated and can reach the temperatures necessary for silicon VPE. As a result, commercial RTP epitaxy systems sometimes use thin susceptors to hold the wafer, providing some compromise between rapid response and good uniformity.

The first part of this chapter emphasized the importance of reducing the residual water and oxygen concentrations in the growth chamber and the relationship to these partial pressures and the minimum allowable growth temperature. To obtain growth at the lowest possible temperatures, it is necessary to obtain a growth ambient with extremely small partial pressures of oxygen and water.

One silicon epitaxial growth process that has been developed to meet this objective is ultrahigh vacuum chemical vapor deposition (UHVCVD) [81]. As shown in Figure 14.27, the system is similar to a load locked LPCVD tube. Up to 35 wafers can be loaded into the chamber at a time in the current generation research reactors. The base vacuum for the system is typically at least 10^{-9} torr.

Growth is carried out at temperatures below 800°C in about 1 mtorr of silane. Under these conditions, it is believed that little decomposition of silane occurs in the gas phase, since this reaction generally requires a collision with another gas molecule [82]. The growth rate is only a few angstroms per minute for undoped or p-type doped layers [83], but the low rate is tolerable due to the batch loading. Because of the very low temperature, heavily doped n-type layers cannot be practically grown using the technique. The process is run well into the reaction rate limited regime. Since the temperature uniformity of a hot wall system is very good, the deposition rate uniformity is often better than 2%.

One of the major concerns with UHVCVD is wafer precleaning. Because of the low pressure in the chamber most of the heating is radiative. Since the wafers are loaded by the batch, heat must flow into or out of the boat load of wafers primarily through radiation exchange between the wafer edges and the tube walls. Temperatures must therefore be slowly ramped to avoid excessive thermoplastic stress in the wafers. This makes the use of a high temperature oxide desorption step rather difficult. If hydrogen terminated surfaces can be made to work in a production environment however, UHVCVD is a promising technology.

Figure 14-27   Schematic of an ultrahigh vacuum CVD epitaxial growth system *(after Mayerson, ©1986 AIP)*.

14.11 Molecular Beam Epitaxy Technology

One of the premier growth techniques for GaAs heterostructures is molecular beam epitaxy (MBE). The technique has the capability of growing device quality layers of semiconductors with atomic resolution of the growth thickness. Furthermore, typical growth temperatures virtually preclude dopant diffusion. MBE has also been applied to silicon epitaxy. Figure 14.28 shows a diagram of a typical MBE system. The basic requirements for such a system are: ultrahigh vacuum (at least 10^{-10} torr base vacuum), in situ sample heating and cleaning, and independently controlled thermal and/or electron beam sources for all materials and dopants. Most systems also include several in situ analysis capabilities, including electron diffraction and Auger spectroscopy. A separate chamber connected through UHV load locks is often provided to evaporate various metallizations without breaking vacuum after the growth is complete.

To achieve these high base vacuums, the MBE chamber (including gaskets) must be suitable for baking at 150 to 250°C. Baking at these elevated temperatures increases the vapor pressure of gasses, such as water adsorbed on the inside walls and surfaces of the system. When the system is pumped hot and then cooled, the pressure in the chamber may drop by an order of magnitude or more. Heating jackets are often provided with the system for this purpose. Particular care must be taken during high temperature bakeouts for the durability of valves and fused silica viewports. These components often will not tolerate repeated high temperature cycling. Critical components such as the evaporation sources may be held at somewhat higher temperatures during bakeout to ensure an even higher level of cleanliness in these areas. Bakeout times vary with the level of exposure. If the entire system must be opened to air such as must be done to refill a source, it is common to require nearly a week of pumping with bakeout before device quality growth can begin.

A variety of pumps can be used to maintain the high vacuum state. Early systems used diffusion pumps, but extreme care must be used to prevent back diffusion of the hot diffusion pump oil into the growth chamber. Some small systems use liquid nitrogen cooled adsorption pumps. These pumps operate by physical adsorption of gasses at the surface of molecular sieves or other adsorption material such as activated Al_2O_3. This material is typically a large surface area solid. The pump is cooled to 77 K, then opened to the area where pumping is desired. After pumping is complete the pump is isolated from the chamber, heated, and evacuated. A number of adsorption pumps may be used to provide an adequate base vacuum. The most popular MBE high vacuum pump is the cryopump. As with adsorption pumps, the cryopump must be isolated from the chamber occasionally and heated to remove the pumped gasses. Newer MBE systems often use turbomolecular pumps for producing high vacuum. This option is particularly attractive if toxic sources are being used, since material from these sources tends to accumulate in the cryopumps between temperature cycling.

In order to achieve routine ultrahigh vacuum, wafers must be introduced into the MBE system using a load lock. The load lock can be evacuated to high vacuum in a short time. Typical pumpdown characteristics for an MBE load lock are 10^{-6} torr in 15 min. Some high volume MBE systems allow a cassette or platen of wafers to be kept under UHV. This increases the wafer throughput since the total pumpdown time per wafer decreases. Many GaAs MBE systems also make use of cryopanels in the growth area. The walls of the chamber have liquid nitrogen jackets, making them large pumps themselves. While capable of maintaining very high vacuums, these systems also consume so much liquid nitrogen (particularly if the substrate must be heated to a high temperature) that it must be plumbed directly to the chamber through double walled vacuum jacketed pipes.

Substrate heating is typically provided through the use of chucks or susceptors that can be either radiantly or resistively heated. Resistive heaters can be refractory metal wire wound on a form or it

Figure 14-28 Schematic and photograph of a molecular beam epitaxy growth system *(after Davies and Williams).*

can be graphite films that are etched into a heating pattern and covered with SiC. Wafers are often attached to the chuck through clips, clamps, or indium based solders, although the latter is becoming less common. The temperature of the wafer can be monitored by measuring the temperature of the chuck with a thermocouple. It can also be measured by sighting the wafer with a pyrometer. Pyrometers can measure the temperature of the wafer directly; however, to be accurate it must be calibrated in the growth chamber. Furthermore, one must be careful to avoid measuring radiation of the heater directly. Radiation from the heater may come from reflections off of the walls of the chamber or it may come directly through the wafer, particularly in the infrared below the bandgap of the semiconductor.

There are three types of sources most commonly used in MBE: Knudsen cells, electron beam sources, and gas sources. Figure 14.29 shows the cross section of a Knudsen cell. A crucible loaded with a source charge is radiantly heated by a foil heater. Thermocouples pressed against the outside of the crucible can be used to control the temperature and therefore the material flux. To avoid contamination, the heater foil is typically made of tantalum and is baffled to prevent a line of sight to the wafer. For III-V growth, most crucibles are made from pyrolytic boron nitride (pBN). Quartz can also be used for silicon MBE. The empty crucible must be annealed at high temperature before use to ensure that the beam, once produced, will be pure. To achieve reasonable fluxes in silicon MBE source, temperatures of about 1700°C are required. At this temperature, silicon is molten and highly reactive. As a result, the films are contaminated with impurities from the crucible. To avoid

Figure 14-29 A simple Knudsen cell for producing thermal beam for MBE *(after Davies and Williams)*.

this problem most silicon MBE systems use electromagnetically focused electron beam sources. This type of source has been discussed as e-beam evaporation (see Section 12.4).

For both e-beam and thermal cells, the source flux is determined by the temperature, which in turn establishes a vapor pressure. Since the charge has a finite thermal mass, this flux cannot be turned off instantaneously. To provide this on/off control most MBE systems use shutters in front of the cells. One difficulty with these shutters is material creep. Since the source is still at temperature, deposition will continue onto the back of the shutter. This material may surface diffuse around the shutter and eventually reevaporate. Shutters may be cooled to reduce this problem, but they are being constantly heated by the exposure to the hot cell. To avoid this problem, as well as to improve uniformity, eliminate chamber venting to replenish the charge, and to improve material quality, MBE has adopted gaseous sources (GSMBE) within the last few years. Figure 14.30 shows a typical gas source MBE.

One of the most popular techniques in research applications for real time growth monitoring is electron diffraction (RHEED). In this method, high energy electrons are diffracted off of the growing surface and imaged on a screen on the opposite side of the chamber. Since the films should grow a layer at a time, the spot intensity should grow, then diminish, then grow again as each monolayer of coverage is achieved (Figure 14.31). For production systems where uniformity is of paramount importance, the substrate must be rotated during growth, making RHEED impractical.

A major source of defects in MBE is particulation from other surfaces that are coated during the deposition process. Particles that fall from the walls are actually pushed by the growth beam

Figure 14-30   GSMBE growth chamber. The solid effusion and gas sources are aimed at the growth position *(courtesy of RIBER S.A.).*

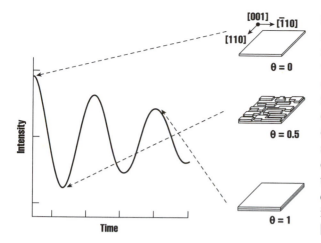

Figure 14-31   Electron diffraction oscillations during MBE growth. The peaks correspond to nearly complete layers.

toward the wafer surface [84]. This is a particularly serious concern in silicon MBE. It has been found that replacing liquid nitrogen cooling of the chamber walls with simple water cooling can dramatically reduce the defect density. Also by using high voltage plates near the evaporation source, silicon dust can be kept from entering the beam, and defect levels as low as 30 cm^{-2} can be achieved [85]. Metal contamination is another serious concern in MBE. In most VPE, chlorinated species are present that react with residual metals to clean the surface. These impurities may reduce the minority carrier lifetime for MBE films unless great care is taken in designing the growth system and preclean cycles [86].

One of the major manufacturing concerns in GaAs MBE is the oval defect: a region of excess gallium concentration [87]. The oval defect density is commonly reported [88] to be in the range 10^3 to 10^4 cm^{-2}. By carefully cleaning the system [89], the oval defect densities can be made as low as 300 cm^{-2}. It was recently observed that oval defects can be completely eliminated by holding the gallium source in a boron nitride crucible that had previously been used for aluminum [90]. The reason for this is not yet known but may be related to oxygen gettering.

The growth rate of an MBE system is simply determined by the flux of atoms leaving the source and the fraction of those atoms that strike the surface of the wafer and stick. This flux of atoms from the source for a thermal source is given by Equation 10.10:

$$I_n = AJ_n = A\sqrt{\frac{n^2kT}{2\pi m}} = A\sqrt{\frac{P_e^2}{2\pi kTm}} \tag{14.16}$$

where A is the cross-sectional area of the source, P_e is the equilibrium vapor pressure that depends on the source temperature, and m is the atomic mass. Knutson cells are designed so that the cross-sectional area remains nearly constant as the source is consumed.

Example 14.3

Assume that aluminum is being evaporated at 1150 K in a 25 cm^2 cell. What is the atomic flux at a distance of 0.5 m if the wafer is directly above the source? What would the growth rate be?

Referring to Chapter 10, the flux at a distance r is

$$J = A\sqrt{\frac{P_e^2}{2\pi kTm}}\frac{\cos\theta\,\cos\phi}{\pi r^2}$$

From Chapter 12, the vapor pressure of Al at 1150 K is about 10^{-6} torr. Inserting and solving with $m = 27*1.67 \times 10^{-27}$ kg, the flux is about 4.8×10^{14} cm^{-2} sec^{-1}. The growth rate can be found using

$$R = \frac{I_n}{\rho}$$

where ρ is the number density of aluminum (6×10^{22} cm^{-3}). Then $R \approx 48$ Å/min.

For compound semiconductor systems, such as GaAs, both gallium and arsenic cells must be used. The solid arsenic source emits As$_4$, which can be converted to As$_2$ if the cell includes a high temperature stage [91]. When growing GaAs, the solid As source is kept at a sufficiently high temperature so that the growth is limited by the arrival of the gallium atoms, typically of order 10^{19} atoms/m^2-sec. The arsenic flux, in the form of As$_2$ and As$_4$, is five to ten times higher. Practically all of the gallium atoms stick to the surface at normal growth temperatures, but only one arsenic atom adheres for each gallium atom.

14.12 BCF Theory[+]

Unlike VPE, a reasonably accurate model of the surface during MBE can be developed. This model starts by examining the opposite of MBE: evaporation. Consider the surface of the wafer to consist of two levels or terraces. The upper terrace is one atomic layer higher than the lower terrace. Kink sites occur where the terrace edge forms an angle with itself. Here, few bond breaking events are required for an atom to liberate itself from the upper level. A lone atom away from the terrace edge, however, is the most loosely bound. The model for evaporation then, is that atoms break loose from kink sites, diffuse across the surface of the wafer, and eventually desorb (Figure 14.32). Assume that the activation energy for liberating an atom from the kink site is W_s. Then the equilibrium density of adsorbed atoms is

$$n_{\text{seq}} = N_s e^{-W_s/kT} \tag{14.17}$$

where N_s is the density of surface positions. Once freed from the kink sites, the adsorbed atoms have only a small energy barrier U_s to move from site to site. If the distance between sites is a, and the frequency of lattice vibrations is f, the surface diffusivity is

$$D_s = a^2 f e^{-U_s/kT} \tag{14.18}$$

Assume that the adsorbed atom desorbs after some mean time T. If the energy barrier for this desorption is E_d,

$$\frac{1}{T} = f' e^{-E_d/kT} \tag{14.19}$$

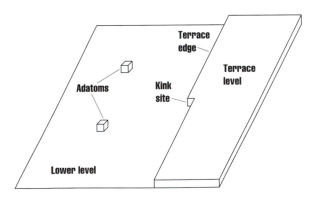

Figure 14-32   A microscopic view of a semiconductor surface during MBE growth or evaporation.

where f' is a pre-exponential factor related to the lattice vibrational frequency f. If the mean migration length before desorption is given by

$$\lambda_s = \sqrt{D_s T} \approx a e^{(E_d - U_s)/2kT}$$

(14.20)

the desorbing flux is given by

$$F_o = \frac{n_{seq}}{T} = N_s f' e^{(E_d + W_s)/kT}$$

(14.21)

Generally W_s and E_d are comparable and much larger than U_s. Due to their exponential temperature dependence, n_{seq}, P_e, and F_o increase strongly with increasing temperature. The surface residence time and the mean migration distance both decrease rapidly with increasing temperature. For typical growth conditions, the surface diffusion length is of order tens or hundreds of microns.

The terraces in this model may be caused by dislocations, previous growth, or by small deviations from perfect crystal orientation. On <111> surfaces, this intentional misorientation is required. If the crystal misorientation is $\Delta\theta$, the terrace separation is given by

$$L_o = \frac{h}{\sin(\Delta\theta)}$$

(14.22)

where h is the step height, typically one atomic layer thickness. Since h is about 1 Å, the mean terrace separation is usually of order microns or less. This separation must be much less than the mean diffusion distance to ensure 2-D growth.

The most well known theory regarding 2-D growth was developed by Burton, Cabrera, and Frank [92]. BCF theory assumes that most adatoms reach the surface between the surface steps and diffuse to the edge of the step where they are captured. The adatoms may then diffuse along the step edge until they reach a kink site, where they are incorporated into the crystal. Terrace edge movement occurs by lateral movement of the kink sites. Crystal growth then occurs as the terrace edges sweep across the surface of the wafer.

If the terrace edge is considered to be a perfect sink for excess adatoms, the adatom concentration near the terrace edge drops to the equilibrium value, n_{seq}. That is, the terrace is not a perfect sink for adatoms but will emit a small number at nonzero temperatures. The net flux of adatoms is given by the difference between the incident flux and the evaporating flux,

$$j_v = j_{inc} - \frac{n_s}{T}$$

(14.23)

The surface current density of adatoms is described by the diffusion equation

$$j_s = -D_s \nabla n_s$$

(14.24)

For steady state growth, the divergence of the surface current density is the net growth flux

$$\nabla \cdot j_s + j_v = 0$$

(14.25)

One can define a supersaturation parameter σ, where

$$\sigma = \frac{j_{inc}}{\frac{n_{seq}}{T}} - 1 \qquad \text{(14.26)}$$

Just as in VPE, when $\sigma > 0$ epitaxial layers will grow. When $\sigma < 0$, the wafer material evaporates. Combining these equations and assuming that the surface diffusivity is independent of position and direction, one can derive the differential equation

$$\lambda_s^2 \, \nabla^2 n_s + n_s = n_{seq}(\sigma + 1) \qquad \text{(14.27)}$$

In reality the diffusivity is strongly directional due to surface reconstruction [93], but we will ignore this refinement of the basic BCF theory.

Example 14.4

Assume that MBE is used to grow a material at a substrate temperature of 500°C. Assume that the material has the following properties: $W_s = 2.0$ eV, $E_d = 1.8$ eV, $U_s = 0.2$ eV, $N_s = 10^{15}$ cm^{-2}, $f = f' = 10^{14}$ sec^{-1}, $a = 2$ Å, $h = 0.5$ Å, and $\Delta\theta = 1°$. Calculate the mean surface diffusion length. Compare this to the mean terrace separation. If the flux on the surface of the wafer is the same as that calculated in Example 14.3, calculate the supersaturation.

Inserting the given parameters into Equations 14.18–14.20 produces $D_s = 2 \times 10^{-3}$ cm^2/sec, $T = 5.7$ ms, and $n_{seq} = 86$ cm^{-2}. Inserting the surface diffusivity and mean desorption time, $\lambda_s = 34$ μm. In comparison, the average terrace separation (given by Equation 16.9) is 28 Å. This separation is much less than the characteristic surface diffusion length. F_o can be found directly as the ratio of n_{seq} and T. It is 1.5×10^4 cm^{-2}-sec^{-1}. Inserting into Equation 14.26, the supersaturation is approximately 8×10^9. These large values of supersaturation are quite typical for MBE. They indicate that the system is run far from equilibrium. As a point of comparison, typical values for growth in a VPE system are less than 2.

Equation 14.27 was solved by BCF, subject to the boundary condition that n_s is equal to n_{seq} at the step edge position. (Step movement can be neglected when the average distance traveled by an atom is much larger than the distance traveled by a step.) Then, assuming that at the step ($y = 0$) $n_s = n_{seq}$ and as $y \rightarrow \infty$, $n = (1 + \sigma)n_{seq}$,

$$n_s = n_{seq}[1 + \sigma(1 - e^{-y/\lambda_s})] \qquad \text{(14.28)}$$

A surface supersaturation can be defined as

$$\sigma_s = \frac{n_s}{n_{seq}} - 1 \qquad \text{(14.29)}$$

For a series of steps separated by L_o,

$$\sigma_s = \sigma \left[1 - \frac{\cosh\left(\dfrac{y}{\lambda_s}\right)}{\cosh\left(\dfrac{L_o}{2\lambda_s}\right)} \right] \qquad (14.30)$$

If one knows the step distribution, then the saturation and the mean diffusion length, the local surface saturation, step velocity, and ultimately the growth rate can be calculated. Finally it should be noted that BCF theory only applies when the degree of saturation is not extremely large. At high surface admolecule concentrations, 3-D growth dominates the process, and this terrace-kink model is no longer valid.

We are now in a position to appreciate the need for a clean environment. Recall that there is a background pressure of impurity gasses in the chamber. When one of these impurity atoms strikes the surface of the wafer, it may adsorb or react chemically so as to remain on the surface. Water for example, is likely to react with the surface of the semiconductor to form an immobile oxide. As the step approaches this oxide, it must grow around it and coalesce on the opposite side. Subsequent layers may not align themselves on this imperfect substrate, resulting in defects. It is also possible that this surface defect may be a getter site for admolecules, leading to island formation. A partial pressure of 10^{-8} torr of water results in an arrival flux of water molecules of 10^{13} atoms/cm^2-sec.

Figure 14-33   A high throughput automated MBE includes load and unload locks, preparation, and growth chambers, all connected under UHV. Three 4 in. wafers are held on circular platens *(courtesy of RIBER S.A.)*.

Figure 14-34 An ion implantation system for introducing impurities during silicon MBE *(after Ota)*.

Assume a sticking coefficient of one. A typical growth rate for MBE is 10000 Å/hr. Then, the arrival rate of growth atoms must be about 10^{15} atoms/cm^2-sec. That is, nearly one in 100 atoms is an impurity. Fortunately, most of the residual gas in the chamber is not water vapor. The semiconductor atoms in the beam flux rapidly react with the residual water and getter it from the chamber (Figure 14.33).

It is also not necessary for defects to be caused by impurities. Anything other than a terrace edge on which admolecules can condense can be a defect site. In VPE, there is a constant flux of chemical reactions, removing and adding the growth species from the surface of the wafer. The degree of supersaturation therefore, is quite modest. In MBE, the primary mechanism for removal of the adatoms once they stick to the surface is evaporation. (In GaAs MBE, the As adatoms may actually form As$_4$, which has a small but not negligible vapor pressure at typical growth temperatures.) Since the substrate temperature during MBE is low, P_e is small and the supersaturation is extremely high. This can also lead to higher defect densities in MBE growth due to self agglomeration. Once nucleated, these islands will grow, producing dislocations that propagate through the epitaxial layer. This effect limits the maximum growth rate that can be achieved in MBE.

For the epitaxial layer to be useful in device structures, they must be doped. In a good MBE system, unintentional doping can be as low as 10^{13} to 10^{15} cm^{-3}. For silicon, n-type dopants include arsenic, phosphorus, and antimony. For heavily doped layers, P and As are preferred due to their high solid solubility and low activation energy. Sb has reasonable vapor pressures at typical source temperatures, so that standard Knutson cells can readily be used for doping. As and P on the other hand, have very high vapor pressures at room temperature. This results in memory effects. For p-type layers in silicon, boron and aluminum have been used.

One of the problems with silicon MBE is that all of the common dopants have large segregation coefficients. The dopant concentration in the film is much less than the concentration on the surface of the wafer [94]. This excess dopant concentration on the surface continues to act like a dopant source even after the actual dopant evaporation source has been shut off. Several techniques have been developed to solve this problem. The most common method is to flash off these surface impurities by raising the temperature high enough to evaporate the dopant atoms. The high temperature required however, is not desirable due to dopant redistribution. In p-type doping, molecular boron sources can also be used. In particular, HBO_3 produces good control and only modest oxygen incorporation at substrate temperatures as low as 650°C [95].

Producing heavily doped layers in either material system is difficult, particularly if high quality epi layers with abrupt junctions are also desired. This problem arises because the sticking coefficients of common dopants decrease as the substrate temperature increases. Moderate or high substrate temperatures are necessary to reduce the defect density. The third method of introducing dopants in MBE is the use of low energy implantation during the growth [96] (Figure 14.34). This may be done using an actual ion gun or by ionizing the dopant atoms and biasing the substrate to imbed the dopant atom into the growing film. Energies for these processes are kept below a few keV to maintain well controlled profiles and greater than 100 eV to minimize sputtering. Alternatively, the growth atoms themselves may be ionized and accelerated toward the growing substrate. Upon reaching the substrate some of them will impact impurities on the surface, driving them into the substrate. This is referred to as knock on implantation. Substrate temperatures for implanted silicon MBE growth must be kept above 650°C to minimize radiation damage [97].

Group IV elements such as Si and Sn are typical dopants for producing n-type layers in GaAs. Silicon is the most common source, since silicon doped layers have the highest mobilities for a given doping level. Optimization of the growth process must be done to minimize the amount of doping compensation for group IV dopants. Zinc and beryllium are commonly used p-type dopants for GaAs. Recently carbon doping has been investigated as an alternate p-type doping source for MBE [98]. Carbon can easily be introduced by using a serpentine graphite filament in the chamber.

14.13 Gas Source MBE and Chemical Beam Epitaxy[+]

The distinction between vapor phase epitaxy and physical techniques begins to blur as VPE crystal growth techniques are being extended to ultrahigh vacuum and MBE crystal growth techniques begin to use gaseous sources. Depending on the details of the procedure, they may be called chemical beam epitaxy [99] (CBE), metal organic molecular beam epitaxy (MOMBE), or gas source molecular beam epitaxy (GSMBE). In each case, gasses are injected into an ultrahigh vacuum environment at very low flow rates. Table 14.3 differentiates the processes.

Table 14.3   Ultrahigh vacuum growth techniques that use gaseous sources

Technique	Group III Source	Group V Source
MBE	Elemental	Elemental
CBE	Gaseous	Gaseous
MOMBE	Gaseous	Elemental
GSMBE	Elemental	Gaseous

See [103].

In early use of GSMBE, it was found that gallium droplets formed at the surface when substrate temperatures were held below 550°C [100]. To solve this problem, most UHV growth systems that use gaseous group V sources have an oven or cracking furnace to decompose the source [101]. In the case of AsH_3, the cracking product is As_2. One of the interesting features of using organometallic sources in UHV is that the molecular sticking coefficients depend strongly on the surface coverage. In growing GaAs for example, it has been shown that the sticking coefficient for trimethyl gallium is almost zero when the surface is gallium terminated but is much higher when it is arsine terminated [102]. This observation suggests that the growth of GaAs can be accomplished in a digital manner by alternate exposures to TMG and an arsine source. These processes are called atomic layer epitaxy (ALE).

14.14 Summary

The epitaxial growth of silicon from the vapor is a well established process for fabricating advanced semiconductor technologies. An understanding of the process requires a detailed knowledge of both the gas phase chemistry and the events at the surface. Cleanliness, both of the chamber and of the growth gasses, is critical to the success of the process. Chlorosilanes are the preferred precursors for silicon epitaxy, which can be done uniformly across the wafer and selectively in windows etched in dielectrics. The growth of GaAs from the vapor however, is complicated by the lack of suitable inorganic gallium source gasses.

The second half of the chapter presented growth techniques that have been developed for controllably producing thin epitaxial layers, often in the form of heterostructures. Some of these processes allow atomic level control of the growth thickness. These methods have been divided into physical (MBE) and chemical (MOCVD, RTCVD, and UHV/CVD) techniques. This distinction is becoming more blurred as UHV processes are developed that use chemical reactions occurring on the surface of the wafer rather than in the gas phase.

Problems

1. Starting from Equation 14.1, derive Equation 14.2.
2. Compare Equation 14.2 to Figure 14.8. For each compound, identify the approximate temperature at which the growth becomes mass transport limited. If

$$k_s = k_o e^{-E_a/kT}$$

   h_g is independent of temperature and C_g is 10^{15} cm^{-3} at all temperatures and for all gasses, find h_g and k_o.
3. An epitaxial growth chamber has a background O_2 pressure of 2×10^{-5} torr. What would the minimum annealing temperature be to ensure a Si stable surface if only the background gasses are present? Alternatively, one can flow H_2 which contains 5 ppb of O_2. If this gas flushes the chamber of its background contaminates, what would the minimum annealing temperature be at 760 torr of H_2?
4. Explain why the growth curve shown in Figure 14.6 becomes negative.
5. Explain why low temperature growth is desirable. To assist you in this explanation, use Figure 14.8 to determine the amount of time necessary to grow 1 μm of epitaxial silicon at 800, 900, 1000, and 1100°C. Also calculate the intrinsic diffusivity of As at these temperatures (see Chapter 3), and from this find the characteristic diffusion length $\sqrt{Dt}$.

6. A 5% mixture of $SiCl_4$ in H_2 is used to grow epitaxial silicon at 1100°C. Calculate the supersaturation (σ) for this condition. Will growth or etching result?

7. Why isn't halide transport used for silicon epitaxial growth?

8. Briefly contrast commensurate, incommensurate, and pseudomorphic growth techniques. Describe the advantages and limitations of each.

9. The growth of GaAs on silicon wafers is able to provide large wafers for IC fabrication. If you wanted to extend this process to InP, HgCdTe, or other compound systems, what would you need to find out to determine whether or not it would be possible?

10. Let us assume that an MOCVD reactor is able to grow 1.0 µm per hour and that the utilization efficiency of the TMG is about 2% (common for run/vent systems). If the V/III ratio is 50:1 and the system grows 2 µm per day, how much arsine (in moles) is required per day? What volume does this correspond to at standard temperature and pressure (Note: 1 mole = 22.8 l at STP). How many runs can be done with a 25 l cylinder containing 10% AsH_3 at 2000 psi?

11. What are the advantages and disadvantages of using AsH_3 in MOCVD of GaAs?

12. Compare the advantages of both RTCVD and UHV/CVD.

13. Repeat Example 14.4 if $E_d = 1.5$ eV, $U_s = 0.1$, and $W_s = 1.7$ eV, all other parameters are unchanged. Estimate n_{seq}, λ_s, and σ.

14. What are some of the driving forces for using gas sources in MBE?

References

1. *Epitaxial Silicon Technology*, B. J. Baliga, ed., Academic Press, Orlando, 1986.

2. S. A. Campbell, J. D. Leighton, G. H. Case, and K. Knutson, ''Very Thin Silicon Epitaxial Layers Grown Using Rapid Thermal Vapor Phase Epitaxy,'' *J. Vac. Sci., Technol.* **B7**:1080 (1989).

3. C. Norris, ''Monitoring Growth with X-ray Diffraction,'' *Phil. Trans. Royal Soc. London Series A: Phys. Sci., Eng.* **344**:1673 (1993).

4. For a comprehensive review of wafer cleaning techniques the reader is referred to *Handbook of Semiconductor Wafer Cleaning Technology: Science, Technology, and Applications*, W. Kern, ed., Noyes, Park Ridge, NJ, 1993.

5. W. Kern and D. A. Puotinen, ''Cleaning Solution Based on Hydrogen Peroxide for Use in Semiconductor Technology,'' *RCA Rev.*, p. 187 (June 1970).

6. W. Kern, ''Purifying Si and SiO_2 Surfaces with Hydrogen Peroxide,'' *Semicond. Int.*, p. 94 (April 1984).

7. W. Kern, ''The Evolution of Silicon Wafer Cleaning Technology,'' *J. Electrochem. Soc.* **137**:1887 (1990).

8. O. J. Antilla and M. V. Tilli, ''Metal Contamination Removal on Silicon Wafers Using Dilute Acidic Solutions,'' *J. Electrochem. Soc.* **139**:1751 (1992).

9. H. R. Chang, in *Defects in Silicon*, W. M. Bullis and L. C. Kimberling, eds., Electrochem. Soc., Pennington, NJ, 1983.

10. G. Ghidini and F. W. Smith, ''Interaction of H_2O with Si(111) and (100),'' *J. Electrochem. Soc.* **131**:2924 (1984).

11. F. W. Smith and G. Ghidini, ''Reaction of Oxygen with Si(111) and (100): Critical Conditions for the Growth of SiO_2,'' *J. Electrochem. Soc.* **129**:1300 (1982).

12. G. W. Rubloff, ''Defect Microchemistry in SiO_2/Si Structures,'' *J. Vac. Sci. Technol.* **A8**:1857 (1990).

13. Y.-K. Sun, D. J. Bonser, and T. Engel, "Spatial Inhomogeneity and Void-Growth Kinetics in the Decomposition of Ultrathin Oxide Overlayers on Si (100)," *Phys. Rev.* **B3**:14,309 (1991).

14. R. Iscoff, "Wafer Cleaning: Wet Methods Still Lead the Pack," *Semicond. Int.*, p. 58 (July 1993).

15. J. H. Comfort, L. M. Garverick, and R. Reif, "Silicon Surface Cleaning by Low Dose Argon-Ion Bombardment for Low Temperature (750°C) Epitaxial Silicon Deposition," *J. Appl. Phys.* **62**:3388 (1988).

16. M. Ishii, K. Nakashima, I. Tajima, and M. Yamamoto, "Properties of Silicon Surface Cleaned by Hydrogen Plasma," *Appl. Phys. Lett.* **58**:1378 (1991).

17. S. Salimian and M. Delfino, "Removal of Native Silicon Oxide with Low-Energy Argon Ions," *J. Appl. Phys.* **70**:3970 (1991).

18. Y. J. Chabal, G. H. Higashi, K. Raghavachari, and V. A. Burrows, "Infrared Spectroscopy of Si(111) and Si(100) Surfaces After HF Treatment: Hydrogen Termination and Surface Morphology," *J. Vac. Sci. Technol.* **A7**:2104 (1989).

19. G. S. Higashi, Y. J. Chabal, G. W. Trucks, and K. Raggavachari, "Ideal Hydrogen Termination of the Si(111) Surface," *Appl. Phys. Lett.* **56**:656 (1990).

20. P. Dumas, Y. J. Chabal, and G. S. Higashi, "Coupling of an Adsorbate Vibration to a Substrate Surface," *Phys. Rev. Lett.* **65**:1124 (1990).

21. H. Ubara, T. Imura, and A. Hiraki, "Formation of Si-H Bonds on the Surface of Microcrystalline Silicon Covered with SiO$_x$ by HF Treatment," *Solid State Commun.* **50**:673 (1984).

22. M. Morita, T. Ohmi, E. Hasegawa, M. Kawakami, and K. Suma, "Control Factor of Native Oxide Growth on Silicon in Air or in Ultrapure Water," *Appl. Phys. Lett.* **55**:562 (1989).

23. D. Gräf, M. Grundner, and R. Schultz, "Reaction of Water with Hydrofluoric Acid Treated Silicon (111) and (100) Surfaces," *J. Vac. Sci. Technol.* **A7**:808 (1989).

24. M. Offenberg, M. Liehr, and G. W. Rubloff, "Ultraclean Integrated Processing of Thermal Oxide Structures," *Appl. Phys. Lett.* **57**:1254 (1990).

25. A. S. Grove, *Physics and Technology of Semiconductor Devices*, Wiley, New York, 1967.

26. J. P. Hirth and G. H. Pond, "Condensation and Evaporation: Nucleation and Growth," *Prog. Mat. Sci.* **11**:1 (1963).

27. R. Pagliaro Jr., J. F. Corboy, L. Jastrzebski, and R. Soydan, "Uniformly Thick Selective Epitaxial Silicon," *J. Electrochem. Soc.* **134**:1235 (1987).

28. P. van der Putte, L. J. Giling, and J. Bloem, "Growth and Etching of Silicon in Chemical Vapor Deposition Systems: The Influence of Thermal Diffusion and Temperature Gradients," *J. Crystal Growth* **31**:299 (1975).

29. J. Bloem and W. A. P. Claassen, "Nucleation and Growth of Silicon Films by Chemical Vapour Deposition," *Philips Technol. Rev.* **41**:60 (1983).

30. J. W. Medernach and P. Ho, *Mat. Res. Soc. Conf. Proc.*, Honolulu, 1987, p. 101.

31. W. A. P. Claassen and J. Bloem, "Rate-Determining Reactions and Surface Species in CVD of Silicon II. The SiH$_2$Cl$_2$-H$_2$-N$_2$-HCl System," *J. Crystal Growth* **50**:807 (1980).

32. H. C. Theuerer, "Epitaxial Silicon Films by the Reduction of SiCl$_4$," *J. Electrochem. Soc.* **108**:649 (1961).

33. T. Arizumi, *Curr. Topics Mat. Sci.* **1**:343 (1975).

34. F. C. Eversteyn, "Chemical Reaction Engineering in the Semiconductor Industry," *Philips Res. Rep.* **19**:45 (1974).

35. P. A. Coon, M. L. Wise, and S. M. George, "Modelling Silicon Epitaxial Growth with SiH$_2$Cl$_2$," *J. Crystal Growth* **130**:162 (1993).

36. T. Aoyama, Y. Inoue, and T. Suzuki, "Gas Phase Reactions and Transport in Silicon Epitaxy," *J. Electrochem. Soc.* **130**:204 (1983).

37. A. A. Chernov, "Growth Kinetics and Capture of Impurities During Gas Phase Crystallization," *J. Crystal Growth* **42**:55 (1977).

38. P. Rai-Choudhury and E. Salkovitz, "Doping of Epitaxial Silicon: The Effect of Dopant Partial Pressure," *J. Crystal Growth* **7**:361 (1970).

39. H. M. Liaw and J. W. Rose, "Silicon Vapor-Phase Epitaxy," in *Epitaxial Silicon Technology*, B. J. Baliga, ed., Academic, Orlando, 1986.

40. C. H. J. van den Breckel, "Characterization of Chemical Vapor Deposition Processing," *Philips Res. Rep.* **32**:118 (1977).

41. M. J. Stowell, in *Epitaxial Growth, Part B*, J. W. Matthews, ed., Academic, New York, 1975, p. 437.

42. B. J. Ginsberg, J. Burghartz, G. B. Bronner, and S. R. Mader, "Selective Epitaxial Growth of Silicon and Some Potential Applications," *IBM J. Res. Dev.* **34**:816 (1990).

43. J. O. Borland, in *Proc. 10th Int. Conf. Chem. Vapor Deposition*, Electrochem. Soc., Pennington, NJ, 1987, p. 307.

44. D. M. Jackson, "Advanced Epitaxial Processes for Monolithic Integrated Circuit Applications," *Trans. Met. Soc. AIME* **233**:596 (1965).

45. K. Tanno, N. Endo, H. Kitajima, Y. Kurogi, and H. Tsuya, "Selective Silicon Epitaxy Using Reduced Pressure Techniques," *Jpn. J. Appl. Phys.* **21**:L564 (1982).

46. C. I. Drowley, G. A. Reid, and R. Hull, "Model for Facet and Sidewall Defect Formation During Selective Epitaxial Growth of (100) Silicon," *Appl. Phys. Lett.* **52**:546 (1988).

47. J. T. McGinn, L. Jastrzebski, and J. F. Corboy, "Defect Characterization in Monocrystalline Silicon Grown over SiO$_2$," *J. Electrochem. Soc.* **136**:398 (1984).

48. L. Jastrzebski, J. F. Corboy, and R. Soydan, "Issues and Problems Involved in Selective Epitaxial Growth of Silicon for SOI Fabrication," *J. Electrochem. Soc.* **136**:3506 (1989).

49. R. Pagliaro, F. Corboy, L. Jastrzebski, J. and R. Soydan, "Uniformly Thick Selective Epitaxial Silicon," *J. Electrochem. Soc.* **134**:1235 (1987).

50. B. P. Jain and R. K. Purohit, "Physics and Technology of Vapour Phase Epitaxial Growth of GaAs—A Review," *Prog. Cryst. Growth, Charact.* **9**:51 (1984).

51. J. L. Gentner, "Vapour Phase Growth of GaAs by the Chloride Process Under Reduced Pressure," *Philips J. Res.* **38**:37 (1983).

52. A. Cadoff and J. Bicknell, "The Epitaxy of Silicon on Alumina—Structural Effects," *Philos. Mag.* **14**:31 (1966).

53. H. Manasevit and R. Simpson, "A Survey of the Heteroepitaxial Growth of Semiconductor Films on Insulating Substrates," *J. Crystal Growth* **22**:125 (1974).

54. P. K. Vasudev, "Silicon-on-Sapphire Heteroepitaxy," in *Epitaxial Silicon Technology*, B. J. Baliga, ed., Academic, Orlando, 1986.

55. W. I. Wang, "Molecular Beam Epitaxial Growth and Materials Properties of GaAs and AlGaAs on Si (100)," *Appl. Phys. Lett.* **44**:1149 (1984).

56. W. T. Masselink, T. Henderson, J. Klem, R. Fischer, P. Pearah, H. Morkóc, M. Hafich, P. D. Wang, and G. Y. Robinson, "Optical Properties of GaAs on (100) Si Using Molecular Beam Epitaxy," *Appl. Phys. Lett.* **45**:1309 (1984).

57. T. Soga and S. Hattori, "Epitaxial Growth and Material Properties of GaAs on Si Grown by MOCVD," *J. Crystal Growth* **77**:498 (1986).

58. T. Soga, S. Hattori, S. Sakai, M. Takeyasu, and M. Umeno, "MOCVD Growth of GaAs on Si Substrates with AlGaP and Strained Layer Superlattice Layers," *J. Appl. Phys.* **57**:4578 (1985).

59. R. People and J. C. Bean, "Calculation of Critical Layer Thickness versus Lattice Mismatch for $Ge_xSi_{1-x}Is$: Strained-Layer Heterointerfaces," *Appl. Phys. Lett.* **47**:322 (1985); ibid **49**:229 (1986).

60. H. M. Manasevit and W. I. Simpson, *J. Appl. Phys. Lett.* **116**:1725 (1969).

61. G. B. Stringfellow, *Organometallic Vapor-Phase Epitaxy*, Academic, Boston, 1989.

62. R. M. Lum, J. K. Klingert, and M. G. Lamont, "Comparison of Alternate As-sources to Arsine in the MOCVD Growth of GaAs," in *Fourth Int. Conf. MOVPE*, 1988, p. P1–3.

63. C. H. Chen, C. A. Larsen, G. B. Stringfellow, D. W. Brown, and A. J. Robertson, "MOVPE Growth of InP Using Isobutylphosphine and tert-Butylphosphine," *J. Crystal Growth* **77**:11 (1986).

64. R. J. Field and S. K. Ghandhi, "Doping of GaAs in a Low Pressure Organometallic CVD System," *J. Crystal Growth* **74**:543 (1986).

65. T. F. Kuech, "Metal-Organic Vapor Phase Epitaxy of Compound Semiconductors," *Mat. Sci. Rep.* **2**:1 (1987).

66. D. J. Schlyer and M. A. Ring, "An Examination of the Product-Catalyzed Reaction of Tri-methylgallium with Arsine," *J. Organometallic Chem.* **114**:9 (1976).

67. D. H. Reep and S. K. Ghandhi, "Deposition of GaAs Epitaxial Layers by Organometallic CVD," *J. Electrochem. Soc.* **130**:675 (1983).

68. P. Rai-Chuadhury, "Epitaxial Gallium Arsenide from Trimethylgallium and Arsine," *J. Electrochem. Soc.* **116**:1745 (1969).

69. Y. Seki, K. Tanno, K. Iida, and E. Ichiki, "Properties of Epitaxial GaAs Layers from a Tri-ethylgallium and Arsine System," *J. Electrochem. Soc.* **122**:1108 (1975).

70. T. F. Kuech, M. A. Tischler, P.-J. Wang, G. Scilla, R. Potemski, and F. Cardone, "Con-trolled Carbon Doping of GaAs by Metalorganic Vapor Phase Epitaxy," *Appl. Phys. Lett.* **53**:1317 (1988).

71. B. T. Cunningham, M. A. Haase, M. J. McCollum, J. E. Baker, and G. E. Stillman, "Heavy Carbon Doping of Metalorganic Chemical Vapor Deposition Grown GaAs Using Carbon Tetrachloride," *Appl. Phys. Lett.* **54**:1905 (1989).

72. B. T. Cunningham, L. J. Guido, J. E. Baker, J. S. Major, Jr., N. Holonyak, Jr., and G. E. Stillman, "Carbon Diffusion in Undoped *n*-type and *p*-type GaAs," *Appl. Phys. Lett.* **55**:687 (1989).

73. M. A. Tischler, "Advances in Metalorganic Vapor-Phase Epitaxy," *IBM J. Res. Dev.* **34**:828 (1990).

74. T. F. Kuech, E. Veuhoff, D. J. Wolford, and J. A. Bradley, "Low Temperature Growth of $Al_xGa_{1-x}As$ by MOCVD," *GaAs, Related Compounds, 11th Int. Symp.*, 1985, p. 181.

75. J. R. Shealey and J. M. Woodall, "A New Technique for Gettering Oxygen and Moisture from Gasses in Semiconductor Processing," *Appl. Phys. Lett.* **68**:157 (1984).

76. J. F. Gibbons, C. M. Gronet, and K. E. Williams, "Limited Reaction Processing: Silicon Epitaxy," *Appl. Phys. Lett.* **47**:721 (1985).

77. S. A. Campbell, J. D. Leighton, G. H. Case, and K. Knutson, "Very Thin Silicon Epitaxial Layers Grown Using Rapid Thermal Vapor Phase Epitaxy," *J. Vac. Sci., Technol.* **B7**:1080 (1989).

78. M. L. Green, D. Brasen, H. Luftman, and V. C. Kannan, "High Quality Homoepitaxial Sili-con Films Deposited by Rapid Thermal Chemical Vapor Deposition," *J. Appl. Phys.* **65**:2558 (1989).

79. T. Y. Hsieh, K. H. Jung, and D. L. Kwong, "Silicon Homoepitaxy by Rapid Thermal Pro-cessing Chemical Vapor Deposition," *J. Electrochem. Soc.* **138**:1188 (1991).

80. K. L. Knutson, S. A. Campbell, and F. Dunn, ''Three Dimensional Temperature Uniformity Modeling of a Rapid Thermal Processing Chamber,'' accepted for publication, *IEEE Trans. Semicond. Manufac.* (1993).

81. B. S. Mayerson, ''Low-Temperature Silicon Epitaxy by Ultrahigh Vacuum/Chemical Vapor Deposition,'' *Appl. Phys. Lett.* **48**:797 (1986).

82. B. S. Mayerson, E. Ganin, D. A. Smith, and T. N. Nguyen, ''Low Temperature Silicon Epitaxy by Hot Wall Ultrahigh Vacuum/Chemical Vapor Deposition Techniques: Surface Optimization,'' *J. Electrochem. Soc.* **133**:1232 (1986).

83. B. S. Meyerson, ''Low Temperature Si and Ge:Si Epitaxy by Ultrahigh Vacuum/Chemical Vapor Deposition: Process Fundamentals,'' *IBM J. Res. Dev.* **34**:806 (1990).

84. D. Bellevance, ''Industrial Application: Perspective and Requirements,'' in *Silicon Molecular Beam Epitaxy* **2**, E. Kasper and J. C. Bean, eds., CRC, Boca Raton, 153, 1985.

85. T. Tatsumi, H. Hirayama, and N. Aizaki, ''Si Particle Density Reduction in Si Molecular Beam Epitaxy Using a Deflection Electrode,'' *Appl. Phys. Lett.* **54**:629 (1989).

86. A. von Gorkum, ''Performance and Processing Line Integration of a Silicon Molecular Beam Epitaxy System,'' *Proc. 3rd Int. Symp. Si MBE, Thin Solid Films* **184**:207 (1990).

87. K. Fujiwara, K. Kanamoto, Y. N. Ohta, Y. Tokuda, and T. Nakayama, ''Classification and Origins of GaAs Oval Defects Grown by Molecular Beam Epitaxy,'' *J. Crystal Growth* **80**:104 (1987).

88. A. Y. Cho, ''Advances in Molecular Beam Epitaxy (MBE),'' *J. Crystal Growth* **111**:1 (1991).

89. J. Saito, K. Nanbu, T. Ishikawa, and K. Kondo, ''In situ Cleaning of GaAs Substrates with HCl Gas and Hydrogen Mixture Prior to MBE Growth,'' *J. Crystal Growth* **95**:322 (1989).

90. N. Chand, ''A Simple Method for Elimination of Gallium-Source Related Oval Defects in Molecular Beam Epitaxy of GaAs,'' *Appl. Phys. Lett.* **56**:466 (1990).

91. J. H. Neave, P. Blood, and B. A. Joyce, ''A Correlation Between Electron Traps and Growth Processes in n-GaAs Prepared by Molecular Beam Epitaxy,'' *Appl. Phys. Lett.* **36**:311 (1980).

92. W. K. Burton, N. Cabrera, and F. C. Franks, *Philos. Trans. R. Soc. London*, Ser. **A 243**:299 (1951).

93. M. G. Legally, ''Atoms in Motion on Surfaces,'' *Phys. Today* **46**:24 (1993).

94. J. C. Bean, ''Silicon Molecular Beam Epitaxy: Highlights of Recent Work,'' *J. Electron. Mat.* **19**:1055 (1990).

95. T. L. Lin, R. W. Fathauer, and P. J. Grunthaner, ''Heavily Boron-doped Si Layers Grown Below 700°C by Molecular Beam Epitaxy Using a HBO_2 Source,'' *Appl. Phys. Lett.* **55**:795 (1989).

96. P. Fons, N. Hirashita, L. C. Markert, Y.-W. Kim, J. E. Green, W.-X. Ni, J. Knall, G. V. Hansson, and J.-E. Sundgren, ''Electrical Properties of Si(100) Films Doped with Low-Energy (<150 eV) Sb Ions During Growth by Molecular Beam Epitaxy,'' *Appl. Phys. Lett.* **53**:1732 (1988).

97. J.-P. Noel, J. E. Greene, N. L. Rowell, S. Kechang, and D. C. Houghton, ''Photoluminescence Studies of Si(100) Doped with Low Energy (<1000 eV) As$^+$ Ions During Molecular Beam Epitaxy,'' *Appl. Phys. Lett.* **55**:1525 (1989).

98. R. J. Malik, R. N. Nottenberg, E. F. Schubert, J. F. Walker, and R. W. Ryan, ''Carbon Doping in Molecular Beam Epitaxy of GaAs from a Heated Graphite Filament,'' *Appl. Phys. Lett.* **53**:2661 (1988).

99. W. T. Tsang, "From Chemical Vapor Epitaxy to Chemical Beam Epitaxy," *J. Crystal Growth* **95**:121 (1989).

100. E. Veuhoff, W. Pletschen, P. Balk, and H. Lüth, "Metallorganic CVD of GaAs in a Molecular Beam Epitaxy System," *J. Crystal Growth* **55**:30 (1981).

101. M. B. Panish, "Molecular Beam Epitaxy of GaAs and InP with Gas Sources for As and P," *J. Electrochem. Soc.* **127**:2729 (1980).

102. M. Uneta, Y. Watanabe, and Y. Ohmachi, "Desorption of Triethylgallium During Metalorganic Molecular Beam Epitaxial Growth of GaAs," *Appl. Phys. Lett.* **54**:2327 (1989).

103. W. T. Tsang, "Current Status Review and Future Prospects of CBE, MOMBE, and GSMBE," *J. Crystal Growth* **107**:960 (1991).

104. J. Davies and D. Williams, "III–V MBE Growth System," in *The Technology and Physics of Molecular Beam Epitaxy*, E. H. C. Parker, ed., Plenum, New York (1985).

Part V

Process Integration

Any IC technology can be divided into three basic components, whether it is FET or bipolar, Si or GaAs. These components are: device fabrication, interconnection, and isolation. When deciding which processes are to be used, the technologist must ensure that all three of these tasks are done. The structure of this section will be to first discuss groups of processes, called *modules*, for isolation and interconnection (including ohmic contacts) for both silicon and GaAs based technologies. This is done both to avoid repetition and so that the student can compare these modules across technologies.

The essence of engineering is knowing what variables you can afford to ignore.[1]

In deciding on the order of the process flow, there are two important considerations. The first is the function of the particular process. As an example, interconnect could be done first by using diffused resistors in the substrate. The problem with such an approach is that the interconnect has a large resistance and a large capacitance. This type of interconnect, while functional, would dramatically reduce the circuit speed. The second consideration is the effect that each process might have on later processes or on the structures that have been fabricated before them. One obvious example of this consideration is thermal budget. Since most technologies involve the use of selectively doped regions of semiconductors, diffusion is a serious concern if high temperature processes are required. It is therefore desirable to do the highest temperature processes first. If aluminum is to be used for interconnection, after aluminum deposition the wafer temperature certainly must not exceed 660°C, the melting temperature of aluminum. These considerations mean that it may not be possible for the three modules to be done sequentially, but that in some cases they may have to be intertwined.

After this introductory chapter, some of the most important individual technologies will be presented. The reader might wonder how the technologies were selected. As a useful reference, refer to the figure. The chart shows the fraction of the IC market for a variety of technologies. Clearly, the dominant technology is complimentary metal oxide semiconductor (CMOS). NMOS and PMOS are simply subsets of the CMOS process. Chapter 16 will therefore begin this section of the book by

---

[1]W. C. Randels, in *Transistors Fundamentals for the Integrated Circuit Engineering*, R. M. Warner and B. L. Grung, eds., Wiley-Interscience, New York, 1983.

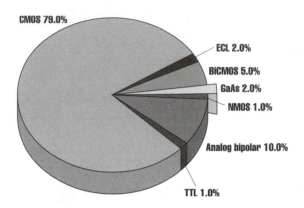

Estimated market share of IC technologies in 1995.

covering CMOS. GaAs, once called the technology of the future, has a small but slowly increasing market share. Chapter 17 will review GaAs field effect transistors (FETs), both because they are representative of GaAs technologies in general and because they present some interesting contrasts with silicon CMOS. The bipolar market share is shrinking, particularly in the areas of low power/moderate performance digital. A large fraction of analog and most of the high end digital however, continues to be done in bipolar, and so Chapter 18 will be devoted to bipolar technologies. A significant share of the market is going toward a combination of CMOS and bipolar called BiCMOS, which will be covered at the end of Chapter 18. Memory technologies will not be covered per se, due to space limitations. References to memory technology requirements are contained primarily in Chapters 15 and 16. Finally, some material common to all technologies will be presented: yield, statistical process control, and design of experiments.

Each of the technology chapters begins with a brief review of the ideal device equations. The emphasis here is not a complete description of the operation of the device. It is assumed that the student has taken a basic introductory semiconductor device course. This device understanding is necessary to be able to point out how the process must be designed to optimize device performance. Next, the series of unit processes used to fabricate a simple version of the technology will be presented. The extension of these simple technologies to higher speed applications is then presented, along with a description of the process modules necessary to obtain the desired performance. Finally, the concerns and process directions for each technology will be discussed.

Chapter 15

Device Isolation, Contacts, and Metallization

This chapter will cover two of the most basic functions of an IC. The first half will discuss device isolation, that is, the ability of the technology to allow each device to operate independently of the state of the other devices. Unless the technology is limited to building discrete devices, this is an essential function. The second half of the chapter will discuss device interconnection including the metal to semiconductor contact. Again, only a brief review of the device physics will be provided to point out the implications for the technology.

15.1 Junction and Oxide Isolation

To fabricate ICs, some sort of isolation module must be developed [1]. The metrics of such a module are: density, process complexity, yield, planarity, and parasitic effects. Tradeoffs exist among these parameters. The result of this simple consideration is that no single isolation module is the most appropriate for all applications. A high density, high complexity module is the most appropriate choice for DRAMs, but a simpler, lower density isolation module can be used for low cost TTL. This section will present several options for device isolation in silicon and GaAs technologies and evaluate each according to the metrics listed above.

The first ICs were bipolar. To understand the necessity for isolation, consider Figure 15.1. From the discussion on unit processes, it is easy to understand how a modest bipolar technology can be constructed. To do this, simply diffuse a deep p layer and a shallower N+ layer into an n-type substrate. The substrate acts as the common collector. Assume that an insulating layer is then deposited on the substrate, contacts are patterned and etched, and the interconnect is applied and patterned. The first question is, how close can the transistors be placed; that is, what is the packing density?

The emitters of two adjacent transistors are automatically isolated from each other by the fact that each emitter is totally enclosed in the base diffusion. In order for the base layers to be isolated from each other, they must be separated by a sufficient distance such that the depletion layers asso-

Figure 15-1 Simple junction isolation in a bipolar transistor technology with a common collector.

ciated with the base collector junctions do not touch. To get an estimate, assume a simple one sided step junction. Then the depletion layer thickness is

$$W_D = \sqrt{\frac{2k_s\varepsilon_o}{qN_D}(V_{bi} + V_{CB})} \qquad (15.1)$$

where V_{bi} is the built in voltage of the junction, which for nondegenerate doping is given by

$$V_{bi} = \frac{kT}{q}\ln\frac{N_A N_D}{n_i^2} \qquad (15.2)$$

and n_i is the intrinsic carrier concentration (about 10^{10} cm^{-3} at room temperature). For a typical collector concentration of 10^{16} cm^{-3}, this reduces to

$$W_D \approx 0.36\ \mu m\sqrt{V_{bi} + V_{CB}} \qquad (15.3)$$

The isolation distance then is just $2W_D$, or about 1.8 μm for a maximum V_{BC} of 5 V, or about 2.4 μm, to ensure isolation for a 10 V V_{BC}. Of course, one must also consider lateral diffusion. The base regions must be at least this far apart at the end of the process, not just on the photomasks.

To extend the approach to completely isolate the transistors, simply add a third diffusion to form the collector and use p-type substrates. To minimize the capacitance associated with the collector and to ensure that the collector doping is independent of variations in the substrate doping, a lightly doped substrate must be employed. Assume that the substrate is now 10^{15} cm^{-3} p-type. Then

$$W_D \approx 1.14\ \mu m\sqrt{V_{bi} + V_{CS}} \qquad (15.4)$$

For a 10 V bias (2× a supply voltage of 5 V), this would require about a 4 μm spacing around each diffusion or a total of 8 μm between devices after processing is complete. Taking into account lateral diffusion, a 12 μm collector to collector spacing might be required. In these processes, it is essential to ensure that the junctions remain reverse biased. This idea of using reverse biased junctions to isolate the devices, first patented in 1959 [2], was the earliest practical method of device isolation.

To get an estimate for the isolation length required for a technology, consider the densest portion of the circuit. Then referring to Figure 15.2, read off the maximum permissible isolation distance. For example, if the densest portion of the circuit has 10^5 transistors/cm^2 and

Figure 15-2 Simple calculation of average isolation distance required between transistors as a function of the device density.

Figure 15-3   Cross section of simple bipolar technology with a metal line crossing the junction isolation region, forming a parasitic MOSFET.

50% of the area is active in that region, the device separation must be 10 μm or less. In many designs, the circuit density is limited not by the isolation distance, but by the metal density. The density goal of the isolation module then, is that the separation necessary to make the interconnect, and not the device isolation, the limiting factor for the circuit density. Generally, for a random logic circuit, the isolation distance should be less than twice the first metal pitch (minimum linewidth + minimum line space). For memories and other highly structured logic, the isolation distance should be no more than the first metal pitch.

The other important point to make about Figure 15.2 is that for isolation limited circuits, the circuit density is a sensitive function of the isolation distance. Reducing the isolation distance by a factor of 3 allows an order of magnitude increase in the circuit density. Consequently, companies that deal in isolation driven circuits such as memories, have developed elaborate isolation techniques.

Density however, is not the only consideration in designing the isolation module. Go back to Figure 15.1, but add an insulating layer and a metal line that happens to pass over, but does not contact both transistors (Figure 15.3). A parasitic MOS transistor now exists. The collectors of the two transistors act as the source and drain of the MOSFET, and the metal line acts as the gate. If a large enough positive bias exists on the line, the surface of the area under the line may invert, turning on the parasitic MOSFET. This effect will short out the two collectors, even if they are sufficiently far apart so that the depletion regions do not touch.

Neglecting oxide charge, the MOSFET threshold for NMOS transistors on p-type substrates is given by (see Section 16.1)

$$V_T = \phi_{ms} + 2\phi_f + \frac{k_s t_{OX}}{k_{OX}} \sqrt{\frac{4qN_A}{k_s \varepsilon_o}} \phi_f \qquad (15.5)$$

where

$$\phi_f = \frac{kT}{q} \ln \left[\frac{N_A}{n_f} \right] \qquad (15.6)$$

and ϕ_{ms} is the metal semiconductor work function that depends on the type of metal used in the interconnect and the doping concentration in the substrate. For aluminum on p-type substrates, ϕ_{ms} varies from about -0.8 to -1.0 V. Since the intrinsic carrier concentration increases with temperature, the parasitic threshold voltage may shift by several volts when operating at high temperatures. Furthermore, the source to drain current of an MOS device increases exponentially with gate voltage in the subthreshold regime. Normal practice is to make the threshold of these parasitic transistors at least 2× (and preferably 3×) the supply voltage. This ensures that they will not turn on, even in the presence of an excess supply voltage and/or voltage spikes on the supply line, nor will the leakage currents be excessive.

Figure 15.4 shows a plot of threshold voltage versus the substrate concentration with oxide thickness as a parameter. The oxide thickness was varied from 0.2 to 1.0 μm in 0.2 μm increments.

Figure 15-4 V_T as a function of the N_A assuming no ϕ_{ms} = 0. Solid lines are a perfect interface, dotted lines are for $N_{it} = 10^{11}$ cm^{-2}.

Figure 15-5 Guard ring isolation for the bipolar technology from Figure 15-3.

Also included is the threshold voltage if a total fixed charge of 10^{11} cm^{-2} is present. To achieve a suitably large parasitic threshold, one must select a thick field oxide and/or a large substrate concentration. The large substrate concentration degrades the performance due to junction capacitance. The thicker oxide improves both performance and parasitic turn on, but obviously due to the effect of oxide charge, some substrate doping is required even for very thick field oxides.

The device separation can be reduced and the parasitic threshold increased by adding a p+ diffused barrier ring around each device (Figure 15.5). Guard rings require an additional mask that must be aligned to the transistor. Furthermore, the guard ring must be deep (at least 2 μm), or the depletion layer will simply extend beneath the guard ring, shorting out the devices. The large thermal cycle needed to produce such a deep junction must be done early in the process. Assuming a 2 μm resolution lithography such as might be used in such a simple technology and recognizing that lateral diffusion that would occur, the final width of the guard ring is over 5 μm. This requirement ultimately limits improvements in packing density.

We can now analyze this result in terms of the metrics described above. Obviously, junction isolation is simple and produces a planar isolation. Because of its simplicity, it would have a high yield. Density however, is not large and must be traded off against the parasitic capacitance of the collector substrate junction. Increasing the substrate concentration increases the density but also increases the capacitance. Guard rings can improve the situation, particularly concerning the prevention of turning on parasitic MOS devices. For these reasons, in spite of its low density, junction isolation is still used in some low cost, low density applications.

15.2 LOCOS Methods

The most straightforward way to produce a thick field oxide is by growing one before device fabrication, then etching holes in the oxide and fabricating the devices in these holes. This approach has two serious shortcomings. The first is the topology that is created. The step coverage for subsequent depositions will be poor and the photolithography will suffer. This is extremely serious if small features are to be printed. The second drawback is less obvious. On lightly doped substrates, a guard ring must be implanted to increase the parasitic threshold voltage. Unless very high energies are used, the implant must be done before the oxidation. Diffusion during oxidation may also be enhanced by point defects released during the oxidation process. Combined with alignment tolerance requirements, this will significantly reduce the density of the IC.

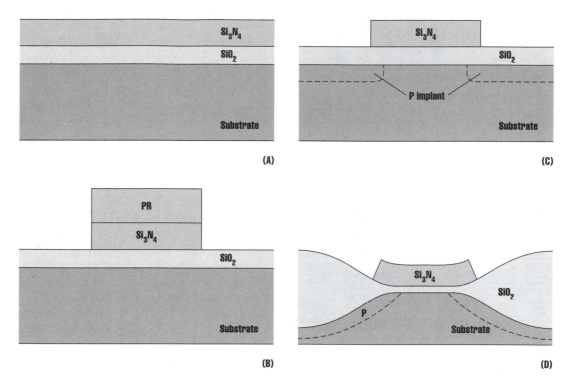

Figure 15-6 Cross-sectional views of a standard local oxidation of silicon (LOCOS) process.

The isolation approach that has become the standard of silicon IC fabrication is local oxidation of silicon or LOCOS [3]. Local oxidation is essentially an outgrowth of junction isolation and addresses both the isolation and the parasitic device formation concerns. A thin oxide is first grown and a layer of Si₃N₄ deposited on the wafer, usually by LPCVD. After the nitride is patterned, a field implant may be done to increase the threshold voltage of the parasitic MOSFET. Then the photoresist is stripped and the wafer is oxidized (Figure 15.6). The nitride acts as a barrier to the diffusion of the oxidant, preventing oxidation in selected regions of the silicon. A thin oxide will also be grown on top of the nitride. This is important because it limits the minimum nitride thickness to about 1000 Å and because the oxide must be removed before the nitride can be stripped after the field oxidation.

Since oxidation consumes 44% as much silicon as it grows, the resultant oxide is partially recessed and has a gradual step onto the field that is easy for subsequent layers to cover. If the silicon is etched before the field implant, the field oxide can be made fully recessed, resulting in a nearly planar surface. Figure 15.7 shows the growth of a local oxide along with a cross-sectional view of a completed LOCOS structure. The process leaves a characteristic bump on the surface, followed by a gradually narrowing oxide tail into the active area. The structure is called a *bird's beak* for obvious reasons. The bump, or bird's head, is particularly pronounced in recessed structures.

The purpose of the thin or pad oxide layer under the nitride is to reduce the stress that occurs in the silicon substrate during oxidation. This stress is due to the mismatch of the thermal expansion coefficients of the substrate and the nitride and due to the volumetric increase of the growing oxide. At high temperature, viscous flow of the oxide greatly reduces the stress. A great deal of work has gone into optimizing the thicknesses of the oxide and nitride layers. If the stress exceeds the yield

A) Standard LOCOS — 800Å Nitride/300Å Oxide

B) LOPOS — 2000Å Nitride/400Å Poly/300Å Oxide

C) LOPOS — 2000Å Nitride/400Å Poly/100Å Oxide

Figure 15-7   Cross-sectional scanning electron micrograph of a typical LOCOS isolation *(after Ghezzo et al., reprinted by permission, The Electrochemical Society).*

strength of silicon, it will generate dislocations in the substrate. A thicker pad oxide will lower the stress in the substrate. The minimum pad oxide thickness that can be tolerated without dislocation formation is about one third the thickness of the nitride [4]. This defect protection must be traded off against increased lateral encroachment of the oxide, which occurs due to lateral diffusion of the oxidizing species through the pad oxide. A nitride to thermal pad oxide thickness ratio of 2.5:1 produces a lateral encroachment or bird's beak, approximately equal to the thickness of the field oxide.

One concern of the LOCOS process is the white ribbon or Kooi nitride effect [5]. In this situation, a thermal oxynitride forms at the surface of the silicon under the edges of the nitride pad. White ribbon is caused by the reaction of Si_3N_4 with the high temperature wet ambient to form NH_3, which diffuses to the silicon/silicon dioxide interface where it dissociates. The surface texture caused by these nitrides can be seen as a white ribbon around the edges of the active area. This defect leads to a reduced breakdown voltage in subsequent thermal oxides (such as gate oxides) in the active region.

The existence of the bird's beak has two importance consequences from a device standpoint. Often the active region defines the edge of the device in at least one direction. Then encroachment reduces the active width of the device, reducing the amount of current that a transistor will drive. A more subtle effect is due to the field doping. The field oxidation causes the field implant to diffuse into the edge of the active region. Figure 15.8 shows a schematic of an MOS transistor along its width, immediately under the gate. If the transistor is narrow enough, the additional dopant from the field diffusion will increase the threshold voltage of the device, reducing its drive current. Known as the narrow channel effect, it is important in extremely dense technologies such as memories.

Various methods have been proposed to modify the LOCOS process to reduce the bird's beak length. The simplest is the use of materials other than a thermal oxide for the pad layer. A sandwich of thermal oxide and polysilicon is extremely effective in reducing lateral encroachment [6] and has become quite popular. The polysilicon absorbs the excessive stress resulting from the use of thin pad oxides and very thick nitrides. Typically about 500 Å of polysilicon is deposited on top of 150 Å of thermal oxide, followed by 1500 Å of nitride. This structure produces a bird's beak that is less than half the thickness of the field oxide [7]. Experiments describing the optimization of the poly buffered process can be found in Ghezzo et al. [8] and Guldi et al. [9]. Although the length of the bird's beak is reduced, the poly buffered process does not solve the problem of lateral diffusion of the field dopant.

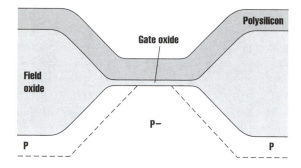

Figure 15-8 Cross-sectional schematic of an MOS transistor cut along its width, illustrating the origin of narrow channel effects.

To gain the full benefit of such a process, the field oxidation should be done at high pressure in a wet ambient to minimize the lateral diffusion. White ribbon effects in poly buffered LOCOS is a concern, however. Care must be taken when etching back the nitride/poly/oxide layers [10] to avoid it.

There are various other improvements to the LOCOS process that attempt to reduce bird's beak by preventing the diffusion of the oxidizing species under the nitride. Two examples are SILO and SWAMI. As shown in Figure 15.9, the sealed interface local oxidation (SILO) method uses a very thin (100 to 200 Å) nitride layer deposited directly on the silicon. Next, layers of silicon dioxide and silicon nitride are deposited. The layers are patterned, and a field oxide is grown. The nitride layer on the surface is thin enough to avoid most of the damage to the surface of

Figure 15-9 Sealed interface local oxidation (SILO) process *(after Molle et al., reprinted by permission, The Electrochemical Society).*

Figure 15-10   Sidewall masked isolation (SWAMI) process.

the silicon under it, but thick enough to serve as an effective diffusion barrier [11]. The upper nitride layer then prevents the oxidation of the lower layer. The encroachment in the SILO process is typically about 0.2 μm per side [12]. As with the poly buffered LOCOS, the step height is large if the silicon is not recessed before oxidation, and the field implant diffusion remains a problem. This process also produces a high density of interface states near the edge of the active regions unless very careful attention is paid to shaping the silicon recess before oxidation.

First proposed in 1982, sidewall masked isolation (SWAMI) has generated a great deal of interest due to its ability to fabricate thick, highly planar field oxides with little encroachment [13]. The SWAMI process is shown in Figure 15.10. A thermal pad oxide is grown, and a nitride layer is deposited. The layers are patterned and the silicon is etched to a depth of about half the desired oxide thickness. This can be done wet chemically with KOH, which produces a 60° slope on <100> substrates, or it can be done in a properly designed plasma etch. A second layer of nitride is then deposited and anisotropically etched. This leaves a sidewall of nitride on the slopes of the silicon, which effectively seals the active region. A thermal oxidation is done for the field oxide. Finally, the nitride and pad oxide are removed. SWAMI can produce thick field oxides with minimal encroachment, but retains the dopant diffusion problems of all LOCOS based modules. Its complexity and the development of alternative isolation techniques have precluded SWAMI's widespread application.

15.3 Trench Isolation

It became obvious during the 1980s that neither LOCOS nor any of its variations might be acceptable for ICs with transistor densities much greater than 10^6 cm^{-2}. It is now believed that the absolute minimum isolation distance for advanced LOCOS processes is about 0.8 μm from the edge of one N+/p junction to another [14]. The ultimate limitation is not surface inversion or simple punchthrough, but a reachthrough effect known as drain induced barrier lowering.

Many new isolation approaches have been developed around the idea of etching away part of the substrate and refilling it with an insulator. These can be

Figure 15-11   Simple shallow trench isolation process.

divided into two classes. Shallow etches with small aspect ratios were considered first [15]. These techniques can be thought of as similar to recessed LOCOS, but a deposited rather than a thermal oxide is used to fill the field regions. After deposition, planarization must be done to remove the unwanted oxide from over the active regions (Figure 15.11). To be effective, an additional photomask step may be required in this oxide removal step. Due to problems of misalignment, this is difficult to achieve reliably. Furthermore, it is difficult to prevent inversion along the etched sidewalls [16], since the field implant is normally done at close to normal incidence. In a CMOS process, this inversion may cause excess leakage between source and drain. Due to their geometry, bipolar transistors are not as prone to this problem as MOSFETs.

Deep trench isolation modules (Figure 15.12) use trenches of fixed width. Typical dimensions are 0.35 to 1.0 μm in width and 2 to 5 μm in depth although trenches with widths of 0.25 μm, and depths of 10 μm have been demonstrated [17] (See Figure 15.13). The smaller trench widths are particularly attractive for memory applications. The process is fabricated by starting from a standard LOCOS structure. After nitride patterning, the trenches are etched. Trench isolation puts extreme demands on the etch process. It must have smooth walls at no more than 85° with respect to the plane of the wafer. More taper is highly desirable [18]. The walls cannot undercut the mask, and should end in a rounded bottom. Sharp corners at the bottom of the trench will result in excess stress during oxidation, and ultimately defects in the oxide [19]. Next a field implant is done. Preventing sidewall inversion becomes even more difficult as the aspect ratio of the structure increases. An important feature of the implant therefore, is that the beam be perpendicular to the surface of the wafer. As discussed in Chapter 5, new implanters have been developed to accomplish this [20]. Another, albeit much less popular, approach to solving this problem is the use of rapid thermal processing combined with planar dopant sources [21] or a doped CVD glass [22].

The implant is followed by a thin local oxidation (typically less than 1000 Å). In applications in which the trench fill is to be used as a charge storage capacitor, it is common practice to use much thinner oxides to increase the capacitance. When the oxide is less than about 100 Å however, the capacitor leakage current increases sharply [23]. Finally, a layer of polysilicon is deposited and etched back. If the poly is thick enough, it will fill the groove. Etching back to the substrate will leave this fill in place. A second thermal oxidation can be used to complete the process by oxidizing the upper part of the polysilicon in the groove. Often this is done as part of a LOCOS process. This minimizes leakage, it allows the fabrication of arbitrary isolation lengths (see below), and it offsets the junctions from the wall of the trench. To minimize lateral diffusion, the channel stop implant associated with the LOCOS may be done after oxidation [24].

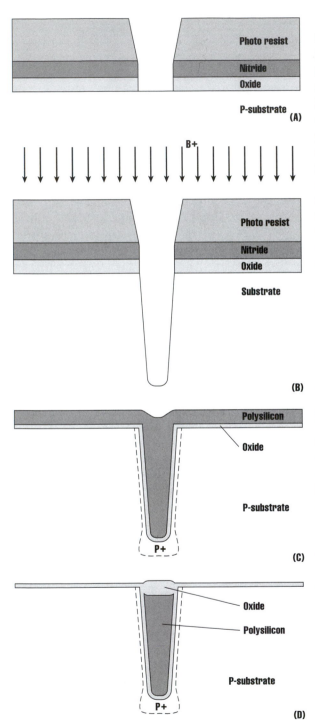

It is important to emphasize that the trench must be of a fixed width for the process to work. This can be made to work with the minimum N+/P+ separation, but if the design also calls for larger separations in some areas, an additional isolation (such as a standard LOCOS) technique must also be used. The result is a complex process that is difficult to control. If the deposition process is not done properly, a void may form in the center of the trench, trapping some material. This may also occur if too steep an angle is etched into the substrate during the silicon RIE. The void may pose a reliability problem. Nevertheless, deep trench isolation technology currently looks attractive for high density CMOS and for many bipolar technologies. N+ to P+ spacings less than 2 μm can be achieved. N+ to N+ spacings can be less than 0.5 μm. Referring to Figure 15.2, such an isolation technique will be adequate for device densities well in excess of 10^6 cm^{-2}.

15.4 Silicon on Insulator Isolation Techniques

The ideal method of device isolation would be to completely encase each device in an insulating material. Several techniques exist for doing this in silicon. Generically they are called silicon on insulator (SOI). All of these methods suffer from problems related to defect density. For that reason SOI technologies in silicon have been relegated to small markets such as radiation hardened devices where extreme isolation is required. For more routine applications, it is found that techniques such as trench isolation not only have fewer defects, but allow closer spacing between devices than most SOI approaches. In some measure, this has changed in the last few years. Several techniques have improved dramatically and LSI to VLSI applications has been demonstrated. One of the most promising methods, SIMOX, has been presented in Chapter 5 and so will not be repeated here. Instead we will concentrate on several other processes that show significant promise.

One of the first SOI methods developed is called dielectric isolation (DI). The DI process sequence, shown in Figure 15.14 [25], was developed to build high voltage telecommunication ICs that required elec-

Figure 15-12   Deep trench isolation process schematic.

Figure 15-13   Cross-sectional scanning electron micrograph of an extreme aspect ratio deep isolation process *(after Rajeevakamur et al., ©1991 IEEE).*

Figure 15-14   Dielectric Isolation (DI) process for forming silicon on insulator.

trically isolated bidirectional switches. It has since gained popularity for other high voltage and radiation hard digital applications. Deep grooves are first etched in the surface of the wafer. The wafer is oxidized and a very thick (>200 μm) layer of polysilicon is deposited. If substrate contact is needed in some islands, windows can be opened in the oxide before deposition. The deposition can be done by conventional CVD or by using a molten silicon spray deposition (MSSD) process [26]. The wafer is turned over and mechanically ground until the grooves penetrate through the wafer. Finally, the wafer is chemically polished and devices are fabricated in the isolated islands.

Dielectric isolation has several severe drawbacks. The wafers are not as planar as normal starting material. This not only impacts further processes such as lithography, but also results in varying silicon island thicknesses across the wafer. Wafers made using the DI process are expensive. Typical costs are well over $100 per 100 mm wafer, before any device fabrication. Finally, if KOH is used to etch the v-grooves, the isolation density is not very large.

Wafer bonding is an alternate approach to producing SOI wafers (Figure 15.15). In this process, two wafers are pressed together at high temperature until they fuse [27]. Alternatively, the wafers can be fused at low temperature by anodic bonding [28],[29]. If the wafers are oxidized before bonding, a layer of oxide remains at the center of the fused wafer. The wafer can be ground back down to thicknesses of 2 to 3 μm using standard grinding and polishing techniques. If thinner layers are required, additional processing can be done to produce submicron semiconducting films on top of the oxide. Devices can be isolated with a simple etch process that produces single crystal islands on top of the insulating oxide. Also, as with DI isolation, wafer bonding suffers from high cost, but can produce high density, well isolated structures.

15.5 Semi-insulating Substrates

There are far fewer papers considering device isolation in GaAs technologies than there are for silicon isolation. Due to the availability of semi-insulating substrates, one can easily achieve high density, planar isolation. The only remaining question is how to separate the conducting islands from each other. The fact that a semi-insulating substrate is used in most GaAs ICs means that the devices are inherently radiation hard. A significant market for GaAs ICs involves defense applications that demand such hardening.

Most of the early semi-insulating GaAs material was made by chromium doping. Both chromium and oxygen have an energy level very close to the center of the GaAs band. If chromium is added in high concentrations (about 10^{17} cm^{-3} in older material, now more commonly in the mid 10^{16} range) to the boule as it is grown, semi-insulating wafers will result. The chromium adds discrete states near the center of the band (Figure 15.16). Consider what happens if a single donor atom is added to a semi-insulating wafer with no other dopants. Normally, the dopant atom will ionize and the electron will go into the conduction band, even though the conduction band of the crystal is a higher energy state. The probability of ionization is nearly 1 at room temperature for most dopants because of the large number of empty states near the conduction band edge as opposed to the single state at the donor atom. For chromium doped GaAs, much lower energy states exist at the chromium sites. The sites will therefore ionize, accepting the donor atom's electron. Since the states are discrete and widely spaced, the electron cannot move. The crystal is therefore semi-insulating. Resistivities of 10^8 Ω-cm have been reported. Not until the dopant concentration approaches that of the chromium will there be significant conduction in the crystal. Conduction will also occur when the injected carrier

Figure 15-15   Wafer bonding method for achieving silicon on insulator *(after Quenzer and Benecke, reprinted with permission, Elsevier Sequoia S.A.).*

Figure 15-16   Deep levels necessary for forming a semi-insulating substrate.

density exceeds the trap concentration. This may occur under illumination, or in GaAs metal semiconductor field effect transistors (MESFETs) in a process known as *sidegating*. The effect is seen as a reduction in the current of a device if an adjacent device is turned on. Sidegating is a problem for high density GaAs ICs.

Many GaAs technologies therefore, begin by producing a thin conducting layer on top of the semi-insulating substrate. Device isolation can be achieved by simply etching through the conducting layer, leaving islands of semiconducting GaAs. This simple method is known as *mesa isolation*. It has the disadvantage that the surface is no longer planar. To avoid this problem, many GaAs technologies use ion implantation. This method implants the field regions of the wafer with hydrogen or another ion. A 10^{13} cm^{-2} dose and 100 keV energy are typical implant parameters. This implant is sufficient to destroy the lattice, producing resistivities as high as 10^7 Ω-cm. The surface is also quite planar. A drawback of proton implantation is that the crystal will be regrown if any further processing is done at temperatures above 350°C. Proton implantation must therefore be done near the end of the process.

Semi-insulating substrates are not produced in silicon. Many of the noble and transition metals have states near the center of the silicon gap, but their diffusivities are very large (10^4 μm^2/hr at 800°C and 10^6 μm^2/hr at 1100°C for gold). As a result, it is very difficult to keep the gold in the substrate from diffusing into the active region during processing. If the gold reaches device junctions, enormous increases in leakage will result since the states that render the wafer semi-insulating are also very efficient recombination centers. In a recent paper, however, semi-insulating silicon has been grown epitaxially using oxygen diluted in argon along with silane. A room temperature resistivity of 10^6 Ω-cm has been achieved [30].

In some applications, chromium doping has been replaced by nonstoichiometric liquid encapsulated Czochralski (see Section 2.7) wafers. Figure 15.17 shows the resistivity of the GaAs produced this way as a function of the arsenic mole fraction in the melt [31]. For compositions close to stoichiometric, the wafers are semi-insulating with bulk resistivities of 10^7 to 10^8 Ω-cm. The semi-insulating property is believed to be due to the presence of mid 10^{16} cm^{-3} deep donor levels called *EL2 sites* [32]. The atomic structure of these sites is not certain, but is believed to be related to arsenic atoms on gallium sites (As$_{Ga}$). Although these defects introduce a large etch pit density in undoped semi-insulating wafers, they do not seem to affect IC yield. The B$_2$O$_3$ plug used in LEC growth also tends to getter silicon impurities that would otherwise dope the GaAs substrate n-type. The EL2 center has the interesting property that when illuminated by visible or near IR light while at low temperature, the defect state is trans-

Figure 15-17   The resistivity of GaAs substrates as a function of the melt stoichiometry *(after Ferry)*.

formed into a higher energy metastable state [33]. This second state has very different properties than the original defect. EL2 centers near the surface of the wafer may also be removed by high temperature hydrogen anneals.

One interesting comparison between GaAs and silicon isolation methods is that GaAs technologies do not generally use low dielectric constant insulators, whereas they are emphasized in most silicon technologies to minimize parasitic capacitance. Many GaAs technologies are designed for microwave applications. Then, it is not the absolute capacitance that is critical but the characteristic impedance of the interconnect. This must be impedance matched to the devices to ensure the maximum power transfer. To achieve this characteristic impedance, the GaAs wafers are often thinned after processing, and a ground plane is deposited on the back of the wafer. Through wafer vias are used to contact the ground plane.

15.6 Schottky Contacts

Later sections of the book will deal with the fabrication of specific transistor structures. To efficiently connect semiconductor devices to the outside world, contacts must be made between the semiconductor and metal lines. Two types of contacts are often used: ohmic contacts and rectifying or Schottky contacts. In ideal ohmic contacts, the current varies linearly with the applied voltage. It is also implicitly assumed that ohmic contacts have low resistance. To transfer as much current as possible from the device to charge the various capacitances of the circuit, the contact resistance must also be a small fraction of the device resistance. Schottky contacts however, should act as perfect diodes. They should have a very low resistance in the forward direction and an infinite resistance in the reverse direction. They should have a well defined and perfectly reproducible ''on voltage.'' The reader should be aware that, while semiconductor technologists often use these descriptions for metal to semiconductor contacts, real contacts are neither perfectly ohmic nor perfectly rectifying. If due care is taken in the fabrication however, good approximations of the ideal can be achieved. This section will deal with the fabrication of Schottky contacts. The next section will discuss process modules for making ohmic contacts.

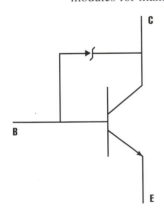

Figure 15-18   Schottky shunted bipolar transistor used for nonsaturating bipolar logic.

Schottky barrier contacts have a variety of applications related to voltage clamping and controlled diode drops. It has the advantage of turning on (and off) more quickly than a p-n diode, and the turn on voltage can be selected during the fabrication sequence. Figure 15.18 shows a typical application of a Schottky diode in a bipolar silicon technology. The base and collector of an NPN bipolar transistor are clamped through the Schottky diode. The turn on voltage of the diode is set to a few tenths of a volt less than the turn on of the base collector diode. If the base is suddenly driven higher than the collector, the base voltage will be limited to the Schottky diode on voltage. This prevents the transistor from being driven into a strong saturation condition from which it would take a long time to recover.

Figure 15.19 shows a band diagram for a Schottky diode on a p-type substrate. ϕ_m is the metal work function, the voltage necessary to remove an electron from the surface of the metal. χ is the electron affinity, the voltage necessary to remove an electron from the conduction band minimum of the semiconductor. ϕ_s is the voltage difference between the Fermi energy of the semiconductor and the vacuum level. When the metal and the semiconductor are brought into contact, charge will flow from one to the other until an electric field builds up that is

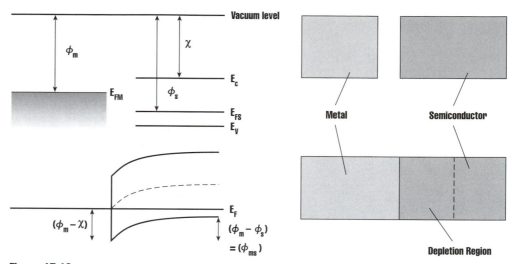

Figure 15-19 Band diagram for an ideal Schottky contact: before contact (right) and after contact (left).

sufficient to prevent further current flow. This occurs when the barrier at the interface reaches ϕ_{ms}. If a Schottky contact is needed, the metal must be chosen to create such a barrier. If for example, this metal were chosen for a Schottky contact to a lightly doped n-type semiconductor, electrons in the metal, which has a higher Fermi energy than the semiconductor, flow readily into the semiconductor. This leads to an electric field that will assist the movement of carriers across the interface. Then a Schottky diode will not be formed. Ideally, if $\phi_m > \phi_s$ and the semiconductor is n-type, or if $\phi_m < \phi_s$ and the semiconductor is p-type, a Schottky diode will result.

The current across a Schottky diode is determined by thermionic emission. That is, it is determined by the fraction of carriers that have sufficient energy to surmount the barrier. It is given by the equation

$$I = I_o \left(\exp \frac{qV}{nkT} - 1 \right) \tag{15.7}$$

where

$$I_o = RT^2 A \, \exp \left[- \frac{\phi_{ms}}{kT} \right] \tag{15.8}$$

In this equation A is the area and n is the ideality factor of the diode. R is Richardson's constant, which has values of 110 and 32 Amp cm^{-2} K^{-2} for n- and p-type silicon, respectively, and 8 and 74 Amp cm^{-2} K^{-2} for n- and p-type gallium arsenide. It is clear from this equation that the current density depends strongly on the barrier height.

Figure 15.20 shows the measured barrier heights for both silicon and gallium arsenide as a function of the metal work function. The experimental points should fall on a line with unity slope.

Figure 15-20 Experimentally measured Schottky barrier heights for silicon and GaAs *(from Sze, reprinted by permission, Wiley).*

The offset would be due to the different electron affinities and (potentially) Fermi levels for the two semiconductors. Instead of a unity slope, a slope of about 0.25 is seen for silicon and about 0.1 is seen for GaAs. From an empirical point of view we need to modify Equation 15.8 to read

$$I_o = RT^2A \exp\left[-\frac{\phi_b}{kT}\right] \tag{15.9}$$

where ϕ_b is the effective barrier height.

To understand this discrepancy, one needs to go back to the basic assumptions in the model. In presenting these equations a perfect interface was implicitly assumed. In most processes that make Schottky diodes, the metal is not deposited under ultrahigh vacuum conditions. Rather, an active region of the semiconductor is first exposed, for example, by etching a hole in an insulating layer. Then the wafer is placed in a vacuum system to deposit the metal layer. During this time the surface of the wafer will oxidize and receive a coating of carbon and other contaminants from the residual pump oil vapor in the chamber and from the air in the room. To improve the process, many sputtering systems can presputter the surface of the wafer before deposition. This however, does not ensure that a perfect surface is achieved. The sputtered species are not volatile; many will redeposit on the surface of the wafer. Furthermore, material will be sputtered wherever the plasma touches a surface. Thus, it would not be surprising to find Fe, Ni, and Co on the surface of the wafer. The plasma also damages the surface of the wafer. Both contamination and damage have an impact on the Schottky diode. Even without these defects, the semiconductor surface itself represents an important ''imperfection'' from the standpoint of the electronic structure. Unless the metal can be epitaxially grown on the surface of the semiconductor, the barrier height and the metal semiconductor work function may be considerably different.

PtSi is commonly used as a Schottky contact for lightly doped n-type silicon. The cross section of a Schottky clamped bipolar transistor is shown in Figure 15.21. The diode is formed by first etching holes down to bare silicon. A dip in dilute HF is normally done just before deposition. A platinum layer is then sputtered onto the surface of the wafer. Typical thicknesses are 300 to 600 Å. It is critically important in obtaining a good surface morphology that the silicon surface is clean before deposition [34], and that little oxygen and

Figure 15-21 PtSi Schottky used to shunt the base and collector of a bipolar transistor.

water vapor are present during the deposition step. Recall that from kinetic theory the flux of residual gasses is given by

$$N = \frac{p}{\sqrt{3kTm}}$$

(15.10)

For nitrogen at 10^{-6} torr, the bombardment rate is 0.17 atoms Å^{-2} $\sec^{-1}$. Thus, if one assumes that all of the atoms that strike the surface of the wafer stick, a monolayer of atoms will deposit within 1 sec. It is therefore impossible to ensure even a reasonably clean surface unless ultrahigh vacuum methods are used. Platinum will reduce a thin oxide however, breaking it up in the anneal. As a result, a standard presputter is sufficient to ensure that a uniform reaction will proceed. It is also critically important to ensure that no oxygen leak exists into the sputtering system. The deposition rate is normally kept low to achieve good process control and uniformity. A small air leak or a significant water partial pressure in the chamber can produce highly oxygen contaminated films. This leads to high resistivity and poor reproducibility in the diode characteristics.

After deposition, the wafer is annealed at about 550°C in a furnace to allow the film to react with the silicon. Toward the end of the process, the furnace is switched to an oxidizing ambient and a thin layer of SiO_2 is grown over the platinum silicide. The platinum itself does not oxidize. After reaction, the wafer is immersed in a dilute solution of aqua regia at 85°C to selectively remove the unreacted platinum. The oxide over the silicide protects it during this etch. Finally, the wafer is dipped in HF to remove the oxide. After silicide formation, the wafer temperature must not be allowed to exceed 800°C, otherwise the morphology of the PtSi layer quickly degrades. Since the silicon surface is consumed during the reaction process, a clean interface results between the semiconductor and the wafer. Typical PtSi to lightly doped n-type Si barrier heights are 0.85 ± 0.05 V.

Schottky diodes are more widely used in GaAs technologies than in silicon. The fundamental reason is that GaAs does not have the exceptional oxide that silicon has. Thus, the basic GaAs field effect device is not the MOSFET but the MESFET. For the MESFET to work well, the gate electrode must make a Schottky contact with the channel. The pinchoff voltage of the FET depends directly on the barrier voltage of the Schottky diode. Furthermore, all of the metals used in GaAs must be fully deposited; they are not reacted with the substrate. Figure 15.22 shows a typical cross section of a GaAs MESFET. WSi and WSi/W are commonly used gate electrodes, although aluminum, chromium, titanium, and molybdenum have also been used. CoAl alloys can also be produced that have good thermal stability and Schottky barrier heights as large as 0.9 V [35]. WSi_x can be deposited using CVD [36], by sputtering from a composite target, or more commonly by cosputtering W and Si [37]. The optimal composition appears to be $WSi_{0.4}$ [38]. Gallium does not diffuse rapidly in WSi unlike many metals, preventing degradation of the interface. WSi is also attractive for this application for the same reason polysilicon is attractive for MOS-FETs in silicon. It is a refractory metal that will tolerate subsequent high temperature processing [39]. This means that the gate can be used as an implant mask for the source and drain, self aligning these diffusions to the gate (see Chapter 17). An 0.8% atomic alloy of Al in W is also an attractive gate material for GaAs MES-FETs. It is thermally stable, has a larger barrier than W [40], and has lower resistance than WSi [41].

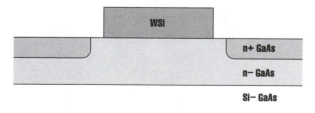

Figure 15-22   Typical GaAs MESFET structure.

As shown in Figure 15.20, the barrier voltage for a Schottky diode in GaAs is about 0.8 V and is nearly independent of the metal. Many compound semiconductors have similar behavior. This represents a serious limitation for MESFET circuits. InP based materials, for example, have several electronic advantages over GaAs, but there is no suitable material that will provide a Schottky barrier larger than 0.4 V [42], virtually precluding their use in MESFETs. Even in GaAs, the gate cannot be forward biased much beyond 1 V or significant gate leakage will result. This leakage increases power consumption and reduces performance. The Schottky barrier also depends strongly on the interface quality. A poor interface results in a larger ideality factor and a reduced barrier height. It has been shown that a sputter cleaning of the interface can greatly reduce the nonuniformity of the pinchoff voltage of GaAs MESFETs [43].

Several methods have been attempted to increase the Schottky barrier height for GaAs. One technique is to use a narrow band of doping called a *charge sheet* or *delta doping* at the surface of the semiconductor [44]. The carriers will be depleted, but the remaining ionic charges are sufficient to shift the barrier height by as much as a few tenths of an eV [45]. Another method is to use a monolayer of alternate materials [46] or heterostructures with a wide bandgap material (AlGaAs) between the GaAs and the WSi. This latter technique produces a barrier height of 1.0 to 1.2 V [47]. Either method requires the use of MBE or MOCVD growth.

15.7 Implanted Ohmic Contacts

Often one is interested in making low resistance ohmic contacts to the semiconductor. The specific contact resistance can be defined as

$$R_c = \left[\frac{\partial J}{\partial V}\right]^{-1}_{V-0} \tag{15.11}$$

From Equation 15.9, for thermionic emission this becomes

$$R_c = \frac{k}{qRT} \exp\left[\frac{\phi_b}{kT}\right] \tag{15.12}$$

Since the barrier height depends logarithmically on the dopant concentration in the substrate, R_c should decrease linearly as the doping concentration increases. This is true up to doping concentrations of 10^{18} to 10^{19} cm^{-3}. For more heavily doped substrates R_c depends strongly on the doping concentration, falling by orders of magnitude. The reason for this behavior is that thermionic emission over the energy barrier is no longer the dominant transport mechanism.

The width of the depletion region at zero applied bias can be calculated using

$$W_D = \sqrt{\frac{2k_s\varepsilon_o\phi_b}{qN_A}} \tag{15.13}$$

Figure 15-23   Two carrier transport mechanisms typically found in metal semiconductor contacts.

For heavily doped substrates this width becomes sufficiently small that carriers can tunnel through the barrier (Figure

15.23) rather than be limited by thermionic emission over the barrier. The specific contact resistance in this region can be approximated by

$$R_c \approx A_o \exp\left[\frac{C_2\phi_b}{\sqrt{N_D}}\right]$$

(15.14)

where

$$C_2 = \frac{4\pi}{h} \sqrt{m_n^* k_{si}\varepsilon_o}$$

(15.15)

h is Plank's constant and m_n^* is the effective mass of the electron in the semiconductor. In contacts to p-type substrates m_n^* is replaced by m_p^*, the effective mass of the hole.

From Equations 15.14 and 15.15, it is apparent that one should make the substrate doping as large as possible to reduce the contact resistance. Alternately, it is possible to make the barrier height as small as possible. This approach has as its limitation the fact that both polarities are contacted in many technologies. One or the other will have a large barrier height. To form ohmic contacts then, very heavily doped junctions must be made in one or both types of semiconductor. Generally, the higher the concentration, the lower the specific contact resistance. Values as low as 10^{-7} Ω-cm^2 are often obtained in semiconductor fabrication.

Most of the work in silicon ohmic contacts has been done with aluminum. Of the few metals that have a lower resistivity than aluminum, each has strong disadvantages from the technologist's viewpoint. Silver and copper are both prone to corrosion and so are reliability concerns. Although interest in copper has reemerged in the last few years, it is still not in common use. Gold, copper, and silver are all difficult to plasma etch since their halides have low vapor pressures. Furthermore, all these materials are extremely efficient recombination sites in silicon (see Section 15.5). For that reason, they are not allowed into many silicon fabrication facilities. Aluminum also has the property that it will react readily with SiO$_2$ to form a thin layer of Al$_2$O$_3$, which promotes adhesion between the silicon dioxide and the aluminum.

To form a low resistance contact with silicon, most aluminum metallization processes include as the last step a low temperature anneal or sinter. Typically done at 450°C, this sinter has several effects. As the aluminum reduces the native oxide, it breaks it up, diffusing the oxygen atoms back into the bulk of the aluminum and allowing fresh Al to diffuse to the metal semiconductor interface. Figure 15.26 shows the diffusion rate of aluminum through aluminum oxide. For a 30 min sinter at

Figure 15-24 Aluminum penetration rate through aluminum oxide as a function of temperature *(after Wolf, reprinted by permission, Lattice Press).*

450°C, the aluminum will penetrate about 10 Å into the aluminum oxide. Just as with Pt deposition, it is important to ensure that the oxide at the contact interface is as thin as possible. Contacts are often dipped in a very dilute HF solution immediately before loading the wafers into the deposition chamber. Once in the chamber, a sputter clean is often carried out immediately before deposition.

While the sinter step is essential to obtaining low resistance contacts, it has undesirable side effects. Figure 15.25 shows a phase diagram for the Al/Si system. As the insert shows, if pure aluminum is heated to 450°C and a source of silicon is present, silicon will begin to dissolve in the aluminum until it reaches a concentration of about 0.5%. If the sample is heated to 525°C, silicon will dissolve to about 1%. The source for this silicon of course, is the wafer. While this does not seem to be a large amount of silicon, the metal line is an enormous sink. Once dissolved in the aluminum, silicon will diffuse rapidly along the grain boundaries, moving silicon away from the contact. The aluminum in turn moves into the holes to fill the voids left by the silicon. Spikes of aluminum can penetrate into the wafer as deep as 1 μm. If the spike penetrates an electrical junction, the result is a short circuit. Figure 15.26 shows a sequence of diagrams that depict junction spiking.

There are several methods that have been used to reducing spiking liability in semiconductors.

Figure 15-25 Phase diagram of Al/Si. Insert shows the low concentration region.

Figure 15-26 Cross-sectional diagrams of Al on Si contact formation process *(after Wolf, reprinted by permission, Lattice Press).*

The simplest is to ensure that deep junctions are employed. This is not always desirable from a device standpoint. The next solution is the use of dilute alloys of aluminum/silicon instead of pure aluminum. If the concentration of silicon in the aluminum exceeds the solid solubility at the anneal temperature, little spiking will occur. For 500°C anneals typical Si alloy concentrations range from 1 to 2%. The use of Al/Si alloys also presents problems. The first is silicon condensation. The fact that the dissolved silicon concentration exceeds the solid solubility at low temperature means that there will be a driving force to condense silicon nodules, which are typically 0.5–1.5 μm in diameter. These nodules form between the aluminum grain boundaries and at the metal/semiconductor interface. Since aluminum is an acceptor in silicon, these nodules are heavily doped p-type. If the contact is between n-type silicon and metal, this p-layer can significantly increase the contact resistance, particularly if the contact is small. Furthermore, the existence of silicon nodules between the grain boundaries of the metal line represents a serious reliability concern. For narrow lines, the nodules approach the full cross-sectional area of the wire. When large currents are passed through the line, significant local heating can occur, eventually leading to a failure of the line.

For very shallow junctions (less than about 0.2 μm), silicon doped aluminum is no longer effective. This can occur due to silicon condensation that lowers the silicon concentration in the metal. It can also be related to electromigration induced contact failures. In this phenomenon, electrons moving under the influence of an imposed electric field, accelerate to high energies and then collide with an aluminum atom. In some cases, the collision transfers enough momentum to the aluminum atom to drive it into the substrate. The effect is known to increase as the junction area decreases. To make reliable ohmic contacts to very shallow junctions, a barrier metallization must be employed. A thin layer of the barrier metal is deposited beneath the aluminum. This metal must have the property that the diffusivity of silicon and aluminum are both low at typical sintering temperatures. It must also be highly conductive and have good adhesion between the semiconductor and the metal. One of the most widely used barrier metals is an alloy of titanium and tungsten. Typically 1000 to 2000 Å of 10 wt% Ti/90 wt% W is first sputter deposited. The aluminum is then deposited without breaking vacuum. Several common improvements to this basic process include the use of PtSi or TiSi$_2$ under the TiW to improve the contact resistance. PtSi is formed as discussed previously. TiSi$_2$ can be formed in a furnace [48] or in a multistep process in a rapid thermal processor [49]. One disadvantage of TiW processes is that the material flakes off the walls of the sputtering system [50]. These flakes can reduce the IC yield. Metal nitrides can also be used as barrier metals. These films can be deposited through reactive sputtering or through rapid thermal annealing of a deposited TiN layer in a nitridizing ambient. In this process, the Ti or TiW is sputtered in the presence of nitrogen. The nitrogen is incorporated in the grain boundaries and further reduces the diffusion rate. It is believed that this reduction in diffusivity is due to a passivation effect that dramatically reduces the grain boundary diffusion coefficient. Contacts made this way are stable up to temperatures of 500°C.

All sputtered barrier metals suffer from a common shortcoming for deep submicron devices. In these devices, the contact holes have large aspect ratios. In such a topology, sputtering will not produce a uniform deposit at the bottom of the contact. The barrier metal will be very thin near the edges. Furthermore, the barrier metal does little to improve the planarity of the contact. As a result, the deposited aluminum will also have poor step coverage going into and out of the contact. One contact method that has received considerable attention in recent years is the use of chemical vapor deposited (CVD) tungsten (Figure 15.27), as discussed in Section 13.8.

Figure 15-27   Tungsten plug structure.

15.8 Alloyed Contacts

Unlike silicon that uses implanted contacts almost exclusively, most ohmic contacts in GaAs are alloyed [51]. To form an ohmic contact to n-type GaAs, the wafers must first be cleaned. A typical process is to rinse the wafer in organic solvents, followed by a DI rinse. Various metallizations have been used for ohmic contacts to n-type GaAs including GeMoW [52] and GeWSi$_2$Au [53]. The most common contact however, is NiAuGe. In this process, the wafer is placed in an evaporator where 1000 to 1500 Å of a eutectic of 88 wt% Au/12 wt% Ge is deposited, followed by 100 to 500 Å of Ni. Layers of a refractory barrier metal and gold may be added to the top of the Ni to reduce the resistance of the metal lines. The ohmic contact is formed by annealing the wafer for 30 min at 450°C in a H$_2$/N$_2$ mixture. Contact resistivities of 10^{-6} Ω-cm^2 have been reported using this technique.

During alloyed contact formation, Au reacts with substrate Ga to form various alloys, leaving behind a large concentration of Ga vacancies. Ge diffuses into the GaAs, occupying the Ga sites and doping the GaAs heavily n-type [54]. The specific contact resistance of ohmic contacts formed in this way depends inversely on the doping concentration in the lightly doped substrate [55] instead of depending on the doping concentration at the contact, which can be as large as 5×10^{19} cm^{-3}. It has been found that the germanium does not penetrate the surface of the contact uniformly. Instead, contact is made in small pockets of hemispherical radius r (Figure 15.28) that have been observed electrically [56] and by using TEM [57]. Then, the specific contact resistance will have two components, one due to the tunneling contact resistance that depends on the Ge doping level in the GaAs (N$_D$) and one due to the spreading resistance of the Ge inclusions. The total specific contact resistance then is

Figure 15-28   Schematic of ohmic contact to n-type GaAs showing the formation of small hemispherical contact regions and subsequent current crowding.

$$R_c \approx A_o \exp\left[\frac{C_2\phi_b}{\sqrt{N_D}}\right] + D^2\frac{\rho}{\pi r} \qquad \textbf{(15.16)}$$

where D is the mean distance between the pockets and ρ is the resistivity of the substrate. This suggests that it is desirable to have a small separation between these inclusions. In practice however, that is difficult to achieve.

Alternatively, the specific contact resistance can be lowered by increasing the dopant concentration near the contact (as opposed to at the contact). The most common method is ion implantation. Implanted alloyed contacts are now commonly used in GaAs technologies. One problem that arises in such technologies is simultaneously maintaining both Schottky and ohmic contacts. Figure 15.29 shows the cross section of a MES-FET with ion implanted contacts. It is important that the implantation is offset from the gate. If it extends under the gate, gate leakage will increase substantially. While a deposited dielectric can be used, followed by an implantation directly into the contact, this approach leaves a high series resistance between the FET and the contact. As will be described in a later section, this seriously reduces the device performance. To achieve this self

Figure 15-29   Cross-sectional schematic of a low resistance MESFET using ion implantation to reduce the contact resistance.

aligned implantation for the contact formation, several schemes have been developed. These will be covered in Chapter 17.

Alloyed ohmic contacts to p-type GaAs has been studied in less detail, in part due to the low hole mobility of GaAs. The same general procedure is used, but the usual metallization employed is Au/Zn [58] with 5 to 15 wt% zinc. The metal structure may be evaporated or sputtered [59]. Since Zn is a Group II material, it will dope the GaAs p-type when it resides on a Ga site. An ohmic contact may be formed in a furnace or in a rapid thermal processor [60]. As with n-type GaAs, specific resistances less than 10^{-6} Ω-cm^2 have been achieved. The thermal stability and the reliability of these contacts are somewhat suspect however, due to the large vapor pressure of Zn.

An alternate approach to forming low resistance ohmic contacts to GaAs is to use indium in the deposited layers, usually close to the GaAs surface. The indium does not dope the GaAs. Rather, it forms $Ga_xIn_{1-x}As$ compounds where x varies smoothly from 1 to 0 through the interface. These compounds have lower barriers and so improved ohmic behavior [61]. This type of contact is also more thermally stable than traditionally formed contacts [62].

15.9 Multilevel Metallization

This section will deal with the problem of making the connection between the contact and the bonding pad. Silicon and GaAs technologies have taken very different approaches in this area. In large part, this is because the tasks for which these technologies have been designed are so different. Most silicon technologies have been designed to achieve high levels of integration. Many GaAs technologies have been optimized for high speed, with only a secondary emphasis on density. These technologies often simply place a single layer of gold on top of the Ni to reduce the interconnect resistance. This layer is deposited directly on top of the wafer after the transistors have been fabricated. As the number of transistors on digital GaAs circuits has increased over the last few years however, GaAs based technologies have had to use multiple layers of interconnect. In doing so, they have sometimes grafted the same metallization approaches that have been used in silicon technologies for many years onto the basic MESFET technology. For that reason, we will approach the metallization process primarily as technology independent.

A few comments about the metrics for the interconnect process module can be made. The most critical for digital circuits is capacitance. Digital switching speed is proportional to the capacitance on each node as

$$\tau \approx \frac{V_{swing}C_{node}}{<I_{drive}>} \tag{15.17}$$

As technologies have improved an increasingly large part of this node capacitance is due to the wire. This can be capacitance between the wire and the substrate or the capacitance between the wires. As the wiring density increases, the wires get narrower. In principle, the wire to substrate capacitance should be proportional to the wire width. In practice, edge effects make the capacitance larger than what is predicted from a simple parallel plate model. This occurs when the linewidth approaches the oxide thickness. Due to fringing capacitance, the reduction of metal linewidth does not reduce the line to substrate capacitance as much as one would expect. Furthermore, the wire to wire capacitance increases as the metal spacing decreases. This line to line capacitance is inversely proportional to the line to line spacing. The result of these considerations is that there is a minimum in capacitance for a particular density. The interconnect resistance is not a major issue in most digital applications, except for extremely long runs, power distribution, and the use of materials other than aluminum. In digital circuits the performance is more often limited by capacitive charging than the RC delay. Planarity, on the other hand, is a major issue, particularly in silicon technology (see Section 15.10). The density in many ICs is limited primarily by the interconnect. As a result, the metal line and space width have often been at or close to the limit of the photolithographic technology. Because of this and because of the amount of wire on a typical IC, interconnect often dominates the yield. As a result, the wafer must be as planar as possible when patterning each layer of interconnect.

There are two ways to accommodate the amount of interconnect required: decreasing the pitch or increasing the number of interconnect layers. Due to crosstalk noise, it is unlikely that metal pitch will decrease much below 2.0 μm [63]. As a result, the number of levels of interconnect on modern IC processing is constantly increasing. The current state of the art for silicon is four layers of metal [64]. There is every indication that this will continue to increase to six or more levels.

Example 15.1

A 0.5 μm digital CMOS technology has 5 μm wide transistors. The minimum wire width is 1 μm and the metallization layer consists of 1 μm thick aluminum. Assume that μ_n is 400 cm²/V-sec, t_{OX} is 100 Å, V_{DD} is 3.3 V, and the threshold voltage is 0.6 V. Finally, assume that the maximum voltage drop that can be tolerated is 0.1 V when a 1 μm² cross section aluminum wire is carrying the maximum current that can be supplied by the NMOS transistor. How long a wire can be allowed? Use a simple square law long channel model to predict the MOS current drive.

The oxide capacitance per unit area is given by

$$C_{OX} = \frac{\varepsilon_{SiO_2}}{t_{OX}} = 3.5 \times 10^{-7} \ F/cm^2$$

and the maximum current supplied by the device is

$$I_{DS} \approx \frac{1}{2} \frac{W}{L} \mu C_{OX} [V_{gs} - V_t]^2 = \frac{1}{2} \frac{5 \ \mu m}{0.5 \ \mu m} 3.5 \times 10^{-7} \ F/cm^2 [V_{gs} - V_T]^2 \approx 5 \ mA$$

and the maximum allowable wire resistance is 0.1 V/5 mA, or 20 Ω. Then, the length of the wire must be

$$L \leq \frac{R \times \text{Area}}{\rho} = \frac{20 \ \Omega \cdot 10^{-8} \ cm^2}{2.7 \times 10^{-8} \ \Omega\text{-}cm} = 0.074 \ cm$$

or 740 μm. This is a long distance compared to most device spacing. When driving signals between widely spaced logic blocks however, minimum feature sized lines would not be appropriate.

In microwave circuits, the important parameter is the wire impedance. To achieve efficient power transfer, the characteristic impedance of the wire must match the input impedance of the device. Two primary methods have been employed to achieve a controlled impedance and a stable ground connection. One common approach is called a *microstrip line*. To do this, the wafer is first thinned

Table 15.1 Properties of commonly used interconnect materials in both Si and GaAs technologies

Material	Bulk Resistivity ($\mu\Omega$-cm)	Melting Point (°C)
Au	2.2	1064
Al	2.7	660
Cu	1.7	1083
W	5.7	3410
PtSi	30	1229
TaSi$_2$	40	2200
TiSi$_2$	15	1540
WSi$_2$	40	2165

to achieve the desired characteristic impedance for the design wire width. Next, a layer of metal is deposited on the back side of the wafer. Holes are etched through the wafer to make contact to this solid ground plane. These through holes are patterned and etched from the back side of the wafer using an infrared aligner. A second method for fabricating controlled impedance lines uses a coplanar waveguide. Here two ground wires are run on either side of the signal wire. This does not require any back side processing, but it decreases the density of the IC and does not provide as stable a ground plane.

Most millimeter microwave ICs (MMICs) often have only a small number of transistors. Many discrete microwave components are also manufactured. In these technologies, wire density is not typically an issue. One layer of metallization is used to interconnect the devices. The Schottky gate metal is also often used as a first layer for parallel plate capacitors and where necessary the Ni/AuGe serves as the second layer of metallization.

In modern silicon technologies a distinction is drawn between global or ''true'' interconnect and local interconnect. While the resistivity of aluminum is low enough that long runs do not degrade performance in most circumstances, the same cannot be said of polysilicon. Its resistivity is typically 10^{-4} Ω-cm. Silicides, which can be run directly on top of the polysilicon to shunt the poly resistance, still have resistivities that are much larger than aluminum (Table 15.1). A rough estimate of the delay can be made using a lumped capacitance model. In this model, the line is thought of as one side of a parallel plate capacitor. Then

$$C = \frac{LWk_{OX}\varepsilon_o}{t_{OX}} \tag{15.18}$$

and

$$R = \rho_{met}\frac{L}{Wt_{met}} \tag{15.19}$$

Therefore,

$$RC \approx \rho_{met}\frac{\varepsilon_{OX}L^2}{t_{met}t_{OX}} \tag{15.20}$$

Figure 15-30   Simple lumped parameter model for the time constant of various interconnect materials.

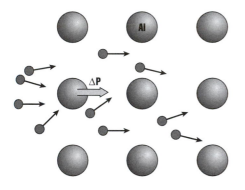

Figure 15-31   Momentum transfer between electrons and matrix atoms in an interconnect material.

Figure 15.30 shows the lumped delay as a function of length for several metals running over 1 μm of oxide. The important feature to understand here is that the delay depends on the square of the length of the interconnect. Designers can therefore use moderate resistance materials to join adjacent transistors or even adjacent cells if they are close together. These materials are often called *local interconnect*. These local interconnects have several advantages over the aluminum that is used on the upper layers of interconnect. They can tolerate high temperature processing. Local polysilicon interconnect is often available for no additional processing cost since it may already be used for the gates of MOSFETs or the emitter of a bipolar structure. Similarly modern devices often use silicide as will be discussed in later chapters. These materials can also be used for local interconnect with little additional process complexity.

Aluminum based alloys have been the metallization of choice for silicon IC technologies. One of its limitations, junction spiking, has already been described. A second limitation for Al and AlSi alloys is electromigration. Electromigration is the movement of conducting atoms as a result of momentum transfer from current carrying electrons (Figure 15.31). This movement gives rise to a net flux of metal atoms. If a nonzero divergence in the atomic flux exists anywhere along the line, metal atoms will be depleted or accumulated. Upon depletion, open circuits form. Hillocks are formed by accumulation. If these hillocks get large enough, adjacent lines or even overlying lines can be shorted together. For aluminum metallization systems in particular, this represents a serious reliability problem. A circuit that works well initially will wear out and eventually fail in the field.

A phenomonological description of the electromigration process is Black's equation [65]:

$$MTF = AJ^{-n} \exp\left[\frac{-E_A}{kT}\right] \qquad \textbf{(15.21)}$$

where *MTF* is the median time to failure, J is the current density, n is a fitting parameter, typically about 2 [66], E_A is the activation energy, and A is a constant. E_A depends on the diffusivity of the metal atom. At high temperatures (above 350°C), the activation energy for aluminum closely matches the self diffusivity of aluminum. At lower temperatures where ICs operate however, the activation energy for electromigration is smaller. At these temperatures grain boundary diffusion along the facets of the aluminum crystals dominates.

A common metal atom flux divergence mechanism is shown in Figure 15.32. At some point in the line, three crystals of aluminum come together. One crystal face leads into this intersection and two lead out. As a result, atoms experiencing grain boundary diffusion will tend to diffuse more easily

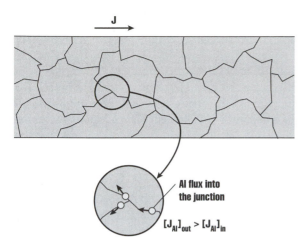

Figure 15-32   Typical point of flux divergence in an interconnect material.

out of this point in the line than into it. Over time a void will form. A second source of flux divergence is the presence of a temperature gradient. Because of the temperature dependence of the diffusivity, there will be a net movement of material away from the hotter portion of the wire. This will ultimately produce a void. Electromigration failures of this sort are often seen near bonding pads or large contacts where a large thermal sink exists. Electromigration in contacts is further aggravated by metal thinning as it goes over the topology.

Accelerated testing for electromigration can be done by holding the wafer at high temperatures and/or high current densities, measuring the MTF, and extrapolating back to use conditions. The result is a rule for the maximum dc current that can be allowed in a line to ensure a certain lifetime. A rather conservative number often used for Al/Si is 10^5 A/cm^2 for a 20 yr life. For pulsed dc stress, Black's equation can be modified by multiplying the current density by the duty factor [67]. Various models have been used to explain ac electromigration results. In the commonly used average current model, the same formula that is used for pulsed dc testing [68] is found appropriate for both unidirectional and bidirectional stressing.

An interesting question in electromigration is what happens when the linewidth approaches the grain size. Typical grains in aluminum metallization after the final sinter are of order 1 μm, so many advanced technologies have linewidths of order the grain size. Early reports suggested a sharp rise in lifetime due to this so called bamboo effect [69]. On closer examination however, it was found that while the median time to failure increased, there was still a sizable fraction of wires that failed in much shorter times. These failures were attributed to the statistical distribution of grain sizes in the wire and ultimately to the existence of an intersection between three crystals.

To increase its current carrying capability, a different metallization system must be used. One commonly employed method is the addition of 1 to 4% copper to the aluminum. The copper atoms reduce grain boundary diffusion effects in aluminum and so greatly increase the activation energy for electromigration. Current densities as high as 10^6 A/cm^2 are considered reliable with many Al/Cu metallurgies. A difficulty with the addition of Cu is plasma etching. Normally, aluminum interconnect is etched in a chlorine plasma. Silicon also etches well in these systems due to the high vapor pressure of the various chlorosilanes. Copper chlorides on the other hand, do not tend to form gasses readily at temperatures below 175°C. Since most photoresists will not tolerate such high temperatures, the wafers cannot be heated sufficiently during the etch to remove all of the copper. As a result, most of the copper in the metallization is left behind after a plasma etch. Chlorinated plasma etch processes tend to corrode the remaining metal due to the presence of AlCl$_2$ etch residue and water vapor in the air. When an Al/Cu alloy is used, AlCu$_2$ is formed in the metal. This compound has a galvanic response with Al that drives the corrosion process. As a result, care must be taken to ensure that all of the chlorine has been removed from AlCu wafers that have been plasma etched. This is sometimes done with an inert plasma after the chlorine plasma is completed.

Another reliability concern for fine pitch metallization systems is stress induced voiding [70]. This effect can open narrow slits [71] or wider features called wedges [72]. The effect arises due to the difference in the thermal expansion coefficients of the metal and the encapsulating dielectric layers and is aggravated at very small metal linewidths. As with electromigration, stress induced voiding is significantly reduced by the introduction of small amounts of Cu in Al [73] or by using refractory metals instead of aluminum.

Several other metal systems have been proposed for use in silicon ICs. Layers of Al and a refractory metal such as Ti or TiN [74] have been suggested to reduce electromigration. The processing is more difficult, however. The most promising material at this point is tungsten. As mentioned in a previous chapter, the increasing contact aspect ratio is making the attainment of an adequate sputtering process extremely difficult. One possible solution is the use of contact plugs made from CVD tungsten [75],[76]. As discussed in Section 13.8, CVD tungsten is typically deposited in a vacuum system from WF_6 or WF_6/SiH_4 mixtures. It is possible to deposit the tungsten either selectively [77], or nonselectively and etch back the film as with trench isolation (Figure 15.27). It is found that tungsten deposition consumes some silicon; however, this can be reduced by the addition of SiH_4 during deposition. The process can be incorporated into a single multichamber system [78]. In this application, an underlayer of TiW or TiN is first sputtered to seal the source/drain surfaces from attack during the tungsten deposition [79] and to serve as an adhesion promoter between the metallization and the oxide [80]. A second possibility is to keep the tungsten layer and to use it as a layer of interconnect. This is very attractive, particularly for local interconnect. Since tungsten has a very low self diffusivity at operating temperatures, electromigration is not a concern. It appears likely that the use of tungsten as the first layer of metallization will become increasingly more common. Selective nickel silicidation of polysilicon can also be used to form these contact plug structures [81].

There has also been a long term interest in the use of copper as an IC interconnect material. Even in ICs whose performance is not resistance limited, the use of Cu would allow thinner conductors at the same resistance. This produces lower line to line capacitance and less crosstalk induced noise. Copper can be sputtered, plated, or deposited through MOCVD using precursors such as bishexafluoro acetylacetonate copper $Cu(HFA)_2$ [82]. Copper has a lower electrical resistance and is strongly electromigration and stress migration resistant [83]. The concerns for copper use that have already been mentioned include difficulty of etch and long term reliability concerns due to corrosion. To address the former concern, it has been shown that copper may be deposited [84] or plated [85] selectively by first patterning a thin conducting base layer. To prevent corrosion the copper lines can be covered on top and bottom before patterning and on the line sidewalls after patterning using Mo, TiN, Ta, or another refractory material [86]. These higher resistance cladding layers however, offset the resistance improvement that can be achieved with copper and add considerable process complexity.

The standard dielectric for ICs has been silicon dioxide. SiO_2 can be grown thermally (on silicon substrates), sputtered, solidified from a spun on solution, or deposited by thermal, plasma enhanced, or photo enhanced chemical vapor deposition. Silicon dioxide has a low dielectric constant (3.9) and does not have strong absorption bands in the microwave region. It is a good insulator (10^{15} Ω-cm bulk) with a large breakdown field (>10 MV/cm). Finally, many films adhere well to it, and it is stable at high temperatures.

Most SiO_2 layers for interconnect isolation are CVD films (see Chapter 13). In a typical process, the wafer is heated to about 400°C and silane or tetraethylorthosilicate (TEOS) and N_2O or O_2 are injected above the surface of the wafer. The addition of phosphine and diborane produces the doped glasses PSG and BPSG [87]. These oxides have the property that the viscosity drops sharply at temperatures ranging from 700 to 900°C. Under these conditions, the as deposited glasses reflow to reduce the severity of steps on the surface of the wafer. These reflow processes were widely used in 2 and 3 μm silicon IC technologies. The arrival of planarization processes and the need for reduced thermal budgets has made them less popular.

Silicon nitride can also be used as an insulator. A disadvantage in most interconnect applications is that nitride has a higher dielectric constant than oxide. This translates into higher capacitance for a given dielectric thickness. The diffusion barrier properties of nitride are attractive however, for implant annealing of GaAs (to be discussed in a later chapter) and for the passivation or scratch protection layer.

Spin on dielectrics such as spin on glasses and various polyimides [88],[89] have been studied as possible replacements for CVD SiO_2. They have the advantage of smoothing out sharp steps. Polyimide is applied in much the same way as photoresist. The wafers are first given a dehydration bake. Polyimide is applied and spun to the desired thickness and a low temperature bake is done to drive off most of the solvent. Some polyimides are soluble in common positive develop solutions. Etching these films can be done by simply extending the resist develop time. Finally, a 350 to 500°C bake is done to imidize the polyamic acid to form the solid polyimide. It also has the effect of driving off the remaining solvents and water in the polyimide. The final bake cycle is important in determining the electrical and long term mechanical properties of the insulator.

Certain forms of polyimide have been shown to have low dielectric constants. Some with a low frequency ε_r as low as 2.0 have been developed; however, the dielectric constant increases sharply near the water absorption bands at a few gigahertz. They also are hygroscopic and tend to outgas water over time. When the subsequent metallization is carried out, this outgassing may adversely affect the deposition process. In depositing selective W for example, from WF_6, the WF_6 attacks the polyimide, which consequently absorbs fluorine. The deposit does not fill the contacts well under these conditions unless the polyimide is coated with a thin inorganic layer [90]. Finally polyimide films are not very etch resistant. As a result, when the metal layer is etched some dielectric is also consumed. This aggravates the step problem for subsequent layers. A thin layer of PECVD SiO_2 or some other ''hard oxide'' deposited on top of the polyimide will also solve this etch stop problem. When polyimide is to be used as the lowest level in an MOS process, a thin layer of nitride may also be used as a sodium diffusion barrier [91].

Spin on glasses are siloxanes or silicates mixed in a solvent. They are also applied like a photoresist, then baked at a low temperature (200°C) to drive off the solvents. The wafer is then placed in a controlled ambient oven and cured at high temperature (400°C) to polymerize the material and to form the ''glass.'' To achieve global planarization, a very thick spin on glass layer must be used. During curing, these films tend to crack due to out diffusion of water formed during the polymerization. New materials such as methyl-phenyl silsesquioxane [92] and new processes such as multiple applications with multiple bakes [93] are claimed to achieve a more global planarization process.

Metal

Oxide

Metal

Oxide

Substrate

Figure 15-33   Nonplanarized plug structure consisting of a coincident contact and via.

15.10 Planarization

Planarized interconnect process modules have become very popular as digital technologies have driven into the submicron, due to the convergence of several factors. The most obvious one is the photolithography. As the imaging lenses have tended to higher numerical apertures, the depth of focus has decreased. Furthermore, nonplanarized interconnect typically has not only the highest steps in the process, but these steps are often nearly vertical. One way of coping with the problem was to impose large design rules for the minimum pitch (linewidth plus line space) that resulted in increasing

Figure 15-34 Etch back planarization process.

the number of required interconnect layers. If the lowest level of interconnect had a pitch of x, the second level might have a pitch of $1.5x$, the third level might have a pitch of $3x$, and so on. If finer pitch can be achieved with planarized interconnect modules, however, their additional process complexity may be more than offset by their ability to reduce the number of interconnect layers that are required.

At the same time, the step height often cannot be reduced. Since the linewidth was shrinking, the resistance was already increasing. To also shrink the metal thickness would invite further problems with step coverage and electromigration in aluminum films, and it would mean a further increase in the resistance of silicides and tungsten. The result is higher aspect ratio structures that are difficult to cover. Not only must the metal have sufficient step coverage into and out of contacts, it must also have good step coverage as it goes over metal lines on lower levels. There can be particularly severe topology as the second level metal exits a via down to the first layer of metal. In an extreme example called a *plug* (Figure 15.33), the via may be coincident with a contact from the first layer metal down to the substrate.

An early, popular planarization approach was the sacrificial oxide etchback method [94]. As shown in Figure 15.34, a thick oxide is first deposited over a patterned metal layer. Next, a layer of photoresist is spun onto the wafer. The wafer is then put into a plasma etch system. The etch ambient is a mixture of O_2 and CF_4 or another fluorinated species. The mixture is set to provide nearly identical etch rates of resist and oxide. Since the resist is thinner over the first metal lines, the oxide will be exposed here first. The etch is done until most of the resist has been removed. A second layer of oxide may then be deposited. Another etchback cycle can also be run if required.

Since these processes follow the resist profile, they are only locally planarizing. That is, they smooth out steps but do not produce a truly planar surface. For feature sizes greater than about 50 μm, the planarizing effect is lost [95], although temperature baking of the resist tends to improve this somewhat. The thickness of the oxide will also vary across the device. This means that subsequent contact etch process must have a high selectivity and must not significantly etch laterally. This second criterion can be difficult to achieve if the etch process must also slope the sidewalls of the oxide.

A method that has been recently introduced to produce more planar interconnect is the oxide trench and refill process [96]. Here, lines are reactive ion etched into the deposited oxide before metal deposition. The etch does not extend all the way through the oxide. With a separate mask, contact holes are opened to the lower level and CVD tungsten is deposited. Finally, the tungsten is etched back. Alternatively, a thin barrier metal can be deposited into the bottom of the trench and contacts, and the tungsten can be selectively deposited [97]. Either way, the result is an oxide trench that has been filled with tungsten. Problems with the approach include the resistance of the tungsten and for the first approach, the fact that the final metal lines may only be a fixed width. Wider lines and pads could not be fabricated with this method.

One way to achieve global planarization is with electromechanical polishing. In these processes, a thick dielectric, commonly a spin on glass, is first applied. The wafer is then mechanically abraded in a slurry containing a silica abrasive and an etching agent [98], such as a dilute HF. The process has successfully been used to fabricate VLSI ICs [99]. Equipment to do electrochemical polishing in a manufacturing environment has now become available. Process concerns include contamination from the etch slurry, deep scratches that will be difficult to cover, residual embedded particles, and the uniformity of the etchback all the way to the edge of the wafer. As of this writing, it appears that this process will become widely accepted for many submicron IC technologies.

15.11 Summary

In this chapter, the process modules of device isolation, contact formation, and interconnection were presented. The simplest isolation techniques involve junction isolation. Various LOCOS based methods have been widely used, but suffer from lateral encroachment and incomplete isolation at small junction separations. Trench based methods have become popular for submicron technologies. For GaAs technologies, nearly all device isolation is accomplished with semi-insulating substrates. Conducting islands can be created via proton implantation or mesa etching.

Contacts are divided into rectifying and ohmic. For rectifying contacts, the barrier height is a sensitive function of the nature of the metal/semiconductor interface. For ohmic contacts, achieving a low contact resistivity requires a large doping concentration at the metal/semiconductor interface. Most silicon technologies achieve heavy doping by implantation. Self aligned silicides (salicides) have been developed to reduce the series resistance of shallow junctions in silicon. Many GaAs technologies use alloyed contacts, sometimes with an implantation, to achieve acceptably low specific contact resistivities.

High performance interconnect requires the use of low resistivity metal on top of low capacitance dielectrics. Aluminum alloys are the most widely used for silicon based technologies. Gold is generally used for GaAs technologies. CVD SiO_2 is the most commonly used dielectric, although lower permitivity films such as polyimide are promising in some applications.

Problems

1. In some microwave applications, the collector of a bipolar transistor is heavily doped to allow the device to be run at large dc-bias currents. Assume that for the simple IC shown in Figure 15.1, the n-type collector concentration is 2×10^{17} cm^{-3}. If the maximum reverse bias ever applied to the base collector junction is 5 V, what is the minimum base to base spacing required to isolate the transistors?
2. If the field oxide is 5000 Å thick, what will the threshold voltage of the parasitic device be? Assume no interface states or metal semiconductor workfunction and a substrate concentration of 2×10^{15} cm^{-3}.

3. Consider the deep trenches shown in Figure 15.13. If the substrate is uniformly doped at 1×10^{16} cm^{-3} and the trenches are 5 μm deep, how large a reverse bias is required to deplete the trench all the way to the bottom?

4. Calculate the specific contact resistance at room temperature for a contact to 10^{17} cm^{-3} n-type GaAs if the metallization has a barrier height of 0.8 eV. (Hint: Differentiate Equation 15.9 and evaluate at $V = 0$.)

5. A triple diffused bipolar technology is fabricated using 10^{14} p-type substrates. The maximum bias that either collector would ever see is 10 V. Assume simple step junctions and calculate the required separation between devices. (Calculate the distance between the collectors after fabrication. The design of the collectors must be further apart than this to allow lateral diffusion during the collector drive.) To improve matters, a p+ guard ring will be added to the process. The guard ring must be at least 3 μm deep to be effective. Assume that the guard ring pattern is 2 μm wide, that the final guard ring diffusion must never be allowed to contact the collector, and the alignment tolerance between the collector and guard ring is 1 μm. What is the new minimum separation between the collector diffusions?

6. Memory makers found that replacing long word lines made from polysilicon with a salicide could decrease the access time of the device. Assuming that the line is 1 cm long, the oxide is 1 μm thick, the poly and silicide are each 0.5 μm thick, and the resistivity of the two films are 10^{-3} and 10^{-4} Ω-cm, respectively. Use the simple lumped RC model to determine the delay associated with each of the two lines. (The dielectric constant of SiO$_2$ is 3.9*8.84E − 14 F/cm.)

7. An ohmic contact is to be made to n-type GaAs using an alloyed Ni/Au/Ge contact on an ion implanted channel layer. The resistivity of the conducting GaAs layer is 0.01 Ω-cm. In the region of the alloyed contact, the surface concentration is 1×10^{19} cm^{-3}. The mass of an electron in GaAs is 0.067*m_o. The barrier height of the metal is 0.75 V, and the constant ''A'' for GaAs in the exponential expression for the ohmic contact resistance is 10^{-8} Ω-cm^2. The contact is found to form pits at the metal/semiconductor interface. These pits have radii of 300 Å with a pit density of 10^9 cm^{-2}. (a) What is the specific contact resistance in ohm-cm^2? (b) What would the specific contact resistance be if the diffused impurity concentration under the contact is raised to 3×10^{19} cm^{-3}? (10 points)

8. An advanced metallization process is proposed for high density silicon based ICs. This process will use several new materials. Identify one advantage and one disadvantage for each new material: (a) CVD tungsten, (b) evaporated copper, and (c) spin on polyimide.

References

1. For additional discussion on silicon process integration the reader is referred to S. Wolf, *Silicon Processing for the VLSI Era—Volume 2, Process Integration*, Lattice Press, Sunset Beach, Calif., 1990.

2. U.S. Patent 3,029,366, K. Lehovec.

3. E. Kooi and J. A. Appels, in *Semiconductor Silicon 1973*, H. R. Huff and R. Burgess, eds., *The Electrochem. Symp. Ser.*, Princeton, NJ, 1973.

4. A. Bogh and A. K. Gaind, ''Influence of Film Stress and Thermal Oxidation on the Generation of Dislocations in Silicon,'' *Appl. Phys. Lett.* **33**:895 (1978).

5. E. Kooi, J. G. van Lierop, and J. A. Appels, ''Formation of Silicon Nitride at an Si/SiO$_2$

Interface During the Local Oxidation of Silicon in NH_3 Gas," *J. Electrochem. Soc.* **123**:1117 (1976).

6. U.S. Patent 4,541,167, R. H. Havemann and G. P. Pollack (1986).

7. Y. Han and B. Ma, "Poly Buffered Layer for Scaled MOS," *VLSI Sci., Technol.*, Electrochem. Soc., 1984, p. 334.

8. M. Ghezzo, E. Kaminski, Y. Nissan-Cohen, P. Frank, and R. Saia, "LOPOS: Advanced Device Isolation for a 0.8 μm CMOS/Bulk Process Technology," *J. Electrochem. Soc.* **136**:1992 (1989).

9. R. L. Guldi, B. McKee, G. M. Damminga, C. Y. Young, and M. A. Beals, "Characterization of Poly-Buffered LOCOS in a Manufacturing Environment," *J. Electrochem. Soc.* **136**:3815 (1989).

10. T.-H. Lin, N.-S. Tsia, and C.-S. Yoo, "Twin White Ribbon Effect and Pit Formation Mechanism in PBLOCOS," *J. Electrochem. Soc.* **138**:2415 (1991).

11. P. Molle, S. Deleonibus, and F. Martin, "Sealed Interface Local Oxidation by Rapid Thermal Nitridization," *J. Electrochem. Soc.* **138**:3732 (1991).

12. J. Hui, et al., "Sealed-Interface Local Oxidation Technology," *IEEE Trans. Electron Dev.* **ED-29**:554 (1982).

13. K. Y. Chiu, J. L. Moll, K. M. Cham, J. Lin, C. Lage, S. Angelos, and R. L. Tillman, "The Sloped Wall SWAMI—A Defect Free Zero Bird's Beak Local Oxidation Process for Scaled VLSI Technology," *IEEE Trans. Electron Dev.* **ED-30**:1506 (1983).

14. J. W. Lutze and J. P. Krusius, "Electrical Limitations of Advanced LOCOS Isolation for Deep Submicrometer CMOS," *IEEE Trans. Electron Dev.* **38**:242 (1991).

15. M. Mikoshiba, T. Homma, and K. Hamano, "A New Trench Isolation Technology as a Replacement of LOCOS," *IEDM Tech. Dig.*, p. 578 (1984).

16. T. Iizuka, K. Y. Chiu, and J. L. Moll, "Double Threshold MOSFETs in Bird's-Beak Free Isolation," *IEDM Tech. Dig.*, p. 380 (1981).

17. T. V. Rajeevakumar, T. Lii, Z. A. Wienberg, G. B. Bronner, P. MacFarland, P. Coane, K. Kwietniak, A. Megdanis, K. J. Stein, and S. Cohen, "Trench Storage Capacitors for High Density DRAMs," *IEDM Tech. Dig.*, p. 835 (1991).

18. K. Shibahara, Y. Fujimoto, M. Hamada, S. Iwao, K. Tokashiki, and T. Kunio, "Trench Isolation with ∇ Shaped Buried Oxide for 256Mega-bit DRAMs," *IEDM Tech. Dig.*, p. 275 (1992).

19. C. W. Teng, C. Slawinski, and W. R. Hunter, "Defect Generation in Trench Isolation," *IEDM Tech. Dig.*, p. 586 (1984).

20. R. Kakoshke, R. E. Kaim, P. F. H. M. van der Meulen, and J. F. M. Westendorp, "Trench Sidewall Implantation with a Scanned Ion Beam," *IEEE Trans. Electron Dev.* **37**:1052 (1990).

21. W. Zagożdżon-Wosik, J. C. Wolfe, and C. W. Teng, "Doping of Trench Capacitors by Rapid Thermal Diffusion," *IEEE Electron Dev. Lett.* **12**:264 (1991).

22. F. S. Becker, H. Treichel, and S. Röhl, "Low Pressure Deposition of Doped SiO_2 by Pyrolysis of Tetraethylorthosilicate (TEOS)," *J. Electrochem. Soc.* **136**:3033 (1989).

23. T. Ozaki, A. Nitayama, T. Hamamoto, K. Sunouchi, and F. Horiguchi, "Analysis of Band-to-Band Tunneling Leakage Current in Trench-Capacitor DRAM Cells," *IEEE Electron Dev. Lett.* **12**:95 (1991).

24. K. Sunouchi, F. Horiguchi, A. Nitayama, K. Hieda, H. Takato, N. Okabe, T. Yamada, T. Ozaki, K. Hashimoto, S. Takedai, A. Yagishita, A. Kumagae, Y. Takahashi, and F. Mas-

uoka, ''Process Integration for 64M DRAM using an Asymmetrical Stacked Trench Capacitor (AST) Cell,'' *IEDM Tech. Dig.*, p. 647 (1990).

25. Y. Sugawara, T. Kamei, Y. Hosokawa, and M. Okamura, ''Practical Size Limits of High Voltage IC's,'' *IEDM Tech. Dig.*, p. 412 (1983).

26. T. Aso, H. Mizuide, T. Usui, K. Akahane, N. Ishikawa, I. Hide, and Y. Maeda, ''An Application of MSSD to Dielectrically Isolated Intelligent Power ICs,'' *1991 IEEE Int. Symp. Power Semicond. Dev., ICs*, M. A. Shibib and B. J. Baliga, eds. IEEE, Piscataway, NJ, 1991.

27. J. B. Lasky, S. R. Stiffler, F. R. White, and J. R. Abernathey, ''Silicon on Insulator (SOI) by Bonding and Etch Back,'' *1985 IEDM Tech. Dig.*, p. 684.

28. H. J. Quenzer and W. Benecke, ''Low-Temperature Silicon Wafer Bonding,'' *Sensors, Actuators* **A32**:340 (1990).

29. J. G. Fleming, E. Roherty-Osmun, and N. A. Godshall, ''Low Temperature, High Strength, Wafer-to-Wafer Bonding,'' *J. Electrochem. Soc.* **139**:3300 (1992).

30. P. V. Schwartz, C. W. Liu, and J. C. Sturm, ''Semi-Insulating Crystalline Silicon Formed by Oxygen Doping During Low-Temperature Chemical Vapor Deposition,'' *Appl. Phys. Lett.* **62**:1102 (1993).

31. D. K. Ferry, *Gallium Arsenide Technology*, Harold W. Sams, Indianapolis, 1985.

32. G. M. Marin, J. P. Farges, G. Jacob, J. P. Hallais, and G. Poublaud, ''Compensation Mechanisms in GaAs,'' *J. Appl. Phys.* **51**:2840 (1980).

33. ''Recent Developments in the Study of the EL2 Defect in GaAs,'' H. J. von Bardeleben and B. Pajot, eds., *Rev. Phys. Appl.* **23**:727 (1988).

34. C. A. Crider et al., ''Platinum Silicide Formation Under Ultra High Vacuum and Controlled Impurity Ambients,'' *J. Appl. Phys.* **52**:2860 (1981).

35. H. C. Cheng, C. Y. Wu, and J. J. Shy, ''Excellent Thermal Stability of Cobalt-Aluminum Alloy Schottky Contacts on GaAs Substrates,'' *Solid-State Electron.* **33**:863 (1990).

36. T. Hara, A. Suga, and R. Ichikawa, ''Properties of CVD WSi$_x$ Films and CVD WSi$_x$/GaAs Schottky Barrier,'' *Phys. Stat. Sol.* **A 113**:459 (1989).

37. J. Willer, M. Heinzle, N. Arnold, and D. Ristow, ''WSi$_x$ Refractory Gate Metal Process for GaAs MESFET's,'' *Solid-State Electron.* **33**:571 (1990).

38. M. Kanamori, K. Nagai, and T. Nozaki, ''Low Resistivity W/WSi$_x$ Bilayer Gates for Self-Aligned GaAs Metal-Semiconductor Field Effect Transistor Large Scale Integrated Circuits,'' *J. Vac. Sci. Technol.* **B6**:1317 (1987).

39. J. Willer, M. Heinzle, L. Schleicher, and D. Ristow, ''Characterization of WSi$_x$ Gate Metal Process for GaAs MESFET's,'' *Appl. Surf. Sci.* **38**:548 (1989).

40. H. Nakamura, Y. Sano, T. Nonaka, T. Ishida, K. Kaminishi, ''A Self-Aligned GaAs MESFET with W-Al Gate,'' *IEEE GaAs IC Symp. Tech. Dig.*, IEEE, Piscataway, NJ, 1983, p. 134.

41. T. Ohnishi, N. Yokoyama, H. Onodera, S. Suzuki, and A. Shibatomi, ''Characterization of WSi$_x$/GaAs Schottky Contacts,'' *Appl. Phys. Lett.* **43**:600 (1983).

42. C. A. Mead and W. G. Spitzer, *Phys. Rev.* **134**:A173 (1964).

43. Y. Sekino, T. Kimura, K. Inokuchi, Y. Sano, and M. Sakuta, ''Effect of Bias Sputtering on W and W-Al Schottky Contact Formation and Its Application to GaAs MESFETs,'' *Jap. J. Appl. Phys.* **27**:L2183 (1988).

44. T.-H. Shen, M. Elliott, R. H. Williams, D. A. Woolf, D. I. Westwood, and A. C. Ford, ''Control of Semiconductor Interface Barriers by Delta-Doping,'' *Appl. Surf. Sci.* **56–58 pt. B**:749 (1992).

45. R. A. Kiehl, S. L. Wright, J. H. Margelin, and D. J. Frank, *Proc. IEEE Cornell Conf.*, 1987, p. 28.

46. S. T. Ali and D. N. Bose, "Improved Au/n-GaAs Schottky Barriers Due to Ru Surface Modification," *Mater. Lett.* **12**:388 (1991).

47. N. C. Cirillo, J. K. Abrokwah, and M. S. Shur, "Self-Aligned Modulation-Doped (Al,Ga)As/GaAs Field-Effect Transistors," *IEEE Electron Dev. Lett.* **EDL-5**:129 (1984).

48. C. Y. Ting, S. S. Iyer, C. M. Osborn, G. J. Hu, and A. M. Schweigart, *Electrochem. Soc. Ext. Abs.* **82-2**:254 (1982).

49. C. Mallardeau, Y. Morand, and E. Abonneau, "Characterization of TiSi$_2$ Ohmic and Schottky Contacts Formed by Rapid Thermal Annealing Technology," *J. Electrochem. Soc.* **136**:238 (1989).

50. E. Waterman, J. Dunlop, and T. Brat, "Tungsten-Titanium Sputtering Target Processing Effects on Particle Generation and Thin Film Properties for VLSI Technology," *IEEE VLSI Multilevel Interconnect Conf.*, 1991, p. 329.

51. T. Sands, "Compound Semiconductor Contact Metallurgy," *Mater. Sci., Eng.* **B1**:289 (1989).

52. C. Dubon-Chevallier, P. Blanconnier, and C. Bescombes, "GeMoW Refractory Ohmic Contact for GaAs/GaAlAs Self-Aligned Heterojunction Bipolar Transistors," *J. Electrochem. Soc.* **137**:1514 (1990).

53. R. P. Gupta, W. S. Khokle, J. Wuerfl, and H. L. Hartnagel, "Design and Characterization of a Thermally Stable Ohmic Contact Metallization on n-GaAs," *J. Electrochem. Soc.* **137**:631 (1990).

54. J. H. Pugh and R. S. Williams, "Boundary Driven Loss of Gas-Phase Group V Sources for Gold/III-V Compound Semiconductor Systems," *J. Mater. Res.* **1**:343 (1986).

55. J. B. B. Oliveira, C. A. Olivieri, and J. C. Galzerani, "Characterization of NiAuGe Ohmic Contacts on n-GaAs Using Electrical Measurements, AUGER Electron Spectroscopy and X-ray Diffractometry," *Vacuum* **41**:807 (1990).

56. M. Kamada, T. Suzuki, T. Taira, and M. Arai, "Electrical Inhomogeneity in Alloyed AuGe-Ni Contact Formed on GaAs," *Solid-State Electron.* **33**:999 (1990).

57. T. S. Kuan, P. E. Baston, T. N. Jackson, H. Rupprecht, and E. L. Wilke, "Electron Microscope Studies of an Alloyed Au/Ni/Au-Ge Ohmic Contacts to GaAs," *J. Appl. Phys.* **54**:6952 (1983).

58. T. Sanada and O. Wada, "Ohmic Contacts to p-GaAs with Au/Zn/Au Structures," *Jpn. J. Appl. Phys.* **19**:L491 (1980).

59. H. J. Gopen and A. Y. C. Yu, "Ohmic Contacts to Epitaxial p-GaAs," *Solid-State Electron.* **14**:515 (1971).

60. Y. Lu, T. S. Kalkur, and C. A. Paz de Araujo, "Reduced Thermal Alloyed Ohmic Contacts to p-type GaAs," *J. Electrochem. Soc.* **136**:3123 (1989).

61. A. A. Lakhani, "The Role of Compound Formation and Heteroepitaxy in Indium-Based Contacts to GaAs," *J. Appl. Phys.* **56**:1888 (1984).

62. T. S. Kalker, "Preliminary Studies on Mo-In-Mn based Ohmic Contacts to p-GaAs," *J. Electrochem. Soc.* **136**:3549 (1989).

63. D. C. Thomas and S. S. Wong, "A Planar Interconnect Technology Utilizing the Selective Deposition of Tungsten—Multilevel Implementation," *IEEE Trans. Electron Dev.* **39**:901 (1992).

64. S. R. Wilson, J. L. Freeman Jr., and C. J. Tracey, "A Four-Metal Layer, High Performance Interconnect System for Bipolar and BiCMOS Circuits," *Solid State Technol.*, p. 67, Nov. 1991.

65. J. R. Black, "Electromigration: A Brief Survey and Some Recent Results," *IEEE Trans. Electron. Dev.* **ED-16**:338 (1969).

66. P. B. Ghate, "Electromigration Induced Failures in VLSI Interconnects," in *Proc. IEEE 20th Int. Rel. Phys. Symp.*, 1982, p. 292.

67. J. M. Towner and E. P. van de Ven, "Aluminum Electromigration Under Pulsed D.C. Conditions," *21st Annu. Proc. Rel. Phys. Symp.*, 1983, p. 36.

68. J. A. Maiz, "Characterization of Electromigration Under Bidirectional (BDC) and Pulsed Unidirectional Currents," *Proc. 27th Int. Rel. Phys. Symp.*, 1989, p. 220.

69. S. Vaidya, T. T. Sheng, and A. K. Sinha, "Line Width Dependence of Electromigration in Evaporated Al-0.5%Cu," *Appl. Phys. Lett.* **36**:464 (1980).

70. T. Turner and K. Wendel, "The Influence of Stress on Aluminum Conductor Life," *Proc. IEEE Int. Rel. Phys. Symp.*, 1985, p. 142.

71. H. Kaneko, M. Hasanuma, A. Sawabe, T. Kawanoue, Y. Kohanawa, S. Komatsu, and M. Miyauchi, "A Newly Developed Model for Stress Induced Slit-like Voiding," *Proc. IEEE Int. Rel. Phys. Symp.*, 1990, p. 194.

72. K. Hinode, N. Owada, T. Nishida, and K. Mukai, "Stress-Induced Grain Boundary Fractures in Al-Si Interconnects," *J. Vac. Sci. Technol.* **B5**:518 (1987).

73. S. Mayumi, T. Umemoto, M. Shishino, H. Nanatsue, S. Ueda, and M. Inoue, "The Effect of Cu Addition To Al-Si Interconnects on Stress Induced Open-Circuit Failures," *Proc. IEEE Int. Rel. Phys. Symp.*, 1987, p. 15.

74. P. Singer, "Double Aluminum Interconnects," *Semicond. Int.* **16**:34 (1993).

75. R. A. Levy and M. L. Green, "Low Pressure Chemical Vapor Deposition of Tungsten and Aluminum for VLSI Applications," *J. Electrochem. Soc.* **134**:37C (1987).

76. R. J. Saia, B. Gorowitz, D. Woodruff, and D. M. Brown, "Plasma Etching Methods for the Formation of Planarized Tungsten Plugs used in Multilevel VLSI Metallization," *J. Electrochem. Soc.* **135**, p. 936 (1988).

77. D. C. Thomas, A. Behfar-Rad, G. L. Comeau, M. J. Skvarla, and S. S. Wong, "A Planar Interconnect Technology Utilizing the Selective Deposition of Tungsten—Process Characterization," *IEEE Trans. Electron Dev.* **39**:893 (1992).

78. T. E. Clark, P. E. Riley, M. Chang, S. G. Ghanayem, C. Leung, and A. Mak, "Integrated Deposition and Etchback in a Multi-Chamber Single-Wafer System," *IEEE VLSI Multilevel Interconnect Conf.*, 1990, p. 478.

79. K. Suguro, Y. Nakasaki, S. Shima, T. Yoshii, T. Moriya, and H. Tango, "High Aspect Ratio Hole Filling by Tungsten Chemical Vapor Deposition Combined with a Silicon Sidewall and Barrier Metal for Multilevel Metallization," *J. Appl. Phys.* **62**:1265 (1987).

80. P. E. Riley, T. E. Clark, E. F. Gleason, and M. M. Garver, "Implementation of Tungsten Metallization in Multilevel Interconnect Technologies," *IEEE Trans. Semicond. Man.* **3**:150 (1990).

81. T. Iijima, A. Nishiyama, Y. Ushiku, T. Ohguro, I. Kunishima, K. Suguro, and H. Iwai, "A New Contact Plug Technique for Deep-Submicrometer ULSI's Employing Selective Nickel Silicidation of Polysilicon with a Titanium Nitride Stopper," *IEEE Trans. Electron Dev.* **40**:371 (1993).

82. F. A. Houle, C. R. Jones, T. Baum, C. Pico, and C. A. Kovac, "Laser Chemical Vapor Deposition of Copper," *Appl. Phys. Lett.* **46**:204 (1985).

83. C. K. Hu, S. Chang, M. B. Small, and J. E. Lewis, "Diffusion Barrier Studies for Copper," *IEEE VLSI Multilevel Interconnect Conf.*, 1986, p. 181.

84. Y. Arita, N. Awaya, T. Amazawa, and T. Matsuda, "Deep Submicron Cu Planar Interconnection Technology Using Cu Selective Chemical Vapor Deposition," *IEDM Tech. Dig.*, 1989, p. 893.

85. P.-L. Pai and C. H. Ting, "Selective Electroless Copper for VLSI Interconnect," *IEEE Electron Dev. Lett.* **10**:423 (1989).

86. D. S. Gardner, J. Onuki, K. Kudoo, and Y. Misawa, "Encapsulated Copper Interconnection Devices Using Sidewall Barriers," *IEEE VLSI Multilevel Interconnect Conf.*, 1991, p. 99.

87. T. Foster, G. Hoeye, and J. Goldman, "A Low Pressure BPSG Deposition Process," *J. Electrochem. Soc.* **132**:505 (1985).

88. C. K. Hu, M. B. Small, F. Kaufman, and D. J. Pearson, "Copper-Polyimide Wiring Technology for VLSI Circuits," *Proc. 1989 Tungsten, Other Advanced Mater.*, S. S. Wong and Furukawa, eds., MRS, 1990, p. 369.

89. K. Mikagi, T. Homma, T. Katoh, K. Tsuenenari, and Y. Murao, "Multilevel Gold Metallization by Use of Selective W-CVD and Polyimide," *IEEE VLSI Multilevel Interconnect Conf.*, 1989, p. 33.

90. R. V. Joshi, L. Hsu, and M. Chow, "Polyimide with Oxide/Nitride Sidewalls for Selective W Deposition," *IBM Tech. Disc. Bull.*, Docket #YO8890137 (1989).

91. R. V. Joshi, L. Hsu, H. Dalal, P. Klymco, M. Jasso, and H. Ng, "A Novel Application of Polyimide-W-Al/Cu for VLSI Interconnect," *IEEE VLSI Multilevel Interconnect Conf.*, 1991, p. 75.

92. D. D. J. Allman, K. P. Fuchs, and D. L. Kwong, "The Material Characteristics of a New Methyl-Phenyl Silsesquioxane Spin-on Glass for Use in a Global Non-Etchback Interconnect Planarization Process," *IEEE VLSI Multilevel Interconnect Conf.*, 1991, p. 373.

93. H. Chung, S. Wong, and S. Lim, "Non-Etchback Silicate Spin-on Glass for Advanced BiCMOS Technology," *IEEE VLSI Multilevel Interconnect Conf.*, 1991, p. 376.

94. A. C. Adams and D. D. Capio, "Planarization of P-Doped Silicon Dioxide," *J. Electrochem. Soc.* **128**:423 (1981).

95. C. Jang et al., *IEEE VLSI Multilevel Interconnect Conf.*, 1984, p. 357.

96. E. K. Broadbent, J. M. Flanner, W. G. M. van den Hoek, and I.-W. H. Connick, "High-Density High-Reliability Tungsten Interconnection by Filled Interconnect Groove Metallization," *IEEE Trans. Electron Dev.* **35**:952 (1988).

97. D. C. Thomas, S. S. Wong, D. R. Dinsmore, and R. J. Soave, "A Multilevel Tungsten Interconnect Technology," *IEDM Tech. Dig.*, 1988, p. 466.

98. B. Davari, C. W. Koburger, R. Schulz, J. D. Warnock, T. Furukawa, M. Jost, Y. Taur, W. G. Schwittek, J. K. DeBrosse, M. L. Kerbaugh, and J. L. Mauer, "A New Planarization Technique Using a Combination of RIE and Chemical Mechanical Polish (CMP)," *IEDM Tech. Dig.*, 1989, p. 61.

99. M. E. Thomas, S. Sekigahama, P. Renteln, and J. Pierce, "The Mechanical Planarization of Interlevel Dielectrics," *IEEE VLSI Multilevel Interconnect Conf.*, 1990, p. 438.

Chapter 16

CMOS Technologies

The last chapter discussed the process modules of device isolation and interconnection. This chapter will apply these modules to the first of the technologies to be considered, CMOS. The discussion begins with a 3 μm CMOS technology. This sort of process was state of the art during the mid 1970s. While it may seem odd to begin with such old techniques, modern CMOS processing is primarily a set of extensions to this basic technology. Device scaling and short channel effects will then be discussed. In these highly scaled devices, parasitics are an important element, so methods of reducing series resistance and source/drain capacitance are covered next. Device reliability and drain engineering will then be covered, and the chapter will finish with a discussion of latchup. For a more extensive review of CMOS devices and technology the reader is referred to Wolf [1].

16.1 Basic Long Channel Device Behavior

This section will review the properties of the ideal long channel MOSFET. It is not intended to serve as a complete description of MOS device physics; that occupies books in itself. (For detailed descriptions of MOS device operation see Sze [2], Nicollian and Brews [3], or Tsividis [4].) Instead, it will review the material necessary to develop an appreciation for the requirements of MOS technologies. To do this we will first review the concept of threshold voltage and then discuss the three operating regimes of the MOS transistor.

Recall that the basic MOS transistor has four terminals: source, drain, gate, and substrate (Figure 16.1). One of the most important parameters in determining the electrical characteristics of the MOSFET is the "turn-on" or threshold voltage, V_t. The threshold voltage is normally defined to be the gate voltage at which the surface of the semiconductor is at the onset of strong inversion. That is, the carrier concentration of the inversion layer at the surface is equal to, but of the opposite type as, the bulk carrier concentration. At this point, the surface potential is $2\phi_f$, where ϕ_f, the Fermi level, is defined as

$$\phi_f = \frac{kT}{q} \ln \frac{N_a}{n_i} \qquad (16.1)$$

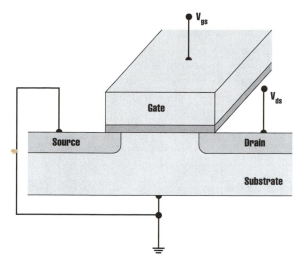

Figure 16-1   Cross-sectional schematic of a four-terminal MOSFET. In this transistor the source and substrate are both grounded.

for a p-type semiconductor and

$$\phi_f = -\frac{kT}{q} \ln \frac{N_d}{n_i} \tag{16.2}$$

for an n-type semiconductor. In these equations, N_a and N_d are the acceptor and donor concentrations in the substrate, and n_i is the intrinsic carrier concentration.

For a p-type semiconductor, the gate voltage can be expressed in terms of the surface potential (ϕ_s) as

$$V_g = \phi_s + \frac{k_s t_{ox}}{k_{ox}} \sqrt{\frac{2qN_a}{k_s \varepsilon_o}} \phi_s \tag{16.3}$$

where k_s and k_{ox} are the relative dielectric constants of silicon and silicon dioxide, and t_{ox} is the gate oxide thickness. Inserting for the surface potential at threshold ($\phi_s = 2\phi_f$), one can obtain for the threshold voltage

$$V_t = 2\phi_f + \frac{k_s t_{ox}}{k_o} \sqrt{\frac{4qN_a}{k_s \varepsilon_o}} \phi_f \tag{16.4}$$

for an NMOS transistor and

$$V_t = 2\phi_f - \frac{k_s t_{ox}}{k_o} \sqrt{\frac{4qN_d}{k_s \varepsilon_o}} (-\phi_f) \tag{16.5}$$

for a PMOS transistor. If one also takes into account the metal semiconductor work function and the presence of oxide and interface charges, an additional term of

$$\Delta V_t = \phi_{ms} - \frac{Q_f + Q_{it}(0) + Q_{MI}\gamma_{MI}}{C_{ox}} \tag{16.6}$$

must be added to the right hand sides of Equations 16.4 and 16.5. In Equation 16.6, Q_f and Q_{MI} represent the fixed and mobile ion charges, γ_{MI} is a constant that depends on the distribution of mobile charge in the oxide, and varies between zero and one, and $Q_{it}(0)$ represents the interface state charge. C_{ox} is the oxide capacitance.

From a process standpoint, it is clear that a wide variety of well controlled threshold voltages are available if one can control the oxide thickness, the charges in the oxide, the metal semiconductor work function, and the doping concentration in the channel. Of these, the doping concentration is most easily manipulated, since reasonable adjustments of this parameter can be accomplished by simply changing an implantation dose. Furthermore, changing the substrate concentration does not have the first order effect on performance and reliability that changing the oxide thickness has.

Detailed equations have been developed for MOS operations that are valid for any set of biasing conditions. The derivation of these equations are beyond the scope of this book. They can be simplified by dividing the operation of the MOSFET (Figure 16.2) into three regions: subthreshold ($V_{gs} < V_t$), linear ($V_t < V_{gs}$ and $0 < V_{ds} < V_{gs} - V_t$), and pinchoff ($V_t < V_{gs}$ and $V_{gs} - V_t < V_{ds}$). Near the transition points, these equations will only be approximate, but they will serve our purpose.

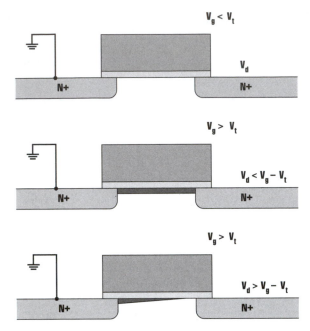

Figure 16-2 Three regions of n-channel MOSFET operation include subthreshold, linear, and saturation.

The linear region is the easiest to derive. In this region the channel forms a thin, nearly uniform layer under the gate. As with any resistor, the conductance is proportional to the width to length ratio, the mobility, and the number of carriers. The number of carriers is proportional to the difference between the gate voltage and the threshold voltage. Collecting these terms, in the simplest (square law) theory one can write the drain current as

$$I_{ds} \approx \frac{W}{L} \mu_{eff} C_{ox} (V_{gs} - V_t) V_{ds} \qquad \textbf{(16.7)}$$

where μ_{eff} is the effective mobility and C_{ox} is the oxide capacitance per unit area. When $V_{ds} > 0$, the carrier concentration shrinks as one approaches the drain edge. Taking this into account,

$$I_{ds} = \frac{W}{L} \mu_{eff} C_{ox} \left[(V_{gs} - V_t) V_{ds} - \frac{V_{ds}^2}{2} \right]$$
$$V_{gs} - V_t > V_{ds} \quad V_{gs} > V_t \qquad \textbf{(16.8)}$$

This expression uses an effective mobility to represent the fact that the carriers in the channel may scatter off each other and the semiconductor/oxide interface in addition to the scattering mechanisms seen in the bulk. The mobility that must be used is substantially less than the bulk value and at high values of vertical electric field, depends on the gate bias.

In the pinchoff region, the channel does not extend fully from source to drain. The current is limited by the movement of carriers from the pinched off channel edge to the drain. As a result, the current in this region is nearly independent of the drain voltage. Since this occurs at the onset of the pinchoff region, we can find the value of the linear current at $V_{ds} = V_{gs} - V_t$ from Equation 16.8. This gives us the drain to source current in pinchoff

$$I_{ds} = \frac{W}{L} \mu_{eff} C_{ox} \frac{(V_{gs} - V_t)^2}{2} \quad 0 < V_{gs} - V_t < V_{ds} \qquad \textbf{(16.9)}$$

The analysis of the subthreshold region is more complicated than either the linear or saturation regions. The drain to source current depends strongly on the surface potential, since the carrier concentration increases exponentially with surface potential. The exact equation for the drain to source current is also beyond the range of this text, but for an ideal device,

$$I_{ds} \propto \exp\left[K_{sub} \frac{qV_{gs}}{kT} \right] \quad V_{gs} < V_t \qquad \textbf{(16.10)}$$

where K_{sub} is a constant that depends on the gate oxide thickness and the substrate doping.

16.2 Early MOS Technologies

The earliest MOS technologies were PMOS. By the early 1970s, they were replaced by NMOS, due to the higher mobility of the electron. Resistor load and enhancement depletion (E/D) NMOS technologies, shown in Figure 16.3a, were widely fabricated through the 1970s and into the early 1980s. As compared to bipolar, the technology was viewed as cheap (it required only five masks for resistor load and seven masks for E/D) and capable of high density (in the range of 1000 to 20,000 gates per chip) but was fairly slow, with digital clock cycles of a few megahertz. As the density increased from 1000 to 10,000 gates, the chip power required grew from hundreds of milliwatts to several watts. Generally, if a chip dissipates about 3 to 5 W, air cooled but relatively expensive ceramic packages can be used. More than 5 W requires water cooling. In order to use inexpensive plastic packages, the chip must dissipate less than a few watts. It was obvious that when another order of magnitude in density was achieved, chip cooling was going to be a major problem.

The idea of complimentary logic (Figure 16.3b) was first proposed by Wanlass and Sah in 1963 [5]. Their devices showed gate delays of over 100 ns per stage. Early work was led by RCA with the first demonstration of an IC in 1966. These technologies were not very dense and were slow. They were also prone to latchup, a self destructive condition will be reviewed later in this chapter. Since LOCOS had not yet been developed, guard rings were used, which had a large area penalty and a large capacitance. During the early 1970s, CMOS was much derided and used primarily in toys, watches, and calculators where such slow speeds could be tolerated and in a few military applications due to its small power dissipation and large noise margins. The invention of the LOCOS process module, the reduction of mobile ionic charge, the introduction of ion implantation, which could provide precise control of the well concentration and the threshold voltage, and improvements in lithography, have dramatically improved the acceptance of CMOS.

CMOS has the basic advantage that, for an inverter structure, only one of the transistors is on when the input is either high or low. Appreciable power is dissipated only during switching transients. Part of this power is capacitive charging current, which can be represented as

$$P \approx \bar{n} * \frac{\text{freq}}{2} * \overline{C_{\text{node}}} * V_{\text{swing}}^2 \qquad \textbf{(16.11)}$$

where $\bar{n}$ is the average number of gates switching at any given time, C_{node} is the average node capacitance, and V_{swing} is the voltage difference between a logical high and a logical low. An additional current arises from the fact that during the switch transient, both transistors are on. Because only a small fraction are likely to switch at any given clock cycle, CMOS power dissipation is much lower than NMOS. The reduced packaging cost and tremendous increase in density completely offsets the cost of the additional process complexity required for CMOS. As a result, it is the dominant semiconductor technology today.

16.3 The Basic 3 μm Technology

In designing a CMOS technology, careful attention must be paid to the concept of thermal budget. Therefore, those processes that involve the formation of the deeply diffused well (also called *tub*) must be done

Figure 16-3   MOS inverters including (a) resistor load, (b) enhancement/ depletion or E/D, and (c) CMOS.

first. The choices for well type are n-well, p-well, and twin well [6]. The substrate material for CMOS is normally lightly doped. This means that the depletion regions of the source and drain junctions for the MOSFET fabricated in the substrate will be large and the parasitic capacitance of these junctions will be small. The well, on the other hand, is typically doped at least an order of magnitude higher than the substrate so that any uncertainty in the substrate concentration will not affect the well concentration. As a result of this higher doping, the device placed in the well is inherently slower than the same device would be if placed in the substrate. If the NMOS devices are placed in a p-well, their degraded performance will more closely match the inherently lower mobility holes in PMOS transistors. On the other hand, if the designer wants to optimize the performance of the NMOS transistor and design portions of the circuit in NMOS only, an n-well may be preferred [7]. The most commonly used tub strategy for submicron technologies is the twin tub. That is, both n-type and p-type ''wells'' are formed in a lightly doped substrate. Typically, this is done with two masks and at least two implants. At very short gate lengths, the performance difference between NMOS and PMOS is reduced. Furthermore, the doping concentration required to obtain acceptable device behavior increases as the gate length shrinks. As a result, neither device really operates in a lightly doped substrate. Although it adds process complexity, the twin tub approach allows the doping profile for each device to be set independently to optimize the performance of both device types [8].

The first step in a simple n-well technology is the growth of a 1000 Å initial oxide. This serves two functions. It prevents photoresist from contacting the silicon, and it will later provide an alignment mark to locate the position of the well. After implantation, the initial oxide is etched, the photoresist is stripped, and the well is driven in a long, high temperature process (Figure 16.4).

The well dose and drive cycle are set by two basic considerations: surface concentration and depth. The surface concentration is determined by the device designer. This is usually not what is required for the threshold voltage, but rather the concentration required to prevent source to drain punchthrough. Thus, the sum of the source and drain depletion widths must be less than the channel length when V_{ds} is at least as large as the supply voltage. To provide a margin of safety, the minimum acceptable punchthrough voltage is often established by biasing the drain at two or three times the supply voltage with the source shorted to the substrate. For this calculation, one must recognize that the channel length is not identical to the drawn gate length. During the gate etch process the gate length may be reduced. The source and drain will also diffuse laterally under the gate. Assume therefore, that the shortest permissible effective channel length devices in this 3 μm technology is 2 μm. The channel doping must be set high enough to ensure that the sum of the depletion regions of the source and drain cannot exceed 2 μm. For a single sided step junction, a substrate concentration of $5 * 10^{15}$ cm^{-3} would provide a total depletion width of 1.8 μm for an unbiased source and a drain biased at 10 V (2 times a supply voltage of 5 V). Common well surface concentrations then are slightly less than 10^{16} cm^{-3}.

The well depth is determined by ensuring that the source and drain do not completely deplete the well vertically. Using the guideline of one tenth of the well doping, the substrate concentration should be no more than 10^{15} cm^{-3}. The full supply bias must be applied between the well and the substrate in order to ensure a proper

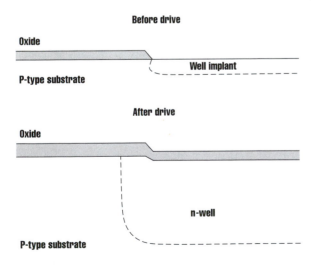

Before drive

Oxide

P-type substrate

Well implant

After drive

Oxide

P-type substrate

n-well

Figure 16-4   Cross-sectional views of an inverter structure after well drive.

Figure 16-5 The minimum well depth must be chosen so as to prevent punchthrough from the reverse biased source/well and source/substrate junctions.

reverse bias on all of the junctions. For the n-well technology discussed above, that was designed to run at 5 V, the depletion layer of the source (at 0 V) in the well is about 1.2 μm. The depletion layer of the well/substrate junction will be primarily in the substrate, owing to its light doping. However, approximately 0.2 μm will be in the well. Thus, the well depth must be at least the source/drain junction depth (which we will take to be 0.5 μm), plus the source/drain depletion width (1.2 μm), plus the width of the well/substrate depletion that lies inside the well (Figure 16.5). The well then must be at least 2.0 μm deep. In reality, the well profile is not uniform. Instead, the carrier concentration decreases near the well substrate junction. Because of this, the depletion width of the substrate junction is considerably larger than 0.2 μm. For a 3 μm process, a typical final well depth then, is about 4 μm. To achieve this deep well, phosphorus is normally chosen as the well dopant due to its high diffusivity.

The well drive must be done at least partially in an oxidizing ambient. This oxide consumes silicon in the exposed well window. When the oxide is later removed, this creates a depression that can be used to align the next level. As an alternative, a separate masking level to etch deep pits in the wafer, which will serve as alignment mark, can be done before well photolithography. Many submicron technologies are done this way to improve registration accuracy. It is important to understand that the well diffuses laterally as well as vertically. The diffusion is not isotropic in this case however, because the oxidation process creates vacancies and interstitials that enhance diffusion. Since the oxidation is proceeding more rapidly in the exposed region, the well diffuses deeper than laterally. A typical ratio of width to depth diffusion is about 0.7.

After the well has been driven, the oxide is stripped, a pad oxide is grown and the nitride layer is deposited for the LOCOS process. The nitride is patterned, and if a recessed LOCOS is desired, so are the oxide and the substrate. Field implants are normally done before oxidation. During the field oxidation phosphorus is rejected and boron is absorbed by the growing oxide. For n-well processes, this means that the threshold voltage of the parasitic PMOS transistors formed in the n-well is large and gets larger due to this dopant redistribution. The lightly doped p-type substrate, however, becomes even more lightly doped due to dopant redistribution. Furthermore, the fixed oxide charge also decreases the threshold voltage of the parasitic NMOS device and increases the threshold voltage of the parasitic PMOS device.

To avoid the formation of parasitic channels in the substrate in an n-well technology, boron implants are often used. (Twin tub approaches have sufficient doping.) The n-well regions are often protected against this implant with an additional photomask. The active regions of the substrate must also be protected from the implant. This can either be done by relying on the thickness of the nitride layer to prevent implantation in the active region or by using a double coat resist process (Figure 16.6). In the former case, BF_2 rather than B is often used as the field implant species so that the implant

Figure 16-6 The field implant mask is the inverse of the well, but the features are bloated to account for lateral diffusion.

can be stopped by the thin nitride layer. In a double coat process, the resist used to pattern the nitride is left on the wafer and hardbaked. A second layer of resist is applied and patterned for the implant mask. The difficult part of the process is to ensure that the first resist is stable upon the application of the second layer. This is primarily a matter of selecting the proper resists and bake cycles. Once the field implant is completed, the resist is stripped and the field oxide is grown (Figure 16.7). Following the oxidation, the residual nitride is stripped. The wafers are first dipped in HF to remove any SiO_2 that has grown on top of the nitride. They are then placed in a hot phosphoric bath to remove the nitride and thereby complete the device isolation.

The pad oxide is removed in HF and a thin oxide is regrown. This layer is used as a screen oxide for the threshold adjust implants. The implant masks are patterned in photoresist, and the implants are done. In the case of n-well technology, multiple implants are often done into the substrate, a deep one to prevent source to drain punchthrough of the NMOS transistor, and a shallower implant to set the threshold voltage. Boron is also typically implanted into the channel region of the PMOS transistor in an n-well. This occurs because the heavily n-doped polysilicon has a work function of about -0.4 V for 10^{16} cm^{-3} n-type layers (PMOS transistors) and about -1.1 V for similarly doped p-type layers (NMOS transistors). Because of this difference, the PMOS device has a large negative threshold voltage that must be raised toward zero. Typical implant doses are about 10^{12} cm^{-2} of boron. Often, part of the shallow NMOS implant is done without a mask, allowing it to implant the PMOS device as well.

Following the channel implants, the sacrificial oxide is stripped and the gate oxide is grown. Next the polysilicon is deposited and patterned. The polysilicon is usually heavily doped with phosphorus. This may be done by the addition of phosphine to silane in the low pressure CVD tube, or it may be done by doping the wafer with solid or liquid sources after undoped polysilicon has been

Figure 16-7   Top and cross-sectional views of a CMOS inverter after field oxidation.

Figure 16-8   Top and cross-sectional views of a CMOS inverter after source/drain implants.

Figure 16-9 Top and cross-sectional views of a CMOS inverter after PSG deposition and reflow and metallization.

deposited. Of course, the fidelity of the pattern transfer during this step is critical to the performance of the circuit. If the final poly gate length is too short, the source and drain may punch through. If the gate length is too long, I_{ds} will be decreased and the IC will operate slowly or not at all.

After polysilicon patterning, the channel masks are reapplied in sequence and the NMOS and PMOS source/drain regions are implanted (Figure 16.8). The polysilicon gate acts as an implant stop, preventing dopant species from reaching the channel. The field oxide defines the other three sides of the source/drain regions. The source/drain implant mask simply protects some of the active regions from receiving the implant.

The p-type impurity that is used almost exclusively is boron. The n-type implant species is normally arsenic since it has a lower diffusion coefficient than phosphorus or antimony. Typical implant doses for the source/drain range from 1 to $5 * 10^{15}$ cm^{-2}. A great deal of work has been done to optimize the anneal cycles used to activate these high concentration implants and repair the implant damage. Typical processes for the 3 μm technology are 1000°C for 30 to 60 min. This may also include a thermal cycle to grow a thin oxide on top of the polysilicon.

After activation, a contact oxide is deposited. This usually contains phosphorus (PSG) and frequently boron as well (BPSG). Concentrations of phosphorus range from 4 to 12%, although high concentrations are generally avoided because of their ability to form phosphoric acid in the presence

Figure 16-10 Top and cross-sectional views of a completed CMOS inverter fabricated with a simple 3 μm process.

of moisture. Following deposition, the oxide is densified through a high temperature annealing process. Densification lowers the oxide etch rate making uniform, reproducible etching much more feasible. Again, temperatures of 1000°C are typical, this time in an oxidizing ambient. As mentioned in the last chapter, this oxide has the capability to reflow at high temperature to reduce the sharpness of vertical steps and so improve the metal step coverage. PSG also serves as an effective gettering agent. Ionic and metallic impurities become trapped in the oxide where they have little effect on the electrical operation of the device.

Next, the contact holes are patterned and etched into the wafer. The contact glass may be reflowed after etch by a final high temperature step. This has the effect of rounding the top edge of the contact, further improving the metal step coverage. The first level of metallization is applied, patterned, and etched (Figure 16.9). If desired, a second set of insulator and metal is applied and patterned. Finally, a sinter is done in H_2 or H_2/N_2 mixtures to form the ohmic contact. This step also reduces the interface state density of the FET devices. A thick layer of insulator, either SiO_2 or Si_3N_4, is then deposited for scratch protection and final passivation. Large vias are opened up so that contact can be made to the bonding pads. Figure 16.10 shows top and cross-sectional views of a completed CMOS inverter.

16.4 Device Scaling

The performance of the baseline 3 μm process can be improved dramatically through device scaling. In this method, an improvement in the speed and packing density can be achieved through a reduction in the various physical lengths in the device. Several schemes can be constructed from the scaling rules shown in Table 16.1. The supply voltage is reduced by a factor of $1/k$. The transistor width, length, oxide thickness, and junction depth are all reduced by the factor $1/\lambda$. The doping in the channel is increased by a factor of λ^2/k. One of the first scaling methodologies that was proposed was constant field scaling [9]. The intent of this approach is to keep the electric fields in the transistor unchanged from their long channel values. To do this k is set equal to λ. Then, to first order the drive current of the MOSFET, as given by Equations 16.8 and 16.9, does not change if W is held fixed. In this method the increase in speed comes from two sources: the reduction in the voltage swing and the reduction in the capacitance that must be driven due to the smaller device sizes.

While intellectually appealing, true constant field scaling has not been widely applied. The approach that has been used in scaling to 1.0 μm is closer to constant voltage scaling [10] in which $k = 1$. In this type of scaling, the voltage swing stays the same, but the drive current increases due to the increase in C_{ox}. Since the drive current increases roughly as the square of the supply voltage, constant voltage scaling produces more speed improvement than constant field scaling. Furthermore, since the supply voltage is held constant there is no need to reduce the threshold voltage. This avoids

Table 16.1   Generalized scaling theory for MOS transistors

Parameters	Variables	Scaling Factor
Dimensions	W, L, t_{ox}, x_j	$1/\lambda$
Potentials	V_{ds}, V_{gs}	$1/k$
Doping Concentration	N_a, N_d	λ^2/k
Electric Field	$\mathcal{E}$	λ/k
Current	I_{ds}	λ/k^2
Gate Delay	T	k/λ^2

$1/\lambda$ is the factor by which the dimensions are scaled and $1/k$ is the factor by which the voltages are scaled.

the increased "off-state" leakage that always occurs when V_t is reduced. The constant voltage scaling approach is ultimately subject to limitations relating to the reliable operation of the devices at very high electric fields. As a result, commercial practice has used constant voltage scaling until adequately reliable operation can no longer be maintained. Commonly, this is considered to be a 20 year life under some sets of worst case operating conditions. The current standard has moved from 5.0 to 3.3 V. It is expected to make additional steps to 2.0 V, 1.5 V, and ultimately 0.9 Volts.

If the gate length of a MOSFET is reduced but these scaling rules are not followed, the MOSFET will operate in the short channel regime. Short channel MOSFETs have several liabilities for IC applications. As shown in Figure 16.11, the output impedance of a short channel MOSFET is much smaller. The subthreshold slope also depends on the drain voltage in a short channel MOSFET. Finally, the threshold voltage of the device drops sharply with the gate length. This effect further increases subthreshold leakage. The result is very poor control of the device characteristics.

Figure 16.12 illustrates the cause of short channel effects. Ideally, the charge under the gate should be controlled only by the gate electrode. For short channel devices, the depletion layer associated with the drain removes a significant fraction of the charge under the gate. Brews has developed an empirical equation [11] that has been used as a guide for the limits of short channel behavior. It has been verified over the range of 1 to 10 μm and has also been applied successfully to deep submicron devices. Brews' rule is given as

$$L_{min} \approx 0.4[x_j t_{ox}(W_d + W_s)^2]^{1/3} \tag{16.12}$$

where x_j is the junction depth in microns, t_{ox} is the oxide thickness in Å, and W_s and W_d are the source and drain depletion widths respectively, in microns. The depletion widths shrink due to increases in the channel doping upon device scaling. The oxide thickness can be reduced to less than 50 Å by changing the oxidation process. In so doing, care must be exercised to minimize the oxide fixed charge and interface state density (Sections 4.7 and 4.8). This is most often accomplished by growing the oxide at high temperature, although recent reports have suggested that very thin oxides can be grown at low temperature without incurring a large interface state density [12]. The reduction of the

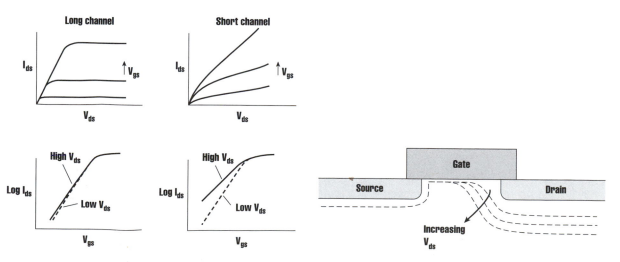

Figure 16-11 I-V characteristics of MOSFETs comparing long and short channel behavior. Typical threshold voltage dependence on gate length.

Figure 16-12 Schematic of device cross section showing enlarged depletion regions that give rise to short channel behavior.

junction depth, however, is a more difficult problem since it usually necessitates increasing the series resistance of the device.

To understand the effect of the parasitic series resistance on MOSFET performance, it is convenient to first introduce the idea of a mobility reduction factor θ [13]. As the vertical electric field increases, the carrier density in the channel increases. This in turn leads to an increase in both the electron-electron and electron-surface scattering rates. To model this effect, the basic equations that describe MOSFET operation are modified by the addition of an empirical mobility reduction parameter. For example, in the linear regime of operation for small V_{ds}, the drain to source current is given by

$$I_{ds} \approx C_{ox} \frac{W}{L} \cdot \frac{\mu_o}{1 + \theta(V_{gs} - V_t)} (V_{gs} - V_t)V_{ds} \qquad (16.13)$$

Now consider the equivalent circuit shown in Figure 16.13. Inserting the effects of the parasitic resistance and rearranging terms,

$$I_{ds} \approx C_{ox} \frac{W}{L} \frac{\mu_o}{(1 + \theta)(V_{gs} - V_t)} [(V_{gs} - R_s I_{ds}) - V_t][V_{ds} - (R_d + R_s)I_{ds}]$$

$$\approx C_{ox} \frac{W}{L} \mu_o \frac{(V_{gs} - V_t)V_{ds}}{1 + \left[\theta + C_{ox} \frac{W}{L} \mu_o(R_s + R_d) \right](V_{gs} - V_t)} \qquad (16.14)$$

Thus, if the parasitic resistance associated with the source/drain diffusions is large, it has the effect of substantially increasing the apparent mobility reduction factor and so reducing the transconductance of the device [14]. The effect is particularly acute at small gate lengths due to the increase in the intrinsic device performance. It has been shown that traditional source/drain structures can reduce I_{ds} by a factor of 2 for gate lengths of 0.2 μm [15], even if the source and drain resistance can be held constant during device scaling.

To achieve shallower junctions, the technologist may decrease the characteristic diffusion length of the activation step, reduce the implant dose, and for boron implantation, preamorphize and reduce the effective implant energy through the use of molecular implants. Of course, the reduction of the implant dose is undesirable since it would increase the parasitic resistance. Furthermore, if the surface concentration decreases as a result of the reduced dose, the contact resistance will increase sharply

$$R_S = R_{series} + R_{crowding} + R_{contact}$$

Figure 16-13   Equivalent circuit for MOSFET with source/drain series resistances displayed.

Figure 16-14 Gate to source/drain shorts caused by misalignment and/or contact blowout during etch.

as discussed in Chapter 15. This resistance has exactly the same effect as the series resistance of the source/drain diffusion.

As discussed in Chapters 3 and 6, a great deal of effort has gone into the development of processes that reduce the thermal cycle during the activation process. Of course, this is subject to the limitation that the implant damage must also be sufficiently annealed. This in turn implies high temperature annealing. Dislocation loops and other extended defects resist annealing unless the temperature is about 1000°C. To allow these high temperature anneals with minimal diffusion, rapid thermal processing techniques have proliferated (Chapter 6).

One way to reduce the series resistance is to minimize the separation between the contact and the gate. Figure 16.14 shows the concern of having a metal contact too close to the gate of the MOSFET. The top edge of the gate is typically a 90° angle that produces a large electric field. If the contact is sufficiently close, the oxide between the gate and the contact can easily be ruptured. The contact may come too close due to misalignment or overetching of the contact. The latter is particularly likely if the contact edges must be tapered, and an oxide planarization process that results in varying oxide thicknesses has been used. Since this breakdown is likely to occur while the device is in use, device designers have been reluctant to push this rule very far in spite of the density and performance consequences.

The second effect that further reduces the performance of the MOSFET for submicron devices is the spreading resistance of the contact. For bipolar transistors and other devices with a vertical current flow, the contact resistance is proportional to the specific contact resistance as

$$R = \frac{R_c}{\text{Area}} \tag{16.15}$$

assuming that the contact size is not very small. In the case of an MOS device, however, if one assumes that the contact extends the entire width of the device, the effective contact resistance is given by

$$R = \frac{\sqrt{R_c \rho_D}}{W} \coth\left[L \sqrt{\frac{\rho_D}{R_c}}\right] \tag{16.16}$$

where ρ_D is the sheet resistance (in $\Omega/\square$) of the diffusion, L is the length of the contact, and W is the width of the transistor and contact. This equation is the 1-D transmission line model. It has two limiting forms that can be used for estimates with about 10% accuracy:

$$R \approx \frac{R_c}{WL} \qquad L < 0.6 \sqrt{\frac{R_c}{\rho_D}}$$

and

$$R \approx \frac{\sqrt{R_c \rho_D}}{W} \qquad L > 1.5 \sqrt{\frac{R_c}{\rho_D}} \tag{16.17}$$

Small contacts that are required to increase the circuit density will therefore increase the parasitic resistance. Several process solutions have been developed for this problem. The most widespread is the use of the self aligned silicidation (salicide) of the contact area. As shown in Figure 16.15, in most salicide processes the source/drain diffusions are formed first. Then, a thin metal film is deposited

(A) Form standard device up to diffusion

(B) Form sidewall oxide spacers

(C) Deposit metal, react to form silicide

(D) Selectively remove unreacted metal

(E) Final structure after glass passivation, reflow, contact opening and metallization

Figure 16-15 Common salicide process sequence (*after Ting, ©1984 IEEE*).

and allowed to react with the exposed active regions of the device. The result is a substantial increase in the metal/semiconductor contact area. The success of this approach depends critically on diffusion and silicon consumption that occur during the silicidation process. For example, one of the most popular silicides, $TiSi_2$, was originally restricted to junctions greater than 0.35 μm [16] to avoid excessive leakage. It was found that to use $TiSi_2$ on more shallow junctions, the Ti thickness must be no more than about one fifth of the original junction depth [17]. Silicides formed over polysilicon may also invert at high temperature. That is, the polysilicon diffuses through the silicide layer to the surface, and the silicide sinks until parts of it come into contact with the gate oxide. Due to the work function difference, these parts of the transistor will have a different threshold voltage, and the device will have very poorly controlled behavior. This problem can be minimized by ensuring a large grain structure in the poly before silicide formation [18].

When this problem is avoided, high temperature steps after $TiSi_2$ formation must still be limited to 900°C because of agglomeration of the silicide layer [19],[20]. Agglomeration effects are aggravated if the thin native silicon oxide, and therefore the silicide reaction, is non-uniform. To solve this problem, the silicide can be formed through a 50 Å thermal oxide. Silicides formed this way are stable against agglomeration up to 1100°C [21]. It is also possible to form the silicide first, then implant the source and drain regions into the silicide [22]. This technique can be used to form very shallow silicided junctions of $CoSi_2$ [23]. Another liability for self aligned silicides is oxygen incorporation during the thermal silicidation, leading to higher than expected resistivity and poor morphology. Specially designed furnaces can be used to minimize oxygen [24]. The more popular technique however, is to react the silicide in a rapid thermal processor. The reduced time at temperature reduces the oxygen incorporation [25].

The use of a salicide process can significantly reduce the parasitic series resistance of the MOSFET. For a salicide, the total resistance between the metal contact and the edge of the reachthrough is given by [26]

$$R = \frac{L_{ms}\rho_D\rho_S}{W\rho_D + \rho_S}$$
$$+ \frac{2\rho_S\rho_D + (\rho_S^2 + \rho_D^2)\cosh\beta L_{ms}}{W\beta(\rho_S + \rho_D)\sinh\beta L_{ms}} \quad \textbf{(16.18)}$$

Table 16.2   Properties of commonly used materials for self aligned silicides

Silicide	Resistivity ($\mu\Omega$-cm)	Anneal Temp (°C)	Dominant Diffuser	n-type Barrier (eV)
$TiSi_2$	13–16	900	Si	0.6
$TaSi_2$	35–45	1000	Si	0.59
$MoSi_2$	90–100	1100	Si	0.55
WSi_2	70	1000	Si	0.65
$CoSi_2$	16.20	900	Co	0.65
PtSi	28–35	600–800	Pt	0.86

where W is the width of the transistor, L_{ms} is the distance between the metal contact and the edge of the silicide, ρ_S and ρ_D are the sheet resistivities of the silicide and diffusion (in $\Omega/\square$), and β is given by

$$\beta = \sqrt{\frac{\rho_D + \rho_S}{R_C}} \tag{16.19}$$

where R_C is the specific contact resistance (in Ω-cm^2) between the silicide and the substrate.

The choice of which silicide to use for a salicide process hinges on the resistivity required, the number of contacts expected to n and p-type substrates, and the subsequent processing that must be done (Table 16.2). In the case of a predominantly NMOS application for example, $MoSi_2$ may be a better choice than $TiSi_2$ since its lower barrier height provides a lower R_C that more than compensates for the higher series resistance. A second consideration is the dominant diffuser. Platinum will readily form rather good contacts since the platinum diffuses to the interface, providing a low contamination contact. Silicides in which Si is the dominant diffuser have a serious problem with the salicide process called *wicking*. Wicking occurs during the silicide formation anneal when the diffusing silicon results in a thread of silicide that short circuits the gate to the source or drain. Furthermore, the oxide spacer itself, particularly if the oxide is heavily damaged during the sidewall spacer formation, can be a source of silicon for silicide formation. For these reasons $PtSi_2$ tends to be one of the most difficult to form reliably, but owing to its low resistance, it is one of the most desirable.

Example 16.1

An NMOS transistor has the dimensions shown in Figure 16.16. The source/drain diffusion has a sheet resistance of 100 $\Omega/\square$, and the surface concentration is 10^{20} cm^{-3} n-type. A 1 μm wide contact had been used on both the source and drain of an MOS transistor. This was replaced by a self aligned 10 $\Omega/\square$ silicide that extends to 0.2 μm from the gate edge. Calculate the resistances before and after the implementation of the silicided process. The contact metal and the silicide both have barriers of 0.72 V to the silicon. The mass of an electron in silicon is 1.1 m_o, and the constant "A_o" in Equation 15.14 is 10^{-12} Ω-cm^2.

The constant C_2 in Equation 15.15 is given by

$$C_2 = \frac{4\pi}{6.6 \times 10^{-34} \text{ J-s}} \sqrt{1.1 * 9.1 \times 10^{-31} \text{kg} * 11.9 * 8.84 \times 10^{-12} \text{ F-m}}$$

$$= 1.95 \times 10^{14} \text{ m}^{-3/2} V^{-1} \tag{16.20}$$

Then

$$R_c = 10^{-12} \text{ }\Omega\text{-cm}^2 \exp\left[\frac{1.95 \times 10^{14} \text{ m}^{-3/2} V^{-1} * 0.72V}{\sqrt{10^{26} \text{ m}^{-3}}}\right]$$

$$= 1.24 \times 10^{-6} \text{ }\Omega\text{-cm}^2 \tag{16.21}$$

For aluminum the contact resistance is given by

$$R = \frac{\sqrt{100 \text{ }\Omega/\text{sq} \ 1.24 \times 10^{-6} \text{ }\Omega\text{-cm}^2}}{1 \times 10^{-3} \text{ cm}} \coth\left[5 \times 10^{-5} \text{ cm} \sqrt{\frac{100 \text{ }\Omega/\text{sq}}{1.24 \times 10^{-6} \text{ }\Omega\text{-cm}^2}}\right]$$

$$= 26.5 \text{ }\Omega \tag{16.22}$$

The total resistance to the edge of the gate is

$$R = 26.5 \text{ }\Omega + 100 \text{ }\Omega/\text{sq} \frac{10^{-4} \text{ cm}}{10^{-3} \text{ cm}} = 36.5 \text{ }\Omega \tag{16.23}$$

For silicided contacts

$$\beta = \sqrt{\frac{100 \text{ }\Omega/\text{sq} + 10 \text{ }\Omega/\text{sq}}{1.2 \times 10^{-6} \text{ }\Omega\text{-cm}^2}} = 9.6 \times 10^3 \text{ cm}^{-1} \tag{16.24}$$

The total resistance for this type of contact is given by

$$R = 100 \text{ }\Omega/\text{sq} \frac{0.2 \text{ }\mu\text{m}}{10 \text{ }\mu\text{m}} + \frac{0.8 \text{ }\mu\text{m}}{10 \text{ }\mu\text{m}} \frac{100 \text{ }\Omega/\text{sq} * 10 \text{ }\Omega/\text{sq}}{100 \text{ }\Omega/\text{sq} + 10 \text{ }\Omega/\text{sq}}$$

$$+ \frac{2 * 100 \text{ }\Omega/\text{sq} * 10 \text{ }\Omega/\text{sq} + [(100 \text{ }\Omega/\text{sq})^2 + (10 \text{ }\Omega/\text{sq})^2] \cosh(9.42 \times 10^3 \text{ cm}^{-1} * 8 \times 10^{-5} \text{ cm})}{1 \times 10^{-3} \text{ cm} * 9.42 \times 10^3 \text{ cm}^{-1} * (100 \text{ }\Omega/\text{sq} + 10 \text{ }\Omega/\text{sq}) * \sinh(9.42 \times 10^3 \text{ cm}^{-1} * 8 \times 10^{-5} \text{ cm})} \tag{16.25}$$

$$R = 2.0 \text{ }\Omega + 0.7 \text{ }\Omega + 17.6 \text{ }\Omega = 20.3 \text{ }\Omega \tag{16.26}$$

The dimensions in this example are compatible with a 0.5 μm device. If one assumes an L_{eff} of 0.4 μm, a gate oxide thickness of 100 Å, and an electron mobility of 400 cm^2/V-sec, one can show that even this modest reduction in resistance can have a significant effect on the device performance. Furthermore, the resistance of the salicide is independent of the size of the metal to silicide contact and so can be scaled to very small feature sizes.

As devices are scaled into the deep submicron, a serious concern is raised about the operation of the p-channel device. If N+ poly continues to be used for the gate electrode, the well concentration necessary to prevent punchthrough produces a threshold voltage of −1.5 to −2.0 V. In order to obtain the desired value of −0.5 to −0.8, boron (a p-type dopant), must be implanted into the surface of the channel. The concentration of boron required normally forms a metallurgical junction with the n-type

10μm

Contact
metal

Poly gate   Oxide
insulator

Silicide

N+

|← 0.8μm →|← 0.5μm →|

0.2
μm

Figure 16-16   Cross section and dimensions of
MOSFET example.

dopant of the well, although at zero gate bias, this region is depleted. As the gate voltage is moved toward V_t, this doping profile causes the channel to form not at the surface, but in the bulk. The PMOS device therefore, has a buried channel for gate voltages close to threshold. While a buried channel has some minor advantages of improved mobility and some types of reliability, the threshold voltage of the device is much more difficult to control, the subthreshold leakage is larger, and the devices are much more susceptible to punchthrough.

While it is possible to achieve some reduction in buried channel behavior by minimizing the range of the boron threshold voltage adjust implant or adding an n-type implant [27], the most effective way to eliminate the problem is to ensure that both NMOS and PMOS are surface channel devices. This is commonly achieved with a symmetric gate, that is, the gate electrode material is chosen so as to have a work function that puts its Fermi energy near the center of the silicon bandgap. Three materials that have been investigated are $MoSi_2$, Mo [28], and W [29]. While MOS fabrication is possible using these materials, stress in the deposited layer tends to reduce the oxide quality. It has also been shown that at very short gate lengths, these materials allow buried channels to form in both NMOS and PMOS devices.

An alternative to the refractory metal common gate structure is the use of dual gate materials. A simple example of this approach is the use of N+ poly for the NMOS gate and P+ poly for the PMOS gate [30]. This can be done by depositing undoped polysilicon and doping it during the source/drain ion implantation step. This approach presents several difficulties. Of course, the P+ poly has a higher resistivity for the same dopant concentration since the hole mobility in silicon is lower than the electron mobility. More importantly however, a problem occurs in polysilicon interconnect. If the N+ poly and the P+ poly contact each other directly, as is commonly done for inverters and most other basic logic functions, a tunnel junction will be formed. Interdiffusion of the dopants can lead to undesirable work function variations. This is particularly a problem when a self aligned silicide is used, as the diffusivity of dopants is extremely rapid in most silicides [31]. The polysilicon on the field oxide must therefore remain undoped and thermal processing minimized after dopant introduction. Of course, the polysilicon layer must then be silicided to shunt the large resistivity of the undoped regions. The other serious problem is boron penetration of the gate dielectric [32]. Grain boundary diffusion in the poly allows the boron to diffuse rapidly to the poly/SiO_2 interface. The thermal cycles, poly thickness, oxide thickness, poly grain size, implant dose, and implant energy must all be optimized to prevent boron punchthrough of the oxide. Fluorine incorporation in the oxide is known to exacerbate this problem [33]; however, nitrided oxides appear to be effective in reducing boron penetration [34].

16.5 Hot Carrier Effects and Drain Engineering

In order to maximize performance, device scaling from 3 μm to about 1 μm has primarily followed a constant voltage rather than a constant field approach. As a result, electric fields in the devices have

● Electron
○ Hole
▣ Channel

Figure 16-17   Impact ionization process in a MOSFET operated in pinchoff.

risen sharply. The elastic mean free path for excited electrons in silicon at room temperature is about 100 Å. When the electric field is high enough such that the energy gained between collisions exceeds the bandgap of the semiconductor by a sufficient amount, impact ionization events can occur (Figure 16.17). Some of the holes so created can recombine with other high energy electrons, creating a population of extremely high energy electrons.

$$e^* \rightarrow e + e + h$$
$$e^* + e + h \rightarrow e^{**}$$

(16.27)

It is also possible for a small number of electrons to gain enough energy directly by traveling a number of mean free paths before suffering an inelastic collision. In either case, these electrons have sufficient energy to change the physical and therefore electrical structure of the device. This is known as hot carrier effects. The remainder of the holes escape into the substrate and are collected in the form of substrate current.

A few of these high energy electrons can surmount the oxide/semiconductor barrier. Once in the oxide they can cause damage directly, or they can create positive charge either by impact ionization in the oxide or by positive hydrogen ion creation at the gate electrode. This charge may become trapped in the oxide, leading the shifts in the threshold voltage and in the transconductance. It is currently believed that at least one important degradation mechanism involves the displacement of hydrogen atoms at the oxide/semiconductor interface [35]. These atoms passivate the interface, tying up unsatisfied interface bonds. Only about 0.3 eV is required to break such a bond. Once released, they leave interface states or other defects behind that may trap charge. The effect is seen primarily at the drain edge [36] and occurs most quickly when the device is held in saturation with a large drain bias and the gate biased near $V_{ds}/2$. The damage is particularly evident when the transistor is used in both forward and reverse configurations and when dealing with small devices at large electric fields. Modeling work has suggested that if constant field scaling is employed, these effects will become less important for gate lengths less than 0.2 μm [37], due to the small supply voltages. However, hot electron effects have been observed in devices as small as 0.07 μm [38]. Due to incomplete scaling of the voltage, this effect, along with power consumption, is expected to limit the maximum allowable supply voltage for deep submicron devices [39]. In recognition of this limit, new scaling methods have been developed that keep the maximum electric field, rather than the average electric field, constant [40].

One of the most effective ways to minimize hot carrier effects is to develop techniques that reduce the maximum electric field in the device. Since the total voltage that must be dropped from drain to source is not to be changed, the distance over which that voltage is dropped must be increased. In particular, at the drain edge when the device is in saturation, the drain/substrate junction operates like a reverse biased p-n diode. The maximum electric field occurs near the metallurgical junction. For a simple n-channel MOSFET with a gate oxide thickness less than 150 Å and gate lengths less than 0.5 μm, one can estimate the peak field as [41],[42]

$$\mathscr{E}_{max} \approx \frac{V_{ds} - V_{sat}}{I} \approx \frac{V_{ds} - \dfrac{\varepsilon_{sat}L[V_{gs} - V_t]}{V_{gs} - V_t + \varepsilon_{sat}L}}{1.7 \times 10^{-2} t_{ox}^{1/8} x_j^{1/3} L^{1/5}}$$

(16.28)

where ε_{sat} is the critical field for velocity saturation ($\sim$40 kV/cm), L is the gate length, and x_j is the junction depth.

Example 16.2

A transistor has $L = 0.5$ μm, $t_{ox} = 125$ Å, and $x_j = 0.2$ μm, and $V_t = 0.7$ V. Calculate the maximum electric field at several gate voltages for V_{ds} of 5 V and 3.3 V. Assume $V_{gs} = V_{ds}/2$.

Inserting the values given into Equation 16.28, the maximum electric field will be

$$\mathscr{E}_{max} = 8.6 \times 10^4 \text{ cm}^{-1}\left[V_{ds} - \frac{2[V_{gs} - V_t]}{V_{gs} - V_t + 2}\right] \tag{16.29}$$

Then the peak fields are approximately 3.6×10^5 V/cm for $V_{ds} = 5$ V and 2.3×10^5 V/cm for $V_{ds} = 3.3$ V.

At large electric fields, a hot electron need only travel about four mean free paths in order to achieve sufficient energy for impact ionization to occur. (Due to momentum conservation the electron must gain more energy than the bandgap to suffer impact ionization. This is usually taken to be about an additional 30%.) In more realistic 2-D simulations, fields approaching 10^6 V/cm have been predicted for this type of structure. The class of process modules used to reduce this peak electric field near the drain/substrate metallurgical junction are referred to as *drain engineering*, although they are usually applied to both sides of the device. These techniques are usually applied only to NMOS devices, since they are far more prone to hot carrier damage than PMOS devices.

The earliest attempts at drain engineering used a process module known as the double diffused drain [43]. In this technique both phosphorus and arsenic are implanted into the substrate to form the source and drain diffusions. Since phosphorus is a faster diffuser than arsenic, near the channel edge the predominant profile is that of the phosphorus and the junction is not as sharp. If the step junction is replaced with the profile shown in Figure 16.18, part of the depletion region will occur on the drain side. In fact, if the right phosphorus concentration is chosen all of the phosphorus may be depleted up to the arsenic diffused edge. A similarly graded profile will extend under the gate electrode as well, due to lateral diffusion. It is possible in this way to reduce the peak field by more than 20%. Even this modest reduction in field has a pronounced effect on the hot carrier injection rate. Unfortunately, it also has the effect of increasing the junction depth. Once again this raises the specter of short channel behavior. For this reason, the double diffused process has seen limited application.

Another problem with the double diffused pro-

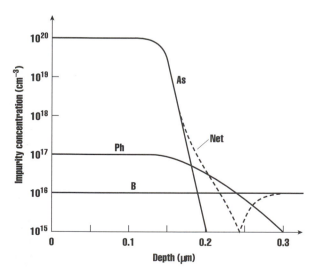

Figure 16-18   Impurity profile in a double diffused process.

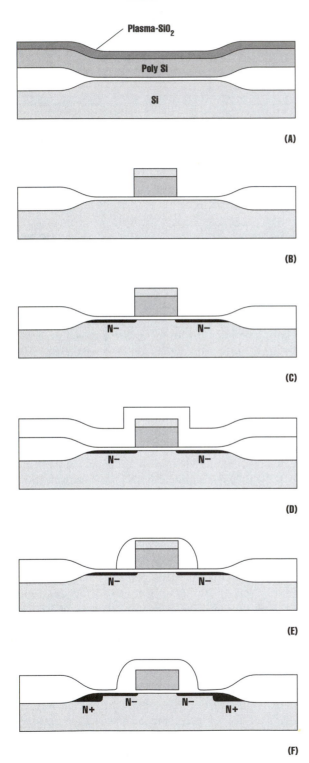

(A)

(B)

(C)

(D)

(E)

(F)

Figure 16-19 Common lightly doped drain (LDD) process sequence.

cess, aside from the junction depth, is that the final impurity profile is still determined by diffusion. The equations of diffusion (i.e., Fick's laws) are such that exponential solutions are almost guaranteed for any realistic set of boundary conditions. By using a material with a higher diffusion coefficient, the argument of the exponential can be adjusted, but it remains an exponential. It is therefore difficult to reliably produce the correct phosphorus profile.

The process module most commonly used to control hot carrier effects is the lightly doped drain (LDD) technique [44]. Since the concept of sidewall spacers was already introduced for salicide formation, it is simple to add a moderate dose implant step before sidewall formation and to reserve the heavier source/drain implant for after the sidewall formation (Figure 16.19). The moderate implant step is called the *reachthrough*. The effect of this process sequence is to put, between the drain and the channel, a region whose length and doping level can be easily controlled so as to reduce the peak electric field and with it hot electron degradation. The LDD structure however, adds additional parasitic resistance. An optimization process for the reachthrough doping, depth, and length must be carried out in which the transistor performance must be balanced against the device reliability. Furthermore, if the doping in the reachthrough is too low, the hot electrons that are trapped in the oxide above the reachthrough can deplete the reachthrough region, substantially increasing the resistance and bringing about a premature failure. To provide the best hot electron resistance, the n-region should be completely under the gate. Notice that this is not the case for the simple LDD process presented here.

Several variations of the basic LDD process, including moderately doped LDD (MLDD) [45], profiled LDD [46] and the gate-drain overlap (GOLD) structure [47] have been proposed to improve the performance of the basic LDD. The latter is designed to provide a large overlap of the gate electrode and the reachthrough. This improvement can also be obtained in a conventional LDD device through the use of high dielectric constant spacers. The larger capacitance causes the fringing fields to modulate the field in the reachthrough as it would if the reachthrough were under the gate in a conventional spacer approach [48]. The most commonly used material is Si_3N_4. Since the interface of nitride and silicon often contains a high density of interface states, a thin thermal oxide is often grown before nitride deposition. One in-

Figure 16-20 In the proposed VLED structure current flows through the n⁻ reachthrough to the elevated n⁺ drain. The peak field and therefore the injected hot electrons occur under the very lightly doped (L⁼) silicon spacer, instead of the gate oxide.

teresting proposed device is the buried LDD. In this process, a heavily doped reachthrough is formed with a high energy implant. This forces the current to flow down away from the Si/SiO$_2$ interface. Although hot carriers are still produced, they lose sufficient energy before reaching the interface so that few can penetrate the oxide. An even larger improvement can be made if the buried LDD is also graded. This is done by performing both high energy arsenic and high energy phosphorus implants [49].

The LDD process has three drawbacks. The first is the added process complexity, although when a self aligned silicide is formed, this is minimal. The second drawback is that the spacer adds a parasitic resistance between the metal contact and the channel. As a result, LDD devices are slower than non-LDD devices operating at the same voltage. The third problem with the LDD process is shadowing. Recall that most implantation is done off axis to avoid the problem of channeling. In the case of the LDD structure, the polysilicon gate acts as an effective shadow that blocks implantation of one side of the device. The result is that the reachthrough may not extend all the way under the gate. This effect results in unusual MOSFET electrical characteristics [50]. When the shadowed side is used as the drain, little performance degradation is seen. When it is used as the source, the drive current in saturation may be reduced by 30 to 40%. This effect is avoided by either implanting on axis or by ensuring that the final poly gate profile has at least a 7° slope to it. Shadowing may also be caused by a polysilicon oxidation process (called a *reox*) that is done after the gate level patterning.

For deep submicron devices, a new type of structure has been proposed. Called a hot carrier suppressed (HCS) elevated structure [51], the device starts with a conventional lightly doped reachthrough. Directly above the reachthrough is a very lightly doped layer, followed by a heavily doped contact layer. Another variation of this device, called the vertically layered elevated drain (VLED) structure [52], uses two spacer layers and an elevated drain. Closest to the gate is a conventional oxide spacer. A second, lightly doped, silicon spacer is formed (Figure 16.20). Finally, a heavily doped elevated source/drain is grown. The high field region occurs under the silicon spacer, dramatically reducing the charge injection into the oxide. While significantly more difficult to fabricate, both structures also reduce the series resistance by allowing the formation of a thick heavily doped region. This allows the fabrication of thick silicide layers without junction leakage concerns. In addition to the added process complexity, the structure's drawbacks include an increased gate/drain and gate/source capacitance.

16.6 Processing for Robust Oxides

Deep submicron MOSFETs require the use of very thin gate oxides, often in the range of 40 to 100 Å. These layers are subject to damage from hot electron effects as well as direct charge injection similar to time dependent dielectric breakdown. The resistance of the oxide to these damaging effects is called the *oxide robustness*. As discussed in Chapter 3, wet thermal oxides have a higher density of nonbridging oxygen sites and a high concentration of hydroxyl groups. Recently it has been shown that even 1 ppm of moisture in the oxygen ambient during oxidation can adversely impact the reliability of these oxides [53],[54]. Moisture may also enter the oxide from TEOS [55] or spun on glass films [56] during a high temperature step.

The other step that can be taken to improve the reliability of these thin gate oxides is the

(A)

(B)

Figure 16-21   (a) Cross section of n-well CMOS inverter with parasitic elements superimposed. (b) Equivalent circuit.

limitation of process induced damage. In particular, it is well known that energetic processes such as reactive ion etching leads to significant damage that cannot be completely removed without anneals at temperatures above 500°C. Glow discharge processes may also produce damage when the plasma leads to charging of the gate electrode or charge accumulation on the surface of the exposed oxide layer. In many cases, the polysilicon and/or metal lines that connect to a particular MOSFET gate electrode can act as an antenna, accumulating charge that makes the problem worse.

Unfortunately the energetic processes required for good resolution are also the processes that tend to damage the oxides and leave them more susceptible to degradation under electrical stress. The problem is most acute for the final steps of the technology when subsequent anneals cannot be done at high temperatures. There is some evidence that a reduction in the plasma excitation frequency [57] or modifications in the plasma chemistry can reduce the effects to some extent.

16.7 Latchup

One of the early limitations of CMOS technology was latchup. Although now well understood, latchup continues to be a concern, and scaled CMOS technologies must be constructed so as to avoid this problem. Latchup is seen as a sudden massive increase in the current draw of an IC. The current may be high enough to rupture metal interconnect lines. As a result, latchup is often destructive. If the supply current is limited, the latchup condition can be eliminated by powering down the circuit, then reapplying power.

The most common form of latchup occurs as a result of parasitic elements in a CMOS IC [58]. Figure 16.21a shows a cross section of an n-well CMOS wafer with the various parasitic elements drawn in. The transistors are connected as an inverter. Explicitly added to this picture are the substrate contacts for both the well and the p-type wafer. A lateral NPN bipolar transistor ($Q_{||}$) is formed from the drain of the NMOS transistor, the p-substrate, and the n-well. A vertical PNP bipolar transistor ($Q_\perp$) is formed from the p+ source of the PMOS transistor, the n-well, and the p-substrate. The collector of the lateral NPN is connected to the base of the PNP through the n-well. The collector of the vertical PNP is connected to the base of the NPN through the p-substrate. The equivalent circuit for the parasitics in this inverter is shown in Figure 16.21b [59].

To understand how this circuit latches, consider that at some time, a sufficiently large current is allowed to flow through the collector of the NPN transistor. This could occur by a voltage spike, an ionizing event, or some other transient [60]. The current through the substrate resistance biases the base of the PNP. If this bias is large enough, transistor action forces current to flow in the collector of the NPN transistor. This current flowing through the n-well resistance can bias the base of the NPN, which forces more current to flow into the substrate resistance. The result is a positive feedback circuit that will not turn off until the power is removed [61].

To analyze the circuit, assume forward active operation of the transistors. Then neglecting leakage,

$$I_c = \alpha I_e \tag{16.30}$$

The total current flowing from the supply is given by

$$
\begin{aligned}
I_t &= I_{c\parallel} + I_{c\perp} \\
&= \alpha_\parallel I_{e\parallel} + \alpha_\perp I_{e\perp} \\
&= \alpha_\parallel (I_t - I_s) + \alpha_\perp (I_t - I_w)
\end{aligned}
\tag{16.31}
$$

where the $\parallel$ and $\perp$ subscripts refer to the transistors parallel and perpendicular to the surface of the wafer, and α is the ratio of the emitter and collector currents of the bipolar transistors. Collecting terms,

$$I_t \left[\alpha_\perp + \alpha_\parallel - 1 - \alpha_\perp \frac{I_w}{I_t} - \alpha_\parallel \frac{I_s}{I_t} \right] = 0 \tag{16.32}$$

For a nontrivial solution,

$$\alpha_\perp + \alpha_\parallel - 1 - \alpha_\perp \frac{I_w}{I_t} - \alpha_\parallel \frac{I_s}{I_t} = 0 \tag{16.33}$$

If we express this equation in terms of β, the ratio of the base and collector currents of the transistors one can show that,

$$\beta_\perp \beta_\parallel = 1 + \frac{I_w}{I_t} \beta_\perp [\beta_\parallel + 1] + \frac{I_s}{I_t} \beta_\parallel [\beta_\perp + 1] \tag{16.34}$$

where $\beta = \alpha/(1 - \alpha)$.

Equation 16.34 points out several important effects. Since the right-hand side of Equation 16.33 must be greater than 1, the product $\beta_\perp \beta_\parallel$ must be greater than 1 for latchup to occur. If this product can be made less than unity, latchup cannot occur. For a uniformly doped ideal bipolar transistor, the gain is given by [62]

$$\beta = \frac{D_B N_E L_E}{D_E N_B W} \tag{16.35}$$

where D_B and D_E are the minority carrier diffusivities in the base and emitter, N_B and N_E are the doping concentrations in the base and emitter, L_E is the characteristic diffusion length of minority carriers in the emitter, and W is the width of the base. More generally this can be rewritten as

$$\beta = \frac{D_B N_E L_E}{D_E \displaystyle\int_0^W N(x)\, dx} \tag{16.36}$$

The lateral transistor typically has a gain less than unity and is sometimes as low as 0.01, but the vertical device often has a gain much larger than one.

The most direct ways to avoid latchup therefore, is to increase the base width, which is the

Figure 16-22 Effects of epitaxial-layer thickness on latchup *(after R. A. Martin, ©1985 IEEE).*

separation between the NMOS and PMOS transistor, to increase the base doping so that the gain product of the two transistors is less than unity, or to use guard rings that can absorb the injected charge, thereby preventing the bipolar action. Any of these approaches incur a penalty in density. There are a number of ways to minimize this penalty. These techniques either tend to reduce the parasitic resistance, and therefore increase I_{sub} and/or I_{well} which allows the technology to avoid latchup with large $\beta_{\perp}\beta_{\parallel}$ products, or they tend to reduce the gain of the parasitic transistors. The two goals sometimes overlap. Increasing the doping concentration in the substrate and the well decreases both the gain and the resistance. Of course, the well concentration at the surface must satisfy the device constraints needed by the MOSFET. This has led to the development of retrograde wells in which the doping concentration rises deep in the well. This is done using high energy implantation of phosphorus for the well shunt and a lower energy implant for the surface profile. While this technique is capable of making some improvement in latchup susceptibility, it does not eliminate it.

Two techniques have come into widespread use that dramatically reduce latchup susceptibility. In the simplest, the devices are fabricated in lightly doped epitaxial layers grown on heavily doped substrates [63]. This technique provides two avenues of relief. The low resistance substrate effectively shunts the substrate resistance, substantially increasing I_{sub} and thereby increasing the maximum allowable gain product. The substrate also serves as an effective sink for the minority carriers in the base. Since the substrate is heavily doped, minority carriers that reach the substrate quickly recombine. This prevents the bipol action of the device and effectively reduces latchup susceptibility. The effectiveness of the epi increases as the thickness of the epi decreases. Typical CMOS epi thicknesses start at about 5 μm. After the well drive, up diffusion may shrink this to about 3 μm. Figure 16.22 shows the result of epitaxial thickness on latchup susceptibility. Problems with the epi approach include the cost of the starting material and defects in the epi that reduce the IC production yield. These considerations preclude the use of epi in very cost sensitive products such as DRAMs.

Trench isolation has also proven effective in improving latchup susceptibility, particularly at small N to P spacings [64]. If the trench depth is considerably greater than the well depth, the carriers are forced to travel around the trench. This increases the effective base length and forces the current to flow in a 2-D manner that reduces the transistor gain [65]. Trench depth is found to have a much stronger impact than trench width. If a P+ ring is implanted into the bottom of the trench, latchup is further suppressed. Proper design can force the voltage necessary to maintain the latched condition to well above the 3.3 V supply used at deep submicron, making the devices latchup free, even when fabricated on bulk substrates without the implanted plug [66]. This is an extremely important result to high density memory ICs, where the trench may be used for charge storage, and the cost of epitaxial substrates is prohibitive.

16.8 Summary

This chapter reviewed the basic long channel device equations for MOSFET operation and qualitatively described the effects of short channel behavior. The description of the technology began with

Table 16.3　Projected characteristics of CMOS technology

	1992	1995	1998	2001	2004	2007
Feature Size (μm)	0.5	0.35	0.25	0.18	0.12	0.10
DRAM Size	16 M	64 M	256 M	1 G	4 G	16 G
Logic Gates/Chip	300 k	800 k	2 M	5 M	10 M	20 M
Chip Size (cm^2)						
Logic/μP	2.5	4.0	6.0	8.0	10.0	12.5
DRAM	1.32	2.0	3.2	5.0	7.0	10.0
Levels Interconnect	3	4–5	5	5–6	6	6–7
Power Supply (V)	5	3.3	2.2	2.2	1.5	1.5
Number of I/Os	500	750	1500	2000	3500	5000
Performance (MHz)						
off chip	60	100	175	250	350	500
on chip	120	200	350	500	700	1000

SIA 1992 Technology Workshop Conclusions.

the basic 3 μm flow and presented scaling options for shrinking the technology. Parasitic effects, in particular series resistance and contact resistance, were discussed and their impact on device behavior reviewed. Salicide process modules were presented as a way to reduce these effects. Large electric fields due to nearly constant voltage scaling have impacted MOS device reliability. Hot electron effects were described semiquantitatively. Double diffused drain and lightly doped drain process modules were presented as ways to improve the device reliability. Finally, CMOS latchup was reviewed, along with options for designing the technology to avoid this problem.

CMOS and BiCMOS are expected to continue to dominate the IC market. Table 16.3 lists the future of CMOS as projected by the Semiconductor Industry Association. By the year 2007, it is anticipated that the industry will be prototyping 0.1 μm CMOS on 8 in. to 16 in. wafers, although the lithographic tool for accomplishing this task is as yet undefined. Random logic will have 20 million transistors per chip, and DRAMs will be at the 16 Gbit level. Clock speeds approaching 1 GHz are expected on chip, and as many as seven levels of interconnect are expected.

Problems

1. For the nonsilicided transistor described at the end of Example 16.1, calculate the current that would flow, ignoring the parasitic resistance effects at $V_{gs} = V_{ds} = 3.3$ V. Assume $V_t = 0.6$ V. From this, calculate the voltage drops across the parasitic resistances and recalculate the current flow in the transistor. Repeat the process until the solution converges. Repeat the problem for the silicided transistor.

2. A particular metallization has a specific contact resistance of 10^{-6} Ω-cm^2. Assume that the contact is 0.5 μm long and 10 μm wide.
   (a) If the current flow is vertical, what is the contact resistance?
   (b) Repeat part (a) if the current flow is horizontal. Assume that R_s is 80 $\Omega/\square$. How much resistance does current crowding add?
   (c) Assume that the contact described in part (b) is used for an MOS transistor. If the contact edge is 1 μm from the edge of the gate, how large is the series resistance between the gate and the contact?

3. To improve the performance of the transistor described in Problem 2(c), it is decided to form a self aligned silicide on the source and drain. Assume that the silicide resistivity is 10 $\Omega/\square$ and

that the silicide extends to 0.2 μm from the gate (0.8 μm from the contact edge). Calculate the series resistance of this contact.

4. For the following process sequences, explain the function of the indicated step. For each of the two sequences, sketch the wafer cross section after the sequence is done. Be sure to show all relevant features.
   N-well implantation
   N-well drive
   Deep trench mask
   Reactive ion etch of silicon
   Field implant
   (a) Thermal oxidation
       Chemical vapor deposition of polysilicon
   (b) Plasma or reactive ion polysilicon etch
       Gate mask
       Polysilicon etch
       PMOS source/drain implant mask
       PMOS source/drain implant
   (c) Reachthrough implant
       Oxide deposition
   (d) Oxide reactive ion etch
       NMOS source/drain implant mask
       NMOS source/drain implant
       Source/drain activation anneal
       Tantalum deposition
   (e) Tantalum reaction/anneal
       Aqua regia strip

5. To reduce hot electron effects, an LDD is implemented in the transistor described in Problem 2. The LDD is doped 10^{17} cm^{-3} n-type. Ignoring depletion regions, calculate the resistance that this LDD adds.

6. Hot electron effects can also be reduced if the power supply is reduced. Discuss the tradeoffs involved in such a decision.

7. The first CMOS devices were not as prone to latch up as some of their scaled counterparts. Discuss the reasons.

References

1. S. Wolf, *Silicon Processing for the VLSI Era*, vol. 3—*The Submicron MOSFET*, Lattice Press, Sunset Beach, Calif., 1995.
2. S. M. Sze, *Physics of Semiconductor Devices*, Wiley, New York, 1981.
3. E. H. Nicollian and J. R. Brews, *Metal Oxide Semiconductor Physics and Technology*, Wiley, New York, 1982.
4. Y. P. Tsividis, *Operation and Modeling of the MOS Transistor*, McGraw-Hill, New York, 1987.
5. F. M. Wanlass and C. T. Sah, ''Nanowatt Logic Using Field-Effect Metal-Oxide-Semiconductor Triodes,'' *IEEE Int. Solid-State Circuits Conf.*, Feb. 1963.
6. J. Y. Chen, *CMOS Devices and Technology for VLSI*, Prentice-Hall, Englewood Cliffs, NJ, 1989.

7. R. Chwang and K. Yu, "CHMOS-An n-Well Bulk CMOS Technology for VLSI," *VLSI Design*, p. 42, (Fourth Quarter 1981).

8. L. C. Parrillo, L. K. Wang, R. D. Swenumson, R. L. Field, R. C. Melin, and R. A. Levy, "Twin-Tub CMOS II," *Tech. Dig. IEDM*, p. 706 (1982).

9. R. H. Dennard, F. H. Gaensslen, H. N. Yu, V. L. Rideout, E. Barsous, and A. R. LeBlanc, "Design of Ion-Implanted MOSFETs with Very Small Physical Dimensions," *IEEE J. Solid-State Circuits* **SC-9**:256 (1974).

10. Y. El-Mansy, "MOS Device and Technology Constraints in VLSI," *IEEE Trans. Electron Dev.* **ED-29**:567 (1982).

11. J. R. Brews, W. Fichtner, E. H. Nicollian, and S. M. Sze, "Generalized Guide for MOSFET Miniaturization," *IEEE Electron Dev. Lett.* **EDL-1**:2 (1980).

12. K. H. Lee, S. A. Campbell, R. Nachman, M. Reilly, and F. Cerrina, "X-ray Damage in Ultrathin Low Temperature Silicon Dioxide," *Appl. Phys. Lett.* **61**:1635 (1992).

13. M. H. White, F. Van de Wiele, and J. P. Lambot, "High-Accuracy Models for Computer-Aided Design," *IEEE Trans. Electron Dev.* **ED-27**:899 (1980).

14. P. L. Suciu and R. I. Johnston, "Experimental Derivation of the Source and Drain Resistance of MOS Transistors," *IEEE Trans. Electron Dev.* **ED-27**:1846 (1980).

15. M.-C. Jeng, J. E. Chung, P.-K. Ko, and C. Hu, "The Effects of Source/Drain Resistance on Deep Submicrometer Device Performance," *IEEE Trans. Electron Dev.* **37**:2408 (1990).

16. C.-Y. Lu, J. M. J. Sung, R. Liu, N.-S. Tsai, R. Singh, S. J. Hillenius, and H. C. Kirsch, "Process Limitation and Device Design Tradeoffs of Self-Aligned $TiSi_2$ Junction Formation in Submicrometer CMOS Devices," *IEEE Trans. Electron Dev.* **38**:246 (1991).

17. B. Davari, W.-H. Chang, K. E. Petrillo, C. Y. Wong, D. Moy, Y. Taur, M. W. Wordeman, J. Y.-C. Sun, C. C.-H. Hsu, M. R. Polcari, "A High Performance 0.25-μm CMOS Technology: II—Technology," *IEEE Trans. Electron Dev.* **39**:967 (1992).

18. S. Nygren and F. d'Heurle, "Morphological Instabilities in Bilayers Incorporating Polycrystalline Silicon," *Solid State Phen.* **23&24**:81 (1992).

19. A. Ohsaki, J. Komori, T. Katayama, M. Shimizu, T. Okamoto, H. Kotani, and S. Nagao, "Thermally Stable $TiSi_2$ Thin Films by Modification in Interface and Surface Structures," *Ext. Abs.* 21st SSDM, 1989, p. 13.

20. C. Y. Ting, F. M. d'Heurle, S. S. Iyer, and P. M. Fryer, "High Temperature Process Limitations on $TiSi_2$," *J. Electrochem. Soc.* **133**:2621 (1986).

21. H. Sumi, T. Nishihara, Y. Sugano, H. Masuya, and M. Takasu, "New Silicidation Technology by SITOX (Silicidation Through Oxide) and its Impact on Sub-Half Micron MOS Devices," *Proc. IEEE Int. Electron Dev. Mtg.* (IEDM), 1990, p. 249.

22. F. C. Shone, K. C. Saraswat, and J. D. Plummer, "Formation of a 0.1 μm n^+/p and p^+/n Junction by Doped Silicide Technology," *IEDM Tech. Dig.*, 1985, p. 407.

23. R. Liu, D. S. Williams, and W. T. Lynch, "A Study of the Leakage Mechanisms of Silicided n^+/p Junctions," *J. Appl. Phys.* **63**:1990 (1988).

24. M. A. Alperin, T. C. Holloway, R. A. Haken, C. D. Gosmeyer, R. V. Karnaugh, and W. D. Parmantie, "Development of the Self-Aligned Titanium Silicide Process for VLSI Applications," *IEEE J. Solid-State Circuits* **SC-20**:61 (1985).

25. R. Pantel, D. Levy, D. Nicholas, and J. P. Ponpon, "Oxygen Behavior During Titanium Silicide Formation by Rapid Thermal Annealing," *J. Appl. Phys.* **62**:4319 (1987).

26. D. B. Scott, W. R. Hunter, and H. Shichijo, "A Transmission Line Model for Silicided Diffusions: Impact on the Performance of VLSI Circuits," *IEEE Trans. Electron Dev.* **ED-29**:651 (1982).

27. K. M. Cham and S. Y. Chiang, "Device Design for the Submicrometer p-channel FET with N+ Polysilicon Gate," *IEEE Trans. Electron Dev.* **ED-31**:964 (1984).

28. R. F. Kwasnick, E. B. Kaminsky, P. A. Frank, G. A. Franz, K. J. Polasko, R. J. Saia, and T. B. Gorczya, "An Investigation of Molybdenum Gate for Submicrometer CMOS," *IEEE Trans. Electron Dev.* **ED-35**:1432 (1988).

29. S. Iwata, N. Yamamoto, N. Kobayashi, T. Terada, and T. Mizutani, "A New Tungsten Gate Process for VLSI Applications," *IEEE Trans. Electron Dev.* **ED-31**:1174 (1984).

30. C. Y. Wong, J. Y.-C. Sun, Y. Tsuar, C. S. Oh, R. Angelucci, and B. Davari, "Doping of N^+ and P^+ Polysilicon in a Dual-Gate CMOS Process," *IEDM Tech. Dig.*, 1988, p. 238.

31. C. L. Chu, C. Saraswat, and S. S. Wong, "Characterization of Lateral Dopant Diffusion in Silicides," *Proc. IEEE Int. Electron Dev. Mtg. (IEDM)*, 1990, p. 245.

32. J. Y. Sun et al., "Study of Boron Penetration Through Thin Oxide with P^+ Polysilicon Gate," *Proc. 1989 Symp. VLSI Technol.*, 1989, p. 17.

33. J. M. Sung, C. Y. Lu, M. L. Chen, S. J. Hillenius, W. S. Lindenberger, L. Manchanda, T. E. Smith, S. J. Wang, "Fluorine Effect on Boron Diffusion of P+ Gate Devices," *Tech. Dig. IEDM*, 1989, p. 447.

34. Y. Sato, K. Ehara, and K. Saito, "Enhanced Boron Diffusion Through Thin Silicon Dioxide in a Wet Oxygen Atmosphere," *J. Electrochem. Soc.* **136**:1777 (1989).

35. C. Hu, S. C. Tam, F.-C. Hsu, P. K. Ko, T.-Y. Chan, and K. W. Terrill, "Hot-Electron-Induced MOSFET Degradation—Model, Monitor, and Improvement," *IEEE Trans. Electron Dev.* **ED-32**:375 (1985).

36. S. Baba, A. Kita, and J. Ueda, "Mechanism of Hot Carrier Induced Degradation in MOS-FET's," *IEDM Tech. Dig.*, 1986, p. 734.

37. L. Hendrickson, Z. Peng, J. Frey, N. Goldsman, "Enhanced Reliability of Si MOSFETs with Channel Lengths Under 0.2 Micron," *Solid-State Electron.* **33**:1275 (1990).

38. H. Kurino, H. Hashimoto, Y. Hiruma, T. Fujimori, and M. Koyanagi, "Photon Emission from 70 nm Gate Length MOSFETs," Tech. Dig. *IEEE Int. Electron. Dev. Mtg. (IEDM)*, 1992, p. 1015.

39. H. Hazama, M. Iwase, and S. Takagi, "Hot Carrier Reliability in Deep Submicrometer MOS-FET's," *Proc. IEEE Int. Electron. Dev. Mtg. (IEDM)*, 1990, p. 569.

40. K. Taniguchi, K. Sonoda, and C. Hamaguchi, "Physical Limitations of Ultrasmall MOSFETs: Constant Energy Scaling and Analytical Device Model," *Proc. 22nd Int. Conf. Solid State Dev., Mater. (SSDM)*, 1990, p. 825.

41. C. Sodini, P. K. Ko, and J. L. Moll, "The Effect of High Fields on MOS Devices and Circuit Performance," *IEEE Trans. Electron Dev.* **ED-31**:1386 (1986).

42. J. Chung, M.-C. Jeng, G. May, P. K. Ko, and C. Hu, "New Insight into Hot-Electron Currents in Deep-Submicrometer MOSFETs," *Tech. Dig. IEDM*, 1988, p. 200.

43. R. H. Dennard, F. H. Gaensslen, H. Yu, V. L. Rideout, E. Bassous, and A. R. LeBlanc, "Design of Ion Implanted MOSFETs with Very Small Physical Dimensions," *IEEE J. Solid State Circuits* **SC-9**:256 (1974).

44. S. Ogura, P. J. Tsang, W. W. Walker, P. L. Critchlow, and J. F. Shepard, "Design and Characteristics of the Lightly Doped Drain-Source (LDD) Insulated Gate Field-Effect Transistor," *IEEE Trans. Electron Dev.* **ED-27**:1359 (1980).

45. M. Kinugawa, M. Kakuma, S. Yokogama, and K. Hashimoto, "Submicron MLDD NMOS-FET's for 5 V Operation," *VLSI Symp. Tech. Dig.*, 1985, p. 116.

46. Y. Toyoshima, N. Nihira, and K. Kanzaki, "Profiled Lightly Doped Drain (PLDD) Structure for High Reliable NMOS-FETs," *VLSI Symp. Tech. Dig.*, 1985, p. 118.

47. R. Izawa, T. Kure, S. Iijima, and E. Takeda, "The Impact of Gate-Drain Overlapped LDD (GOLD) for Deep Submicron VLSI's," *IEDM Tech. Dig.*, 1987, p. 38.

48. T. Mizuno, T. Kobori, Y. Saitoh, S. Sawada, and T. Tanaka, "Gate-Fringing Field Effects on High Performance in High Dielectric LDD Spacer MOSFETs," *IEEE Trans. Electron Dev.* **39**:982 (1992).

49. C. Wei, J. M. Pimbley, and Y. Nissan-Cohen, "Buried and Graded/Buried LDD Structures for Improved Hot-Electron Reliability," *IEEE Electron Dev. Lett.* **EDL-7**:380 (1986).

50. J. Pfiester and F. K. Baker, "Assymmetrical High Field Effects in Submicron MOSFETs," *Tech. Dig. IEDM*, 1987, p. 51.

51. A. F. Tasch, H. Shin, and C. M. Maziar, "A New Vertical Structural Approach for Reducing Hot Carrier Generation in Deep Submicron MOSFET's," *VLSI Tech. Symp. Dig.*, 1990, p. 43.

52. M. Orlowski, C. Mazure, and M. Noell, "A Novel Elevated MOSFET Source/Drain Structure," *IEEE Electron Dev. Lett.* **12**:593 (1991).

53. R. B. Fair and R. C. Sun, "Threshold Voltage Instability in MOSFETs Due to Channel Hot-Hole Emission," *IEEE Trans. Electron Dev.* **ED-28**:83 (1981).

54. Y. Ohji, Y. Nishioka, K. Yokogawa, K. Mukai, Q. Qiu, E. Arai, and T. Sugano, "Effects of Minute Impurities (H, OH, F) on SiO_2/Si Interface as Investigated by Nuclear Resonant Reaction and Electron Spin Resonance," *IEEE Trans. Electron Dev.* **37**:1635 (1990).

55. M. T. Takagi, I. Yoshii, and K. Hashimoto, "Characterization of Hot-Carrier Induced Degradation of MOSFETs Enhanced by H_2O Diffusion for Multilevel Interconnect Processing," *Tech. Dig. IEDM*, 1992, p. 703.

56. N. Lifshitz and G. Smolinsky, "Hot Carrier Aging of the MOS Transistor in the Presence of Sin-On Glass at the Interlevel Dielectric," *IEEE Electron Dev. Lett.* **12**:140 (1991).

57. K. Tsunokuni, K. Nojiri, M. Nawata, "Dependence of RF Frequency on Gate Oxide Breakdown During Microwave Plasma Etching," *Ext. Abs. 38th Mtg. Jpn. Soc. Appl. Phys.*, 1991, p. 499.

58. G. H. Hu, "A Better Understanding of CMOS Latchup," *IEEE Trans. Electron Dev.* **ED-31**:62 (1984).

59. J. E. Hall, J. A. Seitchik, L. A. Arledge, P. Yang, and P. K. Fung, "Analysis of Latchup Susceptibility in CMOS Circuits," *IEDM Tech. Dig.*, 1984, p. 292.

60. T. Ohzone and H. Iwata, "Transient Latchup Characteristics in n-Well CMOS," *IEEE Trans. Electron Dev.* **39**:1870 (1992).

61. A. Herlet and K. Raithel, "Forward Characteristics of Thyristors in the Fired State," *Solid-State Electron.* **9**:1089 (1966).

62. For example, G. W. Neudeck, *The Bipolar Junction Transistor, Modular Series on Solid State Devices*, Addison-Wesley, Reading, Mass., 1989.

63. R. R. Troutman, *Latchup in CMOS Technology*, Kluwer, Norwell, Mass., 1986.

64. R. D. Rung, "Trench Isolation Prospects for Application in CMOS VLSI," *IEDM Tech. Dig.*, 1984, p. 574.

65. S. Bhattacharya, S. Banerjee, J. Lee, A. Tasch, and A. Chaterjee, "Design Issues for Achieving Latchup-Free Deep Trench-Isolated, Bulk, Non-Epitaxial, Submicron CMOS," *IEDM Tech. Dig.*, 1990, p. 185.

66. S. Bhattacharya, S. Banerjee, J. C. Lee, A. F. Tasch, and A. Chatterjee, "Parametric Study of Latchup Immunity of Deep Trench-Isolated, Bulk, Non-Epitaxial CMOS," *IEEE Trans. Electron Dev.* **39**:921 (1992).

Chapter 17

GaAs FET Technologies

This last chapter discussed the dominant silicon FET technology, CMOS. This chapter will review the same topics for GaAs. Since the basic device operation is similar, only the differences in terminology between MOSFETs and MESFETs will be pointed out. The chapter will then review the basic depletion mode MESFET process. GaAs process technologies are much simpler than silicon technologies. This is because the transistor count is less, obviating the need for more than two metal layers because excellent electrical isolation can be achieved with simple processes, and because complementary technologies are almost never used. Once the basic MESFET technology is described, digital applications will be discussed. Enhancement/depletion (E/D) and other performance enhancing technologies, such as self aligned gate processes, will then be introduced. Since microwave analog circuits require very high speed and low noise, they are a good application for GaAs with its large electron mobility and semi-insulating substrate. Microwave technologies typically include a number of unique structures. They will be covered in a separate section. Finally, the chapter will review the application of heterojunctions to GaAs technologies from a processing standpoint.

17.1 Basic MESFET Operation

Figure 17.1 shows the cross section of a typical mesa isolated GaAs MESFET [1]. N-channel devices are used almost exclusively because of the small mobility of holes in GaAs. The threshold voltage of MESFETs is simpler than that of MOSFET transistors, since there is no oxide to support a voltage drop. In a simple depletion mode (d-mode) MESFET, the threshold voltage is given by [2]

$$V_T = V_{bi} - V_{po} \tag{17.1}$$

The pinch off voltage V_{po} is

$$V_{po} = \frac{qN_d d^2}{2k_{\text{GaAs}}\varepsilon_o} \tag{17.2}$$

where d is the thickness of the channel and N_d is the channel doping concentration.

The current voltage characteristics of MESFETs are very similar to those of MOSFETs. As-

Gate electrode

Source ohmic contact

n-channel

Drain ohmic contact

Si GaAS substrate

Figure 17-1   Cross section of a simple mesa isolated MESFET.

suming for the moment that the contact and series resistance are small (not generally a very good assumption in MESFETs), in the linear region the drain to source current is given by

$$
I_{ds} = \frac{q\mu_o N_d W d}{L} \left[V_{ds} - \frac{(V_{ds} + V_{bi} - V_{gs})^{3/2} - (V_{bi} - V_{gs})^{3/2}}{V_{po}^{1/2}} \right] \quad \textbf{(17.3)}
$$

In the saturation region the drain to source current is given by

$$
I_{ds} = \frac{q\mu_o N_d W d}{L} \left[\frac{V_{po}}{3} + \frac{2}{3} \frac{(V_{bi} - V_{gs})^{3/2}}{V_{po}^{1/2}} + V_{gs} - V_{bi} \right] \quad \textbf{(17.4)}
$$

17.2 Basic MESFET Technology

A wide variety of GaAs MESFET technologies have been developed. This section will introduce a basic d-mode technology [3]. The process flow is shown in Figure 17.2. Either three or five masking levels are required, depending on whether one or two levels of metal are needed. A semi-insulating substrate is first coated with a thin layer of Si_3N_4 and then implanted with silicon to form the active layer. The silicon implant must be heavy enough to swamp the Cr or EL2 density that makes the substrate semi-insulating. Typically, this requires a dopant concentration of 2 to 6×10^{17} cm^{-3}. Next, the implant is activated in a high temperature anneal. The activation can be done in a furnace or in a rapid thermal processor [4]. A typical furnace activation process is 850°C for 20 min. The nitride layer protects the wafer from contamination during the implant and serves as a barrier to the out diffusion of arsenic. The state of the semi-insulating substrate, particularly with regard to its boron and carbon impurity concentration and its residual defect density, is critically important in the uniformity of the resistivity of the activated channel [5]. Following the anneal, the nitride is removed and diffused ohmic contacts are formed by evaporating a Ni/AuGe sandwich (see Section 15.7) using a liftoff (see Section 11.8) process, then sintering the contacts at about 450°C. The source to drain contact separation is usually 3 to 4 μm. This is the last high temperature step of the technology.

After ohmic contact formation, the mesas are isolated by wet chemically etching the field regions through the active layer to the semi-insulating substrate. At this point, the pinchoff voltage of the transistor is measured using a mercury probe, or the drain to course ungated current voltage characteristic is measured directly. If the pinchoff voltage must be adjusted, it is done by wet chemically recessing the gate to the desired value. As described in Chapter 11, common wet etchants for GaAs include various proportions of H_2SO_4, H_2O_2, and H_2O [6],[7]. For maximum repeatability, a slow etch is required. Since the ohmic contacts have already been formed, this measurement can be done during the wet etch and the current can be used as an endpoint detection method.

The Schottky gate electrode is then deposited. As discussed in Chapter 15, almost any metal will form a Schottky barrier to moderately doped GaAs. Because of the high diffusivity of Ga in most

Figure 17-2 Process flow for a simple mesa-isolated MESFET. The first step can be replaced with a Si implant and anneal as discussed.

metals however, the metal layer is chosen for its stability with the GaAs channel [8]. Furthermore, the metal must have excellent adhesion to GaAs. Two of the most popular gate metal systems for simple MESFETs are Ti/Pt/Au and Ti/Pd/Au. For these two structures, the Schottky barrier metal (Ti) is very thin, usually 500 to 1000 Å. The barrier metal (Pt or Pd) is also about 500 Å thick. The gold serves as the low resistance interconnect and is typically 2000 to 5000 Å thick. This metal stack may also be applied to the top of the ohmic contacts to further lower their series resistance, if a separate gold deposit was not done on top of the Ni/AuGe layers.

The gate electrode is not necessarily positioned midway between the source and drain electrodes as it is in most MOSFETs. It is essential to minimize the gate to drain capacitance, particularly in microwave technologies, as this parasitic is multiplied by the Miller effect, dramatically degrading performance. It is often a good tradeoff in these designs to opt for the increased series resistance of placing the gate further from the drain than the source. In 1 μm gate length microwave devices, gate to source spacings are typically 1 μm and gate to drain spacings are typically 2 μm.

After gate formation, it is only necessary to add a second layer of interconnect if desired. An insulating layer of 0.5 to 1.0 μm of polyimide, SiO_2, or Si_3N_4 is deposited. SiO_2 and Si_3N_4 must be deposited by plasma enhanced CVD (Section 13.7), since thermal CVD in these materials must be carried out at high temperatures. Finally, contacts are opened up and a second layer of metal is deposited. Ti/Au is typically used for this level. The second level of metal is often ion milled rather than lifted off, so that thick layers (about 1 μm) of metal can be used to lower the resistivity.

17.3 Digital Technologies

This section will review the three design approaches commonly used in GaAs circuits. It will then cover enhancements to the basic FET technology currently being used to fabricate these circuits. The basic process technology that was described in the previous section may be used to build two important classes of GaAs ICs. The fastest, but highest power approach is buffered FET [9] logic (BFL), shown in Figure 17.3a. The two diodes near the output terminal are required to shift the output voltage back to the input level. Due to its large power dissipation (typically more than 5 mW per gate) and the

Figure 17-3 (a) Buffered FET logic (BFL). (b) Schottky diode FET logic (SDFL). (c) Direct coupled FET logic (DCFL).

large area required to lay out a gate, BFL is usually restricted to small and medium levels of integration.

A typical gate designed with Schottky diode FET [10] logic (SDFL) is shown in Figure 17.3b. The level shifting necessitated by the use of d-mode FETs is done by using diodes at the input stage to provide the logical OR input function. Power dissipation of SDFL is typically 1 mW per stage, but the gate delay is about twice as large as BFL. Due to the reduced power consumption, ICs with as many as 10,000 gates can be built in SDFL.

Direct coupled FET logic [11] (DCFL), shown in Figure 17.3c, has the least power consumption of the three approaches, less than 0.5 mW per gate. The name comes from the fact that the output of the basic inverter structure does not involve any diode drops. Because of this, no level shifting is required. Combined with its compact layout, the low power consumption allows ICs with over 65,000 transistors per chip to be built [12], although complexities of current commercial devices are about half that. DCFL requires the fabrication of both enhancement and depletion mode transistors on the same chip, however. Noise margins are small and the gate delay is somewhat larger than SDFL.

One of the most popular design approaches for VLSI GaAs, DCFL requires the simultaneous fabrication of enhancement and depletion mode transistors. In Equations 17.1 and 17.2, it was shown that the threshold voltage depends on the built in voltage, the dopant concentration in the channel, and the channel thickness. In principle, any of these three parameters could be varied in order to achieve an E/D technology. Varying the channel thickness was initially the most popular approach. As shown in Figure 17.4, both devices are recessed to the thickness desired for the d-mode transistors. The d-mode transistors are then masked off with photoresist and the e-mode devices are further etched to obtain the desired V_{po}. The difficulty of this approach is that the pinchoff voltage varies as the square of the thickness. DCFL designs have small noise margins and therefore require very careful control of the pinchoff voltage. Simple wet chemical etching, the most common way to achieve this recess, is notoriously nonuniform and has only moderate reproducibility. Plasma etch recesses have improved uniformity but tend to leave residual damage that shifts the barrier voltage.

An alternate approach to selective recessing is selective implantation. In this approach the initial implant step of the basic FET process is replaced by a lower concentration implant to set the e-mode threshold. The e-mode device areas are masked off, and the d-mode device areas given an additional implant to achieve the desired threshold. Due to its improved uniformity, the selective implantation process is now the most widely used for E/D technologies.

As discussed for MOSFETs, the transistor performance can be degraded by the presence of a series resistance between the source and drain metal and the channel. This can be particularly severe in GaAs technologies, since the traditional MESFET process is not self aligned. Furthermore, the alloyed contact resistance is a strong function of the doping concentration in the channel layer. The primary difficulty in forming self aligned source/drain structures is the lack of an insulating layer between the gate and the channel. If the heavily doped region formed by the alloyed ohmic contact

	Step 1- Epitaxial growth
	Step 2- Ohmic contact
	Step 3- Gate recess
	Step 4- e-mode recess
	Step 5- Gate Schottky
	Step 6- Meso etch

d-mode   e-mode

Figure 17-4 Process flow for an etched-channel E/D technology.

comes into contact with the gate, it will form an ohmic contact to the gate as well. In recent years a great deal of effort has gone into forming advanced gate structures in GaAs. Collectively these devices are sometimes called self aligned gate FETs (SAGFETs).

Two methods of producing self aligned source/ drain contacts to the channel are shown in Figure 17.5. In the self aligned implantation N+ technology (SAINT) [13], the surface is Si_3N_4 passivated, and a trilayer resist sandwich is used to define the N+ implant. The bottom resist is then undercut between 0.1 and 0.2 μm per side and a dielectric deposited at low temperature. The dielectric is lifted off to define the gate region. At this point, a high temperature anneal can be done to activate the N+ implant. Finally, a Schottky metal is deposited for the gate electrode. Since the dielectric isolates the N+ from the gate, excess leakage is avoided.

In the T-gate approach, a procedure more similar to traditional silicon technologies is used [14]. The Schottky metal is applied first. The second layer above the metal may be another metal, or it may be the photoresist itself. As with the SAINT process, the lower level is now intentionally undercut 0.1 to 0.2 μm per side and the N+ implantation done. The upper masking layer is then removed and a thin Si_3N_4 anneal capping layer is deposited. The implant is then activated and the nitride patterned to allow the ohmic contact formation. Finally, the ohmic metal is deposited and annealed. The difficulty of the T-gate process is in finding a suitable Schottky gate material that will tolerate the high temperature implant activation cycle. Most of the materials used for this purpose are refractory metals or their silicides. WSi_2 and PtSi are two of the most popular choices.

A T-gate variation widely used in state of the art GaAs ICs today is a sidewall spacer offset of the source/ drain implants [15]. This process is similar to the silicon salicide LDD. The major advantage of these processes is that the ohmic contact is formed self aligned to the gate electrode, minimizing the parasitic series resistance. When very low resistance gate electrodes are desired, a substitutional gate process can be used. In these processes, the gate electrode spacing is reserved by a photoresist image of the gate [16]. This image is used to lift off an insulating layer, leaving a window for the eventual deposition of the gate electrode. The process has also been combined with the LDD method to produce a self aligned substitutional gate [17].

At deep submicron gate lengths (<0.5 μm), MESFET characteristics are severely degraded by the same short channel effects as those that silicon MOSFET scaling is intended to prevent. Recall that in MOS devices, short channel behavior arose because the depletion region of the reverse biased drain junction modulated the effective charge under the gate electrode. In the case of MESFETs, there is no junction. The ohmic contacts are the same polarity as the channel. The charge under the gate can still be affected by the source/drain diffusions, however. This effect arises because the carriers

Figure 17-5 The self aligned gate process implantation N+ technology (SAINT) *(after Yamasaki, et al., ©1982 IEEE)* and the T-gate process. The order of the last two steps of the T-gate may be reversed to reduce any isolation degradation.

(A)

(B)

(C)

(D)

Figure 17-6   Process flow that employs both a buried p-layer and an LDD *(after Shimura et al., ©1992 IEEE).*

are spread over a characteristic distance known as the Debye length. Thus, even if the ohmic regions are atomically abrupt, the carriers they contribute will spill over into the region under the Schottky gate. As sufficiently short source to drain separations, this effect gives rise to short channel behavior. Since the Debye length is much less than the depletion width, short channel behavior is not usually seen in MESFETs unless the gate length is less than 0.5 μm.

In self aligned gate processes, the minimum alloyed contact to gate spacing results in more pronounced short channel and gate leakage effects. For the smallest devices, the source/drain implant must be done in stages. The LDD like SAGFETs are particularly well suited for this. To further prevent short channel behavior, a buried p-layer can be used under the channel [18]. This layer is commonly formed by high energy implantation of Mg. Figure 17.6 shows a process that combines the buried p-layer and the LDD [19].

17.4 MMIC Technologies

Monolithic microwave IC (MMIC) technologies span a broad range of circuits, from power amplifiers to mixers to transmit/receive modules. Applications include communication such as cellular telephone, direct-broadcast satellite, data links, CATV, radar transmission and detection, and even automobile collision avoidance systems [20]. The technology must often be tuned to the application. This section will therefore describe a "generic" MMIC technology that approximately fits a number of applications. Figure 17.7 shows a cross section of a typical MMIC showing some common analog components [21].

The MMIC technology starts with the basic FET technology described in Section 17.2. As previously mentioned, the gate electrode may be noncentered, although self aligned technologies are gaining more acceptance for MMICs. For power applications, a comb structure may be used for the gate electrode, with alternating sources and drains. Of critical importance is the ability to define very short gate lengths, since the unity gain frequency f_T is proportional to the inverse of the effective channel length. Shorter gate lengths have the advantage of lower noise figures as well. As a result, the current generation of MMICs have gate lengths of less than 0.25 μm. Since the transistor count is low, the gate level of many MMICs is often fabricated by electron beam lithography. In addition to the basic transistor structure, several lumped and distributed passive elements must typically be developed.

Many analog circuits require the use of capacitors and inductors. They may be used to adjust

Figure 17-7 Cross-sectional view of a typical MMIC (*after Decker*).

Figure 17-8 Interdigitated and simple overlay capacitor.

the signal phase, to impedance match a source and load, or to filter a signal. Capacitors may be formed in two ways. Interdigitated capacitors can be formed on a single layer of metal, but typically have capacitances of less than 1 pF. Furthermore, the value of the capacitance is determined by the lithographically defined spacing. This dimension is difficult to control. When a larger or more precisely controlled capacitance is required, an overlay capacitor (Figure 17.8) can be used. The dielectric in overlay capacitors is most commonly Si_3N_4, although SiO_2, Al_2O_3, and polyimide have also been used.

There are three methods for making inductors in MMICs. Metal thickness in all three types is typically several microns to reduce the resistivity and minimize skin losses. Straight line inductors are used at the highest frequencies, but typically have too low an inductance (<1 nH) to be useful in most applications. Single loop ''Ω'' inductors are also easy to form but are limited in inductance to a few nanohenries. The spiral inductor can be made with inductances of up to 50 nH but requires two levels of metal with an underpass. This underpass also represents an unwanted capacitance that must be minimized.

An air bridge process module is often used in forming spiral inductors and elsewhere in the MMIC technology to minimize parasitic capacitance. The air bridge process module is shown in Figure 17.9. A thick polymer such as photoresist is patterned, opening up holes to the substrate. The openings must be large and the polymer sidewalls must be well tapered. A thick metal layer is then deposited and the polymer dissolved. If the step coverage is adequate, the metal will not lift during the dissolution, but rather will be left behind. If long runs of metal are required, a series of these support posts may be used. Air bridges have the lowest possible dielectric coefficient. They have also proven to be rugged and reliable. No final passivation is allowed on MMICs that employ air bridges however, since the CVD process is conformal and would refill the gaps. Figure 17.10 shows a completed spiral inductor using an air bridge crossover [22].

Gold air bridges are often formed for small MMICs because of gold's low resistivity. Due to gold's malleability, these structures require a high density of support posts when the technology is extended to LSI levels of integration. The maximum distance

Figure 17-9   Air bridge process flow.

Figure 17-10   Completed spiral inductor using an air bridge crossover *(after Sciater, reprinted by permission, TAB Books, Inc.).*

between pillars is typically 50 μm. RhAuRh bridges have been demonstrated to allow pillar spacings of up to 300 μm [23], due to the larger Young's modulus of rhodium. It has been shown that U-shaped wires can be used for very long stretches. Even gold can be run up to 3 mm with no loss in yield using this technology [23].

At high frequency it is essential that the metal interconnects have a controlled, reproducible impedance. Lines must also be well shielded from each other to avoid crosstalk. Furthermore, the line loss at frequency must be minimized. Finally, it is essential that a stable ground voltage be provided. At high source inductances, the FETs themselves become less stable. There are two choices for fabricating these interconnects: coplanar and microstrip waveguides (Figure 17.11). Coplanar waveguides terminate the field lines associated with the waveguide with parallel ground lines. These lines should be made as wide as possible and close to the signal line. For low density MMICs, they may be wire bonded directly to a large ground plane surrounding the chip.

Microstrip waveguides use the back of the wafer as the ground plane. The wafer must first be thinned from over 500 μm to about 100 μm. This is commonly done by lapping the wafer in a slurry of water and abrasive materials such as alumina and silicon carbide. After lapping, the backside is commonly polished using a fine abrasive and/or wet chemicals. Wafer thinning also has the advantage of reducing the thermal resistance of the wafer. This is an important consideration, since high f_T's require large dc currents, and GaAs has poor thermal conductivity. A tradeoff must be made between the depth of the through holes and the mechanical strength of the wafer. Control of the final thickness uniformity is critical to obtain a consistent impedance.

After the through holes are patterned and etched, usually wet chemically, the backside is gold coated. Alignment of the frontside and backside must be done using an infrared aligner. The metal is opaque to the IR, while the substrate is transparent. An IR camera and monitor are used to view the image. To increase the device density, a variety of authors have reported anisotropic dry etching of through holes [25],[26].

17.5 MODFETs

An extremely important feature of several of the III-V materials systems is the ability to grow epitaxial heter-

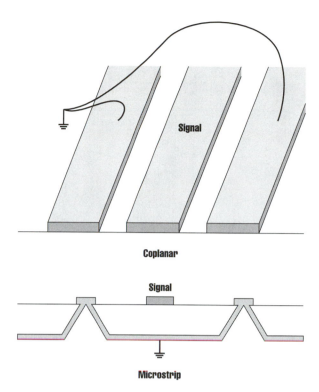

Coplanar

Signal

Microstrip

Fgure 17-11   Coplanar and microstrip waveguides for MMIC interconnect.

ostructures (see Chapter 14). While many devices have sprung from heterostructure technology, this section will very briefly review one such device, the modulation doped field effect transistor (MODFET), also known as the HEMT. We review this not from a device standpoint, about which a great deal has been written, but instead from a processing perspective.

A cross section of a simple MODFET is shown in Figure 17.12. The basic concept of the device is to heavily dope the wide bandgap material (typically $Al_{.25}Ga_{.75}As$) to provide the carriers. Due to the proximity of the narrower bandgap GaAs, the free carriers from the AlGaAs will become trapped by the band discontinuity at the GaAs/AlGaAs interface (Figure 17.12), creating a 2-D electron gas. This has the effect of separating the mobile carriers from their ionic impurity atoms, reducing coulomb scattering and improving the device performance, particularly at low temperature where the scattering contribution from phonons is no longer dominant. To further improve performance, thin (50 Å) undoped AlGaAs spacer layers are frequently grown between the GaAs and AlGaAs layers to further isolate the carriers from the charge centers. It is now common to observe electron mobilities in excess of $100,000 \ cm^2/V\text{-}sec$ at 77 K for the 2-D gas. The ability of the heterointerface to trap carriers close to the gate, along with the enhanced mobility, results in a large transconductance.

MODFETs are not without problems. The defect density of MBE material is a serious issue for manufacturing GaAs ICs based on MODFETs. Deep level traps associated with silicon doping in the AlGaAs can also shift the threshold voltage of the device by several tenths of a volt when they are charged. The effect is most evident at low temperature when the devices are exposed to light. The absorbed optical energy is enough to deplete the traps and increase the carrier concentration. V_t shifts of this order of magnitude are particularly troublesome when operating voltages are limited to less than 2 V as they are for many GaAs digital design approaches. Finally, MODFETs require several process changes from the basic MESFET technology.

The cross-sectional view of the MODFET (Figure 17.12) shows that the structure is quite comparable to a MESFET. The major portion of the effort in developing a MODFET technology is in the growth and in the device design. If a good self aligned MESFET technology is available, only minor modifications are required to form a MODFET technology. Instead of ion implanting the surface or growing a simple conducting epitaxial layer, the technology starts with the epitaxy of the heterostructure on a semi-insulating substrate. The normal MESFET process sequence follows. Device isolation is quite straightforward since the epitaxial layer thickness is typically 2500 Å or less.

In order to extract the optimal performance of the device, low resistance, self aligned source and drain contacts are essential [27] to MODFETs. These may be accomplished using a process such as SAINT. Special care must be used during the implant activation cycle. Typically done at 800 or 850°C, dopant diffusion from the AlGaAs to the GaAs can completely eliminate the desired mobility

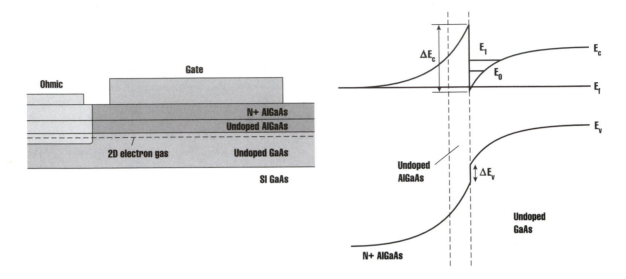

Figure 17-12 Structure and band diagram for a simple MODFET.

enhancement. To avoid this problem, rapid thermal annealing (Section 6.8) processes have been developed. The use of undoped spacer layers also allows some minor diffusion without dramatic performance degradation. Compositional gradients to provide a built in electric field that opposes dopant diffusion have also been tried. These compositional gradients can also be used to lower the effective barrier height for ohmic contact formation.

17.6 Summary

Basic MESFET processes have been presented for both digital and MMIC technologies. To obtain the maximum in performance, self aligned structures are required. To avoid excessive gate leakage, careful attention must be paid with regard to the spacing between the gate edge and the source/drain implant. Several process flows were presented to accomplish this. MMICs also require the use of such unique structures as air bridges, through hole vias, and coil inductors. MODFETs make use of advanced growth processes to improve device performance, particularly at low temperature. In these devices, careful control of dopant diffusion and device parasitics is essential.

Problems

1. Why isn't gold used as the Schottky gate? What would you expect to happen if the Pt barrier metal in a Au/Pt/Ti sandwich is too thin?
2. A spiral inductor is to be made using a metal layer that is 1.5 μm thick. Assume that the minimum metal linewidth and space is 1.0 μm, and the inductor is 50 μm in diameter. Estimate the resultant inductance. (Hint: Consider the inductor to be made up of a series combination of inductors with different radii. Look up a formula for the inductance of a loop of wire.) Com-

pare this to the figure given (50 nH). What could be done to increase the inductance if for example, a 1 μH inductor were needed?

3. Refer back to Chapter 3. If a MODFET with 50 Å spacers is doped with sulphur (S) and annealed at 800°C, how long can it be annealed before the characteristic diffusion length $(\sqrt{Dt})$ is equal to the spacer length? Assume intrinsic diffusivity only.

4. Air bridges have very low capacitance, but are rarely applied to silicon technologies. Why not?

References

1. For a thorough review of GaAs device physics, see M. Shur, *GaAs Devices and Circuits,* Plenum, New York (1987).
2. C. A. Mead, ''Schottky Barrier Gate Field Effect Transistors,'' *Proc. IEEE* **54**:307 (1966).
3. J. Mun, ''GaAs Digital Integrated Circuits,'' in *GaAs for Devices and Integrated Circuits,* H. Thomas, D. V. Morgan, J. E. Aubrey, and G. B. Morgan, eds., IEE—Peregrinus, London (1986).
4. M. Kuzuhara, H. Kohzu, and Y. Takayama, ''Infrared Rapid Thermal Annealing of Si-Implanted GaAs,'' *Appl. Phys. Lett.* **41**:755 (1982).
5. F. Orito, K. Watanabe, Y. Yamada, O. Yamamoto, and F. Yajima, ''Effects of Semi-Insulating GaAs Substrate Properties on Silicon Implanted Layer,'' *Proc. GaAs IC Symp.,* 1990, p. 321.
6. S. Adachi and K. Oe, ''Chemical Etching Characteristics of (001) GaAs,'' *J. Electrochem. Soc.* **130**:2427 (1983).
7. D. N. McFadyen, ''On the Preferential Etching of GaAs by H_2SO_4-H_2O_2-H_2O,'' *J. Electrochem. Soc.* **130**:1934 (1983).
8. R. E. Williams, *Gallium Arsenide Processing Techniques,* Artech, Dedham, Mass., 1984.
9. R. V. Tuyl and C. A. Liechti, ''High Speed Integrated Logic with GaAs MESFET,'' *IEEE J. Solid State Circuits* **SC-9**:269 (1974).
10. R. C. Eden, B. M. Welch and R. Zucca, ''Planar GaAs IC Technology: Applications for Digital IC's,'' *IEEE J. Solid-State Circuits* **SC-13**:419 (1977).
11. T. Mizutani, N. Kato, M. Ida, and M. Ohmori, ''High Speed Enhancement Mode GaAs MESFET Logic,'' *IEEE Trans. Microwave Theory, Techniques* **MTT-28**:479 (1980).
12. R. B. Brown, P. Barker, A. Chandna, T. R. Huff, A. I. Kayssi, R. J. Lomax, T. N. Mudge, D. Nagle, K. A. Sakallah, P. J. Sherhart, R. Uhlig, and M. Upton, ''GaAs RISC Processors,'' *GaAs IC Symp.,* 1992, p. 81.
13. K. Yamasaki, K. Asai, and K. Kurumada, ''MESFET's with a Self Aligned Implantation for n^+ Layer Technology (SAINT),'' *IEEE Trans. Electron Dev.* **ED-29**:1772 (1982).
14. M. Abe, T. Mura, N. Y. Ama, and H. Ishikawa, ''New Technology Towards GaAs LSI/VLSI for Computer Applications,'' *IEEE Trans Electron Dev.* **ED-29**:1088 (1982).
15. S. Asai, N. Goto, M. Kanamori, M. Tanaka, and T. Furutsuka, ''A High Performance LDD GaAs MESFET with a Refractory Gate Metal,'' *Proc. 18th Annu. Conf. Solid State Dev., Mater. (SSDM),* 1986, p. 383.
16. C. F. Wan, H. Schichijo, R. D. Hudgens, D. L. Plumton, and L. T. Tran, ''A Comparison Study of GaAs E/D MESFETs Fabricated with Self-Aligned and Non-Self-Aligned Processes,'' *Proc. GaAs IC Symp.,* 1987, p. 133.
17. S. Shikata, S. Sawada, J. Tsuchimoto, and H. Hayashi, ''A Novel Self-Aligned Gate Process for GaAs LSI Using ECR-CVD,'' *Proc. GaAs IC Symp.,* 1990, p. 257.

18. M. Noda, K. Hosogi, K. Sumitani, H. Nakano, K. Nishitani, M. Otsubo, H. Makino, and A. Tada, ''A GaAs MESFET with a Partially Depleted p Layer for SRAM Applications,'' *IEEE Trans. Electron Dev.* **38**:2590 (1991).

19. T. Shimura, K. Hosogi, Y. Khono, M. Sakai, T. Kuragaki, M. Shimada, T. Kitano, N. Nishitani, M. Otsubo, and S. Mitsui, ''High Performance and Highly Uniform Sub-Quarter Micron BPLDD SAGFET with Reduced Source to Gate Spacing,'' *Proc. GaAs IC Symp.,* 1992, p. 165.

20. T. Noguchi, ''Commercial Applications of GaAs ICs in Japan,'' *Proc. GaAs IC Symp.,* 1990, p. 263.

21. K. Sciater, *Gallium Arsenide IC Technology,* TAB Books, Inc., Blue Ridge Summit, Pennsylvania (1988).

22. D. R. Decker, in *VLSI Electronics: Microstructure Science* **11**, N. G. Einspruch, ed., Academic, New York, 1985.

23. T. Inoue, K. Tomita, Y. Kitaura, T. Terada, and N. Uchitomi, ''A Rh/Au/Rh Rigid Air Bridge Interconnection Technique for Ultra High Speed GaAs LSIs,'' *Proc. GaAs IC Symp.,* 1990, p. 253.

24. M. Hirano, I. Toyada, M. Tokumitsu, and K. Asai, ''Folded U-Shaped Micro-Wire Technology for GaAs IC Interconnections,'' *Proc. GaAs IC Symp.,* 1992, p. 177.

25. L. G. Hipwood and P. N. Wood, ''Dry Etching of Through Substrate Via Holes for GaAs MMICs,'' *J. Vac. Sci. Technol* **B3**:395 (1985).

26. K. Sumitano, M. Komaru, M. Kobiki, Y. Higaki, Y. Mitsui, H. Takano, and K. Nishitani, ''A High Aspect Ratio Via Hole Dry Etching Technology for High Power GaAs MESFET,'' *Proc. GaAs IC Symp.,* 1989, p. 207.

27. R. Dingle, M. D. Feuer, and C. W. Tu, ''The Selectivity Doped Heterostructure Transistor: Materials, Devices, and Circuits,'' *VLSI Electronics: Microstructure Science* **11**, N. G. Einspruch, ed., Academic, New York, 1985.

Chapter 18

Silicon Bipolar Technologies

This chapter will review silicon bipolar technologies. Due to the low hole mobility and the extreme materials requirements, bipolar transistors are much less popular than FETs in III-V material systems. The chapter will begin with a very brief review of bipolar device operation, followed by a discussion of the bipolar IC performance metrics. Next, it will introduce the various bipolar technologies, beginning with the simple triple diffused and ending with highly self aligned double poly technologies.

18.1 Review of Bipolar Devices—Ideal and Quasi-ideal Behavior

The bipolar junction transistor (BJT) consists of two pn junctions in close proximity. The three regions of the device are referred to as emitter, base, and collector (Figure 18.1). Devices can be NPN or PNP. Transistor action refers to the process by which the current flow across one of the pn junctions is modulated by the bias of the other junction. Often a small change in the bias of the first junction can result in a large change in the current through the second junction.

In the forward active region [1], the emitter base junction is forward biased and the base collector junction is reverse biased. In this case, the majority carriers in the emitter diffuse across the emitter base depletion region where they enter the neutral base, becoming minority carriers. These carriers diffuse across the base until they reach the base collector junction, where the electric field sweeps them into the collector. Since the emitter acts as a minority carrier source and the collector acts as a minority carrier sink, a large concentration gradient exists across the base, which leads to the large collector current. For the ideal device, the base current is primarily base majority carriers that are back injected into the emitter. By Kirkoff's law,

$$I_e = I_c + I_b \tag{18.1}$$

Two important figures of merit for the transistor are the common base current gain α, and common emitter current gain β.

$$\alpha = \frac{I_C}{I_E} \text{ and } \beta = \frac{I_C}{I_B} \tag{18.2}$$

Figure 18-1   Cross section view of a simple bipolar transistor showing the active or intrinsic device region and the parasitic or extrinsic regions

The two gains are related by

$$\beta = \frac{\alpha}{1 - \alpha} \tag{18.3}$$

Consider a simple bipolar transistor, which has uniformly doped emitter, base, and collector regions. Ignoring generation/recombination, high current effects, and various 2-D effects, the transistor can be considered ideal. Then

$$\beta = \frac{N_E D_B L_E}{N_B D_E W} \tag{18.4}$$

where D is the diffusivity of the minority carrier in the subscript region, N is the doping concentration in the subscript region, L_E is the characteristic diffusion length of minority carriers in the emitter, and W is the nondepleted base width. Since it is generally desirable to have large gains in bipolar transistors, N_E is typically made much larger than N_B and W is kept as small as practical, as long as the base resistance is not too large and the base is not fully depleted during normal operation.

If the concentration in the base is nonuniform, Equation 18.4 is replaced by

$$\beta = \frac{N_E D_B L_E}{D_E G_B} \tag{18.5}$$

where G_B is the base Gummel number given by

$$G_B = \int_{\text{BASE}} N(x) \, dx \tag{18.6}$$

The next level of approximation is the quasi-ideal device. In this model, a small amount of electron hole recombination is allowed in the base. The recombination rate is represented in this model by the minority carrier diffusion in the base, L_B. The quasi-ideal device is a good approximation of the intrinsic behavior of real devices under moderate current conditions. In this approximation

$$\beta \approx \frac{D_B L_E N_E}{G_B \left[D_E + \dfrac{D_B W_B^3 L_E N_E}{2 G_B L_B^2} \right]} \tag{18.7}$$

If L_B goes to infinity (recombination in the base is not allowed), this form reverts back to the ideal BJT.

18.2 Second Order Effects

There are several high bias/current effects that will only be described qualitatively, since a quantitative description is well outside the scope of this course. The first is the Early effect, also known as base

Figure 18-2 I-V Characteristics BJTs. The dashed line shows the effect of base width modulation and the Early voltage.

width modulation. The gain in Equations 18.2–18.7 all depend on the base width. This width in turn depends on the bias conditions. For a fixed collector current, the base to emitter voltage remains nearly constant. Therefore, as the collector to emitter voltage is raised, the reverse bias across the base collector junction increases, the base width shrinks, and the gain increases. This gives rise to a finite output conductance as shown in the I-V characteristics of Figure 18.2. The family of curves all extrapolate back to a single point known as the Early voltage, V_A.

A large value of the Early voltage is desired, particularly for analog ICs, since the output conductance acts as a parasitic output load which reduces the small signal gain. Analog circuits, therefore, require V_A values of at least 30 V, while BJTs used in digital applications typically exhibit V_A values of 10 to 15 V. V_A can be increased by increasing the base width and/or doping so that the fractional change in the Gummel number with V_{BC} is small. Both of these tend to reduce the gain of the device at low bias. Very short base width BJTs are prone to punchthrough, which is simply the extreme example of the Early effect. The two depletion widths stretch completely across the device, eliminating the neutral base entirely. This has the effect of making the transistor leak in the "off" condition. Punchthrough considerations set an absolute lower bound on the BJT base Gummel number for a desired maximum V_{ce}.

A second type of breakdown known as avalanche, can occur if the bias voltage exceeds the reverse bias breakdown voltage of the diode. The breakdown voltage in turn depends on the doping concentration on each side of the junction. If the field at the junction exceeds some critical value, impact ionization or Zener tunneling will occur within the junction, leading to junction breakdown. Since the collector in most BJTs is lightly doped, the base collector breakdown voltage is typically quite high. The emitter base breakdown voltage is much lower, particularly if the heavily doped base contact region is too close to the emitter. As with punchthrough, the device may not turn off properly if the breakdown voltage is too low. This is an extremely serious concern for high speed self aligned structures.

A number of effects occur under high current conditions that degrade BJT performance. One set of effects are geometry related. BJTs are often fabricated as long narrow stripes. Even in these devices, at least two dimensions must be considered. If the base current is sufficiently high, the base voltage varies laterally across the device, reducing the current drive capability (Figure 18.3a). Similar effects occur at the emitter and collector contacts. Particular care must be taken to provide low resistance contacts to the emitter and collector since I_C and I_E are typically much larger than I_B.

The second effect that occurs under high current conditions is base push out, also called the Kirk effect (Figure 18.3b). This effect arises when the density of the minority carriers injected into the base collector depletion region becomes comparable to the doping level in the collector [2].

Figure 18-3 Second order effects in the bipolar transistor include (a) lateral voltage drop and (b) base push out or Kirk effect.

That is, when

$$J_C \frac{1}{q v_{sat}} \approx N_C \tag{18.8}$$

where N_C is the collector doping concentration, and v_{sat} is the saturation velocity of the minority carriers (about $2*10^7$ cm/sec for electrons in silicon). The effect of these additional carriers is to compensate the ions in the depletion region, reducing the electric field. When the electric field in the depletion region approaches zero, the base begins to encroach into the metallurgical collector, increasing the base width, thereby reducing the current gain. The current density at which this occurs is given by

$$J_C \approx q v_{sat} \left[N_C + \frac{2 \varepsilon_{si} V_{CB}}{q W_C} \right] \tag{18.9}$$

where W_C is the metallurgical collector width. Increasing the collector doping increases the voltage at which the Kirk effect becomes important, but at the cost of increasing the base collector capacitance.

18.3 Performance of BJTs

Figure 18.4a shows a low frequency small signal hybrid-pi model of a high gain BJT. The input resistance is

$$r_n = \frac{\beta}{g_m} \tag{18.10}$$

where g_m is the transconductance, given by

$$g_m = \frac{q I_C}{kT} \tag{18.11}$$

The output impedance due to the Early effect is shown as r_o. To improve the analog performance of the device, one would like a high input impedance and a large transconductance. This requires a large value of I_C and a very large value for β. This implies that for high performance ac applications, the device should be biased so as to conduct as large a dc collector as can be obtained before the onset of Kirk effect or until overall power limitations are a concern.

If high frequency effects and the parasitic resistances are added to the simple model, the equivalent circuit is as shown in Figure 18.4b. Of particular importance in this model is the parasitic capacitance, C_μ. This capacitance can be split into components in parallel with the input resistance and the output resistance using a Miller decomposition. Due to the multiplication effect the input capacitance component will be very large at typical bias conditions. Neglecting the parasitic resistances, the device will have a pole at a frequency of

Figure 18-4   Small signal models of the bipolar transistor at (a) low frequency and (b) high frequency.

$$\omega_p \approx \frac{1}{r_\pi (C_\pi + g_m r_o C_\mu)} \tag{18.12}$$

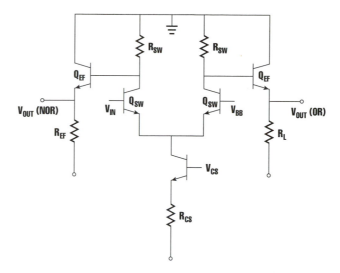

Figure 18-5 Basic ECL logic gate with inverting and non-inverting output.

In practice, both capacitive elements are important. Since the emitter is much more heavily doped than the collector, the emitter base junction capacitance per unit area is much larger than that of the base collector junction. Minimizing both of these capacitances substantially improves the ac performance of the transistor. One commonly used measure of the device performance can be obtained by shorting out the emitter and collector. The frequency at which the gain drops to unity in this configuration is given by

$$f_T \approx \frac{1}{2\pi r_\pi (C_\pi + C_\mu)} \tag{18.13}$$

In a more exact treatment the transit time of carriers across the base must be added to the RC delay

$$f_T \approx \frac{1}{2\pi} \frac{1}{r_\pi (C_\pi + C_\mu) + \dfrac{W_B^2}{\eta D_B}} \tag{18.14}$$

where η is a field enhancement factor, usually between 2 and 20.

Table **18.1**	Typical weighting factors for digital bipolar gate delay contributions	
	Collector base capacitance (k_2)	2.50 + 0.41 * FO
	Collector substrate capacitance (k_3)	1.10
	Wire capacitance (k_4)	0.23
	Emitter follower input capacitance (k_5)	FO
	Diffusion capacitance (k_6)	2.6
	Emitter base capacitance (k_7)	0.24 * FO
	Base collector capacitance (k_8)	1.27

FO is the fanout: the number of gates being driven.

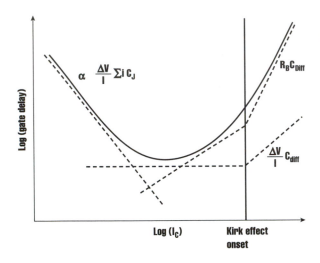

Figure 18-6   Log/log plot of ECL gate delay versus the collector current.

Thus, for high performance of bipolar junction transistors in ac applications, the following transistor characteristics are very desirable:

1. high gain without a low Early voltage or punchthrough,
2. capacitances that are as small as possible,
3. low parasitic resistances,
4. short base widths, and
5. a high Kirk effect onset current.

The requirements for digital applications are somewhat different. Most modern digital BJT applications use ECL and/or CML (Figure 18.5) circuit design approaches. A CML gate is an ECL gate without the emitter followers (Q_{EF}). What is of primary importance here is not just speed, but the speed power product. In these design approaches, additional speed can often be gained for a given technology if the user is willing to pay the penalty of increased power consumption (and heat dissipation). Roughly speaking the gate delay of an ECL circuit is given by

$$t_{pd} \approx R_L C_C + k_1 T_B + R_B C_B \tag{18.15}$$

where

$$C_C = k_2 C_{BC} + k_3 C_{CS} + k_4 C_{\text{WIRE}} + k_5 C_{EF}$$

and

$$C_B = k_6 C_{\text{DIFF}} + k_7 C_{EB} + k_8 C_{BC} \tag{18.16}$$

In these equations the k_x variables are the various weighting factors (Table 18.1), R_L is the pull down resistor, T_B is the base transit time, R_B is the base resistance, C_{BC} is the base collector capacitance, C_{CS} is the collector substrate capacitance, C_{EF} is the input capacitance of the emitter follower, and C_{DIFF} is the diffusion capacitance associated with the stored minority charge in the base.

A general plot of gate delay versus switching current is shown in Figure 18.6. At very high levels of current injection, the diffusion capacitance of the device increases and the transistor gain decreases. For increased current, the diffusion capacitance associated with minority charge storage

Table 18.2 Bipolar scaling methodology	
Emitter stripe width	a
Base doping	$a^{-1.6}$
Collector doping	a^{-2}
Base width	$a^{0.8}$
Collector current density	a^{-2}
Circuit delay	a

After Solomon et al. [3].

actually increases the gate delay with increasing current. Most circuits operate on the left hand side of plot where the gate delay is proportional to the $R_L C_C$ product. Furthermore, most digital circuits are power limited. A particular switching current is budgeted, based on the gate density required for the chip. To improve speed, one must therefore reduce C_C. This in turn means that one must reduce the capacitance associated with the collector base (CB) junction, with the collector substrate (CS) junction, the input of the emitter follower (a combination of the EB and CB capacitances), and the wire capacitance. As these capacitances are reduced, the other components, namely the base transit time and the $R_B C_B$ terms, become increasingly important.

Scaling the bipolar transistor is considerably more complicated than scaling the MOS transistor. In bipolar ECL, the voltage swing is essentially fixed, therefore one must scale the capacitances. Table 18.2 gives a generalized bipolar scaling methodology [3]. In this methodology, the vertical and horizontal dimensions must both be scaled in order to improve the gate delay performance.

There are several limits that one must be aware of in applying these scaling rules. The first two have to do with increasing the base doping. As the base becomes more heavily doped, the mobility in the base decreases due to ion scattering, limiting the reduction of the base resistance. The second effect is bandgap narrowing. Once the doping density is high enough to allow a significant overlap of the dopant atoms, the semiconductor bandgap will begin to shrink. This begins at about 10^{17} cm^{-3} and rises to about 120 meV at 10^{20} cm^{-3}. Bandgap narrowing reduces the gain of the transistor by a factor of $\exp[\Delta E_g / 2kT]$.

18.4 Early Bipolar Processes

One of the earliest bipolar technologies that was used widely in IC manufacturing was the triple diffused (3-D) method. This is still used in some applications due to its low cost and ease of manufacture. A process flow for a simple 3-D technology is shown in Figure 18.7. Only seven masks are required. An initial oxide is grown on a p- substrate and patterned for the guard ring diffusion. The guard rings are driven, the oxide is stripped, and a second oxide is grown. This oxide is patterned for the collector implant. The collector is implanted and driven, the oxide stripped, and a third oxide is grown. This oxide is patterned for the base implant. The base is implanted, the oxide stripped, the base is activated, and a fourth oxide is grown. This oxide is then patterned for the emitter implant. The emitter is implanted, the oxide is stripped, and a final oxide is grown. This oxide is patterned for the base contact (also known as the extrinsic base). A heavy p+ implant is done, and the base contact and the emitter are activated in a final thermal step. A contact glass is deposited, contacts are made, and metallization layers are applied and patterned. As a final note, the extrinsic base implant mask can be

(1) Initial oxidation and guard ring implantation

(2) Guard ring drive, collector oxidation, collector implant

(3) Collector drive, base oxidation, base implant

(4) Base drive, emitter oxidation, emitter implant

(5) Base contact oxidation, base contact implant

Figure 18-7   Process flow for a simple junction isolated triple diffused bipolar technology.

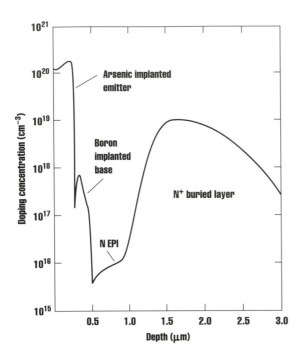

Figure 18-8   Doping profile of a typical buried collector BJT *(after Ko et al., ©1983 IEEE).*

eliminated by implanting the extrinsic base at the same time as the intrinsic base. If this implant is kept shallower and at a lower concentration than the emitter, the emitter diffusion will swamp the extrinsic base in the active region of the device, but will not affect it in the extrinsic device region. The penalty that must be paid is a larger E-B capacitance, a lower breakdown voltage, and a higher base contact resistance.

One of the first improvements in the basic 3-D technology is the addition of a buried collector. This is a heavily doped diffusion under the collector that shorts out the otherwise large collector series resistance. The use of buried collectors implies that the collector must be grown epitaxially on the substrate. This technology has been called *standard buried collector* (SBC). Oxide isolated SBC was the mainstay of the bipolar IC industry through the mid 1970s. A typical SBC bipolar transistor doping profile is shown in Figure 18.8.

As shown in Figure 18.9, the process flow for SBC starts with growing an oxide on a lightly doped p-type substrate. The first mask opens up windows in this oxide for implantation of the buried layer. Implant doses of 2 to $10*10^{15}$ cm^{-2} are typical. Antimony is typically used for this implant for NPN transistors since it tends to diffuse into the epitaxial layer less than arsenic does. Because of the large size mismatch between antimony and silicon, arsenic doped buried layers tend to produce fewer defects in the epi than antimony buried layers. It has also been shown that at reduced pressure, the arsenic updiffusion is greatly diminished. Arsenic buried layers may therefore see a revival. The oxide is then stripped, and an anneal step is done to remove the implant damage and activate the impurities. Another oxide is typically grown in this step. The oxide grows more rapidly over the exposed heavily doped regions. The differential oxidation rate provides a step that can later be used to register the upper layers of the device to the buried collector.

After annealing, the oxide is stripped and the epitaxial layer grown. As discussed in Chapter 14, this is typically done in a radiantly heated barrel reactor with a H_2/SiH_2Cl_2 ambient. The epitaxial growth step is quite critical in an SBC technology. The collector concentration is set during this growth. The collector width is determined by the epitaxial layer thickness less the base diffusion and the updiffusion from the buried collector. Frequently this involves the subtraction of several numbers that are of the same order of magnitude. Control of the collector thickness therefore, requires tight control of the epi growth process and subsequent thermal processes.

Following the epitaxial growth, the LOCOS isolation module is begun by growing a pad oxide and depositing Si_3N_4. Usually the silicon is etched along with the nitride to produce a recessed field oxide. Boron is implanted into the bottom of the recess to prevent parasitic MOS formation, and the field oxide is grown. Next, a sinker implant may be done (not shown) to connect the buried layer and the collector contact. This involves a high dose n-type implant or diffusion of phosphorus, followed by a high temperature drive in diffusion. The processing then continues up through the metallization layers as for the 3-D process.

Figure 18-9 Process flow for an oxide isolated triple diffused bipolar technology without sinkers. Steps include: (a) buried layer formation, (b) epitaxial growth, (c) LOCOS patterning, (d) silicon recessing and channel stop implants, (e) local oxidation, (f) intrinsic base implant, (g) contact mask, (h) extrinsic base implant, (i) emitter and collector contact implant, and (j) metallization *(after Sze, reprinted by permission of John Wiley and Sons).*

18.5 Advanced Bipolar Processes

The preceding discussion suggests that due to the tradeoffs inherent in the BJT, only modest improvements can be made to either the digital or the analog performance of the intrinsic device. Decreasing the base width reduces the base transit time, but requires higher base doping levels to prevent an excessively low Early voltage. This higher doping level increases the E-B capacitance, and reduces the base mobility. Reducing the width of the emitter stripe helps somewhat since it effectively lowers the base resistance, but for most digital circuits, R_B is not a factor of the dominant term in the performance equation. This section will discuss process architectures that have been used to improve the performance of bipolar devices. Many of these improvements are not in the intrinsic transistor, but instead are aimed at reducing the parasitic resistances and capacitances that degrade the performance of the transistor.

One clear improvement in the intrinsic device came with the introduction of poly emitter contacts [4]. When a layer of heavily doped ($>10^{20}$) polysilicon, at least 500 Å thick, is placed between the single crystal emitter and the metal contact, the gain of the device increases. This effect has been studied in considerable detail, but no clear consensus has yet been reached regarding its origin. Any minority carriers in the emitter will immediately recombine at the metal semiconductor contact. Thus, if the emitter thickness is less than L_E, the presence of the contact will increase the minority carrier concentration gradient and thereby increase the base current and so reduce the gain. The addition of the polysilicon moves the metal contact further away from the E-B junction and reduces this effect. Since the minority carrier diffusion length in the emitter is about 0.2 to 0.4 μm [5], increasing the effective emitter width reduces the base current [6]. Even for wide emitter transistors however, a poly emitter increases the gain. Furthermore, the use of a polysilicon emitter contact allows for the use of very shallow emitter base junctions. Typical E-B junctions are now in the range of 200 to 500 Å.

The most popular poly emitter theories invoke an interfacial dielectric between the polysilicon and the substrates. TEM studies have shown that a discontinuous interfacial oxide that is a few atomic layers thick exists near the poly/single crystal interface [7]. This model explains the nonexponential collector and base currents often seen in these devices [8]. This oxide is often due to native oxidation before poly deposition, although a thermal oxide can also be intentionally grown [9]. The gain may be due to the differential tunneling of electrons and holes through this thin oxide if the oxide is thicker than about 10 Å. Such oxides are typically formed during the wet cleaning process. For much thicker oxides however, the emitter resistance becomes excessive [10]. The use of rapid thermal processing during the drive steps has been shown to retain the high gain characteristics while minimizing the emitter resistance [11]. This is believed to be due to oxide thinning for low temperature RTP and agglomeration of the oxide at high temperature RTP [12]. It has also been shown that thermal nitrides can be used as an interfacial layer [13]. Thermal nitrides are self limiting and therefore quite uniform, resulting in good control over the current gain and low emitter resistances if phosphorus doped emitters are used.

The second model for explaining the increased gain of polysilicon emitter BJTs takes into account that a gain enhancement can be observed even if the surface has been carefully prepared to minimize the growth of any interfacial oxide. This model invokes the fact that at high concentrations, arsenic segregates to the grain boundaries in polysilicon films. In poly emitter structures, the poly/single crystal interface is a large grain boundary that will be very heavily doped [14]. Even in the absence of an interfacial oxide, this pile up can be observed in TEM, SIMS, and STEM [15],[16]. This excess n-type dopant leads to a barrier to hole injection but does not affect electron ejection.

This effect has been reinforced by recent observations that phosphorus doped emitters have a larger gain enhancement than arsenic doped emitters. This has been observed in both in situ doped [17] and implanted poly emitters [18]. It has also been speculated that the difference in carrier mobility between the substrate and the poly may also provide this barrier [19],[20].

In addition to the gain enhancement, polysilicon emitters also have considerable technological advantages that alone would justify their application. The emitter contact can be made on the field oxide, minimizing the necessary poly stripe width. Furthermore, as will be described in the next section, the topology of the stripe can be used to self align the transistor. Finally, the emitter and sometimes the base, can be implanted into the polysilicon and diffused into the single crystal substrate. This avoids the problem of implant damage that may serve as recombination sites and degrade device performance.

A variety of process architectures have been developed that exploit the poly emitter structure to further improve bipolar performance. The earliest ones were simple extensions of SBC technologies to include polysilicon contacted emitters, often called *poly emitters*. A common variant is shown in Figure 18.10. A layer of undoped polysilicon is deposited before the base implant. The base and emitter implants are then done into the poly and diffused into the single crystal substrate. If arsenic is used for the N+ implant, the base must be diffused before the introduction of the high dose arsenic, since boron diffusion is severely restricted in the presence of high concentrations of arsenic, owing to the large electric field that it produces. The high temperature step that drives the impurities is quite critical. If it is too high, the benefits of the poly contact are lost. If it is not high enough, the emitter contact resistance will be too large. Of course, the final diffused profile depends on this temperature as well.

After the emitter and base are driven, the poly must be etched back down to the substrate. This is a difficult process to accomplish since a selective plasma (or RIE) is required for polysilicon on silicon. Of course, both are heavily doped n-type silicon and so must etch in the same chemistries. A selective etch process must therefore take advantage of grain boundary erosion. This generally means etching at low power in a chlorine ambient. Selectivities as large as 5:1 can be achieved in this system.

To obtain further improvement in the performance of bipolar technologies, self aligning must be used. That is, the base and emitter contacts must align themselves automatically due to the surface topology, without the need for a critical alignment step. The most successful of these technologies use double poly methods. A typical one, referred to as self aligned (SA) bipolar [21], is shown in Figure 18.11. It has become the standard digital bipolar technology from most of the late 1980s through today. After buried collector and isolation formation, a P+ layer of polysilicon is deposited. This may be

Figure 18-10 In the simple poly emitter technology, the poly is used to diffuse the emitter and intrinsic base layers.

Figure 18-11   Simple process flow for a self aligned double poly bipolar technology.

doped in situ during the deposition or it may receive a blanket implant after deposition. A layer of oxide is then deposited or grown on the p+ poly.

The first poly is then patterned to open a window in an active region. In the SA technology, this alignment is critical. If the overlap of the poly to the active area is large enough to ensure that the poly makes contact with the substrate even with a poor alignment, the device can suffer from an overly large base collector capacitance. If the overlap is too small however, an alignment error will produce a fatal defect since after lateral encroachment of the field oxide, the first poly will terminate on the field oxide and there will be no base contact. After first poly patterning, the extrinsic base is diffused out of the first poly. A second oxide is then deposited or grown and reactive ion etched back down to the substrate, leaving sidewall spacers on the base contacts. The intrinsic base is then implanted through the hole, and the second poly is deposited, doped n-type, and the emitter is driven. Finally, some of the base and emitter areas may be silicided to reduce the series resistance.

There are several advantages of the SA structure. The first is that the base contact to metal is now done through the p+ poly over the field oxide. This reduces the base collector capacitance dramatically. The extrinsic base collector junction area of a double contacted base BJT is typically 10 to 12 μm times the length of the transistor. In the SA process, it may be 1.0 to 1.5 μm times the length of the transistor. The second advantage is that due to the sidewalls, the emitter stripe width is actually less than the minimum lithographic feature size. Typical SA emitter widths are about 0.5 μm for 0.8 μm lithographies.

Aggressive scaling of the SA structure to deep submicron stripe widths along with the use of polysilicon filled deep trench isolation to reduce parasitic capacitances [22] have resulted in ECL gate delays below 20 ps [23] and speed power products of 92 fJ [24]. In order to achieve very high current densities along with high speed, the selectively doped collector can be used [25]. In this process, a high energy n-type implant is done after the first poly to increase the collector concentration only directly under the intrinsic base (Figure 18.12). This allows the extrinsic base collector capacitance to remain low while obtaining a large intrinsic collector concentration. To reduce the collector substrate capacitance, it is possible to grow a thick, undoped epitaxial layer under the buried collector. This has been grown unpatterned on p+ substrates, using a

Figure 18-12 A cross section of a self aligned bipolar technology called Mosaic *(after dela Torre et al., ©1991 IEEE).*

deep trench that penetrates to a heavily doped substrate to provide excellent isolation and high speed [24]. Another concern in scaling the double poly structure is the increased resistance of the extrinsic base. Techniques used to minimize the impact of any reduction in the first poly thickness include using larger grain polysilicon [26] and siliciding as much of the extrinsic base as possible.

Most of the problems with SA technologies relate to making the link between the intrinsic and extrinsic base regions. Just as in CMOS, the reliance on surface topology along with minimal dopant redistribution during the activation means that implant shadowing may present problems [27]. Thus there may be an incomplete linkup between the intrinsic and extrinsic base. If enough shadowing occurs, the linkup region may be completely depleted, resulting in an extremely large base resistance. The simplest method of improving the linkup is to increase the diffusion of the extrinsic base region to extend almost completely under the sidewalls. The problem with this method is that if the heavily doped extrinsic base comes into contact with the heavily doped emitter, not only will the capacitance increase, the base emitter breakdown voltage will decrease. Furthermore, the injection efficiency of the majority carriers in the base into the emitter depends on the doping concentration in the base at the emitter base junction. If the extrinsic base comes into contact with the emitter, the device can be thought of as two bipolar transistors in parallel. The parasitic device has a very low gain due to the back injection of carriers into the emitter.

The linkup problem can be solved with several approaches. The first is to implant the intrinsic base on axis directly into the substrate before the formation of the sidewall spacers. This is even less desirable than in CMOS LDD formation however, in that the channel tail in BJTs will dramatically change the performance of the device. Since the collector concentration is typically about two orders of magnitude less than the base, a small change in the base profile can significantly increase the base width. The second method of improving linkup is to implant the link separately before the formation of the sidewall spacers, then diffuse the intrinsic base from the second poly [28]. In this process, the second poly must be deposited undoped, the intrinsic base implanted and diffused, and then the emitter

Figure 18-13 Walled transistors may have emitter-collector leakage along the isolation edge. *(after Ratnam et al., ©1992 IEEE).*

implanted and diffused. Alternatively, a boron doped glass [29] or P+ doped polysilicon with a thin oxide [30] can be used to form the sidewall spacers and the boron diffused from the spacers themselves.

In addition to the linkup problem, care must be taken in the isolation region of an SA technology. When the BJT stripe is terminated by the isolation, as is commonly done for MOSFETs, the structure is called a *walled transistor*. Walled transistors can have better performance than their unwalled counterparts due to the reduction in the base collector capacitance. They are also more compact, increasing the device density. In simple 3-D structures, the major concern is leakage in junctions terminated along the wall due to boron depletion during oxidation. This is relatively easy to overcome by increasing the doping concentration as evidenced by the density of CMOS currently in manufacture. In SA technologies however, one of the goals is the minimization of the base collector capacitance. This is done by reducing the length of the P+ poly base length that is in contact with the collector. A high concentration of boron under the field oxide, as shown in Figure 18.13, however, can reduce the emitter base breakdown voltage [31]. A tradeoff therefore must be made when selecting the length of the P+-poly contact.

To further improve the performance of BJTs, a series of devices have been proposed that reduce C_{BC} without a critical alignment between the base and isolation levels. The first, shown in Figure

Figure 18-14   The super self aligned or Antipov emitter bipolar transistor.

18.14, was originally called the Antipov emitter [32], but is now more commonly known as the super self aligned transistor [33] (SST). The technology is essentially an extension of the SA transistor. A thin (1000 Å) oxide is grown before the deposition of the first poly. After patterning poly 1, the oxide is undercut by a controlled amount using an isotropic etch such as dilute wet HF. Typical undercuts are 0.1 to 0.3 μm per side. The second poly deposition occurs conformally so that it fills the undercut. The second poly is then oxidized to form the sidewall spacers as in the typical SA process. Care must be taken during this oxidation to ensure that all of the second poly on the planar areas is oxidized, but that the poly plug formed in the undercut region is not. Next, the technology proceeds as a normal SA with the exception that the extrinsic base is driven through the undoped poly plug. This reduces the parasitic C_{BC} junction area to about 0.25 μm times the length of the transistor. The process also removes the necessity of selectively etching the polysilicon down to the substrate. Advances in the SA process and in lithographic tools have reduced the advantages of the SST. As a result, it has not been accepted into widespread manufacture.

Several very high speed bipolar structures have also been fabricated through the use of selective epitaxy. There are two primary advantages of this approach. The first is that all of the layers are grown at the same time. Unlike most BJT technologies where the collector width is determined by the epitaxial collector thickness less the sum of the diffused base collector junction depth and the buried collector updiffusion, the collector width is directly determined by the thickness grown. Similarly, the base width is determined directly by the growth rate. As a result, very thin base transistors can be fabricated. Furthermore, the device parasitics can be reduced sharply by the nature of the structure. Devices with f_T's of over 50 GHz have been fabricated using epitaxial growth.

One heavily studied type of epitaxial BJT, known as the sidewall base contact (SICOS) structure [34], is shown in Figure 18.15. After the formation of the buried collector and LOCOS isolation, a

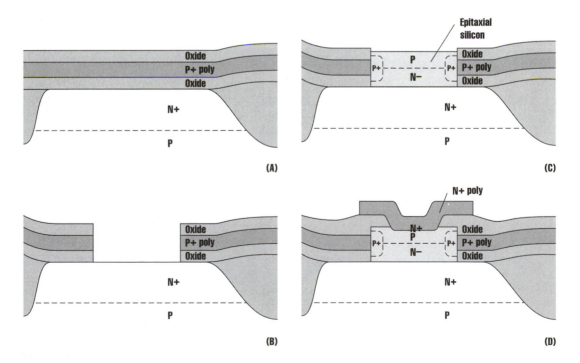

Figure 18-15   The sidewall contacted (SICOS) structure provides the least parasitic B-C capacitance but is difficult to manufacture.

Figure 18-16   Ring oscillator gate delay of various bipolar technologies as a function of the emitter stripe width.

sandwich structure of oxide, P+ polysilicon, and oxide is formed. This structure is then etched back down to the substrate inside of an active area window and an n-collector is selectively grown epitaxially in the window. In this structure, the base contacts are diffused laterally from the P+ poly. As with the other self aligned structures, one problem is the linkup of the intrinsic and extrinsic base regions [35]. An additional difficulty is the epitaxy itself. Growth inside a window tends to produce a high density of defects along the sidewalls [36] (at the epi/oxide interface), facets at the upper surface, and nonuniform growth rates [37]. Growth may also be nucleated at the P+ poly. The orientation of this growth is unlikely to match the growth from the substrate. Unless the boundary between the epi and the poly is contained in the heavily doped region, the device will suffer from high leakage and low gain. Furthermore, it is found that the selective epitaxy process often depends on the amount of exposed silicon in the reactor and so is difficult to control. Although many reports of SICOS transistors have appeared in the literature, it is yet to be used in any mainstream manufacturing process. Figure 18.16 shows the gate delay of various technologies as a function of the gate length [38]. A gross variation can be seen to be due to emitter width, but at a particular lithographic dimension, increasing the amount of self alignment also improves performance. The simple technologies are older and therefore are not reported for narrow emitters.

Perhaps one of the most promising trends for further improvements in silicon bipolar transistors is the use of silicon on insulator (SOI) substrates to further reduce the collector to substrate capacitance. The most commonly discussed substrates are SIMOX and bonded SOI. While still at the demonstration phase [39], as the prices of the material decrease and the quality improves, these technologies are likely to move into the mainstream. One of the disadvantages of this approach is the concern with thermal effects. Silicon dioxide, which is often used as the buried insulator in these structures, has a much lower thermal conductivity than silicon. As a result, BJT fabricated in SOI cannot be operated at the same power density as devices built in bulk silicon, assuming that both devices have the same maximum junction temperature [40].

18.6 Hot Electron Effects in Bipolar Transistors

The effects of hot carriers injected into the gate oxide were discussed in Chapter 16. Several solutions to this problem were presented. Hot carrier degradation phenomena also occur in advanced BJTs. As shown in Figure 18.17, when the emitter base junction is subjected to a large reverse bias, extremely high fields exist near this junction. These fields lead to a degradation of the junction characteristics [41]. One mechanism for this degradation is hot carrier injection. Energetic carriers generated near the junction can be injected into adjacent oxides creating surface states that lead to junction leakage [42]. Higher doping (and hence larger fields) produce larger shifts [43]. Similar effects have also been observed after long time high current forward bias conditions [44]. One effect of the leakage is a gradual reduction of the transistor gain. Since the leakage occurs near the edge of the device, the problem is most pronounced on long and narrow emitter stripes. Fortunately, the current of these leakage states is proportional to $e^{qVbe/2kT}$, while the desired transistor current is proportional to

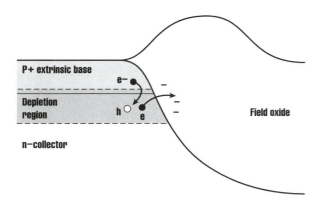

Figure 18-17   Hot carrier charge injection in bipolar transistors.

$e^{qVbe/kT}$. At high dc currents, where most ECL and high performance analog ICs are operated, only small changes are seen.

The reverse bias stress problem can be resolved from a circuit design standpoint by restricting the maximum reverse bias voltage. In typical ECL the logic swing is less than 1 V, so this is easily accommodated. In TTL reducing this voltage is highly desirable since it also reduces the power dissipation and so is already being done. For BiCMOS designs however, the emitter base junction is necessarily subjected to large reverse bias voltages and hot electron degradation of the device is more of a concern. The forward bias problem is not easily solved and restricts the maximum allowable current density in advanced bipolar devices used in many applications.

18.7 BiCMOS

BiCMOS is a combination of both bipolar and CMOS that allows the designer to use both devices on a single chip [45]. This has three primary applications: high speed memory applications such as SRAMs where the bipolar devices are used in sense amps to detect small voltage changes on the lines, high speed digital where the bipolar devices are used to drive large capacitive loads, and merged digital/analog. Bipolar transistors have a very large transconductance and show much less performance degradation upon capacitive loading. In digital circuits, this is particularly useful to drive output pads and long interconnect lines between the circuit elements on a chip that may be separated by as much as several millimeters. Figure 18.18 shows a typical plot of gate delay versus load capacitance for equal area CMOS and BiCMOS circuits. For very lightly loaded gates, CMOS is faster. For large capacitive loads, however, BiCMOS has a significant advantage. The crossover point is typically between fanouts of two and three. The additional process complexity and the additional power consumption however, must be weighed against the improved performance of many ICs. Often these technologies start from a CMOS technology, then graft on bipolar devices. In analog ICs, bipolar is used for its close matching and low noise. Additionally, these circuits frequently require PNPs, precision resistors, and precision capacitors. As a result, BiCMOS technologies for analog can become quite complex and are often developed from bipolar technologies. The section will focus on digital BiCMOS technologies, as these are much more common than analog BiCMOS technologies.

BiCMOS technologies are extremely malleable to a wide variety of design methodologies. It is generally observed that they provide speed and performance somewhere between ECL bipolar and CMOS. At one extreme, it is common to design the chip completely in

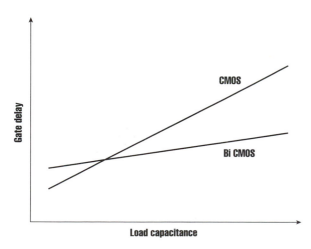

Figure 18-18   Relative gate delays for equal area CMOS and BiCMOS technologies.

CMOS, with only a small number of drivers designed in ECL/CML, or using a merged gate approach. The driver for example, might be used to interconnect blocks of logic, which are internally CMOS. If the transmitted signals are ECL, the reduced voltage swing further minimizes the effects of the large capacitance and can even result in a lower dynamic power dissipation than CMOS [46]. Furthermore, the small device sizes that the bipolar devices require relative to the large CMOS drivers actually reduce the node capacitance. The use of ECL signals requires some level shifting, but is straightforward. This approach has become extremely popular with traditonal CMOS designers as it also allows them to place BiCMOS elements at critical points of the logic path. This can be done after the design is nearly complete, since the BiCMOS gate will often fit into the space required by a moderate to large sized CMOS driver. It is common to have a very small percentage of bipolar devices (<1%) in a digital BiCMOS design, yet these devices make a significant impact on the overall chip performance.

It is also possible to embed the BJTs in a basic CMOS inverter structure and then use these gates as an integral part of the design. Figure 18.19 shows a common example of this type of merged gate structure. This is a push pull or totem pole structure in that BJTs are used to both charge and discharge the output node. The two resistors are used to drain the charge from bases after the MOS-FETs have been turned off. The CMOS front end ensures a high input impedance and a low quiescent power dissipation. This type of design provides a voltage swing on the output from V_{SS} to V_{DD} (full rail). Mixtures of CMOS and BiCMOS gates have been used in gate array or sea of gate logic arrays [47], where the capacitance on each node is unknown.

There are a variety of ways to fabricate BJTs when starting from a CMOS technology. An attempt is usually made to use as much of the existing MOS processes for the bipolar transistor as possible, but the higher the performance required, the greater the additional process complexity. The simplest type of BJT, shown in Figure 18.20a, uses an n-well as the collector, the PMOS source/drain as the base, and the NMOS source/drain as the emitter. It is fortuitous that typical n-well concentrations that are between 1×10^{16} and 1×10^{17} cm^{-3} are about the same as required for digital BJT collectors. Furthermore, the shallow, very heavily doped NMOS source/drain makes a good low resistance emitter. The difficulty with this device is that it has very low gain since the base is heavily doped. If, as shown in Figure 18.20b, an intrinsic base implant mask is added, the PMOS source/drain can be used as the extrinsic base and the intrinsic base profile can be controlled independently of the CMOS transistors.

As with the early BJT 3-D technology, the basic limitation of this device is the collector resistance. To improve these simple devices a buried layer mask must also be added. This implies that a p-layer must be grown on the wafers. The n-well must drive deep enough to merge with the buried layer (Figure 18.20c). This can have the additional advantage of reducing CMOS latchup susceptibility. Thus, with only two additional masking steps one can add a simple, moderate performance NPN device to a CMOS process flow.

High performance BiCMOS technologies have attempted to produce uncompromised BJTs onchip with CMOS [48]. A tradeoff must always be made between the amount of acceptable process complexity and the device performance required. As there is no single way to accomplish this, we will describe a representative BiCMOS technology [49]. The process begins with the high energy implantation of an N+ buried layer in a p- substrate. To minimize outdiffusion

Figure 18-19   A common example of a merged gate BiCMOS structure.

Figure 18-20 Diffused bipolar transistors that can be fabricated in a CMOS technology: (a) the emitter and base are formed by the N+ and P+ source/drain diffusions, respectively, while the n-well serves as a collector, (b) a separate intrinsic base implant is done to set the BJT gain, and (c) a buried layer is added beneath the n-well.

from the buried layer, arsenic may be chosen as the dopant. To further compensate for the outdiffusion, a shallow boron implant may also be done. This blanket implant also forms a moderate P+ buried layer under the epi that reduces latchup susceptibility. Even with this counterdoping, the final depth of the buried layer is only about 0.5 μm.

The n-well/collector is formed next. The concentration of the well must be carefully optimized for obtaining acceptable PMOS device characteristics as well as high performance NPNs. Concentrations of about 5 to 10×10^{16} cm^{-3} allow reasonably high BJT drive currents, and so f_T's, while maintaining a suitably small base collector capacitance and providing adequate punchthrough protection from the PMOS transistors at 0.5 μm gate lengths. After well formation, a conventional LOCOS or trench and LOCOS process is used to isolate the devices. If LOCOS is used, it is recessed about 0.3 μm and is about 0.5 μm thick. To minimize diffusion, this oxidation may be done at high pressure. A field or channel stop implant must be done as with normal CMOS, to prevent inversion of the p- epi. If deep trenches are used, the trenches should extend into, but not through, the buried layers.

To avoid excessive collector resistance, a series of high dose n-type implants are done through a section of the n-well to form a sinker that contacts the buried layer. A trench can also be etched through the epi down to the buried layer. An oxide layer is deposited and reactive ion etched to isolate the sidewalls. Next, a N+ polysilicon layer is deposited and etched back to the substrate. Although somewhat more complex, the trench and refill process offers a lower parasitic collector resistance.

A standard CMOS process then proceeds. A thin screen oxide is grown and a series of channel implants are done to adjust the threshold voltage and punchthrough characteristics of the FETs. Next, the gate oxide is grown and the N+ poly gate electrode material is deposited and patterned. If LDD structures are required, the n-reachthrough is implanted, the dielectric is deposited and etched back to form the sidewall spacers. Finally, the NMOS and PMOS source/drains are implanted. To avoid the need for an extra mask, the intrinsic base implant can be done at this point if the energy of the base implant is kept low enough so that it does not penetrate the gate electrode. Since the intrinsic base concentration is about two orders of magnitude lower than the source/drain regions of the FETs, counterdoping is not significant. Alternatively, the base implant can be done later in the process into the emitter poly layer, followed by an intrinsic base diffusion. This requires more careful attention to the previously discussed linkup problems. Finally, the source/drain regions can be salicided, completing the formation of the MOS devices. The silicide must be selected so as to withstand the high temperature step of the emitter drive. To reduce this problem the emitter may be driven with a rapid thermal process.

The base poly is now deposited and a heavy P+ implant is done to dope the extrinsic base region. The PMOS source/drain implant can also be used to dope the base poly. A thermal and/or deposited oxide is formed on top of the base poly, which is then patterned anisotropically to open the emitter window. Another oxide is grown or deposited, and the oxide is anisotropically etched to form the sidewall spacers. Finally, the emitter poly is deposited, implanted and patterned, and the emitter is driven. If desired, the emitter and base poly layers can be silicided at this point. Interconnect then completes the processing. For three layers of metal, this requires at least 20 masks (compared to 15 masks for the equivalent CMOS) and produces a SA transistor along with high performance CMOS.

18.8 Analog Bipolar Technologies

When bipolar technologies are intended for analog applications, absolute performance often is not as important as matching and careful control of both the active and passive components. Analog designs often require large value capacitors and both high and low value resistors. The availability of PNP devices is also extremely useful for analog designs. As an example, the PNPs are often used as part of the output stage of analog ICs. Complimentary technologies often reduce the power dissipation of the circuit. For that reason, complementary bipolar technology will be reviewed as a prototype analog bipolar technology.

As with BiCMOS, a tradeoff must be made between process complexity and device performance. Figure 18.21 shows some potential PNPs. The simplest type of device, a vertical BJT, uses the extrinsic NPN base region as the emitter. The p-substrate serves as a collector [50]. While it would be desirable to use the NPN collector as the base of the PNP, the doping of the base is insufficient to prevent punchthrough unless a very large W is used. It is possible to include the buried collector of the NPN as part of the base, but of course this device would have very poor performance due to the heavy base doping. As a result, the simplest PNP commonly used is a lateral device. This allows the base length of the device to be chosen lithographically. In addition, a PNP base implant step can be added. Still these devices are quite slow and the critically important Gummel number is subject to variations in the photolithography.

Starting from the SA structure it is possible to diffuse a lateral PNP base from the P+ poly. The buried collector of the NPN can be used to contact this base. This process allows some independent control of the PNP device characteristics. There are large parasitic capacitances associated with this device, and so it has only moderate speed. To obtain the best PNP performance, it is possible to fabricate a full SA PNP onchip with a SA NPN [51],[52], although the process flow is lengthy.

One of the largest additions to the technology required for high performance vertical PNPs is the formation of p-type and n-type buried layers. Obviously, if both types of buried layers are to be isolated, an additional diffusion must be done to form a tub around one

Figure 18-21   Lateral and vertical PNP transistors.

or both buried layers. Alternately, SOI wafers can be used. This is perhaps the most natural way to obtain true high performance complimentary bipolar technologies [53].

18.9 Summary

Historically, the performance of bipolar transistors has been limited to a significant degree by device parasitics. Primary among these is the junction capacitance associated with the contact or extrinsic regions. Simple bipolar technologies, such as the triple diffused or standard buried collector have very large extrinsic capacitances. More advanced structures uses self aligned polysilicon structures to contact the emitter, the base, or both. The metal to polysilicon contact in these structures is formed on top of the thick field oxide, allowing the junction areas to be shrunk considerably. Furthermore, the use of polysilicon to contact the emitter produces intrinsically larger device gains. Recently the combination of bipolar and CMOS has become popular for a wide variety of applications, but is more expensive than conventional CMOS due to the increased process complexity.

Problems

1. In the following NPN triple diffused (3-D) bipolar technology, identify the purpose of the starred (*) steps:
   Initial oxidation
   Pattern and etch oxide
   *Boron Implant
   Strip and regrow oxide
   Pattern and etch oxide
   *Implant phosphorus
   (and so on)

2. In the following NPN standard buried collector bipolar technology, identify the purpose of the starred (*) steps:
   Initial oxidation
   Pattern and etch oxide
   *Antimony Implant
   Oxidation and anneal
   Strip oxide
   Epitaxial growth
   Oxidation
   Pattern and etch oxide
   Boron Implant
   Strip and regrow oxide
   Pattern and etch oxide
   *Medium and high energy phosphorus implant
   Anneal
   (and so on)

3. In the following NPN single poly bipolar technology, identify the purpose of the starred (*) steps:
   (Skip many steps . . . start in the middle of the process)
   Undoped polysilicon deposition

*Boron implantation
Anneal
Arsenic implantation
Polysilicon etch
*Anneal
(and so on)

4.   (a) The delay time of an ECL circuit is found to depend primarily on the base collector capacitance. Which of the following parameters would you also expect to be very important to the delay equation: base resistance, load resistance, emitter resistance, collector resistance? (b) If this ECL circuit had been implemented in a standard buried collector (SBC) technology, what type of technology improvement could be done to increase the speed of the circuit?

5.   The base concentration of a 3-D technology was 1×10^{18} cm^{-3}, but when the technology was changed to add a poly emitter, the base doping was increased to 5×10^{18} cm^{-3}. What benefits are gained by this increase and why does the poly emitter allow it?

6.   Explain why the base transit time is more important in a SICOS transistor than it is in an SBC transistor. The base doping in very high performance bipolars may be graded to form an internal field that reduces the base transit time.

7.   Design a process flow that would produce an oxide isolated complementary bipolar technology (both NPN and PNP), where both transistors are double poly self aligned. Try to minimize the number of masks required.

References

1.   More complete descriptions of bipolar devices at an undergraduate level can be found in: *Modular Series on Solid State Devices. Vol III: The Bipolar Junction Transistor,* G. W. Neudeck, Addison-Wesley; and *Solid State Electronic Devices,* B. Streetman, Prentice-Hall (1990). Advanced level bipolar texts include: *Bipolar Semiconductor Devices,* D. J. Roulston, McGraw-Hill (1990); and *Transistors: Fundamentals for the Integrated-Circuit Engineer,* R. M. Warner, Jr. and B. L. Grung, Wiley-Interscience (1983).

2.   C. T. Kirk, ''A Theory of Transistor Cutoff Frequency (f_T) Falloff at Large Current Densities,'' *IRE Trans. Electron Dev.* **EDL-9**:164 (1962).

3.   P. M. Solomon and D. D. Tang, ''Bipolar Circuit Scaling,'' *Int. Solid State Circuits Conf. Tech. Dig.,* 1979, p. 96.

4.   J. Graul, A. Glasl, and H. Murrmann, ''High Performance Transistors with Arsenic Implanted Polysil Emitters,'' *IEEE J. Solid-State Circuits* **SC-11**:291 (1976).

5.   W. C. Ko, T. C. Gwo, P. H. Yeung, S. J. Radigan, ''A Simplified Fully Implanted Bipolar VLSI Technology,'' *IEEE Trans. Electron Dev.* **ED-30**:236 (1983).

6.   P. Ashburn, ''Polysilicon Emitter Technology,'' *Proc. IEEE Bipolar Circuits, Technol. Mtg.,* 1989, p. 90.

7.   H. C. deGraaf and J. G. de Groot, ''The SIS Tunnel Emitter: A Theory for Emitters with Thin Interface Layers,'' *IEEE Trans. Electron Dev.* **ED26**:1771 (1979).

8.   H. Schaber and T. F. Meister, ''Technology and Physics of Polysilicon Emitters,'' *Proc. IEEE Bipolar Circuits, Technol Mtg.,* 1989, p. 75.

9.   G. R. Wolstenholma, N. Jorgensen, P. Ashburn, and G. R. Booker, ''An Investigation of the Thermal Stability of the Interfacial Oxide in Polycrystalline Silicon Emitter Bipolar Transistors by Comparing Device Results with High Resolution Transmission Electron Microscopy Observations,'' *J. Appl. Phys.* **61**:225 (1986).

10. H. Schaber, B. Benna, L. Treitinger, and A. Weider, "Conduction Mechanisms of Polysilicon Emitters with Thin Interfacial Oxide Layers," *IEDM Tech. Dig.,* 1984, p. 738.

11. J. E. Turner, D. Coen, G. Burton, A. Kapoor, and S. J. Rosner, "Interface Control in Double Diffused Polysilicon Bipolar Transistors," *Proc. IEEE Bipolar Circuits, Technol Mtg.,* 1989, p. 33.

12. T. M. Liu, Y. O. Kim, K. F. Lee, D. Y. Jeon, A. Ourmazd, "The Control of Polysilicon/ Silicon Interface Processed by Rapid Thermal Processing," *Proc. IEEE Bipolar Circuits, Technol. Mtg.,* 1991, p. 263.

13. F. Nouri and B. Scharf, "Polysilicon-Emitter Bipolar Transistors with Interfacial Nitride," *Proc. IEEE Bipolar Circuits Technol. Mtg.,* 1992, p. 88.

14. C. C. Ng and E. S. Yang, "A Thermionic Diffusion Model of Polysilicon Emitter," *IEDM Tech. Dig.,* 1986, p. 32.

15. C. Y. Wong, C. R. M. Grovenor, P. E. Batson, and D. A. Smith, "Effects of Arsenic Segregation on the Electrical Properties of Grain Boundaries in Polycrystalline Silicon," *J. Appl. Phys.* **57**:438 (1985).

16. H. Schaber, R. V. Criegern, and J. Weitzel, "Analysis of Polycrystalline Silicon Diffusion Sources by Secondary Ion Mass Spectrometry," *J. Appl. Phys.* **58**:4036 (1985).

17. M. Nanba, T. Kobayashi, T. Uchino, T. Nakamura, M. Kondo, Y. Tamaki, S. Iijima, T. Kure, and M. Tanabe, "A 64 GHz Si Bipolar Transistor Using in-situ Phosphorus Doped Polysilicon Emitters," *IEDM Tech. Dig.* 1991, p. 443.

18. G. Streutker, A. Pruijmboom, D. B. M. Klaasen, and J. W. Slotboom, "Thermionic Emission Limited Recombination in Phosphorus-Implanted Polysilicon Emitters," *Proc. IEEE Bipolar Circuits, Technol. Mtg.,* 1992, p. 50.

19. A. A. Eltoukhy and D. J. Roulston, "Minority Carrier Injection into Polycrystalline Emitters," *IEEE Trans. Electron Dev.* **ED-29**:961 (1982).

20. Z. Yu, B. Ricco, and R. W. Dutton, "A Comprehensive Analytical and Numerical Model of Polysilicon Emitter Contacts in Bipolar Transistors," *IEEE Trans. Electron Dev.* **ED-31**:773 (1984).

21. D. D. Tang, P. M. Solomon, T. H. Ning, R. D. Isaac, and R. E. Burger, "1.25 μm Deep-Groove-Isolated Self-Aligned ECL Circuits," *Proc. ISSCC,* 1982, p. 242.

22. T. M. Liu, G. M. Chin, D. Y. Jeon, M. D. Morris, V. D. Archer, H. H. Kim, M. Cerullo, K. F. Lee, J. M. Sung, K. Lau, T. Y. Chiu, A. M. Voshchenkov, and R. G. Swartz, "A Half-Micron Super Self-Aligned BiCMOS Technology for High Speed Applications," *Tech. Dig. IEDM,* 1992, p. 23.

23. J. D. Warnock, "Silicon Bipolar Device Structures for Digital Applications: Technology Trends and Future Directions," *IEEE Trans. Electron Dev.* **42**:377 (1995).

24. V. dela Torre, J. Foerstner, B. Lojek, K. Sakamoto, S. L. Sundaram, N. Tracht, B. Vasquez, and P. Zdebel, "MOSAIC V—A Very High Performance Bipolar Technology," *IEEE Bipolar Circuits, Technol. Mtg.,* 1991, p. 21.

25. A. Felder, R. Stengl, J. Hauenschild, H. M. Rein, and T. F. Meister, "25 to 40 Gb/s Si IC in Selective Bipolar Technology," *Int. Solid State Circuits Conf. Tech. Dig.,* 1993, p. 156.

26. T. Shiba, Y. Tamaki, T. Kure, T. Kobayashi, and T. Nakamure, "A 0.5 μm Very-High-Speed Silicon Devices Technology U-Groove-Isolated SICOS," *IEEE Trans. Electron Dev.* **38**:2505 (1991).

27. C.-T. Chuang, G. P. Li, and T. H. Ning, "Effect of Off-Axis Implant on the Characteristics of Advanced Self-Aligned Bipolar Transistors," *IEEE Electron Dev. Lett.* **EDL-8**:321 (1987).

28. T. Yuzuhira, T. Yamaguchi, and J. Lee, "Submicron Bipolar-CMOS Technology for 16 GHz f_T Double Poly-Si Bipolar Devices," *Tech. Dig. IEDM,* 1988, p. 748.

29. M. Sugiyama, H. Takemura, C. Ogawa, T. Tashiro, T. Morikawa, and M. Nakamae, "A 40 GHz f_T Silicon Bipolar Transistor LSI Technology," *IEDM Tech. Dig.,* 1989, p. 221.

30. J. D. Hayden, J. D. Burnett, J. R. Pfiester, and M. P. Woo, "An Ultra-Shallow Link Base for a Double Polysilicon Bipolar Transistor," *Proc. IEEE Bipolar Circuits, Technol. Mtg.,* 1992, p. 96.

31. P. Ratnam, M. Grubisich, B. Mehrotra, A. Iranmanesch, C. Blair, and M. Biswal, "The Effect of Isolation Edge Profile on the Leakage and Breakdown Characteristics of Advanced Bipolar Transistors," *Proc. IEEE Bipolar Circuits, Technol. Mtg.,* 1992, p. 117.

32. I. Antipov, "Bipolar Transistor with Minimized Collector-to-Base Junction Area," *IEEE Trans. Electron Dev.* **ED-30**:723 (1983).

33. Y. Tamaki, T. Shiba, I. Ogiwara, T. Kure, K. Ohyu, and T. Nakamura, "Advanced Device Process Technology for 0.3 μm Self-Aligned Bipolar LSIs," *IEEE Bipolar Cicuits, Tech. Mtg.,* 1990, p. 166.

34. T. Nakamura, T. Miyazaki, S. Takahashi, T. Kure, T. Okabe, and M. Nagata, "Self-Aligned Transistor with Sidewall Base Electrodes," *IEEE Trans. Electron Dev.* **ED-29**:596 (1982).

35. J. Van der Veldedn, R. Dekker, R. van Es, S. Jansen, M. Koolen, H. Maas, and A. Pruijmbuch, "Basic: An Advanced High Performance Bipolar Process," *Tech. Dig. IEDM,* 1989, p. 233.

36. J. O. Borland and C. I. Drowley, "Advanced Dielectric Isolation Through Selective Epitaxial Growth Techniques," *Solid State Technol.* **28**:141 (1985).

37. A. Ishitani, N. Endo, and H. Tsuya, "Local Loading Effects in Selective Silicon Epitaxy," *Jpn. J. Appl. Phys.* **23**:L391 (1984).

38. P. J. Zbedel, "Current Status of High Performance Silicon Bipolar Technology," *Proc. GaAs IC Symp.,* 1992, p. 15.

39. E. Bertagnolli, H. Klose, R. Mahnkopf, A. Felder, M. Kerber, M. Stolz, G. Schutte, H.-M. Rein, and R. Kopl, "An SOI-Based High Performance Self-Aligned Bipolar Technology Featuring 20 psec Gate Delay and a 8.6 fJ Power-Delay Product," *1993 Symp. VLSI Tech. Dig.,* 1993, p. 63.

40. P. R. Ganci, J.-J. J. Hajjar, T. Clark, P. Humphries, J. Iapham, and D. Buss, "Self-Heating in High-Performance Bipolar Transistors Fabricated on SOI Substrates," *Tech. Dig. IEDM,* 1992, p. 417.

41. F. Verwey and B. J. de Maagt, "Avalanche Injection Electron Current in SiO_2 at High Injection Densities," *Solid-State Electron* **17**:963 (1974).

42. D. D.-L. Tang and E. Hackbarth, "Junction Degradation in Bipolar Transistors and the Reliability Imposed Constraints to Scaling and Design," *IEEE Trans. Electron Dev.* **35**:2101 (1988).

43. E. Hackbarth and D. D.-L. Tang, "Inherent and Stress Induced Leakage in Heavily Doped Silicon Junctions," *IEEE Trans. Electron Dev.* **35**:2108 (1988).

44. T. C. Chen, C. Kaya, M. B. Ketchen, and T. H. Ning, "Reliability Analysis of Self-Aligned Bipolar Transistor Under Forward Active Current Stress," *IEDM Tech. Dig.,* 1986, p. 650.

45. A. R. Alvarez, *BiCMOS Technology and Applications,* Kluwer, Norwell, Mass. (1989).

46. T, Oguri and T. Kimura, "A New 0.8V Logic Swing, 1.6V Operational High Speed BiCMOS Circut," *IEEE Bipolar Circuits, Tech. Mtg.,* 1992, p. 187.

47. "BiCMOS, Is it the Next Technology Driver?," *Electronics,* Feb. 4, 1988.

48. G. Shahidi, J. Warnock, B. Davari, B. Wu, Y. Taur, C. Wong, C. Chen, M. Rodriquez, D. Tang, K. Jenkins, P. McFarland, R. Schulz, D. Zicherman, P. Coane, D. Klaus, J. Sun, M. Polcari, and T. Ning, ''A High Performance BiCMOS Technology using 0.25 μm CMOS and Double Poly 47 GHz Bipolar,'' *Symp. VLSI Technol.* 1992, p. 28.

49. C. K. Lau, C.-H. Lin, and D. L. Packwood, ''Sub-Micron BiCMOS Process Design for Manufacturing,'' *IEEE Bipolar Circuits, Technol. Mtg.,* 1992, p. 76.

50. D. Tang *et al.,* ''73 psec Bipolar ECL Circuits,'' *Tech. Dig. ISSCC,* 1986.

51. J. Warnock, P.-F. Liu, T.-C. Chen, K. Y. Toh, J. D. Cressler, J. A. Jenkins, D. D. Tang, J. Burghartz, J. Y. C. Sun, C. T. Chang, G. P. Li, and T. H. Ning, ''A 27 GHz 20 psec PNP Technology,'' *Tech. Dig. IEDM,* 1989, p. 903.

52. C. Davis, G. Bajor, J. Butler, T. Crandell, J. Delgado, T. Jung, Y. Khajeh-Noori, B. Lomenick, V. Milam, H. Nicolay, S. Richmond, and T. Rivoli, ''UHF-1: A High Speed Complementary Bipolar Analog Process on SOI,'' *IEEE Bipolar Circuits, Technol. Mtg.,* 1992, p. 260.

53. C. Davis, G. Bajor, J. Butler, T. Crandell, J. Delgado, T. Jung, Y. Khajeh-Noori, B. Lomenick, V. Milam, H. Nicolay, S. Richmond, and T. Rivoli, ''A High Speed Complementary Bipolar Analog Process on SOI,'' *IEEE Bipolar Circuits, Technol. Mtg.,* 1992, p. 260.

Chapter 19

Integrated Circuit Manufacturing

The previous chapters have reviewed the processing steps that make up several popular technologies. The ultimate goal of these technologies is to be able to manufacture functional components at high volume and low cost. As discussed in Chapter 1, the amount of work required to complete a batch of wafers does not depend on the amount of information that the photomasks contain. Clearly, one way to improve the manufacture of ICs is to increase the number of functional die per wafer. To do this, there is a great deal of emphasis on the design of an IC to make it as small as possible without violating the design rules. IC manufacturing has also pursued increasing the number of die per wafer by increasing the wafer size. Thus, silicon wafer diameters have steadily increased from 100 mm, which was common in the mid 1980s, to 150 and 200 mm wafers commonly used today. A few facilities are set up to run 250 mm wafers. Due to edge losses and scribe line effects, the usable die count increases slightly faster than the area of the wafer. For example, when a manufacturer increases the wafer diameter from 125 mm to 200 mm, the number of 7 mm × 7 mm die per wafer increases by a factor of 2.8 while the area increases by a factor of 2.6. GaAs wafers have also increased in diameter, although more slowly. Currently, the most common wafer diameters are 75 and 100 mm. The increase in wafer diameter sometimes reduces the number of wafers that a unit process can run simultaneously. Some tradeoff must therefore be made to optimize the line throughput.

Perhaps the most fruitful method of increasing functional die count however, is the focus of this chapter: increasing the die yield, the percentage of possible sites that are successfully fabricated, packaged, and tested. This is a very attractive avenue since, unlike increasing wafer size, considerable improvement can often be made without making large capital equipment investments. The die yield is the product of several factors, including the wafer yield (the fraction of wafers that complete processing), the process yield (the fraction of die on the wafers that complete processing and are considered to be functional), the assembly yield (the fraction of good die that are successfully packaged), and burn in yield (the fraction of good packaged parts that are still functional after an initial stress test):

$$Y_d = Y_f * Y_p * Y_a * Y_{bi} \tag{19.1}$$

This book does not consider packaging and other back end processes. For most technologies however, these yield terms are so close to unity that to a first approximation only the process yield, which will now be called simply the yield, must be considered.

490

The improvement of process yield has serious implications for both processing and design. The area of design for manufacturability is an active and fruitful one; however, this chapter will concentrate on processing for manufacturability. That is, what considerations must be taken into account when applying the technologies that have been described so that they can be manufactured with a high probability of success? By now the reader should be familiar with the survey nature of this text. Nowhere is this more true than in this chapter. Each of the sections (yield modelling, particle control, statistical process control, and design of experiments) introduce materials that are the subjects of books themselves. The reader should be aware that only the briefest introductions are given to material that is far more rich than can be presented here.

19.1 Yield Prediction and Yield Tracking

A typical chronological yield plot, called a learning curve, is shown in Figure 19.1. When a new technology is first introduced into manufacturing, it is not uncommon for the IC yields to be as low as 20%. With 80% of the product being discarded, the ability to quickly understand and correct the origin of yield losses is absolutely critical to a plant's economic viability. Not only does this understanding allow the manufacturer to concentrate resources at the source of the problem, it also allows an accurate projection of the supply of good die, so that the proper number of wafers can be kept in the fabrication line. As a simple example of the economics of yield, consider a facility that produces 1000 wafers per week. Each wafer contains 100 die, which, when packaged, sell for $50 each. When the facility is yielding 30%, it is producing a gross income of $75,000,000 per year. If the yield can be increased to 50%, the plant's income increases to $125,000,000 per year. This net gain of $50,000,000 is almost pure profit since very little additional work is involved in packaging and selling the good die. (Remember, the same number of die are produced whether or not they are good.)

Furthermore, since it is impossible to completely test every possible mode of operation of many complicated ICs, it is generally believed that the quality of the product improves with increasing yield.

Integrated circuit yield is normally limited by defects. Figure 19.2 shows the defect density trend with generation of the dynamic random access memory or DRAM [1]. A number of defect classification schemes exist. The next two sections will discuss the nature of two common types of defects, but this is not critical to the discussion of yield and yield models. Perhaps the most fundamental distinction is the one between benign and killing defects, although the difference between the two is sometimes difficult to detect. A large diameter spot of polysilicon left in the field region far from any other lines may be benign (assuming that the topology it causes for subsequent layers is not too severe). A much smaller defect of exactly the same type is killing if it is present in a region of minimum size lines and spaces. Since killing defects are by definition the only ones contributing to yield loss, the term *defect* will be used as shorthand for *killing defect*.

Defects can also be classified as large area and

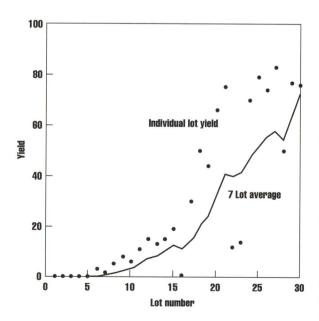

Figure 19-1 A plot of yield versus lot number for a particular technology. Also shown is the running average for the last seven lots.

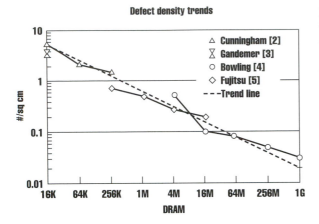

Defect density trends

Figure 19-2   Defect-density trend with generation of DRAMs *(after Dance and Gildersleeve).*

point defects. Large area defects such as scratches are often caused by poor wafer handling. Other large area defects can be the result of nonuniform processes, such as an incomplete etch that leaves some undesired material after patterning. Nonuniform deposition rates in the preceding process can contribute to this type of large area defect. Figure 19.3 shows a yield pattern that might be caused by nonuniformities in a single wafer process. In some processes, all of the wafers are aligned to the flat. Nonuniformities in these processes will result in a distinct yield loss region. If the wafers are randomly oriented in the process in question, the yield loss averaged over a number of wafers would be more difficult to observe, perhaps showing up as an unusually large loss with radial distance. It is more obvious however, in the yield of individual wafers.

Point defects are often assumed to be randomly distributed with a well defined size distribution. This is not necessarily true. Substrate defects like stacking faults, slip, and dislocations are often related to poorly designed thermal processes, incomplete annealing, or low quality epitaxial growth steps. They will lead to highly nonuniform defect distributions in the substrate. A good IC technology will be designed so as to avoid most of these problems. Point defects associated with deposited layers may be visible and often are major yield limitations. These defects can be categorized as voids (also called pinholes or intrusions), protrusions (also called bridges) and random spot defects. Their size distribution and density can be probed with test structures consisting of meander lines and interdigitated combs with varying sizes, spaces, and areas (Figure 19.4) [2].

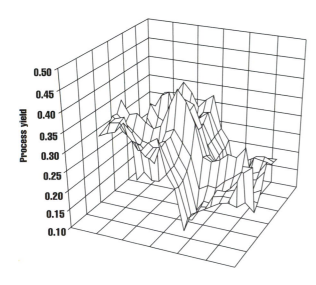

Figure 19-3   Yield plot as a function of die position on the wafer. Yield is often higher near the center of the wafer, but the right side of the wafer shows a strong yield fall off that bears investigation.

Point defect occurrence is generally found to increase sharply with decreasing size as shown in Figure 19.5 [3]. Ferris-Prabhu [4] has proposed that the distribution of defect sizes can be modeled as

$$D_{size} = c\,\frac{\delta^q}{\delta_o^{q+1}} \qquad 0 \le \delta \le \delta_o$$

$$D_{size} = c\,\frac{\delta_o^{p-1}}{\delta^p} \qquad \delta_0 \le \delta \le \delta_{max}$$

$$D_{size} = 0 \qquad \delta > \delta_{max} \qquad \textbf{(19.2)}$$

where D_{size} is the defect density at a particular defect size (often quoted in defects/cm^2), δ_{max} is the largest observed defect size, δ_o is the defect size with the maximum density, c is a normalization constant, and p and q are fitting parameters. In reality, the part of the curve describing very small defects is often very difficult or even impossible to measure.

The simplest type of yield model assumes Poisson statistics and independent, randomly distributed killer

Figure 19-4   Meander lines and interdigitated comb structures used to determine a layer defect density.

Figure 19-5   Theoretical and measured particle density as a function of particle diameter *(after deGyvez).*

defects. This model uses two parameters to define the yield: the defect density (D) and gross fail area (G) where G is the fraction of the wafer in which all of the circuits fail. Then the yield is given by

$$Y = (1 - G)e^{-AD} \tag{19.3}$$

where A is the area of the chip. This simple model is often used to project the yield of a new IC that will be fabricated in the same facility as a previous chip. The implicit assumption is that the defect density probably will not change significantly from one chip to the next as long as the facility and the technology are both fixed. Of course, this approximation is only valid if the chips are both designed so as to be equally sensitive to the types and sizes of defects found in that facility. For example, a low density IC with only a very small fraction of the features at the design rule limit will tend to yield better than the model predicts if the defect density is extracted from a chip that more aggressively pursues minimum feature size geometries.

Example 19.1   Simple Yield Prediction

A particular facility is producing several sizes of the same SRAM chip. The chips are identical except for the size of the array and therefore the die size. Calculate the defect density and gross fail area:

Array Size	Die Size	Yield
256 k	0.4 cm × 0.4 cm	75%
1 M	1.4 cm × 0.4 cm	55%
4 M	1.4 cm × 1.4 cm	20%

According to Equation 19.3, plotting the ln of the yield versus the area produces the defect density (from the slope) and the gross fail area (from the intercept). For this data $D = 0.75 \text{ cm}^{-2}$ and $G = 0.15$.

The predictive capability of this model is limited since it treats all defects as independent and equal. Relaxing the independence requirement produces another commonly used yield model, the negative binomial:

$$Y = \frac{1}{\left[1 + \dfrac{DA}{C}\right]^2}$$

(19.4)

where D is the defect density, A is the critical area of the chip, and C is the clustering factor [5],[6]. One can show that as C becomes large, the defects are once again independent, and the negative binomial model predictions approach those of the Poisson model.

Thus far, it has been assumed that all defects are equally likely to be killing defects. It is more reasonable to assume that a 10 μm boulder due to some errant CVD process is much more likely to result in a killing defect than is a 0.2 μm notch in a 1.0 μm metal line. It also must be pointed out that the 0.2 μm defect may be far more likely to occur. To take this difference into account, a quantity ϕ is defined to be the averaged probability that the mean size defect will be fatal. In this model

$$Y = (1 - G)e^{-D\phi A}$$

(19.5)

The averaged probability is given by

$$\phi = \int_0^\infty D_{\text{size}}(x)K(x)\ dx$$

(19.6)

where $D_{\text{size}}(x)$ is the defect size distribution and $K(x)$ is the fault probability kernel. $K(x)$ is the fraction of the total die area for which a particular size defect would be fatal. The general shape of a fault kernel curve for defects is shown in Figure 19.6. These curves have three regions of interest. For very large defects, the fault kernel approaches unity. For very small defects, the kernel may or may not go to zero depending on the exact nature of the defect. As an example, gate dielectric defects as small as half the gate oxide thickness may prevent complete oxidation and therefore lead to a short between the gate and the substrate. We refer to this zone as zero size or point defects. Between the two extremes, when the defect size is of order the feature size, the kernel increases in some manner that is often approximated as linear.

Defect size distributions depend to a great extent on the process itself. Various distribution shapes have been theorized, but as already discussed, it is generally believed that the defect density has a maximum at some size and decreases for both larger and smaller defects. Figure 19.7 shows an overlay of $K(x)$ and a reasonable curve for $D_{\text{size}}(x)$. Clearly, the integral of the product

Figure 19-6   Two typical fault kernel curves.

Figure 19-7   An overlay of the fault kernel and the defect distribution. Small shifts in the fault kernel curve can lead to dramatic changes in the yield.

depends sensitively on the position of the peak of $D_{size}(x)$ relative to the $K(x)$ curve. As the linewidth is decreased, the $K(x)$ curve shifts to the left, while the $D_{size}(x)$ curve remains relatively constant. Smaller linewidths generally therefore, imply smaller yields, unless the defect size distribution is also shrunk correspondingly. Since many defects are particle related, a great deal of attention has been paid to reducing the particle counts in the facility and in the processing equipment. It is also of great interest to be able to reliably detect smaller and smaller particles as the linewidths shrink.

The yield equation is sometimes broken down to the product of the individual process level yields. Then

$$Y = \prod_{i=1}^{levels} Y_i = \prod (1 - G_i)e^{D_i \phi_i A}$$

$$= \left[\prod (1 - G_i) \right] e^{A \Sigma \phi_i D_i} \quad \textbf{(19.7)}$$

Some levels, such as implant layers, have very low values of ϕ since they involve only gross features. It would be a good approximation for example, in a simple laboratory single level metal CMOS process, to assume that defects are present primarily in the polysilicon gate, metal, and contact levels, since these are done at or near the photolithographic resolution limit. As such they tend to dominate the yield equation. A model might also include the active area layer since relatively small defects in the active mask could lead to killing defects. These four are then called *critical layers*. If only a small number of critical layers are present, a special yield test chip can be made up to vary the size of a critical layer slightly, thereby shifting its $K(x)$ curve, to determine the relative sensitivity of Equation 19.7 to the various critical levels. Once it is known for example, that the IC yield has a strong correlation to a small shift in the metal linewidth, work can be done on reducing the defects at this level. This may involve improving the control of the metal deposition or etching processes, or changing the photolithography process at this level, improving the training of the technicians, or a variety of other variables. The facility and the process tools generally determine the defect number and size distribution in that these components generate the particles that become defects. The process and the design determine the yield kernel since they determine the sensitivity of the process to a defect. If a design uses a great deal of minimum feature size components, it will have a lower yield than one that is less aggressive. If the process has little latitude, smaller defects will prevent circuit operation. An extensive bibliography of yield models is given in Cheek and O'Donoghue [7].

19.2 Particle Control

Some of the most common defects are created by particles. The particle may remain on the wafer and be visible as a short circuit between lines or a boulder that the upper layers cannot cover, or it may be removed from the wafer by a clean after blocking an implant or locally disrupting pattern development during a photolithography step. One estimate suggests that particles are responsible for 75%

of the yield loss [8] in volume manufactured VLSI ICs. Because of the importance of defects on IC yield, most facilities now pay strict attention to particle detection, control, and reduction. The facility itself must be designed and constantly monitored to maintain minimum particle levels in the air. Clean rooms have gone from class 100 to class 10 and are now approaching class 1 in an effort to reduce contamination (Table 19.1). Maintenance of this level of particle control requires a strict discipline of the clean room occupants. Because of the success in controlling airborne particles, the emphasis has now largely shifted to reducing particle counts in the various processes and process equipment. Not only must particle generation be minimized when purchasing new equipment or developing new processes, when process equipment is taken off line for maintenance, it must be particle qualified before it is returned to service. Periodic particle surveys are also commonly done to ensure that low defect levels are maintained.

There are two common types of particle detection: on the wafer and in the ambient. Both methods use optical scattering of a laser beam by a particle. On wafer detection is done by sending a blank monitor (witness) wafer through the system to be checked. Blank wafers are usually used in current generation systems to avoid scattering of light by the pattern, which can obscure particle counts. The witness wafer is measured before and after processing, using a surface particle counter to determine the number of particles added by the process. Typical surface particle systems provide particle maps by measuring the scattered light as a function of position by scanning a laser beam across the surface of a wafer [9]. In some cases, the witness wafer may be run through the process a number of times to improve the signal to noise. Then the total number of particles added is divided by the number of runs to obtain the particles added per pass [10]. The location of the particles on the wafer and the position of the wafer in the processing equipment (in batch processes) can in some cases be used to help diagnose the source of the contamination. While the minimum particle size that can be accurately detected depends on the type of surface scanner and the shape and refractive index of the particles, it is usually quite difficult to use these systems to reliably detect particles smaller than 0.2 μm. Patterned wafers, when used, have much larger minimum detection sizes and produce a high rate of false alarms.

As an alternative to laser scanners, optical inspection systems are used with actual product wafers. These systems compare the optical pattern on the wafer to an expected pattern. The reference image can be either an adjacent die or a database containing the mask information. Any areas with significant differences are shown to the operator for a visual inspection. Not only do these systems detect particles, they can also be used to detect resist defects and for etch line monitoring of high contrast surfaces. While much slower (6 to 10 wafers per hour) than simple surface scanners, the fact that they can be used on product wafers make them an important tool for some critical levels [11]. A disadvantage of any type of witness wafer particle detection system is that it provides information

Table 19.1   Maximum number of particles per cubic foot of air greater than or equal to various sizes in four classes of clean rooms

	0.1 μm	0.2 μm	0.3 μm	0.5 μm	5.0 μm
Class 1	35	7.5	3	1	NA
Class 10	350	75	30	10	NA
Class 100	NA	750	300	100	NA
Class 1000	NA	NA	NA	1000	7

Federal Standard 209D, U.S. GSA 1989.

Figure 19-8   A schematic view of the optical path in the Model HVLIS detector made by Particle Measuring Systems, Inc. *(used by permission).*

only about the total amount of particles deposited during a process. It cannot provide real time information about the particle sources.

Gas and liquid particle detection is done by drawing a particle laden stream through a laser beam. This type of system can be used to measure particle counts in the clean room, or they can be connected to the exhaust of a piece of process equipment to determine the particle creation rate in the process. Figure 19.8 shows a schematic of an optical detector. The amount of light scattered depends on the intensity, wavelength, and polarization of the optical source, the size, shape, orientation, and refractive index of the particle, and the geometry of the optical detection method [12]. As commonly applied to microelectronics however, simple optical detectors are not reliable for particles much less than 0.3 μm in diameter. Condensation nucleus counters (CNCs) can be used to detect smaller particles in a gas ambient [13]. In these systems, the particle laden gas is first drawn through a saturated vapor. Various alcohols are often used to generate the vapor. The vapor condenses on the particles, making them large enough to be detected by optical scattering. Unlike simple optical systems, CNCs are limited to atmospheric pressure gasses.

19.3 Statistical Process Control

Thus far, defects have been thought of as particles or other randomly distributed contamination. If the definition of a defect is a process irregularity that leads to a circuit malfunction, defects may be caused by simple variations in process parameters. As an example, consider the gate level of a MOSFET. Assume that the mean linewidth is 0.35 μm and that the standard deviation of the linewidth is 0.03 μm. Finally, assume that the MOSFET will punch through from source to drain if the gate length is less than 0.20 μm. This is five standard deviations from the mean, which would seem to be a rather comfortable distance. Assuming a normal distribution of linewidths, the probability of any transistor having a linewidth of 0.20 or less, is only about 3×10^{-7}. If the circuit contains 10^7 transistors, however, this represents a significant source of "defects." Furthermore, the effective channel length is not only determined by the photolithography itself, but also on the etch process, the photomask that is used, the source/drain implant and diffusion steps, including the possibility of anomalous diffusion into the channel region due to defects, and the channel implant and anneal processes. Each of these processes have characteristic standard deviations. The combination of multiple variances can significantly increase the likelihood of process variation defects.

The effect of these process variations on yield is a very sensitive function of design. Ideally the IC would be designed so that it could operate with the full range of device parameters produced by the fabrication facility. Statistical process simulators exist to help in this exercise. Disturbances in a process can be used to project the range of transistor performance that can be expected. The output can then be fed into a circuit simulator like SPICE to determine the effects of the perturbation on the circuit performance. Since the performance must fall within certain limits in order to be regarded as good, one can then construct a multidimensional box in parameter space within which the IC must be fabricated. The design can be modified in order to increase the size of the box, but often this tradeoff requires a compromise in either performance or yield. Designing the circuit to guarantee operation for any possible set of device parameters would almost certainly require extremely poor performance.

A great deal of emphasis is therefore placed on controlling the individual process steps. A simple model for this control is shown in Figure 19.9. In this diagram, wafers are introduced into the process to accomplish some task. The quality of the result is measured against a series of desired outcomes. As an example, assume that the process step is intended to etch contacts down to silicon through an oxide. The process variables under the direct control of the engineer might include the residual gas pressure in the etch chamber, the process gas flows, the plasma power, the electrode temperature, and the chamber pressure during the etch. Some of the variables that are not controlled during the process include the composition of the material to be etched, the photoresist profile, the gasses adsorbed onto the surface of the wafer, the films coating the walls of the chamber and the electrodes deposited during previous etching, and the residual gas composition. The desired outcomes are the completeness of the etch, the amount of erosion of the underlying silicon, and the size and profile of the etched contact.

The ideal solution to obtaining a reliable contact etch process would be real time adaptive control. This requires the construction of a process tool that measures the process outcomes and continuously adjusts the process variables to obtain the desired outcomes [14]. There is an important difference between this type of control and simply manually controlling the process variables. In the etcher example, as long as the values of the uncontrolled variables are not too far from the norm, the control parameters of power, gas flows, pressure, and electrode temperature could be adjusted automatically to achieve the desired etch outcome. Because the process is adjusted in real time to compensate for the condition of the wafers as they are loaded into the etch chamber, process variations do not accumulate, resulting in a far better process.

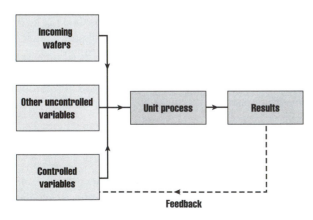

Figure 19-9   A schematic of a semiconductor process. Both controlled and uncontrolled variables must be considered.

For this type of control to be implemented however, it is necessary to be able to measure all of the process outcomes in real time and in the process ambient. Returning to the etch example one more time, one can construct a system for measuring the remaining film thickness on the wafer. Various endpoint detectors, which shut off the etch when some signal indicates etch completion (plus some percent overetch), are a simple and very useful form of this sort of control. Imagine however, using the etch rate information to allow the etch system to adjust itself to maintain the etch rate at some predetermined value. One demonstration of such a system has shown the ability to etch nitride layers down to 503 Å $\pm$ 12 Å over 23 trials when the targeted final thickness was 500 Å. The initial nitride thickness was about 2300 Å [15]. In a similar application in situ ellipsometry was to compensate for long term etch rate drift in a multistep polysilicon gate etching process [16]. While these systems are currently single point detectors, it might also be possible to extend them to multipoint sensors, thereby giving the system the ability to adjust the process to improve uniformity. On the other hand, it is very hard to envision any type of system that could continuously measure the etch profiles. As a result, one can expect to see some progress in this area but for the foreseeable future real time adaptive control is not a panacea.

Since the adaptive control techniques described above are still in the demonstration phase, current manufacturing must use another approach. The method commonly used in IC manufacturing to deal with uncontrolled variables has two components. The first of course, is to control as many variables as possible. Even if the parameter cannot be adjusted to optimize the process result, control of the parameter reduces the process variability. Because of the enormous number of uncontrolled parameters however, this approach alone cannot solve the problem. The most widely used technique for dealing with poorly controlled process variables is to design a robust process, that is, a process that is insensitive to the uncontrolled parameters.

Since uncontrolled variables are by definition impossible to adjust, a valuable measure of process robustness is its stability or performance over time. To improve reproducibility and process control, statistical process control (SPC) has widely been adopted throughout the IC industry. SPC is a method to provide immediate feedback at each process step. It is often used to try to drive each process step to be as consistent as possible. To institute an SPC program trend data must first be collected that shows the behavior of a process as a function of time. Once enough data have been accumulated they can be used to control the process. In a typical CVD application for example, a film thickness is measured on several samples from a batch of wafers immediately after the deposition. The result is stored in a computer along with all previous data from the same process. Plots of thickness, wafer to wafer and across the wafer thickness variation, refractive index and refractive index variation, and other important thin film parameters versus batch number are updated and displayed on the screen. By presenting trend data it is immediately obvious if the lot data is significantly different than previous runs. It is also apparent if a long term drift is present in the process.

Of primary interest is the process variance. From elementary statistics, the variance of a set of data that represents a larger population is given by

$$\sigma^2 = \frac{\Sigma[x - \bar{x}]^2}{N} = \frac{N\Sigma x^2 - [\Sigma x]^2}{N^2} \qquad (19.8)$$

The graph of the variance is sometimes called an *s-chart*. If the variance is large, any information regarding the process mean is of very limited value. Increases in the variance calls for immediate action to improve the process uniformity and repeatability. If the variation is sufficiently small, the mean (the x-chart) is then examined. Alert conditions exist if the process mean is outside some predetermined limits, if two consecutive readings are near these limits, or if a linear regression of a number of readings (for example the last seven) show a suitably large slope with a high regression coefficient. Generally speaking, any of these conditions require immediate corrective action, although the statistical nature of random process variations suggest that occasionally such an excursion will occur even when nothing has changed. As more robust processes are developed, the process limits for the x-chart and the acceptable upper limit for the s-chart can be successively tightened. Experience has shown that this approach will generally lead in turn to higher IC yields.

The ramifications of improving process control are sometimes subtle and quite often indirect. In one example, Hewlett-Packard instituted SPC for the polysilicon deposition process in a CMOS technology. The poly thickness variation improved from a standard deviation of about 5% to a variation of about 1%. The next step in the process was gate patterning, which was done at the limit of the lithography tool. The change in the reflected light due to the polysilicon thickness variation was found to be enough to increase the variation of the poly linewidth measurably. This variation resulted in a yield loss. As a result, reducing the variability in the deposition process reduced the excursions of one of the most important uncontrolled variables for the lithography process and in so doing produced more repeatable lithography results. Implementing a full SPC program is a valuable, but time consuming enterprise. One facility reported requiring 600 charts on line at any time. These charts require not only constant updating, but wholesale modification whenever the technology changes [17].

19.4 Full Factorial Experiments and ANOVA

One of the goals of the process engineer is to reduce the variability of the process outcome as evidenced by control charts and to make minor adjustments to a process to compensate for long-term trends. To do this, it is desirable to understand the effect of each of the control variables on the process outcome and possible interactions between variables. Most often these relationships are unknown. To develop a process then, the engineer can experimentally map out the response of a process to a set of variables. Figure 19.10 shows a simple example showing an etch rate as a function of RF power and etch gas pressure. All other considerations aside, the etch rate appears to be the least sensitive to the pressure at low power and high pressure. Strictly from a manufacturing control standpoint therefore, it would be desirable to run the etch in this regime, but clearly tradeoffs must be evaluated with regard to other important parameters such as etch rate, selectivity, and etch profile.

When there are only two variables it is possible to try a wide range of both variables including all possible combinations. Generating the data in Figure 19.10 would require measurements at 5 powers times 6 pressures or 30 experiments. Such an experiment does not allow for the effects of uncontrolled variables. To get some idea of the variability, multiple measurements must be taken for each of the 30 pressure/power combinations. Thus, at least 90 and perhaps as much as 270 measurements must be done. This type of experiment is called a *full factorial*.

The full factorial analysis is clearly inappropriate when the number of process variables exceeds three or four. The simplest technique for analyzing data more

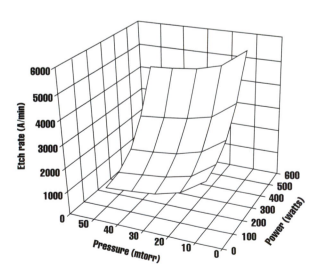

Figure 19-10   Etch rate of a plasma system as a function of power and pressure.

systematically is the analysis of variance (ANOVA) technique. The method, developed by Sir Ronald Fisher in the 1930s to analyze agricultural data, is a statistical tool to determine the effects of several variables. ANOVA experiments are classified by the number of variables that are controlled. As an example consider a two-way ANOVA. Return to the etching experiment once again where the experiment varies pressure and power. Each variable is run at a high and a low value, and each combination of parameters is run twice to get some idea about the variability of the data. The results are summarized in Table 19.2.

The total variance for the experiment is given by

$$SS_{Total} = \sum_{i=1}^{N} y_i^2 - N * \bar{y}^2 \tag{19.9}$$

where y_i are the individual measurements, $\bar{y}$ is the average of the measured etch rates, and N is the number of observations (eight in this case). Applying the formula SS_{Total} is 0.5075 [kÅ/min]^2, and the mean etch rate is 1.10 kÅ/min. Adding all of the results in each column gives a total of the etch rates at power 1 of 4.50 kÅ/min and a total of the etch rates for power 2 of 4.30 kÅ/min. The variance due to the power parameter is given by

$$SS_{Power} = \frac{[\text{total of results at power 1}]^2}{\text{number of observations at power 1}}$$
$$+ \frac{[\text{total of results at power 2}]^2}{\text{number of observations at power 2}} - N * \bar{y}^2 \tag{19.10}$$

For this data, SS_{Power} is 0.005 [kÅ/min]^2, a very small part of the total variance. Similarly, by summing the data for each row, the $SS_{Pressure}$ can be calculated as 0.263 [kÅ/min]^2. Finally, to determine the variance of the etch rate due to the interaction power $\times$ pressure, one must calculate

$$SS_{Pow \times Press} = \sum_{i=1}^{n_{cells}} \frac{(Cell)_i^2}{n_{cells}} - SS_{Pow} - SS_{Press} - N * \bar{y}^2 \tag{19.11}$$

where $(Cell)_i$ is the total of the observed etch rates in each of the cells in Table 19.2. In this case, $SS_{Pow \times Press}$ is

$$SS_{Pow \times Press} = \frac{2.225^2}{2} + \frac{1.45^2}{2} + \frac{2.275^2}{2} + \frac{2.85^2}{2} - 0.005 - 0.263 - 8 \times 1.1^2 \tag{19.12}$$

Table 19.2   The raw data for etch rate in kÅ/min for a simple two-way experiment

	Power 1	Power 2
Pressure 1	1.15, 1.075	0.675, 0.775
Pressure 2	1.175, 1.10	1.40, 1.45

Table 19.3   ANOVA results for a two-way experiment

	SS	v	V	F
Power	0.005	1	0.005	1.67
Pressure	0.263	1	0.263	87.67
Pow×Press	0.228	1	0.228	76
Error	0.012	4	0.003	
Total	0.508	7		

or 0.228 [kÅ/min]^2. The final variability is the error that can be readily determined as the difference between SS_{Total} and the sum of all of the other variances. For the data listed in Table 19.2, SS_e is 0.0115 [kÅ/min]^2.

For this experiment pressure appears to have the largest effect, followed by the *power×pressure* interaction, but it hasn't been determined whether or not the effects are statistically significant. That is, is the variation caused by the change in the control variables significantly larger than the noise level in the experiment? First, one must find the number of degrees of freedom for each component of the variation. In statistics, a degree of freedom (v) means the number of fair (i.e., independent) comparisons that can be made from the data. This ANOVA is calculating mean etch rates. A mean etch rate can only be referenced to one number, 0. Each variable therefore, has one degree of freedom. The cross variable *power×pressure* has the product of the degrees of freedom of the two component variables (*Power* and *Pressure*). In this case, this is also 1 since $1 \times 1 = 1$. The total number of degrees of freedom is given by the number of observations minus 1 or $N - 1 = 7$. The number of degrees of freedom associated with the error then, is the difference between the total number of degrees of freedom and the degrees of freedom associated with the variables. In this case, the degrees of freedom associated with the error is $7 - 1 - 1 - 1 = 4$. Dividing the variations by the degrees of freedom determines a V score as summarized in Table 19.3. By dividing the V scores by the error V score, one calculates the F score. The purpose of this score is to determine whether or not the effect of the variable is statistically significant. In this case, the V score of the error is very small. As a result, some of the F scores are quite large. Statistical significance is always stated at some level of confidence. The F score must be larger than the number tabulated in Appendix VI in order to say that the variable is significant to that confidence level. Each of the upper three parameters have one degree of freedom while the denominator (the error) has four. According to the appendix, $F_{0.05;1;4}$, and $F_{0.01;1;4}$, which correspond to the 95% and 99% confidence are, 7.71 and 21.2, respectively. Thus both *Pressure* and *power×pressure* are significant at the 99% level, while *Power* alone is not significant at the 95% confidence level.

19.5 Design of Experiments

The primary disadvantage of any full factorial method is the number of measurements required for a realistic system. One alternative to the full factorial experiment is to run two sets of one at a time experiments. For example, a set of experiments could be done by varying pressure with power fixed near the middle of the range to be investigated and then fixing the pressure similarly and varying the power. The primary advantage of one at a time experiments are the relatively small number of data points required. It is common for realistic processes to have 6 to 12 variables not including interactions. A full factorial experiment with three levels for each of six parameters would require 3^6 or 728

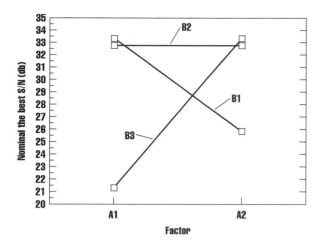

Figure 19-11   Particle counts as a function of silane flow and chamber pressure for the low pressure deposition of polysilicon *(after DePinto).*

experiments without repetition. Clearly, running full factorial experiments is not practical. Running a one at a time experiment would require only $6 * 3 - 1 = 17$ experiments, but will not provide complete information about the parameter space.

Consider performing two one at a time experiments instead of a full factorial experiment for the simple etch experiment described in the last section. The power could first be held constant while the pressure is varied, and then the pressure could be fixed while the power is varied. This method would work fairly well for the data shown in Figure 19.10, but consider the data in Figure 19.11. This experiment tried to minimize the particle counts in an LPCVD polysilicon furnace by trying two silane flows (A1 = 45% of full flow and A2 = 35% of full flow) at three levels of pressure (B1 = nominal, B2 = nominal − 40 mtorr, B3 = nominal − 80 mtorr). At the nominal pressure, increasing the silane flow decreases particle production while the opposite is true when operating at a pressure 80 mtorr below nominal [18]. The two parameters are said to be confounded. The effect of one depends on the value of the other. In the data of Table 19.3, power and pressure are confounded because of the large significance of *power×pressure*. The presence of confounded variables make the conclusions of any one at a time experiments extremely uncertain.

Clearly, some less haphazard technique must be developed in order to economically optimize a process with a large number of variables. This can be done using design of experiments (DOE). It must be reemphasized that this is an extremely useful topic that is really a course in itself. The intention of this section is only to describe an example of a simple designed experiment to introduce the reader to the possibilities of DOE. Readers are strongly encouraged to consult more specialized texts on the topic. In addition, a number of software products exist that can lead the engineer through the process of setting up a designed experiment.

The first step in DOE is to identify all of the possible variables as well as all possible combinations of variables. These variables and combinations of variables are called *factors*. At this point, it is important to identify potential variables without regard to whether or not they are expected to be important. It is not unusual for the first round of experiments to contain 10 or more factors. A reasonable range for each variable must also be identified. The first round of experiments, called *screening experiments*, are intended to be able to eliminate the factors that have little effect on the desired process outcome. Once this is done, a second and possibly third round of experiments are done to focus on the important variables.

Table 19.4 shows an example of a set of eight experiments that can be run to investigate the effects of seven factors. The set of experiments are called *orthogonal*. Each factor is set to a low level in four tests and high in the other four tests. Orthogonality means that factors can be evaluated independently, that the total experiments are designed so that one factor does not affect another. The matrix described in Table 19.4 is therefore called an eight level orthogonal array (L8 OA). One now has to assign the variables and interactions between variables to the factor columns. One method would be to assign seven variables. For an etching experiment, one might choose RF power, pressure, platen temperature, %CF$_4$ in the gas flow, chamber base pressure, number of wafers in the chamber, and total gas flow. Once the eight measurements are taken, the results can be analyzed using the

Table 19.4   An L8 OA for a two level experiment. The H and L indicate high and low values of the factors in the columns

	1	2	3	4	5	6	7
Trial 1	L	L	L	L	L	L	L
Trial 2	L	L	L	H	H	H	H
Trial 3	L	H	H	L	L	H	H
Trial 4	L	H	H	H	H	L	L
Trial 5	H	L	H	L	H	L	H
Trial 6	H	L	H	H	L	H	L
Trial 7	H	H	L	L	H	H	L
Trial 8	H	H	L	H	L	L	H

ANOVA method. If an estimate of the error is desired, 16 or 24 measurements can be done by repeating each row two or three times. A wide variety of OAs have been tabulated for these applications.

This arrangement would seem to be incredibly efficient. The effects of seven variables can be determined in only eight measurements. The tradeoffs here is that no information can be obtained about the interaction between variables. More importantly however, if these interactions exist, they will effect the outcome of the experiment. In this example, the third column of the OA is related to the first two. Using the Boolean laws

$$L \times L = L$$
$$L \times H = H \times L = H$$
$$H \times H = L \tag{19.13}$$

Column 3 is exactly the product of the first two columns. Similarly, Column 5 is the product of Columns 1 and 4, and Column 6 is the product of Columns 2 and 4. Extending these rules to three terms, one can show that Column 7 is the product of Columns 1, 2, and 4. Thus, if the seven variables indicated above are assigned, the experiments run, and Column 3 is found to be significant, it may be that the variable assigned to this column has a significant effect, or it may be that the interaction between the first two variables is important. One can avoid this ambiguity by assigning these columns to the interactions described. Then only three of the factors are variables (Columns 1, 2, and 4). Performing eight trials on two levels of three variables is once again a full factorial experiment. An advantage of DOE, then, is that it allows the engineer to allocate the experiments to investigate the variables that are considered important.

Example 19.2   ANOVA Analysis of L8 OA

Assume that the etching experiment described above was run. The column assignments and levels were as in Table 19.5.

Table 19.5   Experiment parameters for the L8 OA etch experiment

	Variable	H Level	L Level
1	RF power (W)	500	100
2	Etch pressure (mtorr)	50	10
3	Platen temperature (°C)	80	40
4	%CF$_4$	75	50
5	Base pressure (torr)	1×10^{-4}	1×10^{-5}
6	Number of wafers	4	1
7	Total gas flow (slpm)	2.5	1.0

One can now calculate the net effect of the first factor by adding all of the results at low power and comparing them to the sum of all of the results at high power. From Table 19.6 these sums are 2.81 and 6.06 kÅ/min. Since there are eight observations, the sum of the squares for power is

$$SS_{Power} = \frac{[6.06 - 2.81]^2}{8} = 1.32 \qquad \textbf{(19.14)}$$

The results of the analysis for each column is given in Table 19.7.

Notice that the largest effects seem to be caused by the variables assigned to Columns 1, 2, 4, and perhaps 6. Column 3 tests as having a small effect; however, one must also bear in mind that the data in this column can also be affected by the Column 1-Column 2 product. The small response may mean that there is no significant effect of the platen temperature or of the Power-Pressure cross term, or it may mean that the two effects cancel each other. Since power has by far the largest effect, columns that may involve cross products of power must be viewed with the most suspicion.

There are several ways available in an OA to get an error estimate. Sometimes a column may be unassigned and so reserved for error estimation. Otherwise, the lowest response columns may be pooled. In doing so, the assumption is made that these parameters have little effect on the etch rate and the results are roughly indicative of the random error in the experiment. In our case, variables 3, 5, and 7 are clearly candidates for error estimation. Adding the sum of the squares for these three entries and dividing by 3 (the denominator in Equation 19.12 now is 24 instead of 8) gives an SS_e of 0.0022. Then, dividing the error estimation to get the F scores and comparing these scores to the F scores in Appendix VI assuming one degree of freedom in the

Table 19.6   Etch results (in kÅ/min) for the L8 OA experiment using the factor assignments listed above

	1	2	3	4	5	6	7	Etch Rate (kÅ/min)
Trial 1	L	L	L	L	L	L	L	0.760
Trial 2	L	L	L	H	H	H	H	0.895
Trial 3	L	H	H	L	L	H	H	0.400
Trial 4	L	H	H	H	H	L	L	0.755
Trial 5	H	L	H	L	H	L	H	1.575
Trial 6	H	L	H	H	L	H	L	1.800
Trial 7	H	H	L	L	H	H	L	1.170
Trial 8	H	H	L	H	L	L	H	1.515

Table 19.7   The raw data of the L8 OA described in Tables 19.5 and 19.6

	Variable	*H* Total	*L* Total	*SS*
1	RF power (W)	6.06	2.81	1.320
2	Etch pressure (mtorr)	3.84	5.03	0.177
3	Platen temperature (°C)	4.53	4.34	0.005
4	%CF$_4$	4.965	3.905	0.140
5	Base pressure (torr)	4.39^5	4.47^5	0.001
6	Number of wafers	4.265	4.605	0.014
7	Total gas flow (slpm)	4.385	4.485	0.001

Table 19.8   Analysis of the data given in Tables 19.7

	Variable	*SS*	*df*	*V*	*F*
1	RF power (W)	1.32	1	1.32	603.4
2	Etch pressure (mtorr)	0.177	1	0.177	80.9
3	Platen temperature (°C)	0.005	*	*	*
4	%CF$_4$	0.140	1	0.140	64.0
5	Base pressure (torr)	0.00^1	*	*	*
6	Number of wafers	0.014	1	0.014	6.4
7	Total gas flow (slpm)	0.001	*	*	*

numerator and three in the denominator, the power, pressure, and %CF$_4$ are all significant at the 99% level, while the number of wafers is significant at the 90% level.

A final word of caution is in order regarding the use of OAs. The analysis described in this section, often called the Taguchi approach, is a powerful tool that allows one to quickly optimize a process in a complicated parameter space even in the presence of noise. This section has only scratched the surface of the technique. A wide variety of orthogonal arrays exist to fit almost any set of variables that you may wish to investigate. An accurate interpretation of the results however requires some skill with the technique. It is easy to be misled and confirming experiments are extremely important before the results can be relied upon.

19.6 Computer Integrated Manufacturing

Computer integrated manufacturing (CIM) is an umbrella term that encompasses a number of ways to use computers to improve the manufacturing enterprise. For IC manufacturing, CIM includes the following major components [19]:

Planning and scheduling of factory operations and resources
Modeling of factory operation
Process and product specifications and recipes
Tracking of work in progress (WIP)
Factory performance monitoring
Machine and process monitoring, control, and diagnosis

All of these functions are not present in every CIM system, but the list includes the most common components. Whatever its form, CIM generally involves the installation of a local area network throughout the factory including the clean room. Before each process step, the operator enters the batch ID into the computer. Where possible, the process tools themselves report the process results. Ion implanters for example, now have in situ sensors that determine the implant results and can output the results to a LAN interface via the semiconductor equipment communication standard (SECS), which was developed by SEMI [20]. Newer characterization and process tools, such as ellipsometers, linewidth measuring instruments, and etchers similarly are equipped with SECS or SECS-II interfaces

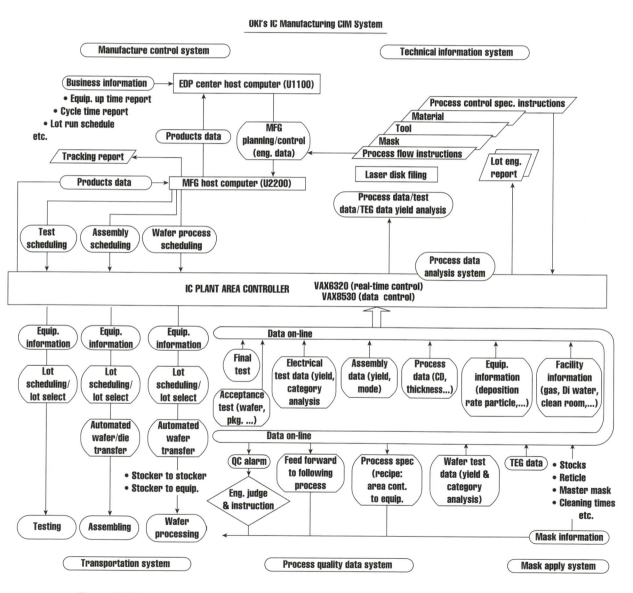

Figure 19-12 A full CIM system implemented by one major semiconductor manufacturer *(from Mizokami)*.

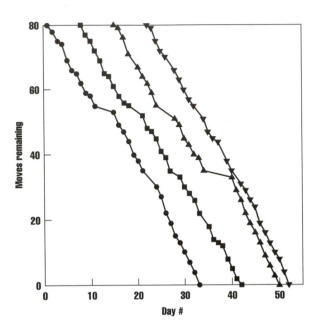

Figure 19-13   Charts showing the remaining lot moves as a function of time are often used to point out bottlenecks in the line. These diagrams are sometimes called chicken scratches.

for CIM as well as other developing SEMI standards [21]. Figure 19.12 shows a schematic for a CIM system [22].

A full CIM system represents an enormous amount of information potentially available for use. The issue is not data collection, but data distribution and utilization. The use of relational, distributed databases allows the users easy query of the database without requiring a very large central computer system. In spite of the emerging availability of the hardware and software to operate CIM systems, intelligent applications of the information that they produce are not as forthcoming.

One of the earliest and perhaps most important applications of CIM has been the scheduling of wafers through particular pieces of equipment in the fab line. It is generally accepted that the longer it takes to fabricate a batch of wafers, the lower the yield. It is desirable then to schedule the fabrication line to maximize throughput without incurring undo delays. Since the process is sequential, one needs to model the throughput of the various pieces of equipment and then examine the projected effects of a particular wafer move. By projecting ahead to the expected inventory at particular steps, long delays for particular process equipment can be avoided. Perhaps even more importantly, the wafer inventory can be controlled so as to minimize cycle time and/or maximize tool utilization.

Charts like those shown in Figure 19.13 are often used to point out bottlenecks in the line. The number of steps remaining until the lot is completed is plotted for each batch of wafers as a function of time. Each line is sometimes called a *scratch*. Process steps at which the scratch frequently is horizontal represent fabrication line bottlenecks that limit the overall throughput of the factory. Cycle time is also often measured against the theoretical cycle time, the minimum time that it would take to move a batch of wafers through the line if they were the only wafers in the factory. A well managed CIM system can produce routine turn times of twice the theoretical as compared to about seven times theoretical for a poorly controlled line [23].

19.7 Summary

Semiconductor IC manufacturing has as one of its primary goals the maximization of the process yield. This involves the reduction of particles both in the clean room ambient and in the process equipment, and the reduction of variability in the various unit processes. Statistical process control is an extremely valuable tool for reducing this variability. ANOVA was presented as a method to analyze experiments that determine the effects of various process parameters on a process. For realistic processes, the number of variables that must be optimized is excessive, and so orthogonal arrays have been developed as a more efficient method for exploring complex parameter spaces.

References

1. D. Dance and K. Gildersleeve, "Estimating Semiconductor Yield from Equipment Particle Measurements," *Proc. 1992 IEEE/SEMI Int. Semiconductor Manufacturing Sci. Symp.,* p. 18.
2. D. Carpenter, "Defect Monitor and Characterization Standard," *Proc. IEEE Int. Workshop Defect, Fault Tolerance,* Nov. 1980, p. 70.
3. J. P. deGyvez, *Integrated Circuit Defect-Sensitivity: Theory and Computational Models,* Kluwer, Boston, 1993.
4. A. V. Ferris-Prabhu, "Role of Defect Size Distribution in Yield Modeling," *IEEE Trans. Electron Dev.* **ED-32**:1727 (1985).
5. C. H. Stapper, "Fact and Fiction in Yield Modeling," *Microelectron. J.* **20**:129 (1989).
6. C. Kooperberg, "Circuit Layout and Yield," *IEEE J. Solid-State Circuits* **23**:887 (1988).
7. G. Cheek and G. O'Donoghue, "Yield Models in a Design for Manufacturability Environment: A Bibliography," *Proc. 1993 IEEE/SEMI Int. Semiconductor Manufacturing Sci. Symp.,* p. 133.
8. T. Hattori, "Contamination Control: Problems and Prospects," *Solid State Technol.* **33**(7):S1 (1990).
9. P. Lilienfeld, "Optical Detection of Particles on Surfaces: A Review," *Aerosol Sci., Technol.* **5**:145 (1986).
10. B. Tullis, "Particle Contamination by Process Equipment," in *Handbook of Contamination Control in Microelectronics,* D. L. Tollivier, ed., Noyes, Park Ridge, NJ, 1988.
11. P. Burggraaf, "Patterned Wafer Inspection: Now Required!," *Semicondvector Int.,* p. 57, Dec. 1994.
12. D. W. Cooper, "Contamination Monitoring," in *Microelectronics Manufacturing Diagnostics Handbook,* A. H. Landzberg, ed., Van Nostrand Reinhold, New York, 1993.
13. K. H. Ahn and B. Y. Liu, "Particle Activation and Growth Processes in Condensation Nucleus Counter—I. Theoretical Background, II Experimental Study," *J. Aerosol Sci.* **21**:249, 263 (1990).
14. H. F. Guo, C. J. Spanos, and A. J. Miller, "Real-Time Statistical Process Control for Plasma Etching," *1991 IEEE/SEMI Int. Semiconductor Manufacturing Sci. Symp.,* p. 113.
15. S. Maung, S. Banerjee, D. Draheim, S. Henck, and S. W. Butler, "Integration of *In-situ* spectral Ellipsometry with MMST Machine Control," *IEEE Trans. Semiconductor Manufacturing* **7**:184 (1994).
16. S. W. Watts and J. A. Stefani, "Supervisory Run-to-Run Control of Polysilicon Gate Etch Using *InSitu* Ellipsometry," *IEEE Trans. Semiconductor Manufacturing* **7**:193 (1994).
17. R. Abugov, "From Trend Charts to Control Charts: Setup Tests for Making the Leap," *1992 IEEE/SEMI Int. Semiconductor Manufacturing Sci. Technol.,* p. 3.
18. G. DePinto, "A Methodology for Optimizing a Constant Temperature Polysilicon Deposition Process," *IEEE Trans. Semiconductor Manufacturing* **6**:332 (1993).
19. J. McGehee, J. Hebley, and J. Mahaffey, "The MMST Computer-Integrated Manufacturing Framework," *IEEE Trans. Semiconductor Manufacturing* **7**:107 (1994).
20. SEMI Standards "SEMI Equipment Communications Standards 1 (Message Transfer: SECS-I, E4-917, 1992) and 2 (Message Content: SECS-II, E5-93, 1993), Semiconductor Equipment and Materials Institute.
21. R. Beaver, A. Coleman, D. Draheim, and A. Hoffman, "Architecture and Overview of MMST Machine Control," *IEEE Trans. Semiconductor Manufacturing* **7**:127 (1994).

22. Y. Mizokami, ''The Total CIM System for Semiconductor Plants,'' *Proc. 1990 Int. Semiconductor Manufacturing Sci. Symp.,* IEEE, Piscataway, NJ, p. 24.

23. M. Adams and B. Smoak, ''Managing Manufacturing Improvement Using Computer Integrated Manufacturing Methods,'' *Proc. 1990 Int. Semiconductor Manufacturing Sci. Symp.,* IEEE, Piscataway, NJ, p. 9.

Appendix I

Acronyms and Common Symbols

The cgs units listed in the parentheses are not necessarily the most commonly used units for this variable.

A	Area (cm^2)
	Coefficient of Deal-Grove equation (cm)
	Absorbance (dimensionless)
APCVD	Atmospheric pressure chemical vapor deposition
ANOVA	Analysis of variance
AR	Aspect ratio
B	Magnetic field (Gauss)
	Coefficient of Deal-Grove equation (cm^2/sec)
BHF	Buffered hydrofluoric acid
BiCMOS	Bipolar/complimentary metal oxide semiconductor
BJT	Bipolar junction transistor
BPSG	Borophosphosilicate glass
C	Concentration (cm^{-3})
	Conductance (cm^3/sec)
CIM	Computer integrated manufacturing
CMOS	Complimentary metal oxide semiconductor
CMP	Chemical mechanical polishing
CMTF	Critical modulation transfer function
CNC	Condensation nucleus counter
C_{ox}	Oxide capacitance per unit area (F/cm^2)
	Solubility of oxygen in silicon (cm^{-3})
CV	Capacitance voltage curve
CVD	Chemical vapor deposition
CZ	Czochralski growth
D	Diffusion coefficient (cm^2/sec)
	Defect density (cm^{-2})
DCFL	Direct coupled FET logic

DCS	Dichlorosilane
DI	Dielectric isolation
D_o	Prexponential for diffusion coefficient (cm^2/sec)
DOE	Design of experiments
DQN	Diazoquinone novalac resist
DRAM	Dynamic random access memory
DUV	Deep ultraviolet
e	Charge of an electron (coul)
E	Energy (ergs)
E_a	Activation energy (erg)
E_G	Energy gap (erg)
EBL	Electron beam lithography
ECL	Emitter coupled logic
ECR	Electron cyclotron resonance
EEPROM	Electrically erasable programmable read only memory
F	View factor (dimensionless)
FET	Field effect transistor
f_T	Unity gain frequency (Hz)
FZ	Float zone growth
g	Gap between mask and wafer (cm)
G	Gibbs free energy (erg)
	Gross fail area (dimensionless)
G_B	Gummel number of the base (cm^{-2})
g_m	Device transconductance (amp/volt)
GOLD	Gate overlap of drain
GSMBE	Gas source molecular beam epitaxy
HEMT	High electron mobility transistor
h_g	Mass transport coefficient (cm/sec)
HMDS	Hexamethyldioxysilazane
IC	Integrated circuit
IR	Infrared
I_x	Atomic current (sec^{-1})
J	Flux density (cm^{-2}-sec^{-1})
k	Relative dielectric constant (dimensionless)
	Thermal conductivity (erg/°C-cm)
	Segregation coefficient (dimensionless)
K_p	Reaction equilibrium constant (Units depend on reaction)
k_x	Reaction rate constant (units depend on reaction)
LDD	Lightly doped drain
LEC	Liquid encapsulated Czochralski
LOCOS	Local oxidation of silicon
LPCVD	Low pressure chemical vapor deposition

m	Atomic mass (gm)
	Segregation coefficient (dimensionless)
M	Molar mass (gm)
MBE	Molecular beam epitaxy
MERIE	Magnetically enhanced reactive ion etching
MESFET	Metal semiconductor field effect transistor
MMIC	Millimeter microwave integrated circuit
MOCVD	Metallorganic chemical vapor deposition
MODFET	Modulation doped field effect transistor
MOS	Metal oxide semiconductor
MTF	Median time to failure
	Modulation transfer function
$M_\lambda(T)$	Spectral radiant exitance
n	electron concentration (cm^{-3})
N	Number density (cm^{-3})
n_i	Intrinsic carrier concentration (cm^{-3})
N_{it}	Interface state density (cm^{-2})
NMOS	N-channel metal oxide semiconductor
N_{re}	Reynolds number (dimensionless)
N_v	Vacancy concentration (cm^{-3})
OA	Orthogonal array
OM	Organometallic
OSF	Oxidation induced stacking faults
p	Hole concentration (cm^{-3})
P	Total pressure (dyne/cm^2)
	Probability function (dimensionless)
PAC	Photoactive compound
PECVD	Plasma enhanced chemical vapor deposition
PMMA	Polymethyl methacrylate
PMOS	P-channel metal oxide semiconductor
PSG	Phosphosilicate glass
p_x	Partial pressure of species x (dyne/cm^2)
q	Charge on an electron (Coul)
Q	Gas throughput (dyne-cm/sec)
Q_f	Fixed charge density (C/cm^2)
R	Growth or deposition rate (cm/sec)
	Resistance (Ω)
	Reflectivity
R_c	Specific contact resistance (Ω-cm^2)
RCA	Popular wet chemical clean
RF	Radio frequency
RIE	Reactive ion etch
R_{ME}	Mass evaporation rate (gm/cm^2-sec)
R_p	Projected range (cm)

R_s	Sheet resistance (Ω/square)
RTCVD	Rapid thermal chemical vapor deposition
RTA	Rapid thermal annealing
RTP	Rapid thermal processing
S	Sputter yield (dimensionless)
SA	Self aligned bipolar technology
SAGFET	Self aligned gate field effect transistor
SAINT	Self aligned implantation N+ technology
SBC	Standard buried collector
SDFL	Schottky diode FET logic
SEM	Secondary electron microscopy
SICOS	Sidewall contacted structure
SILO	Sealed interface local oxidation
SIMOX	Separation by implanted oxygen
SIMS	Secondary ion mass spectroscopy
SOI	Silicon on insulator
S_P	Pumping speed (cm^3/sec)
SPC	Statistical process control
SPE	Solid phase epitaxial regrowth
SPICE	Stanford program for integrated circuit emulation
SRAM	Static random access memory
SS	Sum of squares
SST	Super self aligned bipolar transistor
SUPREM	Stanford University process emulation module
SWAMI	Sidewall masked isolation
t	Time (sec)
T	Temperature (Centigrade)
TCA	Trichloroethane
TCE	Trichloroethylene
TDDB	Time dependent dielectric breakdown
TEG	Triethyl gallium
TEM	Transmission electron microscopy
TEOS	Tetraethyloxysilane
TMG	Trimethyl gallium
TOC	Total organic content
t_{ox}	Oxide thickness (cm)
u	Surface free energy (erg/cm^2)
UHVCVD	Ultrahigh vacuum chemical vapor deposition
ULSI	Ultra large scale integration
UV	Ultraviolet
U_∞	Gas velocity far from a surface (cm/sec)
v	Velocity (cm/sec)
V_A	Early voltage of bipolar transistor (volts)
V_{bi}	Built in voltage (volts)
VLSI	Very large scale integration

VPE	Vapor phase epitaxy
V_{po}	Pinchoff voltage of MESFET (volts)
V_t	Threshold voltage (volts)
W	Base width of bipolar transistor (cm)
W_D	Depletion width (cm)
WIP	Work in progress
x_j	Junction depth (cm)
Y	Yield (dimensionless)
Z	Atomic number
α	Inverse absorption length (cm^{-1})
	Ratio of collector and emitter currents in a bipolar transistor (dimensionless)
β	Ratio of collector and base currents in a bipolar transistor (dimensionless)
γ	Resist contrast (dimensionless)
δ	Boundary layer thickness (cm)
ε	Emissivity (dimensionless)
	Relative dielectric constant (dimensionless)
η	Dynamic viscosity (cm^{-1}-sec^{-1})
λ	Wavelength of radiation (cm)
	Mean free path (cm)
μ	Kinematic viscosity (cm^2/sec)
	Mobility (cm^2/volt-sec)
μ_{eff}	Effective mobility (cm^2/volt-sec)
ν	Frequency of radiation (sec^{-1})
ρ	Resistivity (Ω-cm)
	Mass density (gm/cm^3)
ρ_D	Sheet resistance (Ω/square)
σ	Cross section (cm^2)
	Supersaturation (dimensionless)
	Stress in wafer ($dyne/cm^2$)
σ_o	Saturation (dimensionless)
θ	Fraction of free sites on a surface (dimensionless)
ϕ_{ms}	Metal semiconductor work function (erg)
ϕ_b	Barrier height (erg)
ψ	Critical implant angle (degrees)
$\mathscr{E}$	Electric field (volt/cm)
3D	Triple diffused bipolar technology

Appendix II

Properties of Selected Semiconductor Materials

Table II.1 Properties of Various Materials I: Semiconductors and Insulators

	Si	GaAs	Ge	αSiC	SiO$_2$	Si$_3$N$_4$
Density (gm/cm^3)	2.33	5.32	5.32	3.21	2.2	3.1
Breakdown field (MV/cm)	0.3	0.5	0.1	2.3	10	10
Dielectric constant	11.7	12.9	16.2	6.52	3.9	7.5
Energy gap (eV)	1.12	1.42	0.66	2.86	9	5
Electron affinity (eV)	4.05	4.07	4		0.9	
Index of refraction	3.42	3.3	3.98	2.55	1.46	2.05
Melting point (°C)	1412	1240	937	2830	~1700	~1900
Specific heat (J/gm-°C)	0.7	0.35	0.31		1	
Thermal conductivity (W/cm-°C)	1.31	0.46	0.6		0.014	
Thermal diffusivity (cm^2/sec)	0.9	0.44	0.36		0.006	
Thermal expansion ($\times 10^{-6}$)	2.6	6.86	2.2	2.9	0.5	2.7

Table II.2 Properties of Various Materials II: Metals

	Al	Cu	Au	TiSi$_2$	PtSi
Density (gm/cm^3)	2.7	8.89	19.3	4.043	12.394
Resistivity ($\mu\Omega$-cm)	2.82	1.72	2.44	14	30
Temperature coefficient	0.0039	0.0039	0.0034	4.63	
Barrier to n-type Si (eV)	0.55	0.60	0.75	0.60	0.85
Thermal conductivity (W/cm-°C)	2.37	3.98	3.15		
Melting point (°C)	659	1083	1063	1540	1229
Specific heat (J/gm-°C)	0.90	0.39	0.13		
Thermal expansion ($\times 10^{-6}$)	25	16.6	14.2	12.5	

Appendix III

Physical Constants

Quantity	Symbol	Value
Gravitational constant	G	6.67×10^{-11} N m^2/kg^2 (or m^3/kg s^2)
Avogadro's number	N_0	6.0222×10^{23} particles/mole (or amu/gm)
Boltzmann's constant (microscopic gas constant)	k	1.3806×10^{-23} J/K
	$\dfrac{1}{k}$	8.617×10^{-5} eV/K
		$11{,}605$ K/eV
Macroscopic gas constant	$R (= N_0 k)$	8.314 J/mole K
		1.9872 kcal/kmole K
Quantum unit of charge	e	1.60219×10^{-19} C
		4.8033×10^{-10} esu
Faraday constant (1 mole of electricity)	$F (= N_0 e)$	9.6487×10^4 C/mole
		2.8926×10^{14} esu/mole
Permittivity of space	$\varepsilon_0 \left(= \dfrac{1}{\mu_0 c^2} \right)$	8.85419×10^{-12} C^2/N m^2
	$4\pi\varepsilon_0 \left(= \dfrac{4\pi}{\mu_0 c^2} \right)$	1.112650×10^{-10} C^2/N m^2
	$\dfrac{1}{4\pi\varepsilon_0} \left(= \dfrac{\mu_0 c^2}{4\pi} \right)$	8.98755×10^9 N m^2/C^2
Permeability constant	μ_0	$4\pi \times 10^{-7}$ N/A^2 *exact, by definition* or 1.256637×10^{-6} N/A^2
	$\dfrac{\mu_0}{4\pi}$	*exactly* 10^{-7} N/A^2
Speed of light	c	2.997925×10^8 m/s
Planck's constant	h	6.6262×10^{-34} J s
		4.1357×10^{-15} eV s
		4.1357×10^{-21} MeV s
	$\hbar \left(= \dfrac{h}{2\pi} \right)$	1.05459×10^{-34} Js
		6.58822×10^{-16} eV s
		6.5822×10^{-22} MeV s
Charge to mass ratio or electron	$\dfrac{e}{m_e}$	1.75880×10^{11} C/kg
		5.2728×10^{17} esu/gm

continued

Quantity	Symbol	Value
Mass of electron	m_e	9.1096×10^{-31} kg
		5.4859×10^{-4} amu
Mass of proton	m_p	1.67261×10^{-27} kg
		1.0072766 amu
		$1836.11 m_e$
Mass of neutron	m_n	1.67492×10^{-27} kg
		1.0086652 amu
		$1838.64 m_e$
Intrinsic energy of electron	$m_e c^2$	0.51100 MeV
Intrinsic energy of proton	$m_p c^2$	938.27 MeV
Intrinsic energy of neutron	$m_n c^2$	939.55 MeV
Rydberg constant for infinitely massive nucleus	$\mathscr{R}_\infty \left[= \left(\dfrac{1}{4\pi\varepsilon_0} \right)^2 \dfrac{m_e e^4}{4\pi\hbar^3 c} \right]$	1.0973731×10^7 m^{-1}
Rydberg constant for hydrogen 1	$\mathscr{R}_H$	1.0967758×10^7 m^{-1}
Fine structure constant	$\alpha \left(= \dfrac{1}{4\pi\varepsilon_0} \dfrac{e^2}{\hbar c} \right)$	7.29735×10^{-3} or $1/137.036$

(This table is adapted from B. N. Taylor, W. H. Parker, and D. N. Langenberg, *The Fundamental Constants and Quantum Electrodynamics* [New York: Academic Press, 1969]. A good popular article on the fundamental constants, by the same authors, is to be found in the October, 1970, issue of *Scientific American*. The numbers recorded here have been truncated so that the uncertainty in each is at most ± 1 in the last digit.)

Appendix IV

Conversion Factors

Conversion factors preceded by a dot (●) are exact and serve to define one unit in terms of another. For example, the factor 0.3048 m/ft defines the foot as exactly 0.3048 m.

1. Length
- 10^2 cm/m
- 10^3 m/km

- 2.54 cm/in.
- 12 in./ft
- 5,280 ft/mi

- 0.3048 m/ft
- 1.609344×10^3 m/mi
- 1.609344 km/mi

  1.49598×10^{11} m/A.U.
  9.461×10^{15} m/light-yr
  3.084×10^{16} m/pars

- 10^{-6} m/μm (or m/micron)
- 10^{-10} m/Å
- 10^{-15} m/fm

2. Volume
- 10^{-3} m^3/L
- 10^3 cm^3/L
  0.94635 L/qt
  3.7854×10^{-3} m^3/gal

3. Time
(The day is a mean solar day; the year is a sidereal year.)
- 3,600 s/hr
- 8.64×10^4 s/d

365.26 d/yr

3.1558×10^7 s/yr

4. Speed

- 0.3048 (m/s)/(ft/s)

1.609×10^3 (m/s)/(mi/s)

0.4470 (m/s)(mi/hr)

1.609 (km/hr)/(mi/hr)

5. Acceleration

- 0.3048 $(\text{m/s}^2)/(\text{ft/s}^2)$

6. Angle

- 60 s of arc($''$)/min of arc($'$)
- 60 min of arc($'$)/deg
- $180/\pi(\cong 57.30)$ deg/rad
- $2\pi(\cong 6.283)$ rad/r

7. Mass

- 10^3 gm/kg

453.59 gm/lb

0.45359 kg/lb

2.2046 lb/kg

1.66053×10^{-27} kg/amu

6.0222×10^{26} amu/kg

6.0222×10^{23} amu/gm

8. Density

- 10^3 (kg/m^3)/(gm/cm^3)

16.018 (kg/m^3)/(lb/ft^3)

1.6018×10^{-2} (gm/cm^3)/(lb/ft^3)

9. Force

- 10^5 dyne/N
- 10^{-5} N/dyne

4.4482 N/lbf

(1 lbf = weight of 1 pound at standard
gravity [$g = 9.80665$ m/s^2])

10. Pressure

- 0.1 (N/m^2)/(dyne/cm^2)
- 10^5 (N/m^2)/bar
- 1.01325×10^5 (N/m^2)/atm
- 1.01325×10^6 (dyne/cm^2)/atm
- 1.01325 bar/atm

133.32 (N/m^2)/mm of Hg (0°C)

3.386×10^3 (N/m^2)/in. of Hg (0°C)

6.895×10^3 (N/m^2)/(lbf/in.^{2}, or psi)

11. **Energy**
    (For mass to energy conversion, see the values of c^2 at the end of Appendix IIIA.)

- 10^7 erg/J
- 10^{-7} J/erg

- 4.184 J/cal
- 4.184 J/kcal
- 10^3 cal/kcal
    (The kilocalorie [kcal] is also known as the food calorie, the large calorie, or the Calorie.)

    1.60219×10^{-19} J/eV
    1.60219×10^{-13} J/MeV
- 10^6 eV/MeV

- 3.60×10^6 J/kW hr

    4.20×10^{12} J/kton
    4.20×10^{15} J/mton

    0.04336 (eV/molecule)/(kcal/mole)
    23.06 (kcal/mole)/(eV/molecule)

12. **Power**
- 746 W/hp

13. **Temperature**
- 1.00 °F/°R
- 1.00 °C/K
- 1.80 °F/°C
- 1.80 °F/K
- $T(K) = T(°C) + 273.15$
- $T(°C) = [T(°F) - 32]/1.80$
- $T(K) = T(°R)/1.80$

14. **Electrical quantities**
    (Note that 2.9979 is well approximated by 3.00.)
    Charge: 2.9979×10^9 esu/C
    Current: 2.9979×10^9 (esu/s)/A
    Potential: 299.79 V/statvolt
    Electric field: 2.9979×10^4 (V/m)/(statvolt/cm)
- Magnetic field: 10^4 G/T
- Magnetic flux: 10^8 G cm^2/Wb
- Pole strength: 10 cgs unit/michel
    (cgs unit = $\sqrt{\text{ergcm}}$; michel = Am)

Appendix V

Some Properties
of the Error Function

$$\text{erf } u = \frac{2}{\sqrt{\pi}} \int_0^u e^{-z^2}\, dz = \frac{2}{\sqrt{\pi}} \left(u - \frac{u^3}{3 \times 1!} + \frac{u^5}{5 \times 2!} - \cdots \right)$$

Therefore

$$\text{erf}(-u) = -\text{erf } u$$

$$\text{erfc } u = 1 - \text{erf } u = \frac{2}{\sqrt{\pi}} \int_u^\infty e^{-z^2}\, dz$$

$$\text{erf } u \approx \frac{2u}{\sqrt{\pi}} \qquad \text{for} \qquad u \ll 1$$

$$\text{erfc } u \approx \frac{1}{\sqrt{\pi}} \frac{e^{-u^2}}{u} \qquad \text{for} \qquad u \gg 1$$

$$\text{erf}(\infty) = 1, \qquad \text{erf}(0) = 0$$

$$\text{erfc}(0) = 1, \qquad \text{erfc}(\infty) = 0$$

$$\frac{d \text{ erf } u}{du} = \frac{2}{\sqrt{\pi}} e^{-u^2}$$

$$\int_0^u \text{erfc } z\, dz = u \text{ erfc } u + \frac{1}{\sqrt{\pi}} (1 - e^{-u^2})$$

$$\int_0^\infty \text{erfc } z\, dz = \frac{1}{\sqrt{\pi}}$$

$$\int_0^\infty e^{-u^2}\, du = \frac{\sqrt{\pi}}{2}, \qquad \int_0^u e^{-z^2}\, dz = \frac{\sqrt{\pi}}{2} \text{ erf } u$$

Error Function erf(*w*)

w	erf(*w*)	*w*	erf(*w*)	*w*	erf(*w*)	*w*	erf(*w*)
0.00	0.000 000	0.43	0.456 887	0.86	0.776 110	1.29	0.931 899
0.01	0.011 283	0.44	0.466 225	0.87	0.781 440	1.30	0.934 008
0.02	0.022 565	0.45	0.475 482	0.88	0.786 687	1.31	0.936 063
0.03	0.033 841	0.46	0.484 655	0.89	0.791 843	1.32	0.938 065
0.04	0.045 111	0.47	0.493 745	0.90	0.796 908	1.33	0.940 015
0.05	0.056 372	0.48	0.502 750	0.91	0.801 883	1.34	0.941 914
0.06	0.067 622	0.49	0.511 668	0.92	0.806 768	1.35	0.943 762
0.07	0.078 858	0.50	0.520 500	0.93	0.811 564	1.36	0.945 561
0.08	0.090 078	0.51	0.529 244	0.94	0.816 271	1.37	0.947 312
0.09	0.101 281	0.52	0.537 899	0.95	0.820 891	1.38	0.949 016
0.10	0.112 463	0.53	0.546 464	0.96	0.825 424	1.39	0.950 673
0.11	0.123 623	0.54	0.554 939	0.97	0.829 870	1.40	0.952 285
0.12	0.134 758	0.55	0.563 323	0.98	0.834 232	1.41	0.953 852
0.13	0.145 867	0.56	0.571 616	0.99	0.838 508	1.42	0.955 376
0.14	0.156 947	0.57	0.579 816	1.00	0.842 701	1.43	0.956 857
0.15	0.167 996	0.58	0.587 923	1.01	0.846 810	1.44	0.958 297
0.16	0.179 012	0.59	0.595 936	1.02	0.850 838	1.45	0.959 695
0.17	0.189 992	0.60	0.603 856	1.03	0.854 784	1.46	0.961 054
0.18	0.200 936	0.61	0.611 681	1.04	0.858 650	1.47	0.962 373
0.19	0.211 840	0.62	0.619 411	1.05	0.862 436	1.48	0.963 654
0.20	0.222 703	0.63	0.627 046	1.06	0.866 144	1.49	0.964 898
0.21	0.233 522	0.64	0.634 586	1.07	0.869 773	1.50	0.966 105
0.22	0.244 296	0.65	0.642 029	1.08	0.873 326	1.51	0.967 277
0.23	0.255 023	0.66	0.649 377	1.09	0.876 803	1.52	0.968 413
0.24	0.265 700	0.67	0.656 628	1.10	0.880 205	1.53	0.969 516
0.25	0.276 326	0.68	0.663 782	1.11	0.883 533	1.54	0.970 586
0.26	0.286 900	0.69	0.670 840	1.12	0.886 788	1.55	0.971 623
0.27	0.297 418	0.70	0.677 801	1.13	0.889 971	1.56	0.972 628
0.28	0.307 880	0.71	0.684 666	1.14	0.893 082	1.57	0.973 603
0.29	0.318 283	0.72	0.691 433	1.15	0.896 124	1.58	0.974 547
0.30	0.328 627	0.73	0.698 104	1.16	0.899 096	1.59	0.975 462
0.31	0.338 908	0.74	0.704 678	1.17	0.902 000	1.60	0.976 348
0.32	0.349 126	0.75	0.711 156	1.18	0.904 837	1.61	0.977 207

continued

w	erf(*w*)	*w*	erf(*w*)	*w*	erf(*w*)	*w*	erf(*w*)
0.33	0.359 279	0.76	0.717 537	1.19	0.907 608	1.62	0.978 038
0.34	0.369 365	0.77	0.723 822	1.20	0.910 314	1.63	0.978 843
0.35	0.379 382	0.78	0.730 010	1.21	0.912 956	1.64	0.979 622
0.36	0.389 330	0.79	0.736 103	1.22	0.915 534	1.65	0.980 376
0.37	0.399 206	0.80	0.742 101	1.23	0.918 050	1.66	0.981 105
0.38	0.409 009	0.81	0.748 003	1.24	0.920 505	1.67	0.981 810
0.39	0.418 739	0.82	0.753 811	1.25	0.922 900	1.68	0.982 493
0.40	0.428 392	0.83	0.759 524	1.26	0.925 236	1.69	0.983 153
0.41	0.437 969	0.84	0.765 143	1.27	0.927 514	1.70	0.983 790
0.42	0.447 468	0.85	0.770 668	1.28	0.929 734	1.71	0.984 407
1.72	0.985 003	2.15	0.997 639	2.58	0.999 736	3.01	0.999 979 26
1.73	0.985 578	2.16	0.997 747	2.59	0.999 751	3.02	0.999 980 53
1.74	0.986 135	2.17	0.997 851	2.60	0.999 764	3.03	0.999 981 73
1.75	0.986 672	2.18	0.997 951	2.61	0.999 777	3.04	0.999 982 86
1.76	0.987 190	2.19	0.998 046	2.62	0.999 789	3.05	0.999 983 92
1.77	0.987 691	2.20	0.998 137	2.63	0.999 800	3.06	0.999 984 92
1.78	0.988 174	2.21	0.998 224	2.64	0.999 811	3.07	0.999 985 86
1.79	0.988 641	2.22	0.998 308	2.65	0.999 822	3.08	0.999 986 74
1.80	0.989 091	2.23	0.998 388	2.66	0.999 831	3.09	0.999 987 57
1.81	0.989 525	2.24	0.998 464	2.67	0.999 841	3.10	0.999 988 35
1.82	0.989 943	2.25	0.998 537	2.68	0.999 849	3.11	0.999 989 08
1.83	0.990 347	2.26	0.998 607	2.69	0.999 858	3.12	0.999 989 77
1.84	0.990 736	2.27	0.998 674	2.70	0.999 866	3.13	0.999 990 42
1.85	0.991 111	2.28	0.998 738	2.71	0.999 873	3.14	0.999 991 03
1.86	0.991 472	2.29	0.998 799	2.72	0.999 880	3.14	0.999 991 60
1.87	0.991 821	2.30	0.998 857	2.73	0.999 887	3.16	0.999 992 14
1.88	0.992 156	2.31	0.998 912	2.74	0.999 893	3.17	0.999 992 64
1.89	0.992 479	2.32	0.998 966	2.75	0.999 899	3.18	0.999 993 11
1.90	0.992 790	2.33	0.999 016	2.76	0.999 905	3.19	0.999 993 56
1.91	0.993 090	2.34	0.999 065	2.77	0.999 910	3.20	0.999 993 97
1.92	0.993 378	2.35	0.999 111	2.78	0.999 916	3.21	0.999 994 36
1.93	0.993 656	2.36	0.999 155	2.79	0.999 920	3.22	0.999 994 73
1.94	0.993 923	2.37	0.999 197	2.80	0.999 925	3.23	0.999 995 07
1.95	0.994 179	2.38	0.999 237	2.81	0.999 929	3.24	0.999 995 40
1.96	0.994 426	2.39	0.999 275	2.82	0.999 933	3.25	0.999 995 70
1.97	0.994 664	2.40	0.999 311	2.83	0.999 937	3.26	0.999 995 98
1.98	0.994 892	2.41	0.999 346	2.84	0.999 941	3.27	0.999 996 24
1.99	0.995 111	2.42	0.999 379	2.85	0.999 944	3.28	0.999 996 49
2.00	0.995 322	2.43	0.999 411	2.86	0.999 948	3.29	0.999 996 72
2.01	0.995 525	2.44	0.999 441	2.87	0.999 951	3.30	0.999 996 94
2.02	0.995 719	2.45	0.999 469	2.88	0.999 954	3.31	0.999 997 15
2.03	0.995 906	2.46	0.999 497	2.89	0.999 956	3.32	0.999 997 34
2.04	0.996 086	2.47	0.999 523	2.90	0.999 959	3.33	0.999 997 51
2.05	0.996 258	2.48	0.999 547	2.91	0.999 961	3.34	0.999 997 68
2.06	0.996 423	2.49	0.999 571	2.92	0.999 964	3.35	0.999 997 838
2.07	0.996 582	2.50	0.999 593	2.93	0.999 966	3.36	0.999 997 983
2.08	0.996 734	2.51	0.999 614	2.94	0.999 968	3.37	0.999 998 120
2.09	0.996 880	2.52	0.999 634	2.95	0.999 970	3.38	0.999 998 247
2.10	0.997 021	2.53	0.999 654	2.96	0.999 972	3.39	0.999 998 367
2.11	0.997 155	2.54	0.999 672	2.97	0.999 973	3.40	0.999 998 478
2.12	0.997 284	2.55	0.999 689	2.98	0.999 975	3.41	0.999 998 582
2.13	0.997 407	2.56	0.999 706	2.99	0.999 976	3.42	0.999 998 679
2.14	0.997 525	2.57	0.999 722	3.00	0.999 977 91	3.43	0.999 998 770
3.44	0.999 998 855	3.58	0.999 999 587	3.72	0.999 999 857	3.86	0.999 999 952
3.45	0.999 998 934	3.59	0.999 999 617	3.73	0.999 999 867	3.87	0.999 999 956

w	**erf(*w*)**	*w*	**erf(*w*)**	*w*	**erf(*w*)**	*w*	**erf(*w*)**
3.46	0.999 999 008	3.60	0.999 999 644	3.74	0.999 999 877	3.88	0.999 999 959
3.47	0.999 999 077	3.61	0.999 999 670	3.75	0.999 999 886	3.89	0.999 999 962
3.48	0.999 999 141	3.62	0.999 999 694	3.76	0.999 999 895	3.90	0.999 999 965
3.49	0.999 999 201	3.63	0.999 999 716	3.77	0.999 999 903	3.91	0.999 999 968
3.50	0.999 999 257	3.64	0.999 999 736	3.78	0.999 999 910	3.92	0.999 999 970
3.51	0.999 999 309	3.65	0.999 999 756	3.79	0.999 999 917	3.93	0.999 999 973
3.52	0.999 999 358	3.66	0.999 999 773	3.80	0.999 999 923	3.94	0.999 999 975
3.53	0.999 999 403	3.67	0.999 999 790	3.81	0.999 999 929	3.95	0.999 999 977
3.54	0.999 999 445	3.68	0.999 999 805	3.82	0.999 999 934	3.96	0.999 999 979
3.55	0.999 999 485	3.69	0.999 999 820	3.83	0.999 999 939	3.97	0.999 999 980
3.56	0.999 999 521	3.70	0.999 999 833	3.84	0.999 999 944	3.98	0.999 999 982
3.57	0.999 999 555	3.71	0.999 999 845	3.85	0.999 999 948	3.99	0.999 999 983

Appendix VI

F Values

Table VI.1 95th percentile values for the F distribution
n_1 = degrees of freedom for numerator
n_2 = degrees of freedom for denominator (shaded area = 0.95)

n_2 \ n_1	1	2	3	4	5	6	8	12	16	20	30	40	50	100	∞
1	161.4	199.5	215.7	224.6	230.2	234.0	238.9	243.9	246.3	248.0	250.1	251.1	252.2	253.0	254.3
2	18.51	19.00	19.16	19.25	19.30	19.33	19.37	19.41	19.43	19.45	19.46	19.46	19.47	19.49	19.50
3	10.13	9.55	9.28	9.12	9.01	8.94	8.85	8.74	8.69	8.66	8.62	8.60	8.58	8.56	8.53
4	7.71	6.94	6.59	6.39	6.26	6.16	6.04	5.91	5.84	5.80	5.75	5.71	5.70	5.66	5.63
5	6.61	5.79	5.41	5.19	5.05	4.95	4.82	4.68	4.60	4.56	4.50	4.46	4.44	4.40	4.36
6	5.99	5.14	4.76	4.53	4.39	4.28	4.15	4.00	3.92	3.87	3.81	3.77	3.75	3.71	3.67
7	5.59	4.74	4.35	4.12	3.97	3.87	3.73	3.57	3.49	3.44	3.38	3.34	3.32	3.28	3.23
8	5.32	4.46	4.07	3.84	3.69	3.58	3.44	3.28	3.20	3.15	3.08	3.05	3.03	2.98	2.93
9	5.12	4.26	3.86	3.63	3.48	3.37	3.23	3.07	2.98	2.93	2.86	2.82	2.80	2.76	2.71
10	4.96	4.10	3.71	3.48	3.33	3.22	3.07	2.91	2.82	2.77	2.70	2.67	2.64	2.59	2.54
11	4.84	3.98	3.59	3.36	3.20	3.09	2.95	2.79	2.70	2.65	2.57	2.53	2.50	2.45	2.40
12	4.75	3.89	3.49	3.26	3.11	3.00	2.85	2.69	2.60	2.54	2.46	2.42	2.40	2.35	2.30
13	4.67	3.81	3.41	3.18	3.03	2.92	2.77	2.60	2.51	2.46	2.38	2.34	2.32	2.26	2.21
14	4.60	3.74	3.34	3.11	2.96	2.85	2.70	2.53	2.44	2.39	2.31	2.27	2.24	2.19	2.13
15	4.54	3.68	3.29	3.06	2.90	2.79	2.64	2.48	2.39	2.33	2.25	2.21	2.18	2.12	2.07
16	4.49	3.63	3.24	3.01	2.85	2.74	2.59	2.42	2.33	2.28	2.20	2.16	2.13	2.07	2.01
17	4.45	3.59	3.20	2.96	2.81	2.70	2.55	2.38	2.29	2.23	2.15	2.11	2.08	2.02	1.96
18	4.41	3.55	3.16	2.93	2.77	2.66	2.51	2.34	2.25	2.19	2.11	2.07	2.04	1.98	1.92
19	4.38	3.52	3.13	2.90	2.74	2.63	2.48	2.31	2.21	2.15	2.07	2.02	2.00	1.94	1.88
20	4.35	3.49	3.10	2.87	2.71	2.60	2.45	2.28	2.18	2.12	2.04	1.99	1.96	1.90	1.84
22	4.30	3.44	3.05	2.82	2.66	2.55	2.40	2.23	2.13	2.07	1.98	1.93	1.91	1.84	1.78
24	4.26	3.40	3.01	2.78	2.62	2.51	2.36	2.18	2.09	2.03	1.94	1.89	1.86	1.80	1.73
26	4.23	3.37	2.98	2.74	2.59	2.47	2.32	2.15	2.05	1.99	1.90	1.85	1.82	1.76	1.69
28	4.20	3.34	2.95	2.71	2.56	2.45	2.29	2.12	2.02	1.96	1.87	1.81	1.78	1.72	1.65
30	4.17	3.32	2.92	2.69	2.53	2.42	2.27	2.09	1.99	1.93	1.84	1.79	1.76	1.69	1.62
40	4.08	3.23	2.84	2.61	2.45	2.34	2.18	2.00	1.90	1.84	1.74	1.69	1.66	1.59	1.51
50	4.03	3.18	2.79	2.56	2.40	2.29	2.13	1.95	1.85	1.78	1.69	1.63	1.60	1.52	1.44
60	4.00	3.15	2.76	2.53	2.37	2.25	2.10	1.92	1.81	1.75	1.65	1.59	1.56	1.48	1.39
70	3.98	3.13	2.74	2.50	2.35	2.23	2.07	1.89	1.79	1.72	1.62	1.56	1.53	1.45	1.35
80	3.96	3.11	2.72	2.48	2.33	2.21	2.05	1.88	1.77	1.70	1.60	1.54	1.51	1.42	1.32
100	3.94	3.09	2.70	2.46	2.30	2.19	2.03	1.85	1.75	1.68	1.57	1.51	1.48	1.39	1.28
150	3.91	3.06	2.67	2.43	2.27	2.16	2.00	1.82	1.71	1.64	1.54	1.47	1.44	1.34	1.22
200	3.89	3.04	2.65	2.41	2.26	2.14	1.98	1.80	1.69	1.62	1.52	1.45	1.42	1.32	1.19
400	3.86	3.02	2.62	2.39	2.23	2.12	1.96	1.78	1.67	1.60	1.49	1.42	1.38	1.28	1.13
∞	3.84	2.99	2.60	2.37	2.21	2.09	1.94	1.75	1.64	1.57	1.46	1.40	1.32	1.24	1.00

Table VI.2   99th percentile values for the *F* distribution
n_1 = degrees of freedom for numerator
n_2 = degrees of freedom for denominator (shaded area = .99)

n_2 \ n_1	1	2	3	4	5	6	8	12	16	20	30	40	50	100	∞
1	4052	4999	5403	5625	5764	5859	5981	6106	6169	6208	6258	6286	6302	6334	6366
2	98.49	99.01	99.17	99.25	99.30	99.33	99.36	99.42	99.44	99.45	99.47	99.48	99.48	99.49	99.50
3	34.12	30.81	29.46	28.71	28.24	27.41	27.49	27.05	28.63	26.69	26.50	26.41	26.35	26.23	26.12
4	21.20	18.00	16.69	15.98	15.52	15.21	14.80	14.37	14.15	14.02	13.83	13.74	13.69	13.57	13.46
5	16.26	13.27	12.06	11.39	10.97	10.67	10.27	9.89	9.68	9.55	9.38	9.29	9.24	9.13	9.02
6	13.74	10.92	9.78	9.15	8.75	8.47	8.10	7.72	7.52	7.39	7.23	7.14	7.09	6.99	6.88
7	12.25	9.55	8.45	7.85	7.46	7.19	6.84	6.47	6.27	6.15	5.98	5.90	5.85	5.75	5.65
8	11.26	8.65	7.59	7.01	6.63	6.37	6.03	5.67	5.48	5.36	5.20	5.11	5.06	4.96	4.86
9	10.56	8.02	6.99	6.42	6.06	5.80	5.47	5.11	4.92	4.80	4.64	4.56	4.51	4.41	4.31
10	10.04	7.56	6.55	5.99	5.64	5.39	5.06	4.71	4.52	4.41	4.25	4.17	4.12	4.01	3.91
11	9.05	7.20	6.22	5.67	5.32	5.07	4.74	4.40	4.21	4.10	3.94	3.86	3.80	3.70	3.60
12	9.33	6.93	5.95	5.41	5.06	4.82	4.50	4.16	3.98	3.86	3.70	3.61	3.56	3.46	3.36
13	9.07	6.70	5.74	5.20	4.86	4.62	4.30	3.96	3.78	3.67	3.51	3.42	3.37	3.27	3.16
14	8.86	6.51	5.56	5.03	4.69	4.46	4.14	3.80	3.62	3.51	3.34	3.26	3.21	3.11	3.00
15	8.68	6.36	5.42	4.89	4.56	4.32	4.00	3.67	3.48	3.36	3.20	3.12	3.07	2.97	2.87
16	8.53	6.23	5.29	4.77	4.44	4.20	3.89	3.55	3.37	3.25	3.10	3.01	2.96	2.86	2.75
17	8.40	6.11	5.18	4.67	4.34	4.10	3.79	3.45	3.27	3.16	3.00	2.92	2.86	2.76	2.65
18	8.28	6.01	5.09	4.58	4.25	4.01	3.71	3.37	3.19	3.07	2.91	2.83	2.78	2.68	2.57
19	8.18	5.93	5.01	4.50	4.17	3.94	3.63	3.30	3.12	3.00	2.84	2.76	2.70	2.60	2.49
20	8.10	5.85	4.94	4.43	4.10	3.87	3.56	3.23	3.05	2.94	2.77	2.69	2.63	2.53	2.42
22	7.94	5.72	4.82	4.31	3.99	3.76	3.45	3.12	2.94	2.83	2.67	2.58	2.53	2.42	2.31
24	7.82	5.61	4.72	4.22	3.90	3.67	3.36	3.03	2.85	2.74	2.58	2.49	2.44	2.33	2.21
26	7.72	5.53	4.64	4.14	3.82	3.59	3.29	2.96	2.77	2.66	2.50	2.41	2.36	2.25	2.13
28	7.64	5.45	4.57	4.07	3.76	3.53	3.23	2.90	2.71	2.60	2.44	2.35	2.30	2.18	2.06
30	7.56	5.39	4.51	4.02	3.70	3.47	3.17	2.84	2.66	2.55	2.38	2.29	2.24	2.13	2.01
40	7.31	5.18	4.31	3.83	3.51	3.29	2.99	2.66	2.49	2.37	2.20	2.11	2.05	1.94	1.81
50	7.17	5.06	4.20	3.72	3.41	3.18	2.88	2.56	2.39	2.26	2.10	2.00	1.94	1.82	1.68
60	7.08	4.98	4.13	3.65	3.34	3.12	2.82	2.50	2.32	2.20	2.03	1.93	1.87	1.74	1.60
70	7.01	4.92	4.08	3.60	3.29	3.07	2.77	2.45	2.28	2.15	1.98	1.98	1.82	1.69	1.53
80	6.96	4.88	4.04	3.56	3.25	3.04	2.74	2.41	2.24	2.11	1.94	1.84	1.78	1.65	1.49
100	6.90	4.82	3.98	3.51	3.20	2.99	2.69	2.36	2.19	2.06	1.89	1.79	1.73	1.59	1.43
150	6.81	4.75	3.91	3.44	3.14	2.92	2.62	2.30	2.12	2.00	1.83	1.72	1.66	1.51	1.33
200	6.76	4.71	3.88	3.41	3.11	2.90	2.60	2.28	2.09	1.97	1.79	1.69	1.62	1.48	1.28
400	6.70	4.66	3.83	3.36	3.06	2.85	2.55	2.23	2.04	1.92	1.74	1.64	1.57	1.42	1.19
∞	6.64	4.60	3.78	3.32	3.02	2.80	2.51	2.18	1.99	1.87	1.69	1.59	1.52	1.36	1.00

Appendix VII

SUPREM Commands

A short summary of common SUPREM III commands. You are encouraged to obtain a more complete set of command documentation from your software provider or system administrator. SUPREM III is a trademark of the Board of Trustees of the Leland Stanford Junior University.

Step	Logical Flags	Variables
Title/Comment	none	none
Initialize	antimony/arsenic/boron/phosphorus silicon <100>/<110>/<111>	concentration (cm^{-3}) spaces (#)—number of spaces to be used in the calculation thickness (μm) dx (μm)—grid spacing at xdx xdx (μm)—distance from the top of the first layer at which the grid spacing is dx
Diffusion	antimony/arsenic/boron/phosphorus dryO$_2$/nitrogen/wetO$_2$ solidsol—sets surface conc to solid solubility	HCl% (%) pressure (atm) temp (°C) time (min)
Deposition	aluminum/nitride/oxide/ polysilicon/silicon ⟨100⟩/⟨110⟩/⟨111⟩ antimony/arsenic/boron/phosphorus	concentration (cm^{-3}) pressure (atm) spaces (#)—number of spaces to be used in the deposited layer temp (°C) thickness (μm) time (min)
Etch	all/aluminum/nitride/oxide/ polysilicon/silicon	thickness (μm)
Implant	antimony/arsenic/boron/BF2/ fluorine/phosphorus Gaussian/2-Gaussian/Pearson	dose (cm^{-2}) energy (keV)
Print	active/chemical/all/total/net antimony/arsenic/boron/phosphorus layers	xmin (μm) xmax (μm)
Plot	active/chemical/total/net antimony/arsenic/boron/phosphorus	xmin (μm) xmax (μm) cmin (cm^{-3}) cmax (cm^{-3})

Index